NAIL SURGERY

A TEXT AND ATLAS

NAIL SURGERY

A TEXT AND ATLAS

EDWARD A. KRULL, M.D.

Department of Dermatology
Henry Ford Hospital
Detroit, Michigan

ELVIN G. ZOOK, M.D.

Division of Plastic Surgery
Southern Illinois University
Springfield, Illinois

ROBERT BARAN, M.D.

Nail Disease Centre
Cannes, France

ECKART HANEKE, M.D.

Institute of Dermatology
Klinikk Bunæs
Sandvika, Norway

LIPPINCOTT WILLIAMS & WILKINS
A **Wolters Kluwer** Company
Philadelphia • Baltimore • New York • London
Buenos Aires • Hong Kong • Sydney • Tokyo

Acquisitions Editor: Beth Barry
Developmental Editor: Sonya L. Seigafuse
Production Editor: Cassie Moore
Manufacturing Manager: Benjamin Rivera
Cover Designer: Mark Lerner
Compositor: Lippincott Williams & Wilkins Desktop Division

530 Walnut Street
Philadelphia, PA 19106 USA
LWW.com

Printed in China

Library of Congress Cataloging-in-Publication Data

Nail surgery : a text and atlas / editors, Edward Krull . . . [et al.].
p. ; cm.
Includes bibliographical references and index.
ISBN 0-7817-0154-6
1. Nails (Anatomy)—Surgery. 2. Nails (Anatomy)—Surgery—Atlases. I. Krull, Edward A.
[DNLM: 1. Nails—surgery. WR 475 N1557 2000]
RL 165 .N35 2000
617.4′77—dc21
00-042435

---

10 9 8 7 6 5 4 3 2 1

CONTENTS

CONTRIBUTING AUTHORS

Robert Baran, M.D. Nail Disease Centre, 42, Rue des Serbes, 06400 Cannes, France

Jacques Bittoun, M.D., Ph.D. Centre Interétablissement de Résonance Magnétique (C.I.E.R.M.), Hôpital Bicêtre, Le Kremlin-Bicêtre, France

Rodney P. R. Dawber, M.D. Department of Dermatology, Churchill Hospital, Headington, Oxford, OX3 7LJ (England)

Jean-Luc Drapé, M.D., Ph.D. Professor of Radiology, CHU Cochin-AP-HP, Université Paris 5, 24, rue du Faubourg Saint-Jacques, 75014 Paris, France; and Department of Radiology, Hôpital Cochin, 27, rue du Faubourg Saint-Jacques, 75014 Paris, France

Sophie Goettmann, M.D. Department of Dermatology, Hôpital Bichat, Paris, France

Eckart Haneke, M.D., Ph.D. Professor, Institute of Dermatology, Klinikk Bunæs, Løkkeåsveien 3, Sandvika, Norway, N-1337

Edward A. Krull, M.D. Chairman Emeritus, Department of Dermatology, Henry Ford Hospital, 2799 West Grand Boulevard, Detroit, Michigan 48202

Andrew M. Morgan, M.D. (Colonel, USAF, Retired) Former Assistant Chairman, Department of Dermatopathology, Armed Forces Institute of Pathology (AFIP), Washington, DC 20306; and Dermatologist, Department of Medicine, Williamsburg Community Hospital, Williamsburg, Virginia 23185

Timothy Rosio, M.D. Department of Dermatology, University of California at Davis Medical School 4860 Y Street, Suite 3400, Sacramento, California 98517; and Fulton Skin Institute of Newport-Beach, 1617 Westckuff Drive, Suite 100, Newport Beach, California 92660

Elvin G. Zook, M.D. Professor and Chairman, Division of Plastic Surgery, Southern Illinois University, 800 North Rutledge Street, PO Box 3926, Springfield, Illinois 62708

PREFACE

Interest in diseases of the nails is increasing, as reflected in the growing number of publications on nails. Part of this may be related to the new, more successful systemic antifungal treatments for onychomycosis. Besides KOH (potassium hydroxide) preparations and fungal culture methods, treatment with antifungals alone has in some instances become a standard to determine whether the disease is caused by a fungus. Failure to respond to antifungal agents has led physicians to look to other possible etiologies. Further, because of the promotion of antifungal agents directly to the public, more patients are seeking treatment for nail diseases. All of these factors have contributed to a heightened interest in nail surgery in both sectors.

Interest in nail surgery involves many physician groups, especially dermatologists, plastic surgeons, orthopedists, hand surgeons who may be subspecialists of plastic or orthopedic surgery, podiatrists, and others who deal with common problems such as ingrown toenails. Nail surgery became part of the official dermatology residency special requirements in 1990, which has led to more training in nail surgery for dermatology residents. Investigations by Zaias into the etiology and pathologic findings in nail disease were a stimulus to many of us in expanding our knowledge and skills in this area.

Textbooks on nail diseases have traditionally included information on nail surgery, while textbooks on dermatologic surgery and dermatology also usually cover some aspects of nail surgery. However, the editors felt there was a need for a different type of book which emphasizes the specific techniques of nail surgery, and our goal was to create a detailed yet practical surgical atlas-textbook that would be a cooperative effort and share knowledge between different specialists.

This book is intended to be used as a working manual or atlas as well as a textbook. It is not meant to be a complete review of all nail surgical procedures and nail conditions. There was a purposeful selection of the procedures that were included. It builds on a strong foundation of anatomy and basic nail surgery, progressing to more advanced techniques. Many specific surgical details are illustrated, especially in basic nail surgery, that will assist the surgeon in acquiring the necessary skills.

There is a close biologic-anatomic functional relationship in nail diseases, planning surgery, and predictable surgical outcome. The appearance of the nail plate and surrounding tissues offer significant evidence to the site and nature of the pathologic process. A far better understanding of these relationships can be gained from the extraordinary three dimensional illustrations of the nail tissue in this book. The in-depth section on imaging of tumors of the nail unit adds to this understanding.

The text to a great extent reflects each author's selected and preferred surgical decisions and methods. But other surgical approaches are described, and comparison with other methods and rationale for the selection of the surgical approach are discussed.

The organization and format of the book are designed to benefit not only those beginning to develop interest and experience in nail surgery, but also those who have advanced knowledge and skills. We hope that whatever your level or expertise, you will find what you are looking for in these pages.

We are deeply indebted to our contributing authors: Jacques Bittoun, Rodney P. R. Dawber, Jean-Luc Drapé, Sophie Goettmann, Andrew Morgan, and Timothy Rosio. They have added invaluable information to this work.

Edward A. Krull
Elvin G. Zook
Robert Baran
Eckart Haneke

NAIL SURGERY

A TEXT AND ATLAS

1

ANATOMY OF THE NAIL UNIT IN RELATION TO THE DISTAL DIGIT

ANDREW M. MORGAN
ROBERT BARAN
ECKART HANEKE

Optimal nail surgery begins with a complete knowledge of the anatomy and physiology of the nail unit. The structural and functional features of the nail unit are unique, and the skillful specialist must understand them thoroughly. Moreover, the surgeon needs to appreciate the nail unit in relation to the whole anatomy of the distal digit because of concern for blood supply, innervation, and precise relationships to supportive structures. It is important to understand how the basic structural components of the distal digit interrelate, because abnormality of one of these structures has a profound impact on the other components.

Although this chapter discusses each component individually, correlation is the key to intimate familiarity. To this end, the nail unit and distal digit are methodically and graphically reconstructed, component by component, to give a composite view of their interrelationships. When examined in this systematic manner, the exquisite design of the distal digit and interrelationships of its components become evident.

Through a close look at its structure, the extraordinary utility of this anatomic unit, which so effectively serves each of us, can be more fully appreciated. Because it is seemingly mundane, often the distal digit receives only cursory and simplistic attention in medical textbooks. However, it is a complex structure that is truly vital to daily life and civilized existence.

The six basic components of the distal digit to be discussed are as follows:

1. *Skeletal* (distal phalanx and joint)
2. *Fibrous framework* (ligaments, tendons, and deep connective tissue)
3. *Nail unit* (matrix, nail plate, sheaths, nail bed, epithelial folds, and grooves)
4. *Blood vessels* (with their special architectural arrangements)
5. *Nerves* (especially sensory nerves and receptors)
6. *Tip and pad* (or "pulp," comprised of epidermis, dermis, subcutis, and eccrine glands, in addition to the special neural and vascular components).

The opinions or assertions contained herein are the private views of the author and are not to be construed as official or as reflecting the views of the United States Air Force or the Department of Defense.

A. M. Morgan: Formerly at the Department of Dermatopathology, Armed Forces Institute of Pathology, Washington, D.C. 20306-6000. *Current:* Private practice (Dermatology and Dermatopathology, Williamsburg, Virginia).
R. Baran: Nail Disease Centre, 42, Rue des Serbes, 06400 Cannes, France.
E. Haneke: Klinikk Bunæs, Løkkeåsveien 3, 1337 Sandvika, Norway.

Brief histologic correlations are provided for the surgeon's complete understanding of the anatomy and for use by the surgical pathologist who is asked to evaluate the specimens resulting from nail surgery. Finally, the internal structure is correlated with topography to allow the specialist to fully appreciate and mentally visualize the interrelating components within the patient.

Additional general information on the nail unit and distal digit may be obtained from the following sources: standard anatomy and histology textbooks (1–3), several excellent general reviews on nail unit anatomy (4–8), and classic papers on the subject (9–11).

SKELETAL

The foundation of the nail unit, and thus the starting point of this discussion, is the *distal phalanx*. Each phalanx (Fig. 1.1) has a head, tapered shaft, and a broadened proximal base. Its dorsal surface is convex on cross section, whereas its volar surface is transversely nearly flat.

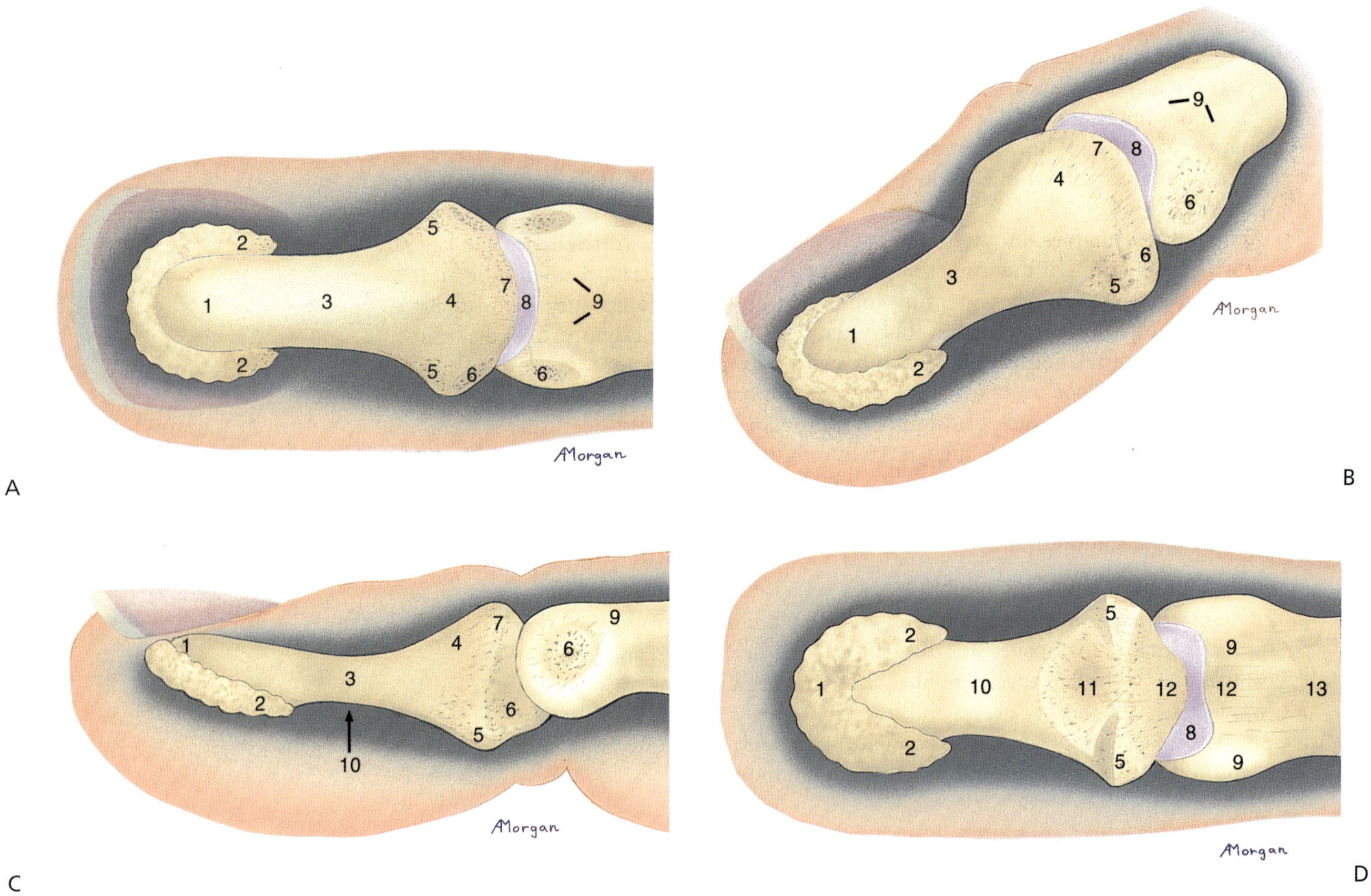

FIGURE 1.1. Distal phalanx and distal interphalangeal joint. **A:** Dorsal view. **B:** Dorsal oblique view. **C:** Lateral view. **D:** Ventral view. 1, ungual process; 2, spines of ungual processes; 3, waist; 4, base of the distal phalanx; 5, lateral tuberosities for the proximal attachment of the lateral interosseous ligaments and fibers from the collateral ligaments and extensor tendon; 6, fossae for attachment of collateral ligaments; 7, dorsal ridge for attachment of the extensor tendon; 8, articular cartilage; 9, bicondylar head of the middle phalanx; 10, ungual fossa; 11, area of attachment of flexor (profundus) tendon; 12, area of attachment of volar plate which blends into the dorsal wall of the fibrous flexor sheath; 13, area of attachment of dorsal wall of the fibrous flexor sheath.

The distal phalanges also are known as *terminal* or *ungual phalanges* (Latin *ungula* hoof or claw; thus, pertaining to the nail). They are slightly concave ventrally along the long axis forming the ungual fossa. The distal phalanges of the toes generally resemble those of the hand, but they are shorter, their shafts flatter, bases broader, and distal ends more expanded.

The *distal interphalangeal joint* (DIPJ) (12,13) is a hinged (or "ginglymus" in anatomic terminology) synovial joint. The head of the middle phalanx shows a bicondylar articular surface separated by a median groove to form a trochlear, or pulley, shape. The base of the distal phalanx shows an articular surface with two shallow concavities separated by an intervening ridge to match the middle phalangeal head. The base of the distal phalanx has prominent lateral tubercles for the attachment of ligaments.

The terminal portion of the distal phalanx exhibits a flattened, horseshoe-shaped tuberosity called the *ungual process*. The distal portion of the nail bed attaches to it dorsally. Its curved distal edge and, even more so, its ventral aspect are roughened where dense connective tissue of the tip of the digit anchors to it. Prominent bilateral *spines* serve to anchor lateral interosseous ligaments (LILs). These lateral ligaments form bilateral protected passages, along with waisting of the terminal phalanx, for neurovascular supply to the nail bed.

The distal phalanx differs from other phalanges in its method and timing of ossification. A cap of bone, which becomes the ungual process, is laid down in membrane as the initial center of ossification for the digit. The base of the distal phalanx shows the more usual enchondral ossification at a later time.

Clinicopathologic Correlations: Relationship of Bone to Nail Tissue

The terminal phalanx largely defines the gross shape of the nail. "One cannot conceive of a normal nail over an abnormal bony phalanx" (14). A broad terminal phalanx will cause a wide nail, as is seen in racket nails, and a narrow phalanx develops a narrow nail. Shortening of the terminal phalanx may result in a hook nail, a wide or broad nail, or a pseudoracket nail. Racket nails are due to premature epiphyseal ossification of the terminal phalanx that prevents longitudinal growth but allows thickening of the base of the terminal phalanx. A variety of bone anomalies with resultant nail changes characterize certain syndromes.

Acquired bone changes also cause corresponding nail alterations. Fractures of the terminal phalanx may result in nail dystrophy, depending on the degree of dislocation of the bone fragments. Distal osteolysis in hyperparathyroidism results in acquired racket nails. Lateral subperiosteal appositions produce macronychia. Degenerative osteoarthritis and Heberden's nodes manifest as transverse overcurvature of fingernails.

Pincer toenails result from osteophytes formed laterally on the base of the terminal phalanx (15). Because the normal matrix curves in a semicircle around the base of the phalanx, the once evenly curved nail becomes unrolled or flattened proximally and overcurved distally. The distal nail plate portion heaps up from the nail bed. This action results in a small traction osteophyte on the tip of the terminal phalanx.

FIBROUS FRAMEWORK

A remarkable fibrous framework gives the distal digit its critical structural support. It includes the tendons, ligaments, joint capsule, fibrous flexor sheath, fasciae, and retinaculae (Fig. 1.2) (16–25). The fibrous framework provides for stable motion and strength of the DIPJ, attachment of the nail unit, resistance to trauma, and facilitation of prehension. In addition to ligaments and tendons involved with the DIPJ, strong ligamentous and fascial support firmly anchors the nail bed and skin to the underlying bone. All these structures interrelate and work together to give stability to the joint and the entire distal digit.

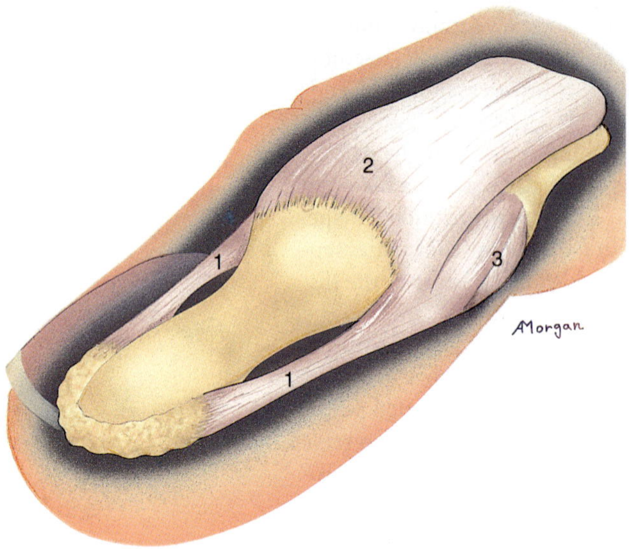

A

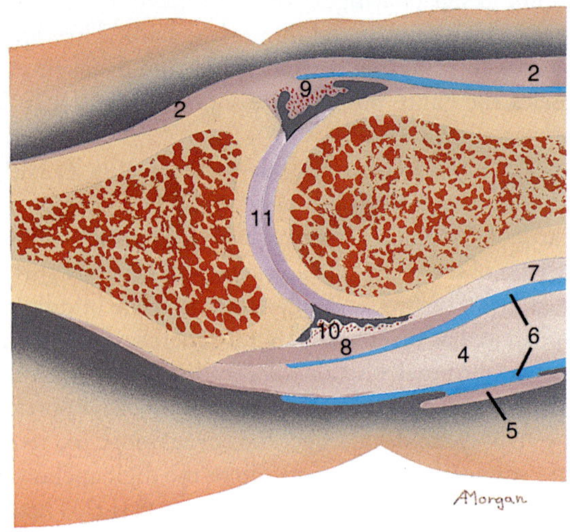

B

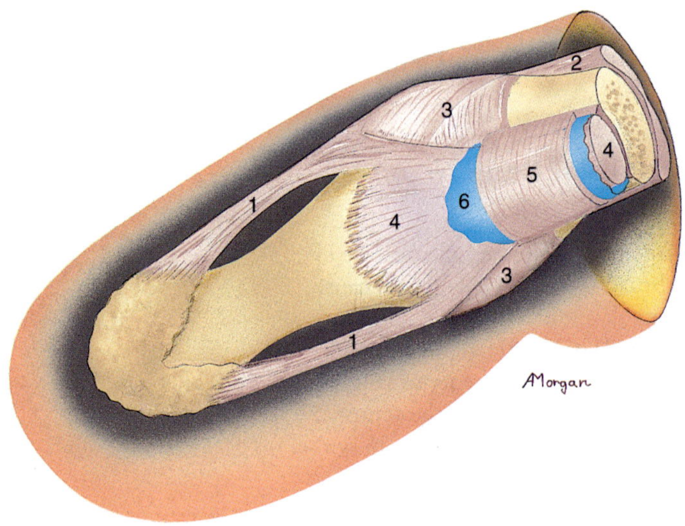

C

FIGURE 1.2. Ligaments and tendons of distal digit relating to the distal interphalangeal joint (DIPJ) and anchoring of the nail unit. **A:** Dorsal oblique view. **B:** Sagittal section through the DIPJ. **C:** Ventral oblique view. 1, lateral interosseous ligament; 2, extensor tendon; 3, collateral ligaments; 4, flexor tendon; 5, fibrous flexor sheath with annular pulley; 6, synovium of flexor sheath; 7, dorsal wall of the fibrous flexor sheath, attached to middle phalanx; 8, volar plate joins fibrous flexor sheath proximally and the flexor tendon distally; 9, dorsal capsule and synovium of DIPJ blending with extensor tendon; 10, ventral capsule and synovium of DIPJ blending with the volar plate; 11, articular cartilage joining the base of the distal phalanx on the left and the head of the middle phalanx on the right.

The connective tissue of the nail bed and pulp are discussed elsewhere in this chapter. However, one needs to be aware of their structural interrelationship to some the structures discussed here. Specifically, the reader should note that the connective tissue supporting and anchoring the nail matrix relates largely to the collateral and, to some extent, the LILs at the lateral tubercles.

Lateral interosseous ligaments ("paraterminal ligaments") extend from the lateral tuberosities at the base of the distal phalanx to the spines of the ungual process (16,24). These ligaments provide support for the posterolateral nail bed fibroconnective tissue, lateral strength to the digit, and a protected neurovascular conduit. The fibers of the ligament are continuous with the collateral ligament and attachments of the flexor and extensor tendons.

The *extensor tendon* attaches to the prominent lip of the dorsal base of the distal phalanx, flares to the lateral tuberosities, and blends with the LILs. The tendon attaches with the dorsal aspect of the *joint capsule* onto the distal phalanx, but is otherwise separated from the capsule by a *dorsal bursa*.

Ventrally, the *joint synovium* fuses with a thickened fibrocartilaginous structure known as the *volar plate* to form the *articular capsule*. The volar plate is roughly square and thickest distally, where it inserts across the width of the volar base of the distal phalanx. The volar plate also is referred to as the palmar or plantar ligament.

Thick *collateral ligaments* extend from pitlike areas on the lateral aspect of the condyles of the middle phalangeal head to the posterior aspect of the lateral tuberosities of the distal phalanx. Running bilaterally from the middle phalangeal head to the sides of the volar plate are *accessory collateral ligaments*. These ligaments also serve as lateral joint capsule.

In addition, the collateral ligaments send *dorsal expansions* that blend into the proximal nail bed connective tissue and serve to anchor the proximal matrix laterally (25). In this manner, a "bone-nail" ligament forms to attach the nail root to the lateral tuberosities. As a consequence of this relationship, deformity of the bone or joint results in nail dystrophy, by way of effects on the ligament and thereby the matrix.

The dorsal expansion of the collateral ligament and the LIL serve as special distal expressions of Cleland's ligament. Cleland's ligament extends from the sides of the phalanges to insert in the skin, thus forming a "lateral digital band" of skin retinaculae (12,16,25).

On the ventral aspect of the joint, the volar plate blends with the dorsal wall of the *fibrous flexor (tendon) sheath*. The dorsal wall of the fibrous flexor sheath is thick and firmly attached to the volar surfaces of the proximal and middle phalanges. The strong attachment of the fibrous flexor sheath to the phalanges provides a strong conduit, known as the "osseo-aponeurotic canal," for the flexor tendons.

The *flexor profundus tendon* inserts broadly along the volar aspect of the base of the distal digit and blends with the LIL at the lateral tuberosities. A system of *annular* and *cruciate retinaculae* known as "pulleys" keep the flexor tendon from "bow stringing." These retinaculae, or pulleys, compose the ventral aspect of the fibrous flexor tendon sheath. The inner lining of the sheath is a double-walled (parieto-visceral) synovial membrane.

NAIL UNIT

The nail unit, or nail organ, consists of several interdependent structural components (Figs. 1.3–1.5) (3–11,26). Because the structural and functional boundaries of these interrelated components can overlap at certain junctures, they are best presented in terms of their primary roles (5). For purposes of discussion and conceptual convenience, the nail unit is divided here in the current conventional manner. Thus, the nail unit is composed of the *generative* portion (matrix and nail bed epithelium [NBE]), the *product* (nail plate), its *framing* (nail folds and grooves), its *sheathing* (cuticles; eponychium and hyponychium), and their *supportive* tissues (nail unit fibrocollagen). The framing and sheathing of the nail are referred to as *paronychium*. Even more inclusive, all the components forming, relating to, and surrounding the nail plate are known collectively as the *perionychium*. However, the use of these terms adds nothing to referring to the collection of components simply as "nail unit."

Generative (Matrix and Nail Bed Epithelium)

The germinative matrix is the nail-forming epithelial structure (Figs. 1.3, 1.6, and 1.7). Terminology can be confusing depending on one's viewpoint as to whether the matrix has two or three portions. Here, it is viewed as being composed of three portions: dorsal, intermediate, and ventral. Their putative contributions are discussed. The *intermediate* (or *conventional*) matrix is the primary keratogenous zone of the nail unit. The term "matrix" generally refers to this portion. It is the principal source of nail plate and histologically distinctive. The conventional matrix has a well-demarcated, distally convex juncture with the nail bed.

The lunula is the visible distal portion of the conventional matrix. It appears as a pale blue-gray half-moon or fusiform structure emerging from under the proximal nail fold

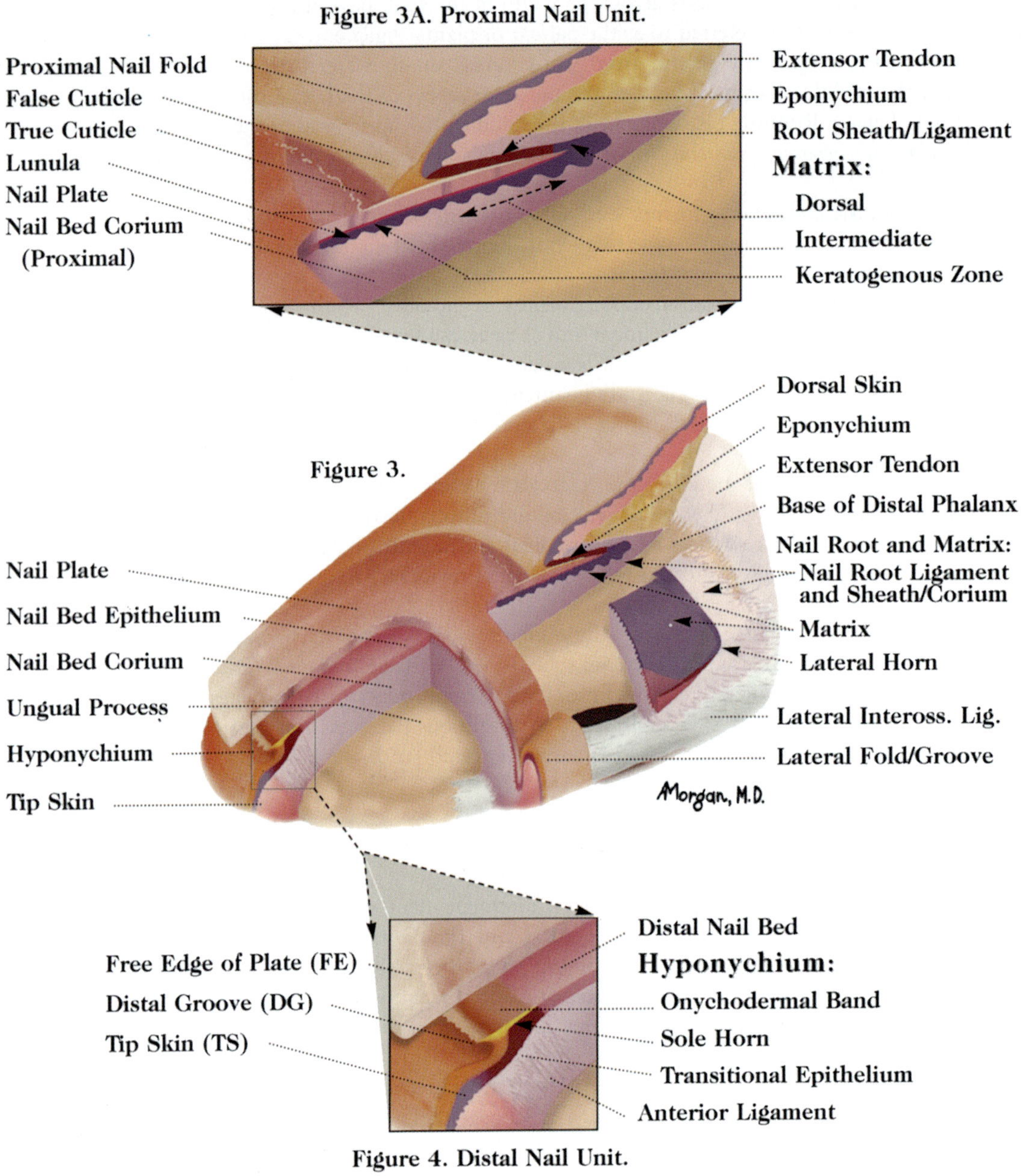

Nail Unit

FIGURE 1.3. Illustration of sagittal section of the matrix area and proximal nail fold showing eponychial false cuticle, nail vest or true cuticle, eponychial epithelium, dorsal matrix, conventional or intermediate matrix, nail bed epithelium or ventral matrix, nail bed fibrocollagen or corium, nail plate, and lunula.

FIGURE 1.4. Illustration of sagittal section of the hyponychial area showing onychodermal band, distal groove, nail bed, nail bed fibrocollagen, skin of the tip, nail plate, and area of anterior ligament connecting the periosteum of the ungual process to the hyponychial area.

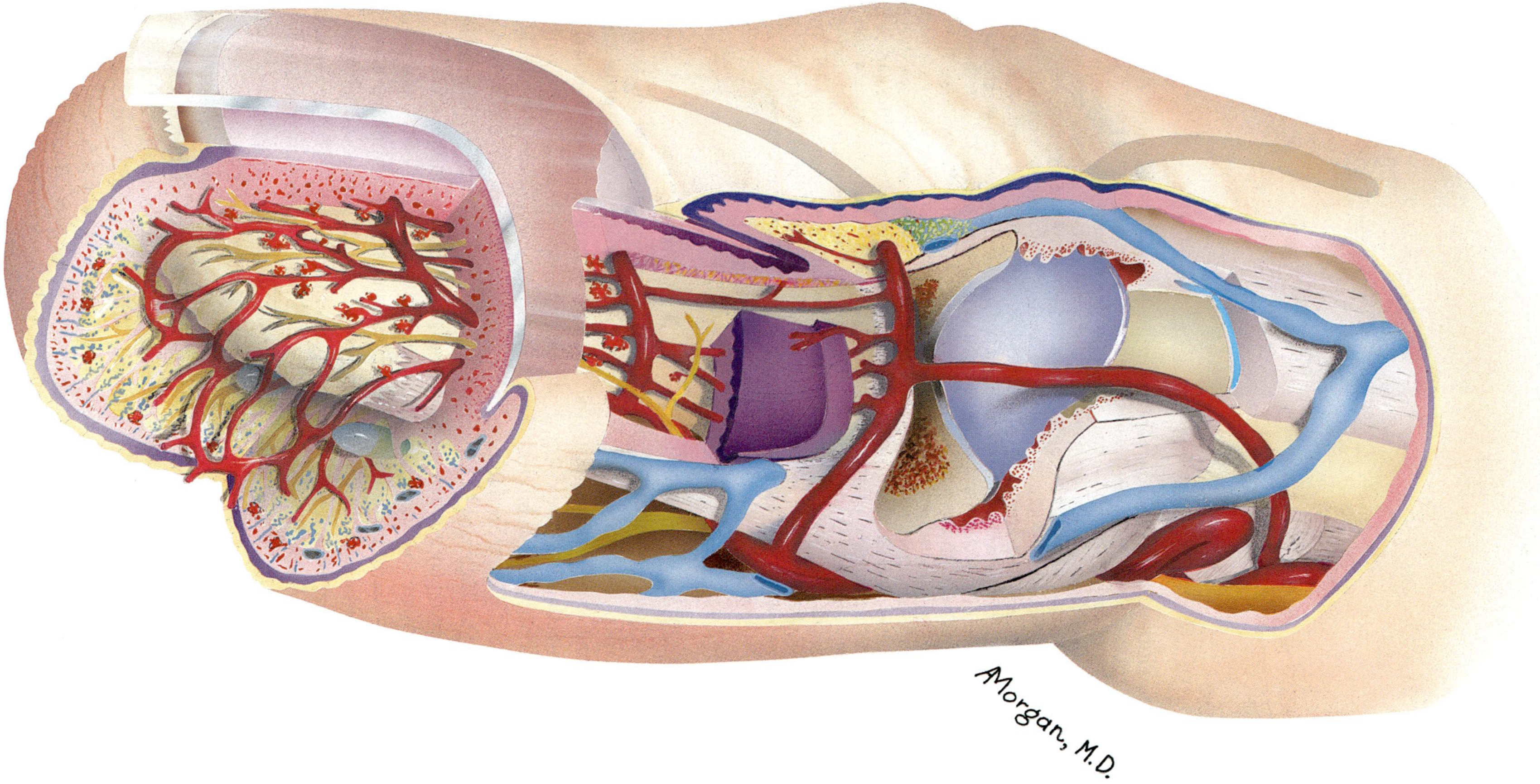

FIGURE 1.5. Overall composite illustration of the distal digit showing the nail unit in relation to its surrounding foundational structures, vascular supply, nerves, and skin. Specific structures can be identified by consulting other accompanying illustrations detailing the components.

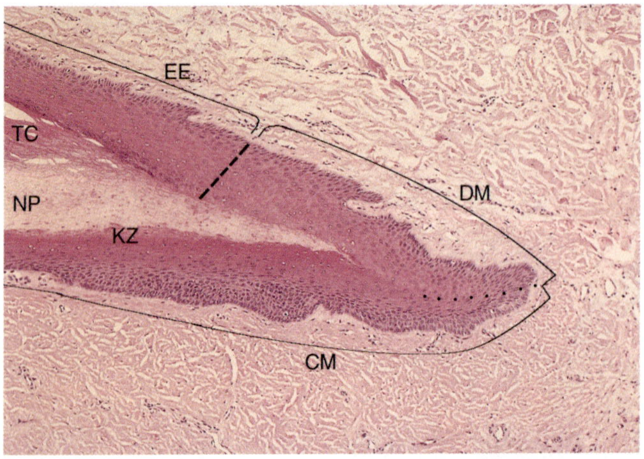

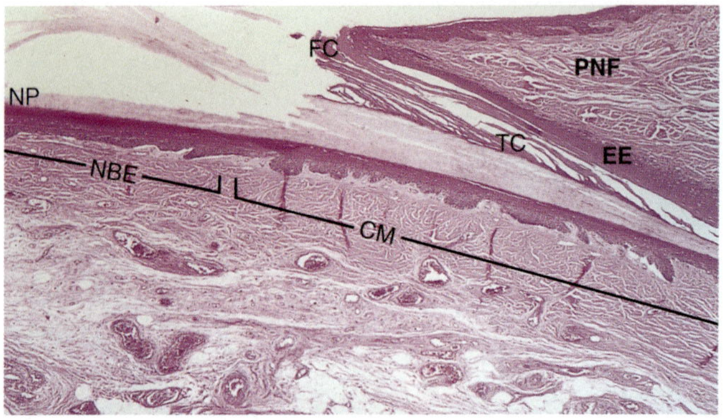

FIGURE 1.6. Photomicrographs of matrix in longitudinal section. **A:** Proximal matrix composed of the dorsal matrix (DM) plus conventional matrix (CM) (also known as the intermediate matrix) producing the proximal nail plate (NP). The conventional germinative matrix beneath the NP shows rounded rete or mamelons that slant distally and are firmly attached to the corium. Basophilic, basaloid cells differentiate to form flattened, eosinophilic cells in the prominent keratogenous zone (KZ). The cells lose most of their nuclear material and become onychocytes. There is no keratohyaline granular layer. Superiorly, the epithelium of the eponychium (EE), which can be differentiated from the DM by the presence of a keratohyaline granular cell layer, produces the thin, hard, adherent nail vest or true cuticle (TC). **B:** Slightly more distal area of matrix beneath the proximal nail plate and proximal nail fold (PNF). The matrix (CM) shows its characteristic mamelons with basaloid proliferation and prominent eosinophilic KZ. Distally, the matrix transitions to nail bed epithelium (NBE). The dorsal skin shows its typical distally slanting papillae. The epidermis and, to a small extent, the ventral epithelium of the PNF produce the softer epidermoid keratin of the false cuticle (FC). A more glassy, eosinophilic true cuticle (TC) is firmly adherent to the dorsal NP. The eponychial epithelium (EE) producing this sheath does not show its usual granular cell layer due to decalcification of the specimen. The nail plate is artifactually fractured. The features of the proximal nail unit fibrocollagenous support system and the loose vascular and adipose tissue, detailed in the text, can be seen.

(PNF). Its color, actually the color of the whole conventional matrix, probably is due largely to the effect of light scattered by nucleated cells of the keratogenous zone of the matrix. Observation of some opacity in the avulsed nail plate supports this opinion. Other factors may explain the different appearance of the (lunular) matrix. The thicker epithelium and the distinctly different patterns of rete in the matrix versus ridges in the nail bed are possibilities. These microscopic features result in the different gross appearance of these areas. Deeper connective tissue and vascular differences, although real, probably are irrelevant in explaining the blue-gray color of the lunula. These differences do, however, play a role in making clearly visible the postlunular vascular flush that serves to accentuate the lunular margin. Compressing the nail bed, by pressing on the nail plate, greatly decreases the color distinction, but does not completely eliminate it.

Proximally, the intermediate (conventional) matrix connects with *dorsal matrix* that is the deepest and most proximal part of the ventral PNF. The smaller dorsal matrix produces the superficial portion of the nail plate, whereas the intermediate matrix produces the bulk of the nail plate. Together the dorsal and intermediate matrices are known as "proximal matrix" or simply as "germinal matrix."

More controversial is the *"distal matrix"* or *ventral matrix* (27,28), which is synonymous with the *NBE,* or at least the most proximal portion of the NBE. The NBE (often simply called "nail bed" if one disregards the nail bed corium [NBC]) sometimes is called the "sterile matrix." It is traditionally thought that NBE adds only a minor amount to the nail plate, just enough to allow it to slide along the nail bed as it grows. However, this view is questioned, and the controversy is discussed in the "Product" section. In essence, there appears to be a gradient of growth along the NBE that is matrix-like proximally but essentially non-proliferative distally.

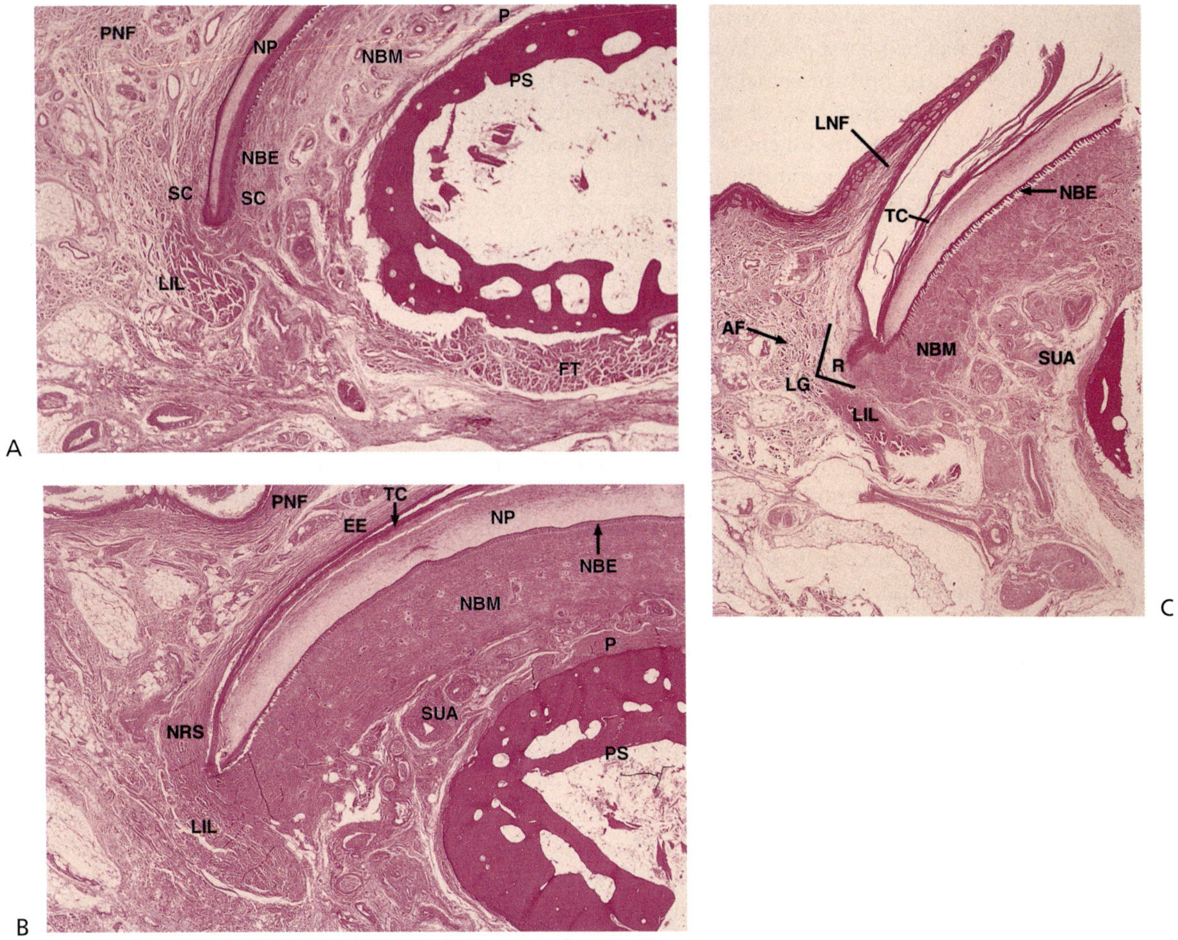

FIGURE 1.7. Photomicrographs of the proximal nail unit in transverse section. **A:** Cross-sectional photomicrograph of a lateral aspect of the proximal nail unit at the level of the distal matrix. The phalangeal forms the shaft bone (PS). It is oval with a large cancellous cavity, indicating it is near the base, as does the presence of the flexor tendon (FT). Proximal nail fold (PNF) overlies the germinal root with its shallow ridges and relatively thin keratogenous zone consistent with distal lunular area. Laterally, the matrix transitions to the nail bed epithelium (NBE). The nail root is largely surrounded by sheathlike collagen fibers (nail root sheath, NRS) which anchor to the longitudinal fibers of the lateral interosseous ligament (LIL). Nail bed connective tissue is relatively delicate and loose (nail bed mesenchyme, NBM), especially deeper where larger vessels are seen. At this point the periosteum (P) is still distinct. **B:** Cross-section, slightly more distal than **A**, but still largely covered by PNF and at the waist of the phalanx. The nail unit fibrocollagenous support system is changing from a sheath of collagen to a thicker, denser intermediate one or nail bed corium, which is attached to the lateral interosseous ligament (LIL) and periosteum medially. At this level, the proximal subungual arcade of arteries (SUA) comes up from the volar cruciate anastomosis. Glomus bodies are not yet prominent. The phalangeal shaft (PS) is narrowing. **C:** The lateral nail fold (LNF), shown in cross section, approaches the proximal fold as evidenced by the presence of false (FC) and true cuticles (TC), yet it is just beyond the lunula because shallow nail bed epithelium (NBE) ridges are present. They show characteristic radiation toward the bone. Thus, we can guess that this is probably a more medial finger or a lesser toe. Note that the lateral interosseous ligament (LIL) at this level snugly supports the lateral nail bed fibrocollagen (nail bed mesenchyme, NBM) along with the root (R) of the lateral groove (LG) showing apical epithelium, which curiously resembles both the matrix and eponychial area. It is indeed an area of epithelial transition, as discussed in the text. Note how the vessels of the subungual arcade are routing under and up from the area of the LIL to enter the nail bed. This, along with the lateral winglike flaring of the NBM over the looser vascularized tissue, locates this section at the waist of the phalanx, proximal to the ungual process. Note also the delicate oblique ligamentous anchoring fibers (AF) of the dermis of the LNF to the LIL and the distinct layering of the nail plate (NP) related to production of the dorsal and conventional matrices.

The *nail bed* extends from the distal lunula to the onychodermal band (ODB). The nail bed has a pink flush in healthy persons, due to vessels in the NBC, which is accentuated both proximally and distally. The intervening area of pallor corresponds to the intermediate zone of nail bed fibrocollagen with its dense, vertically-oriented fibers and its strong attachment to the ungual process.

It is important to recognize that the NBE is "completely derived from the matrix epithelium and is not a self regenerating and independent structure" (5). Also, the NBE moves distally along with the nail plate. Nail bed epithelium shows a low rate of proliferation in contrast to matrix cells, and keratinization of the NBE is far less active than the matrix. A thin parakeratotic keratinizing layer is produced and is moved distally along with the growing nail plate. The NBE and its thin, semihard keratin layer concomitantly provide for strong adherence and distal outward growth of the nail plate, in a manner that is not well understood.

Because the NBE is firmly bound to the nail plate, nail avulsion causes laceration of the epithelium from the deeper connective tissue, which results in bleeding of the nail bed. This is not seen in the proximal matrix, where only the superficial layer of the matrix epithelium remains attached to the avulsed nail plate.

In cross section, the matrix curves around the terminal phalanx (Fig. 1.7) and nearly forms a semicircle in the hallux. The proximal border of the matrix is convex distally. It bows toward the tip of the digit as seen in the shape of the lunula and ODB, although the amount varies. It is particularly pronounced in the great toe, giving rise to prominent *lateral matrix horns,* or extensions of matrix, which are located at a deeper level than those of the fingernails. The surgeon needs to appreciate some degree of proximo-lateral protrusion of the matrix, or "horns," beyond the plate to avoid isolation of persistent germinal tissue that could produce residual nail spicules (25). As will be discussed further in the "Supportive" section, dense, parallel (that is, ligamentous) connective tissue fibers hold the proximal matrix and lateral horns in place. These fibers are continuous with the dorsal expansions of the collateral ligaments (25), the proximal portion of the LILs (16), and periosteum of the dorsum of the phalanx (29)

The distance between the most proximal portion of the matrix and the insertion of the extensor tendon is small (0.8 mm; E. Haneke, *personal data*, 1995) (Figs. 1.2, 1.3, and 1.5). Ligamentous fibers from the nail root sheath blend into the periosteum near the distal aspect of the extensor tendon insertion. In addition, the DIPJ is nearby. The surgeon needs to consider these close proximities and relationships when performing diagnostic matrix biopsies and other procedures.

The newly forming nail plate, situated under the PNF, is very soft, thin, and easily cut and punched. Because it is not firmly attached to the matrix epithelium, it rapidly shears off when one cuts or punches the matrix through the nail plate.

Histology

Although germinative matrix and NBE both lack a granular cell layer, the histology of matrix epithelium is distinct from NBE. Conventional matrix epithelium is relatively thick and interdigitates in budlike projections with the nail bed fibrocollagen (Fig. 1.6). On longitudinal sections the basaloid buds, called mamelons, are more prominent proximally and incline toward the tip of the digit. This orientation is a major factor in the direction of growth of cells that make up the nail plate. Laterally, the mamelons are less prominent. The connective tissue of the matrix region is described in the section on nail unit fibrocollagenous support ("Supportive").

The nail plate forms by *onychokeratinization,* which is characterized by a prominent keratogenous zone without formation of a granular layer (Fig. 1.6) and coexpression of both soft and hard keratin. As they grow outward from the matrix epithelium, the maturing cells lose their nuclei and organelles. They flatten, cornify, and arrange themselves into compact sheets of onychocytes to form the nail plate. Abruptly cornifying eosinophilic squamous cells with cytoplasm that condenses and hardens produce the plate.

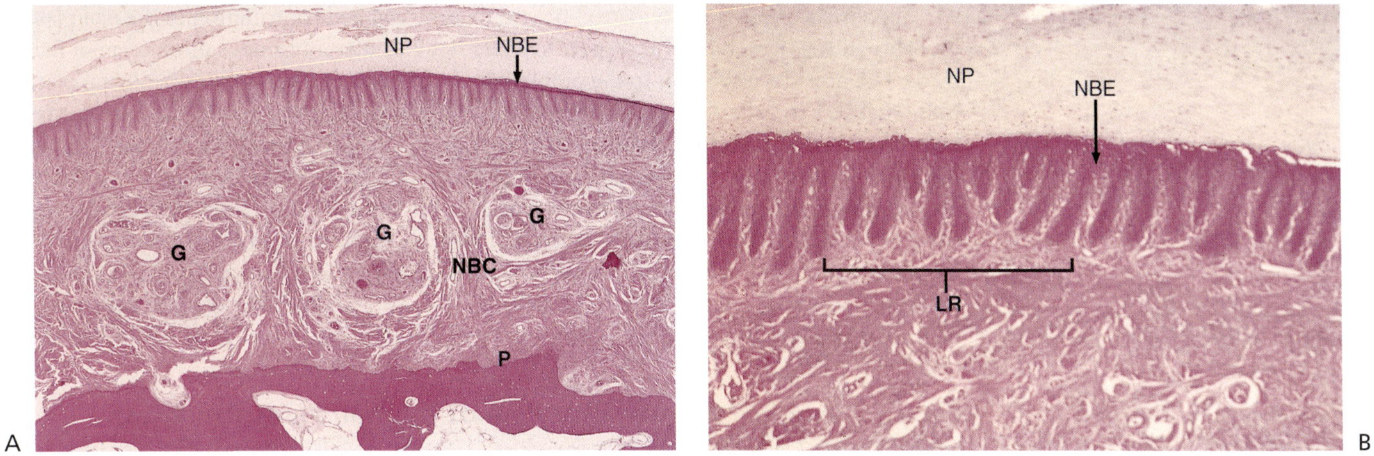

FIGURE 1.8. Cross-sectional photomicrographs at the middle level of the nail bed. The elongated longitudinal ridges (LR) of the thin nail bed epithelium (NBE) interdigitate in their characteristic manner with the nail bed fibrocollagen, which contains thick, coarse collagen bundles and prominent vascularity. **A:** Glomus bodies (G) with their cellular, thick-walled shunts, tortuous and ectatic collecting venous channels, and separate investment by a more delicate stroma. Note the vertically and longitudinally oriented, thick, ligamentous collagen of the NBC blending into the periosteum (P). **B:** Closer view of NBE, cut in cross section, showing LR with a thin, eosinophilic layer of keratinizing cells attached to the nail plate (NP). The ventral NP shows undulating or serrated interdigitation with the abruptly (onycholemmal) keratinizing NBE and shallow grooving of the plate.

Microscopically, NBE is unique in its regular, longitudinal, parallel epithelial ridges (not "rete," which show bridging) that interdigitate, tongue-and-groove fashion, with corresponding corium ridges (not "papillae," which are fingerlike projections associated with rete) when viewed on transverse section (Fig. 1.8). Similarly, but to a lesser degree, NBE shows serrated interdigitation with the ventral surface of the nail plate to form grossly visible longitudinal striations characteristic of the ventral surface of the nail plate. On longitudinal section, the NBE appears flat and thin, being as few as three to four cells thick. Significant proliferative activity and mitoses are not observed. Laterally, NBE transitions into epithelum showing prominent rete ridges in association with lateral nail folds (NLFs) where the lateral edges of the nail plate are anchored. This forms a sort of "lateral nail root" that can be seen on Figure 1.7.

Because matrix and NBE do not normally show a granular cell layer, its appearance in either of these locations is pathologic. Clinically, whenever a granular layer develops, due to trauma or nail disease, the nail plate structure is disturbed and could appear as lamellar horizontal splitting of the plate.

Product (Nail Plate)

The product, indeed purpose, of the nail unit is the *nail plate* that serves as armor for the distal digit and as an extraordinary self-replenishing tool. The nail plate is a durable, partially translucent, keratinous structure that gently curves in longitudinal and transverse directions.

The nail plate shows three distinct layers histologically: a thin dorsal layer that gives it a smooth surface, a thick intermediate layer, and a ventral layer with an irregular surface featuring longitudinal striations. The trilaminate nail plate is predominantly the product of the dorsal and intermediate germinative matrices, and its layering, to some extent, reflects varying degrees of maturation of onychocytes. However, some observers believe that the postlunular NBE produces a significant portion (20%) of the plate, known as the ventral nail plate (9,10,27,28). In this view, the nail plate forms in a gradient fashion along both the matrix and nail bed, and it gradually increases in thickness as it grows outward. Hence, the term "sterile matrix" for the nail bed is inappropriate.

Supporters of the "ventral matrix" believe that the nail bed contribution to the nail plate probably adds to flattening of nail growth. More significantly, continuous production of nail by the bed provides a solution to the problem of attachment of a continuously moving plate (28). Nail bed contribution to the nail plate also helps explain why subungual hematomas become smaller in their longitudinal axis as they grow out, being incorporated into the nail plate. Also, koilonychia may result from more active proliferation of the distal matrix portion.

However, nail growth is still controversial, and some authors hold that NBE contributes nothing or only minimally to the nail plate (8,26,30–34). In this more traditional viewpoint, substance added to the ventral nail plate as it moves forward is not nail plate tissue but onycholemmal material. As such, it serves in the processes of plate adherence, migration, sealing, and eventual separation, rather than primary nail plate structure.

Clinicopathologic Correlations: Role of the Nail Plate, Thickness, Growth, and Shape

The nail plate generally is regarded as "the nail." It is important for the protection of the finger and toe tips. It is a powerful tool for defense and attack. It markedly increases tactile and sensory functions of the fingers by providing counterpressure to the pulp. The nail aids considerably in manual dexterity, helps in grasping of fine objects, and permits scratching and scraping. Furthermore, it is of utmost cosmetic importance. The nail plate provides a protective covering for the wounded nail bed, enables accurate repositioning of damaged subungual tissue, and acts as a splint for fractures of the phalangeal tuft. Conversely, a smooth, even nail bed is crucial for the regenerating nail plate.

Nail plate thickness, which is normally about 0.5 mm in fingers and 1.0 mm in toes of adults, depends on the length of the matrix, the speed of growth of the nail layers, and the size of the onychocytes (35). The superficial portion of the nail plate grows faster than the deeper portion because the proximal matrix proliferates more actively than the distal one (33).

There are few studies on nail growth kinetics. Flow cytometric studies of the lateral matrix horns of great toes from young people with ingrown toenails have shown that 5.4% of matrix epithelium cells are in S and G_2+M phases; this rate is double that of surrounding skin but far less than the proliferation of hair bulbs (E. Haneke, *personal data,* 1995).

In young adults, the fingernail grows about 0.1 mm per day. In toenails, linear nail growth is 30% to 50% that of the fingernail. Impaired arterial blood supply, smoking, cold, malnutrition, and cytostatic drugs decrease nail growth, whereas psoriasis is said to increase nail growth rate.

In addition to growth rate, another characteristic difference between fingernails and toenails is relative proportions. Fingernails are longer than wide. Toenails are rather wider than long, with the nails of both the thumb and the great toe being the widest.

The longitudinal and transverse curvatures of the nails, when equally pronounced, cause the nail to look like a watch glass. This is particularly evident in finger clubbing. The degree of convexibility of the matrix is defined by its firm attachment to the underlying base of the terminal phalanx. When the bone's shape changes, the nail plate's transverse curvature will change correspondingly, as is seen characteristically in pincer nails. Normally, fingernails are less convex than toenails.

Framing (Folds and Grooves)

Nail folds surround the normal nail plate on three sides (Figs. 1.3, 1.5, and 1.7C). The two lateral nail folds merge with the posterior nail fold (synonym: dorsal nail fold), which covers most of the germinal matrix along with the newly forming nail plate. In this way, a pocket is formed, termed the *proximal nail groove,* for the matrix and nail root (the proximal 25% of the nail plate). The PNF is composed of dorsal and ventral portions. The dorsal portion is a continuation of the dorsal skin that has a thin dermis and subcutis and lacks pilosebaceous structures. The ventral portion has a thin, flattened epithelium and lacks adnexae. The proximal nail groove is an invagination that has a roof (the ventral

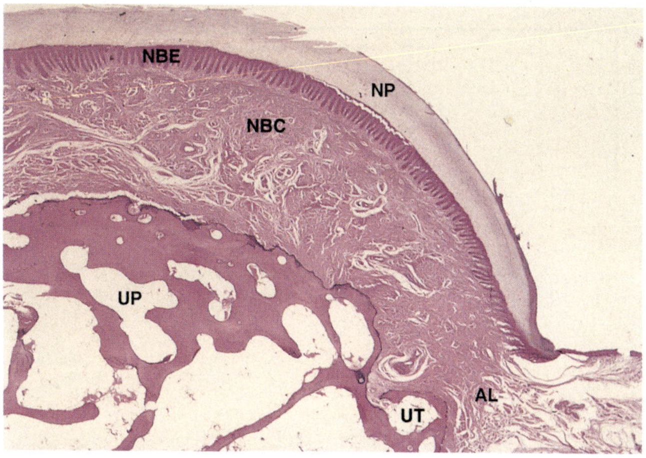

FIGURE 1.9. Distal nail unit in transverse section. Distally, the nail bed corium (NBC) is more broadly attached to the broader ungual process (UP) of phalanx, NBE ridges are more elongated, NBC is coarser with distinct vertical ligamentous collagen and glomus bodies, and LNF disappears as the NP becomes externalized. LIL is no longer present because the section is at the ungual tuberosity (UT). At this point, the corium blends into the LNF dermis forming the lateral aspect of the anterior ligament (AL).

aspect of the PNF) and a floor (the matrix). The PNF and groove form at an acute angle that nearly parallels the skin.

Lateral nail folds and grooves have features analogous to proximal fold and groove (Fig. 1.7C). The LNFs, which are rolls that flatten distally, can be envisioned as distal continuations of the PNF. Lateral folds are not very pronounced on the lesser toes, but they are well developed on the fingers and on the great toe. The latter may exhibit very marked LNFs in subjects prone to develop ingrowing toenails, in tall young persons with very large feet, and in pincer nails.

The lateral nail grooves are the indentation into which the nail plate is embedded. They are most marked where the nail folds are most pronounced. Lateral nail grooves sometimes are lacking in racket nails. The lateral nail grooves provide an abutment for the nail plate and exhibit a specialized connective tissue structure to meet these requirements. Hence, the lateral nail grooves should not be wantonly sacrificed for diagnostic biopsies or for treatment of ingrowing nails.

Under the free edge of the nail plate, the so-called *distal groove* is actually a band of skin that transitionally relates the hyponychium to the skin of the tip of the digit (Fig. 1.4 and 1.10A). Unlike its proximal and lateral counterparts, the distal groove does not normally envelop the nail plate, but potentially may do so when a distal nail fold forms.

At first, as traditionally viewed, these components seem to be separate structures. They are actually closely related sub-structures of the sheathing system of the nail unit. A distal nail fold may be seen in immature newborns and especially at the great toes after nail avulsion or after consistently cutting the nails too short. The distal nail fold in neonates always resolves spontaneously. A hypertrophic lateral fold, or lip, is another anomaly of neonatal great toes that usually remits spontaneously.

The nail folds and grooves provide support for the nail plate and cushioning against physical stress. They constitute a physical seal against the entry of infectious agents and foreign material, especially by way of the sheathing mechanism.

Sheathing (Cuticles, Eponychium, Hyponychium, and Onychodermal Band)

The distal free margin of the PNF forms the eponychial hyperkeratotic rim commonly known as the "cuticle," although it would be more proper to refer to it as a *"false cuticle."*

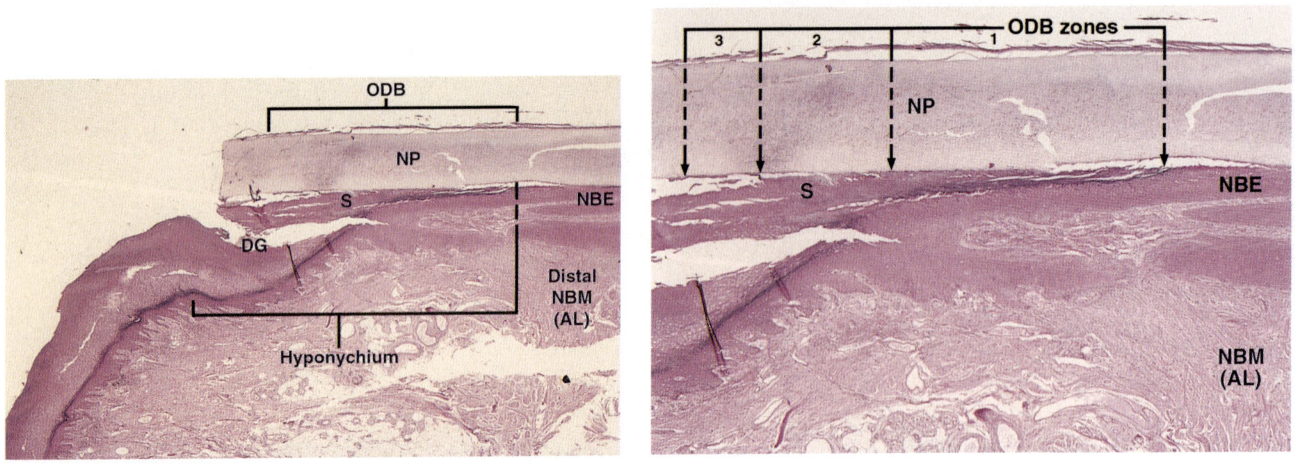

FIGURE 1.10. Photomicrographs of hyponychium in longitudinal section. **A:** Hyponychium with keratinization via a granular cell layer, thickening of epithelium, and a more bulbous and varied pattern of rete ridges compared to adjacent digital tip skin and NBE. In the portion under the attached nail plate (NP), the thicker epithelium appears as the dull pink, milky proximal part (zone 1) of the onychodermal band (ODB). The more distal epidermoid keratin of stratum corneum is seen more tangentially, almost on end, to form the darker distal rim of the onychodermal band (ODB) (zone 2) just before shedding of the soft keratin which forms a thin yellow-white rim (zone 3) beneath the nail. Deeper, the distal NBM appears as anterior ligament (AL). DG is the so called "distal groove." **B:** Closer view of the hyponychial area of ODB demonstrating keratinization by formation of a thin granular cell layer and its juncture with the more eosinophilic, thin, onycholemmal keratin of the nail bed epithelium (NBE) *(right)*. The thin rim of keratin adheres to the ventral plate. *Left:* More typical epidermal keratinization with distinct granular cell layer and looser or softer appearing keratin. The area shows typical epidermal rete ridges and papillae. The keratin of the proximal hyponychial area, beneath the NP and corresponding to the ODB clinically, has a denser, more laminated appearance and can be interpreted as solehorn (S). On the right, longitudinal ridges of NBE are tangentially sectioned, appearing as islands—a sectioning artifact.

The false cuticle is a soft keratinous structure formed by the horny layers of the dorsal and, to a lesser extent, ventral surfaces of the PNF. It is nonadherent to the nail plate. In order to form, the false cuticle needs a growing nail plate, to which its horny layer loosely attaches, and the acute angle of the PNF's free margin (Figs. 1.3–1.5, 1.6B and 1.7C). Hence, it follows that the cuticle is lacking when the nail stops growing, as in scleronychia and yellow nail syndrome, as well as in chronic paronychia when the PNF is thickened and its free margin is rounded.

Eponychium

The cutaneous structure overlying the proximal nail plate is called *eponychium* (Greek *epi* on + *onyx* nail). Because the term is somewhat general, there is some imprecision in the use of the term, specifically in relation to the cuticle, or what was referred to earlier as the "false cuticle." Sometimes the terms are used interchangeably. It would be better to distinguish the epithelium producing epidermoid keratin from that of the larger part of the ventral PNF that has a sheathing (onycholemmal) function. The situation is analogous to differentiating the keratinization of follicular root sheath (trichilemma) from the joining infundibulum with its epidermoid keratinization. We favor the following viewpoint.

The term *eponychium* more accurately refers to the undersurface of the distal three-fourths of the PNF (3). The eponychium is a transitional zone of epithelium that connects the matrix to the epidermis surrounding the proximal unit. It has an epidermis-like epithelium with *onycholemmal keratinization* producing a thin layer of semihard keratin (Figs. 1.3, and 1.5–1.7) (5). This horny layer attaches strongly to the nail plate's dorsal surface to form the colorless, tough, adherent *"true cuticle"* or "nail vest." Using a loupe, this thin layer can be seen disintegrating below and just distal to the false cuticle extending from the rim of the PNF. Because it attaches firmly to the nail plate, the true cuticle protects the matrix by sealing it off from harmful environmental agents (7). Interestingly, some consider this portion of the eponychium that produces the true cuticle to be derived from the matrix (5).

Proximal to the eponychium, the dorsal matrix forms the dorsal nail plate, which is particularly firm, resistant, and responsible for nail sheen. The border between the eponychium and the proximal matrix is demarcated by loss of the granular layer in the latter (Fig. 1.6A). When this particular portion of the matrix is lost due to trauma or altered (e.g., by ungual lichen planus), the nail will appear dull and rough.

Hyponychium and Onychodermal Band

The *hyponychium* is thought to be derived from the embryonic distal ridge and is the first site of keratinization of the embryonic nail unit (10). The hyponychium has features that parallel those of the eponychium. They can be thought of, respectively, as ventral and dorsal counterparts of sheathing and transformational epithelia that are both analogous and homologous. As with eponychium, use of terminology sometimes is imprecise and vague.

Hyponychium and eponychium represent transition zones from nail unit to digital skin. They can be envisioned as a continuous annular seal of the generative areas and nail plate from the outside world. The point of transition laterally from eponychium to hyponychium, which also represents the inversion from dorsal to ventral sheath, corresponds to the point of externalization of the nail plate. This pivotal microanatomic *nail unit confluence* is described fully in the Topography section under "Observations from the Lateral Aspect."

The hyponychium begins where the distal NBE terminates and blends into the epithelium of the distal groove (Figs. 1.4 and 1.10). Thus, it is a transitional zone of epithelium that connects the NBE to the epidermis of the digital tip. It allows for physiologic separation of the nail plate and at the same time seals the nail unit from environmental agents.

A special onycholemmal cuticular product of the proximal hyponychium is the *solehorn.* This terminology refers to its presumed vestigial relationship to the formation of animal hooves. However, it may have a more pertinent function in the human nail unit. It probably is the actual subungual epidermal keratinous product that allows for the concurrent processes of separation of the ventral nail plate from the strong attachment of the nail bed while maintaining a physiologic seal.

Although the solehorn may be considered the ventral counterpart of the true cuticle (5,7), there is no consensus in the medical literature on the precise origin, nature, extent, and function of the solehorn. As understood here, the solehorn is physiologically important rather than being merely vestigial in humans. Some authors contend its importance relates to possible roles in pathologic states giving rise to hyperkeratosis and parakeratosis of pachyonychia congenita and pityriasis rubra pilaris. Also, a defect in the mechanism by which the stratum corneum remains adherent at this site could produce pterygium inversum unguis in which the distal nail bed remains adherent to the nail plate (36).

Under the microscope, the hyponychium shows, like the eponychium, keratinization via a granular cell layer. The solehorn of the hyponychium appears as a thin, compact keratin layer in proximity to the nail plate. Distally, the hyponychial keratin at the free edge of the plate transitions into epidermoid keratinization, which joins that of the digital tip epidermis (Fig. 1.10).

The hyponychial attachment shows through the nail plate as the onychodermal band (Figs. 1.4, 1.5, and 1.10), and the distal end of the nail plate parallels this band. Because the nail plate does not come into direct contact with the dermis and the band corresponds to the most proximal point of attachment of the fingertip stratum corneum to the plate, it has been suggested that the term "onychodermal band" be replaced by "onychocorneal band" or "junction" (36) though conventional use favors ODB.

On close inspection, the ODB is actually three subbands, which show slight individual variation in width and racial variation in color. The two more distal bands also may vary with certain environmental conditions (manicuring habits and exposures) because they represent expressions of the stratum corneum. The proximal and relatively more constant subband of the ODB is a broad (1 to 2 mm), faint, dull pink, and slightly milky band of subungual epithelium that abuts the NBE. It is an expression of the thicker transformational squamous epithelium that is histologically distinct from the NBE and correlates with the region of the solehorn (5). The middle sub-band is a narrow, dark amber (to brown in darker races) band that corresponds to compact orthokeratin seen more or less on edge through the nail. This band abruptly changes distally into a barely perceptible, thin, yellow-white band of externalized, desquamating soft keratin that, in turn, blends into the more gray-white free edge of the nail plate. Because the colors of the ODB are somewhat variable and subjective, it might be better to see the three sub-bands as tonal values: intermediate, dark, and light, from proximal to distal.

The hyponychium has a vascular bed that is separate from that of the adjacent nail bed. This is seen grossly (better with magnification) and histologically. By lightly compressing the free edge of the nail plate, the junction of the proximal ODB and the nail bed, with its unique vascular bed described later in the chapter, can be appreciated. The normal slight rim of vascular flush of the distal nail bed shows accentuation of its fine longitudinal striations with compression. This corresponds to termination of a capillary bed that is separate from that of the hyponychium. In addition, the connective tissue in this transitional area is looser and less bound to the ungual process. Because of this, with compression of the nail plate, the vascular rim stands out from the more dense ligamentous nail bed fibrocollagen associated with the ungual process. Compression of the tip alone will blanch back to the firmer, bound-down fibroconnective tissue of the ungual process. Simultaneous compression of the free edge and tip, or pulp, of the digit with its relatively looser connective tissue will readily blanch all vasculature.

To reiterate, the *proximal* hyponychium, which corresponds to the ODB, is the principal zone enabling physiologic separation of the nail plate from its the strong adherence to the NBE. This occurs by a process that is still not well understood. At the same time, the attachment is replaced by an effective sealing from the external environment. Histologically, the thin layer of eosinophilic subungual keratin of the nail bed slides up over a narrow wedge of hyponychial keratin (solehorn) as the nail plate separates from its bed (Fig. 1.10).

From a larger perspective, the juncture of the NBE and hyponychium represents an interface of opposing, but balanced, forces relating to the directions of growth and differing functions of two distinct squamous epithelia. Not only are the derivation and modes of keratinization different in these epithelia, but the three-dimensional pattern of rete ridges versus the longitudinal ridges of NBE also changes at this point. Deeper, the corresponding connective tissue and vascular patterns also are in transition. Altogether, these observations support the concept of the hyponychium as a broad, complex, and multifaceted zone. This zone has an indispensable role as the distal transitional interface of the nail unit, enabling effective exteriorization of the nail plate.

The *distal* hyponychium, corresponding to the *distal groove,* is a narrow rim of post-separation transformational skin. The skin of the distal groove is included in this zone of transition because it is slightly different from the skin of the tip of the digit. This narrow zone lacks epidermal ridges, or fingerprints, and shows denser deep connective tissue. This area, and that of the whole hyponychial zone, shows a band of anchoring connective tissue termed the *anterior ligament,* although it would be more appropriate to call it "distal ligament" of the nail unit. It attaches the dermis of the hyponychial region to the periosteum of the ungual process at the rim and, thereby, acts as the distal anchor of the nail unit.

In summary, the hyponychium is the distinctive and complex transitional distal part of the nail unit. It provides a vital counterpoint to the growth and adherence of the proximal nail unit and allows for completion of the nail forming process.

Supportive (Nail Unit Support System)

The epithelial components of the nail unit rest on a fibrocollagenous structural support system that is unique. First, it is mostly situated between two hard tissues, the nail plate and bone. Second, it directly connects squamous epithelium to bone. Last, it is a unique amalgamation of fascia and periosteum that forms ligamentous attachments and anchors the nail unit. It ties the nail unit to surrounding dermal connective tissue as well as the deeper fibrous framework of the distal digit, described earlier in the chapter. The nail unit support system (NUSS) has a distinctive and orderly structural organization in relationship to other parts of the nail unit. It should not be thought of as merely indifferent or featureless connective tissue.

A satisfactory comprehensive term for the NUSS has been lacking, and various terms have been applied to it, generally with regard to the nail bed alone. Although nail unit epithelia are epidermal derivatives, it is inappropriate to refer to their subepithelial support tissue as "dermis." It lacks the characteristic features of the papillary and reticular connective tissue relating to epidermis and adnexae and, in actuality, is made up largely of modified periosteum. Although basically synonymous with dermis, the term "corium," referring to hide-like connective tissue, may be more appropriate for the dense fibrous tissue joining the matrix and NBE to the ungual process. In this chapter, we use the term "nail bed corium" to refer specifically to the fibrocollagenous connective tissue in direct relationship to the NBE. This distinguishes it from other nail unit fibroconnective tissue.

Often the nail unit connective tissue is referred to generically as "nail unit mesenchyme." This is not a satisfactory term because it properly refers to loose, primitive, embryologic connective tissue, which is quite unlike that of the nail unit. In addition, for a general term, it fails to recognize the orderly, integral structure of nail unit connective tissue. However, the term "nail unit mesenchyme" enjoys wide current usage and is a pragmatic synonym.

A comprehensive term that collectively refers to the connective tissue surrounding the nail unit epithelial structures as an integrated structural arrangement is "nail unit fibrocollagenous support system." Although histologically accurate, because it is composed of collagen producing fibrous connective tissue that varies from loose to dense, it is referred to here simply as the *NUSS.* This term recognizes a complex structural arrangement of connective tissue components, each with a specific role and relationship to adjacent structures, tied together for a common purpose.

For convenience of description, the NUSS is divided into proximal, intermediate, and distal zones. In each zone the connective tissue pattern varies in relation to the adjacent nail unit component, that is, the matrix, NBE, and hyponychium, respectively. Thus, the NUSS includes a form of *fibrous sheath* around the matrix (analogous to the fibrous sheath of the hair follicle) that extends to form *nail root ligaments,* the *NBC,* and the *anterior ligament.* Although these areas gradually blend together, they have sufficiently distinctive gross and microscopic features to warrant separate consideration.

Longitudinally oriented bands of collagen, comprising the proximal NUSS, surround the matrical root in the manner of a fibrous sheath (Figs. 1.6 and 1.7). Some of these fibers blend into deeper ligamentous connective tissue near the point of attachment of the extensor tendon where it blends with the periosteum. However, the major attachment of the matrical root is by way of fibers that pass obliquely to blend with the dorsal expansions of the collateral ligaments as well as fibers of the LILs at their attachments to the lateral tuberosities.

Laterally, the NUSS attaches to the LILs (Fig. 1.7). The proximal NUSS, in the area of the proximal matrix, is more delicate, looser, and less bound to bone. It contains finer vessels and fewer glomus bodies than the intermediate zone of NUSS. Deeper, the proximal NUSS shows larger vessels with loose adventitial stroma and collections of fat cells. The proximal NUSS gradually blends into the intermediate zone of NUSS, or NBC.

In the large intermediate zone of NUSS, relating to NBE, broad vertical bands of collagen connecting periosteum to NBE are arrayed in almost parallel, slightly fanned, longitudinal rows (Figs. 1.8A and 1.9). The term *"nail bed corium"* refers specifically to the NUSS underlying the NBE. The NBC is richly vascularized, contains many glomus bodies, and shows longitudinally oriented microvasculature (Fig. 1.8). Its capillary bed ends abruptly at the ODB.

The NBC shows proximal and distal portions that vary slightly in relative proportion of vessels and density of connective tissue, and this is apparent even grossly. The proximal portion of this zone is looser and more vascular, and clinically corresponds to the postlunular vascular flush. In contrast, the distal portion is more firmly bound to the ungual process and grossly appears through the nail plate as a paler band.

In the distal NUSS the vertical collagen fibers become inclined toward the hyponychial zone to about 45 degrees and are relatively loosely arranged. In this region the collagen fibers blend into subcutaneous and dermal connective tissue of tip skin (Figs. 1.4 and 1.10). This transition of distal NUSS into subcutis and dermis in the area of the hyponychium extends the concept of a zone of transition beyond the epidermis to include deeper connective tissue. In other words, the distal NUSS corresponds to the deeper part of the hyponychial transition zone. The cresentic band of collagen fibers radiating from the rim of the distal ungual process forms the *anterior ligament* as described above. The relatively looser connective tissue of the hyponychium is reflected as a normal faint vascular flush in this transition zone of the distal nail unit. This is analogous to the flush just distal to the lunula.

Clinical Correlations: Nail Plate Changes due to Alterations in the Nail Unit

Loss of the cuticle commonly results in irregular transverse furrows of the nail plate due to penetration of bacteria, fungi, and other substances harmful to the matrix and nail bed. Furthermore, discoloration of nail plate margins often is seen. Habitual pushing back of the PNF is the most frequent cause of "wash board" nails and pseudodystrophia mediana canaliformis. Whereas most of the distal aspect of the PNF may be resected and reforms within a few months without adverse sequelae, the nail plate's surface will be rough and lusterless whenever its most proximal part is damaged. Scarring from longitudinal cuts and severe lichen planus involving the matrix may cause pterygium unguis. Damage to the hyponychium commonly results in onycholysis that may be permanent or lead to pterygium inversum unguis.

BLOOD VESSELS

There is more orderliness and consistency to the vessels of the distal digit than is generally appreciated. Several excellent general sources for further information on vasculature of the distal digit are available (37–45). Figure 1.11 illustrates the arterial supply to the distal digit and serves as a reference for the following discussion.

To begin with, the major arteries of the fingers and toes are volar. The dorsal digital arteries supply only the dorsal aspect of the proximal and intermediate phalanges. The paired *proper palmar* (and *plantar*) *arteries* (also known as the *common volar digital arteries*), and their equivalents in the thumb and great toe, nourish the distal digit. Although the arteries are paired, one normally is dominant (38,42,46).

Tortuosity of the proper digital arteries, with turns of up to 270 degrees, is characteristic of the digital blood supply, especially around the joints. This provides the vessel protection, resilience (springlike elasticity), and patency under sustained flexion (44).

The nail unit and dorsal digit get their blood supply via a series of arcades coming from the paired proper digital arteries. The first arcade is the *superficial arcade* (also known as the "dorsal nail fold arch" or "proximal matrix arch") at the base of the terminal phalanx. This arcade is located just proximal to the nail matrix and over the attachment of the extensor tendon. The arcade receives contributions from both the middle and distal digital phalangeal segments. In this manner, the superficial arcade has a double arterial supply. Thus, the nail

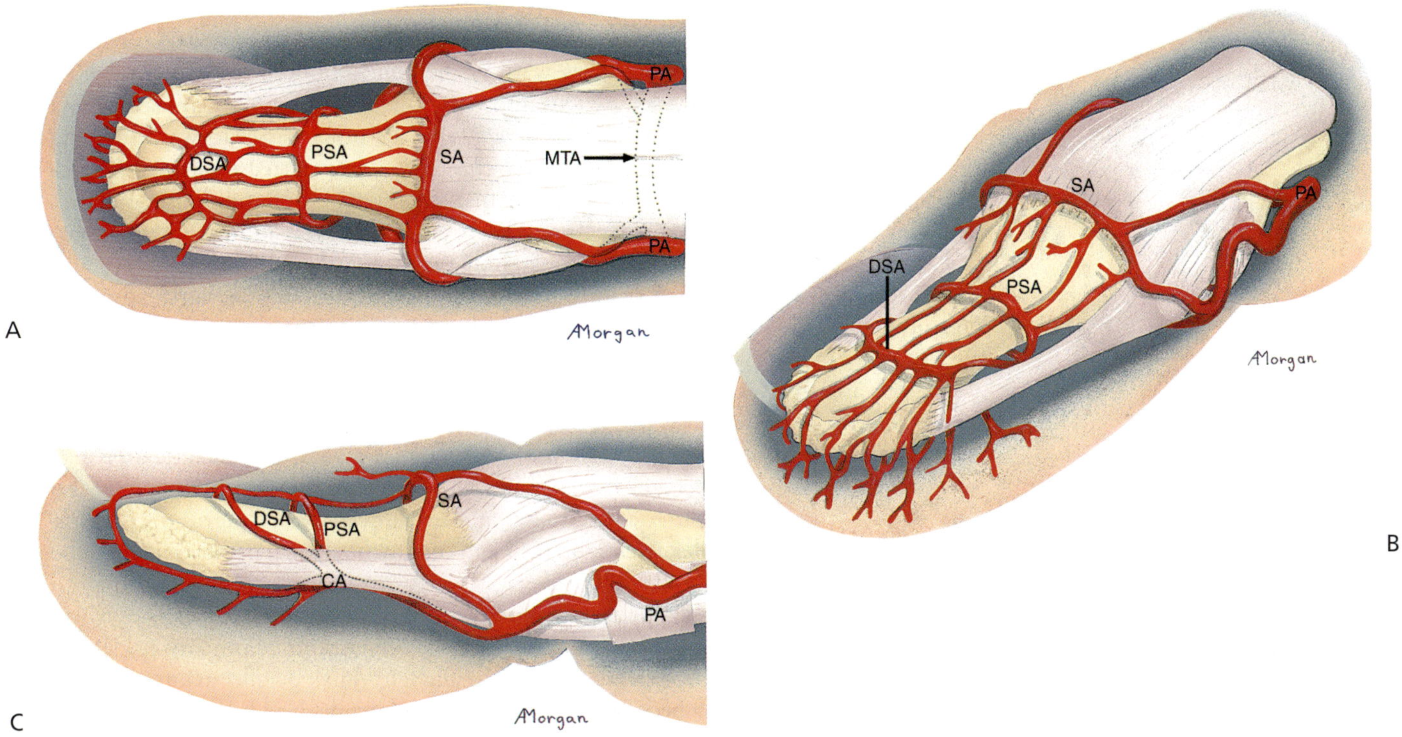

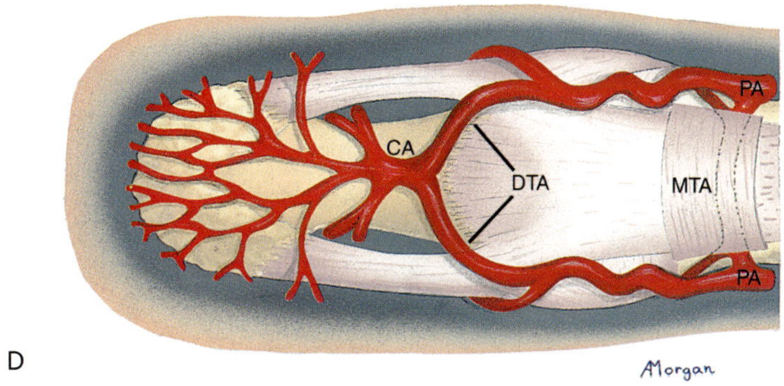

FIGURE 1.11. Illustrations showing the arterial supply of the distal digit with its network of arcades. **A:** Dorsal view. **B:** Dorsal oblique view. **C:** Lateral view. **D:** Ventral view. CA, cruciate anastomosis; DSA, distal subungual arcade; DTA, distal transverse arch; MTA, middle transverse arch (deep to flexor tendon); PA, proper artery; PSA, proximal subungual arcade; SA, superficial arcade. Perforating and anastomosing arteries are seen in the pulp. Note how the nail unit receives contributions from both the middle phalangeal segment as well as dorsal and ventral branches beyond the distal interphalangeal joint (DIPJ). Further, the cruciate anastomosis assures blood supply to both lateral and medial aspects of the nail unit.

unit and joint can survive extensive damage to the blood supply of the volar aspect of the distal digit. From the middle phalanx come the dorsal skin branches that originate from the *middle transverse (palmar/plantar)* arch near their branching from the proper arteries. Branching of the proper arteries to form the middle transverse arch occurs near the head of the middle phalanx and in proximity to the distal cruciate ligament. All "transverse arches" are volar.

Branches from the superficial arcade pass deep to the nail root to enter the nail bed and connect with the *proximal subungual arcade.* Other branches supply the PNF and the dorsum of the digit, such as the extensor tendon insertion and DIPJ.

The pair of palmar (plantar) arteries converge deep in the pulp of the distal digit to form the *distal transverse (palmar or plantar) arch* or "digital pulp arch." It forms distal to the level of the dorsal superficial arcade, just past the attachment of the flexor tendon, and in a protected area. The protected area is located near the point of maximal concavity of the distal phalanx and corresponding thickest amount of pulp.

The confluence of distal palmar or plantar arteries forms a *cruciate anastomosis.* From this point two vessels arise, one to each side of the digit. These vessels pass laterally and then dorsally around the narrow waist of the phalanx to enter from below the proximal nail bed. Branches from these arteries are the source of the *dorsal arcades* of the nail bed. As the branches pass around the narrow waist of the bone and deep to the LILs, which give lateral protection, each vessel divides into two. One limb passes transversely across the waist to form the *proximal subungual arcade,* and the other passes forward over the ungual process to form the *distal subungual arcade.* These are referred to also as the second and third arcades, respectively. The proximal subungual arcade (or "middle matrix arch") forms beneath the intermediate nail matrix or lunular area, and the distal subungual arcade is in the distal nail bed. Although only the principal arcades are discussed and illustrated in the accompanying figures, lesser intermediate arcades also may be formed. The subungual arcades give rise to many longitudinally running vessels which in turn give rise to a rich vascular plexus.

Vessels arising from the distal subungual arcade extend distally to anastomose with vessels from the pulp that pass over the tip of the phalanx. The anastomotic pulp arteries form a three-dimensional, lacelike pattern of tiered arches that send branches, known as *perforating arteries* and *anastomosing arteries,* in a volar direction to supply the pulp of the distal digit. The perforating arteries end in the subdermal/deep dermal plexus from which terminal arteries and arterioles ascend, branch, and supply the dermal papillae via the subpapillary vascular plexus.

The most terminal vascular patterns within the distal digit vary from site to site. Volar skin contains terminal capillary loops aligned with the fingerprints or primary epidermal ridges. In contrast, nail bed vessels are arranged in a longitudinal, parallel manner back to the lunula where the pattern becomes more tortuous proximally. Nail bed capillary architecture is simpler than digital skin (40). Lymphatic vessels also vary with the site. Lymphatics of the nail bed are more numerous than those of the dermis (11).

There are two systems of digital veins, deep and superficial, which connect by oblique communicating veins largely along the sides of the digits (41). The deep system is mostly volar and parallels the arteries as venae concomitantes. The superficial venous system of the distal digit is composed of dorsal, lateral, and palmar (or plantar) drainage systems

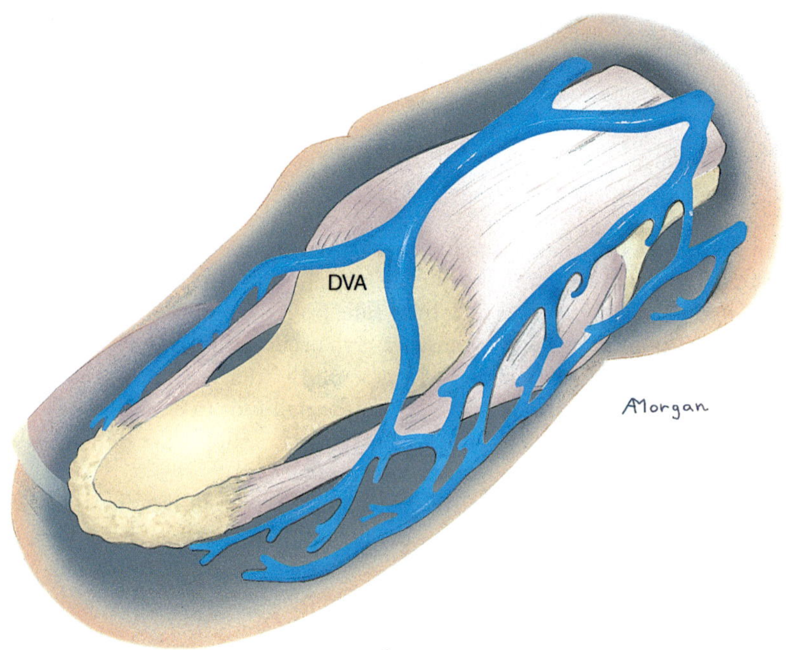

FIGURE 1.12. Superficial venous network. The distal venous arch (DVA) forms a common channel that lies between the condyles of the head of the middle phalanx.

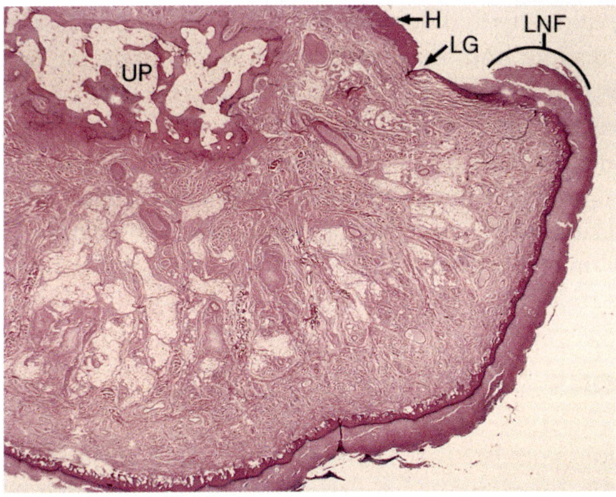

FIGURE 1.14. Cross section of the tip of the distal digit showing the irregular surface of the tip of the ungual process (UP) from which spokelike, radial, trabecular, ligamentous extensions pass through the subcutaneous tissue of the pulp to connect intimately with the dermis. In the tip, as opposed to the pad, fibrous tissue is more plentiful than adipose tissue. Pacinian corpuscles (*arrowheads*), nerves, eccrine glands, and vascular network are distributed in the deep pulp. At this level the nail bed epithelium has given way to the more irregular rete of the hyponychium (H). Here, the lateral aspect of the hyponychium is blending into the distal lateral groove (LG) along with the lateral nail fold (LNF). This LNF joins the skin of the tip, which shows an epidermis with a more conventional rete pattern and thick epidermoid keratinization.

gertips greatly impairs all function of the entire hand. Because tactile sensory perception (along with its companion sensations of pressure, temperature, vibration, and pain) is the only one of the "five senses" not confined to the head, loss of the sensory function of the fingertips is "to render the eyes of the finger blind" (60).

TIP AND PAD ("PULP")

There are actually two distal digit pulp spaces with different structure and functions (19). We refer to the distal pulp region as the *tip* and the proximal pulp region as the *pad.* With the nail plate they work together to make a unique and elegant tool.

Subcutaneous tissue of the tip is separated into wedge-shaped or conical fibrofatty compartments, with their apices at the ungual tuberosity, by well-defined radiating fascial bands (Fig. 1.14). These ligamentous connections firmly attach the dermis to the periosteum of the leading and inferior edge of the ungual tuberosity and restrict the mobility of the finger tip. They also allow dissipation of pressure forces. The tip functions more as a probe and contact point of pressure and abrasion than does the pad. The area used preferentially for fine touch appears to be where tip and pad meet, rather than the extreme tip.

The pad is bulkier and more mobile than the tip. The pad shows a less pronounced pattern of fat lobulations encased by radiating fibrous tissue than does the tip. Fat lobules are more spherical. The pad contains prominent arteries and nerves, along with many pacinian corpuscles deep in the subcutis. The pad is used primarily for gripping because its soft tissue firmly molds about an object. This padded pliers-like function is backed up by the concave configuration of the distal phalanx deep to the pad.

Eccrine glands of the pulp are prominent and deep. They are found mostly pushed to the side of the fat lobules and next to the spokelike, radiating, interconnected fibrous trabeculae of the tip and pad. Because eccrine sweat serves to enhance grip, fine manipulation, and touch sensation, it is instrumental in the chief functions of the distal digit.

The tip and pad are covered by epidermis and dermis with the special characteristics of palmar and plantar skin that are distinct from that immediately surrounding the nail unit. In addition to the lack of hair follicles, the presence of many eccrine glands, and prominent Meissner's corpuscles, volar digital skin is characterized by thick stratum corneum, prominent stratum lucidum, and a complex pattern of rete ridges and papillae (61). A pattern of primary epidermal ridges, or fingerprints, surmounted by regular rows of eccrine ostia reflects an orderly relationship of rete ridges and papillae. The dermis is thick and connected to the deeper connective tissues such as periosteum, ligaments, and retinaculae via subcutaneous fibrous trabeculae (16,17,19).

TOPOGRAPHIC CORRELATIONS

Essentially all topographic observations correlate with the internal anatomy of the distal digit. Thus, on examining the patient, the clinician can derive a more complete mental picture of the internal structure of the distal digit as an anatomic unit. The ability to, in effect, "see into" the nail unit in relation to the distal digit by correlating observed external structure, and careful palpation, with knowledge of internal detail has useful diagnostic and surgical value. With practice, it enables the surgeon to precisely orient and relate all the structural components. To that end, here is a list of useful topographic correlations.

Observations from the dorsal aspect (Fig. 1.15):

1. The LILs are palpable between the ungual spine and lateral tuberosity of the base of the terminal phalanx.
2. The tip of the ungual process is located beneath the juncture of the onychodermal band and distal nail bed.
3. The distance from the distal onychodermal band to the PNF at midline is approximately equal to the distance from the PNF to the primary dorsal crease. It is quite constant from digit to digit in an individual.
4. The primary transverse dorsal skin crease over the DIPJ is located directly over the point of articulation of that joint.
5. The superficial arcade or dorsal nail fold arch lies in a palpable groove formed by the proximal matrix and the insertion of the flexor tendon. The groove can be felt by gently rocking the thumb or index finger while applying firm pressure to the area or with the free edge of the examiner's nail plate.
6. The dusty rose flush of the nail bed just distal to the lunula corresponds to the area of NBC that is relatively less dense and less firmly attached to the phalangeal ungual process. It is especially loose laterally where the subungual arcades enter the deep aspect of looser nail unit fibrocollagen.

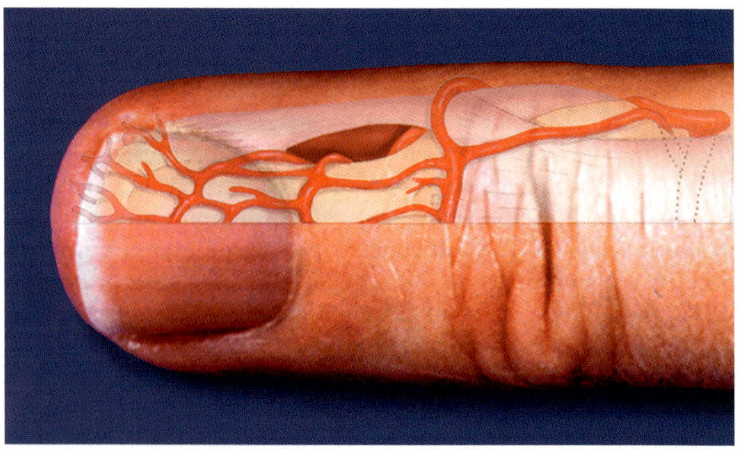

FIGURE 1.15. Topography, dorsal aspect. See text for details, and correlate with Fig. 1.5.

7. The area of central nail bed pallor corresponds to the area of intermediate NBC with its dense, vertically oriented collagen and strong attachment to the ungual process.
8. The slight rim of vascular flush just proximal to the ODB corresponds to a distinct change in the vascular bed from that of the NBC to the looser connective tissue that underlies the area of the hyponychium.
9. The extensor tendon and confluent dorsal vein run between the bilaterally palpable condyles of the head of the middle phalanx. The veins can generally be observed directly in light-colored skin.
10. With flexure of the DIPJ, the middle phalangeal condyles become prominent at the knuckle and should not be mistaken for the base and lateral tuberosities of the distal phalanx.

Observations from the lateral aspect (Fig. 1.16):

1. The most lateral point of emergence of the free edge of nail plate corresponds to a point of confluence, the *nail unit confluence*, of several histologic transitions, including (a) the distal extremity of the LNF and groove transition into the most proximolateral aspect of the distal groove, which marks a point where dorsal skin meets volar skin; (b) the transition point of the extremities of eponychium and hyponychium—a dorsal-to-ventral sheath inversion that commences the onychodermal band; (c) the lateral nail bed fibrocollagen becomes the most proximolateral aspect of the anterior ligament, which is actually a "distal ligament"; (d) deeper, it marks the beginning of the transition of NUSS directly to dermis of the tip, without a subcutis; and (e) deeper still, the LIL attaches at this point to the ungual spines. Fundamentally, this confluence of histologic transitions is coincident with the emergence of the nail plate, from its internal to its external environment.
2. The DIPJ lies between the primary dorsal and volar creases.
3. The subtle erythematous flush of the tip and pad skin reflects the rich vascular anastomotic terminal network of the volar distal digit. Note how this flush is more prominent on the distal digit than other palmar/plantar skin, reflecting the unique vascular arrangements of the distal digit.
4. In a distribution corresponding to the pulp vascular flush, fingerprints are especially prominent distally. Print markings are oriented diagonally transverse in a distal-volar to proximal-dorsal direction, roughly paralleling the LNF.
5. The fingerprints are countered, or crossed, diagonally by faint lines of stress in a zone where the volar skin of the pulp meets the dorsal skin of the nail folds. This zone generally overlies the area of the LIL.
6. The zone of transition from volar to dorsal skin may reflect a difference in epidermal pigmentation in individuals with darker skin.

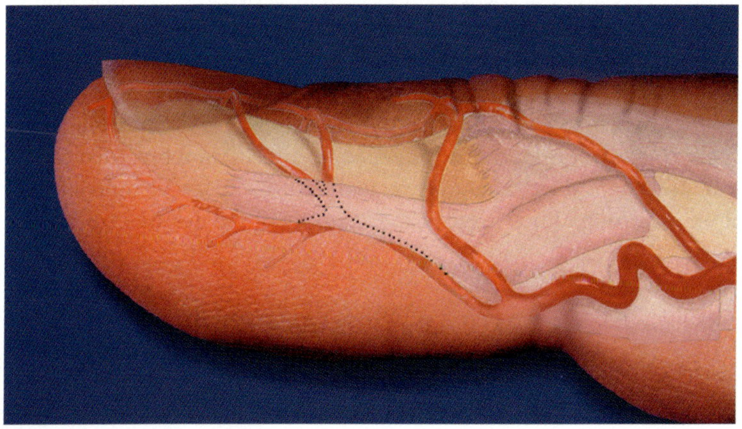

FIGURE 1.16. Topography, lateral aspect. See text for details, and correlate with Fig. 1.5.

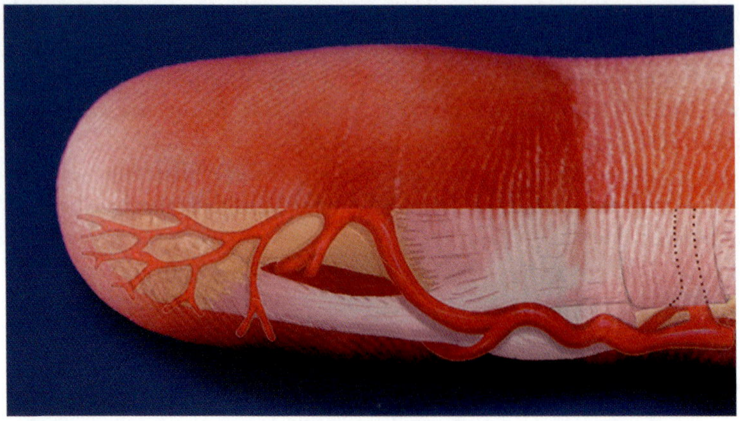

FIGURE 1.17. Topography, volar aspect. See text for details, and correlate with Fig. 1.5.

Observations from the volar aspect (Fig. 1.17):

1. The central whorl or focus of the fingerprints marks the center of the pad area.
2. The confluence of the transverse arterial arch is located midway between the tip of the digit and the DIPJ. Thereby, it is situated at the deepest area of the central pad. Externally, this area corresponds with the central whorl or focus of the fingerprints. Internally, it is located at the most concave and protected volar aspect of the distal phalanx. From the lateral aspect this point is deep to the midportion of the LIL.
3. Transverse epidermal ridges, or fingerprints, reflect the amount of subdermal fibrous connective tissue. Therefore, they are more prominent at the tip and anterior pad.
4. Correspondingly, the same area contains more Meissner's corpuscles and is the area of maximal sensitivity to touch. Paradoxically, this sensitive area is tough and resistant to pressure and abrasion.
5. The tip shows a central, firmer, subtle wedge-shaped ridge that points toward the central, whorled confluence of fingerprints. The base of this wedge represents the "tip" of the finger.
6. The flexor tendon attaches approximately two-thirds of the way between the central whorl of fingerprints and the primary flexor crease.
7. The primary flexural crease demarcates the DIPJ. Distally, the base of the distal phalanx is palpable. The regions of the bilateral neurovascular bundles and principal arterial tortuosities are located ventromedially to the lateral tuberosities of the base.

By systematic examination of nail unit anatomy, particularly in relation to the distal digit, the reader gains an appreciation of the elegant design, integrity, and practical beauty of this unique anatomic unit. There is clearly a consistent order and interdependence of the structural components. This chapter seeks to examine and correlate in detail macroscopic and microscopic anatomy and to surmount the lack of knowledge concerning the nail unit in relation to the whole distal digit. In doing so, we attempt to define terms more precisely and clarify concepts. Most importantly, the specific information provided can be useful for the diagnosis and surgical treatment of the nail unit disorders. Competent nail surgery is contingent on a thorough knowledge of the anatomy of the nail unit in relation to the distal digit.

REFERENCES

1. Williams PL, Warwick R, Dyson M, et al., eds. *Gray's anatomy,* 38th ed. Edinburgh: Churchill Livingstone, 1997.

2. Romanes GJ, ed. *Cunningham's textbook of anatomy,* 12th ed. London: Oxford University Press, 1981.
3. Conejo-Mir JS, Ortega MN. Nail. In: Sternberg SS, ed. *Histology for pathologists,* 2nd ed. Philadelphia: Lippincott-Raven Publishers, 1997:47–68.
4. Dawber RPR, deBerker D, Baran R. The science of the nail apparatus. In: Baran R, Dawber RPR, eds. *Diseases of the nails and their management,* 2nd ed. London: Blackwell Science, 1994:1–34.
5. Gonzalez-Serva A. Structure and function. In: Scher RK, Daniel CR III, eds. *Nails:* therapy, *diagnosis, surgery,* 2nd ed. Philadelphia: WB Saunders, 1997:11–30.
6. Zaias N. *The Nail in Health and Disease,* 2nd ed. Norwalk: Appleton & Lange, 1990:3–14.
7. Ditre CM, Howe NR. Surgical anatomy of the nail unit. *J Dermatol Surg Oncol* 1992;18:665–671.
8. Fleckman P. Anatomy and physiology of the nail. *Dermatol Clin* 1985;3:337–381.
9. Lewin K. The normal finger nail. *Br J Dermatol* 1965;77:421–430.
10. Lewis BL. Microscopic studies of fetal and mature nail and surrounding soft tissue. *Arch Dermatol Syphilol* 1954;70:732–747.
11. Zook EG, Van Beek AL, Russell RC, et al. Anatomy and physiology of the perionychium: a review of the literature and anatomic study. *J Hand Surg* 1980;5:528–536.
12. Gigis PI, Kuczynski K. The distal interphalangeal joints of human fingers. *J Hand Surg* 1982;7: 76–182.
13. Craig S. Anatomy of the joints of the fingers. *Hand Clin* 1992;8:693–700.
14. Kelikian H. *Congenital deformities of the hand and forearm.* Philadelphia: WB Saunders, 1974.
15. Haneke E. Etiopathogenie et traitment de l'hypercourbure transversale de l'ongle du gros orteil. *J Med Esth Chir Dermatol* 1992;19:123–127.
16. Shrewsbury MM, Johnson RK. The fascia of the distal phalanx. *J Bone Joint Surg* 1975;57:784–788.
17. Shrewsbury MM, Johnson RK. A systematic study of the oblique retinacular ligament of the human finger: its structure and function. *J Hand Surg* 1977;2:194–199.
18. Bogumill GP. A morphologic study of the relationship of collateral ligaments to growth plates in the digits. *J Hand Surg* 1983;8:74–79.
19. Shrewsbury MM, Johnson RK. Form, function, and evolution of the distal phalanx. *J Hand Surg* 1983;8:475–479.
20. Jones MM, Amis AA. The fibrous flexor sheaths of the fingers. *J Anat* 1988;156:185–196.
21. Strauch B, de Moura W. Digital flexor tendon sheath: an anatomic study. *J Hand Surg* 1985; 10A:785–789.
22. Hunter JM. The vincular system and its variations. In: Hunter JM, Schneider LH, Mackin EJ, eds. *Tendon surgery of the hand.* St. Louis: CV Mosby, 1987:24–36.
23. Doyle JR. Anatomy of the finger flexor tendon sheath and pulley system. *J Hand Surg* 1988;13A: 473–484.
24. Soon PSH, Arnold MA, Tracey DJ. Paraterminal ligaments of the distal phalanx. *Acta Anat* 1991; 142:339–346.
25. Gure S, Guichard S, Fraitag SR. Ligamentary structure of the base of the nail. *Surg Radiol Anat* 1994;16:47–52.
26. Zaias N, Alvarez J. The formation of the primate nail plate. An autoradiographic study in squirrel monkey. *J Invest Dermatol* 1968;51:120–136.
27. Johnson M, Comaish JS, Shuster S. Nail is produced by the normal nail bed: a controversy resolved. *Br J Dermatol* 1991;125:27–29.
28. Johnson M, Shuster S. Continuous formation of nail along the bed. *Br J Dermatol* 1993;128:277–280.
29. Austin RT. A method of excision of the germinal matrix. *Proc R Soc Med* 1970;63:33–34.
30. Achten G, Heenen M. Pathology of the nails. In: Flarer F, Serri F, eds. *Dermatology Proceedings of the International Congress, Padua-Venice,* Series Number 289. Amsterdam: Elsevier, 1972:890–891.
31. Samman PD. The human toe nail: its genesis and blood supply. *Br J Dermatol* 1959;71:296–302.
32. Samman PD. The ventral nail. *Arch Dermatol* 1961;84:192–195.
33. Achten G, Parent D. The normal and pathologic nail. *Int J Dermatol* 1983;22:556–564.
34. Jarrett A, Spearman RIC. The histochemistry of the human nail. *Arch Dermatol* 1966:94:652–657.
35. Germann H, Barran W, Plewig G. Morphology of corneocytes from human nail plates. *J Invest Dermatol* 1980;74:115.
36. Sonnex TS, Griffiths WAD, Nichol WJ. The nature and significance of the transverse white band of human nails. *Semin Dermatol* 1991;10:12–16.
37. Flint MH. Some observations on the vascular supply of the nail bed and terminal segments of the finger. *Br J Plast Surg* 1955;8:186–195.
38. Edwards EA. Organization of the small arteries of the hand and digits. *Am J Surg* 1960,99:837–846.
39. Hale AR, Burch GE. The arteriovenous anastomoses and blood vessels of the human finger. *Medicine* 1960;39:191–240.
40. Inoue H. Three-dimensional observations of microvasculature of human finger skin. *Hand* 1978;10: 144–149.
41. Moss SH, Schwartz KS, Grace von Drasek-Archer AMI, et al. Digital venous anatomy. *J Hand Surg* 1985;10A:473–450.

42. Chaudakshetrin P, Kumar VP, Satku K, et al. The arteriovenous pattern of the distal digital segment. *J Hand Surg* 1988;13B:164–166.
43. Strauch B, de Moura W. Arterial system of the fingers. *J Hand Surg* 1990;15A:148–154.
44. Smith DO, Oura C, Kimur C, et al. Artery anatomy and tortuosity in the distal finger. *J Hand Surg* 1991;16A:297–302.
45. Smith DO, Oura C, Kimur C, et al. The distal venous anatomy of the finger. *J Hand Surg* 1991; 16A:303–307.
46. Leslie BM, Ruby LK, Madell SJ, et al. Digital artery diameters: an anatomic and clinical study. *J Hand Surg* 1987;12A:740–743.
47. Barbolini G, Tischendorf F, Curri SB. Histology, histochemistry, and function of the human digital arteriovenous anastomoses (Hoyer-Grosser's organs, Masson's glomera). I. The possible relationship with Pacinian corpuscles. *Microvasc Res* 1971;3:142–153.
48. Tischendorf F, Barbolini G, Curri SB. Histology, histochemistry, and function of the human digital arteriovenous anastomoses (Hoyer-Grosser's organs, Masson's glomera). II. The possible relationship with sweat glands. *Microvasc Res* 1971;3:323–336.
49. Samman PD, Strickland B. Abnormalities of the fingernails associated with impaired peripheral blood supply. *Br J Dermatol* 1962;74:163.
50. Lynn RB, Steiner, van Wyck FAF. Arteriographic appearances of the digital arteries of the hands in Raynaud's disease. *Lancet* 1955;1:471–474.
51. Currie AE, Gallagher PJ. The pathology of clubbing: vascular changes in the nail bed. *Br J Chest* 1988;82:845–847.
52. Wilgis EFS, Maxwell GP. Distal digital nerve graft: clinical and anatomical studies. *J Hand Surg* 1979;4:439–443.
53. Zook EG. Injuries of the fingernail. In: Green DP. *Operative hand surgery*, 2nd ed. New York: Churchill Livingstone, 1988.
54. Barron JN. The structure and function of the skin of the hand. *Hand* 1970;2:93–96.
55. Montagna W. Morphology of cutaneous sensory receptors. *J Invest Dermatol* 1977;69:4–7.
56. Munger BL, Ide C. The structure and function of cutaneous sensory receptors. *Arch Histol Cytol* 1988;51:1–34.
57. Winklemann RK. Cutaneous sensory nerves. *Semin Dermatol* 1988;7:236–268.
58. Resnik M. Human skin innervation, and some related disorders. In: Pierard GE, Pierard-Franchimont C, eds. *The dermis:* from biology to diseases. *Monographies dermatopathologiques liegeoises.* Liege, Belgium: CHU du Sart Tilman, 1989:36–50.
59. Dankmeijer J, Waltman JM. Sur l'innervation de la face dorsale des doigts humains. *Acta Anat* 1950;10:377–384.
60. Bunnel S. *Surgery of the hand.* Philadelphia: JB Lippincott Co., 1944:222.
61. Harty M. The dermal papillae in the fingertip. *Plast Reconstr Surg* 1970;45:141–145.

2

PREOPERATIVE AND POSTOPERATIVE MANAGEMENT

ELVIN G. ZOOK

For preoperative evaluation, it is important that the anatomy and developmental physiology of the nail be well known. Using this knowledge, one can postulate from where in the nail bed the defect originates. Using information obtained from the literature, experience, and consultations, a plan of approach to the problem can be formulated.

GENERAL AREAS OF NAIL DEFORMITY CAUSATION

Trauma

Trauma is one of the most common causes of nail deformity. Macrotrauma deformity is easily diagnosed. The effects of trauma consist of nonadherence, splitting, angulated growth, or absence of all or portions of the nail. Microtrauma, such as consistent and repeated picking at the eponychium, detaches the dorsal roof from the nail. Inflammation and invasion by microbes separate the dorsal roof from the nail and create a rough irregular nail surface. Vigorous pushing back of the eponychium or freeing of the eponychial border from the nail to insert artificial nails may cause nail surface deformity. If these actions were not of such duration to cause permanent damage, cessation usually allows growth of a normal nail.

Disease

Systemic

There are many systemic diseases that cause abnormality of the nail. Cardiac and pulmonary problems can decrease oxygenation of the nail bed, and such conditions must be considered and evaluated with appropriate tests.

Local

Local diseases such as psoriasis, chronic infection, and osteoarthritis of the distal interphalangeal joint may cause nail abnormalities.

EQUIPMENT AND PERIOPERATIVE CARE

Most nail surgery of the hand or foot can be performed using a local anesthetic and digital tourniquet, with very little risk to the patient.

E. G. Zook: Division of Plastic Surgery, Southern Illinois University, Springfield, Illinois 62708.

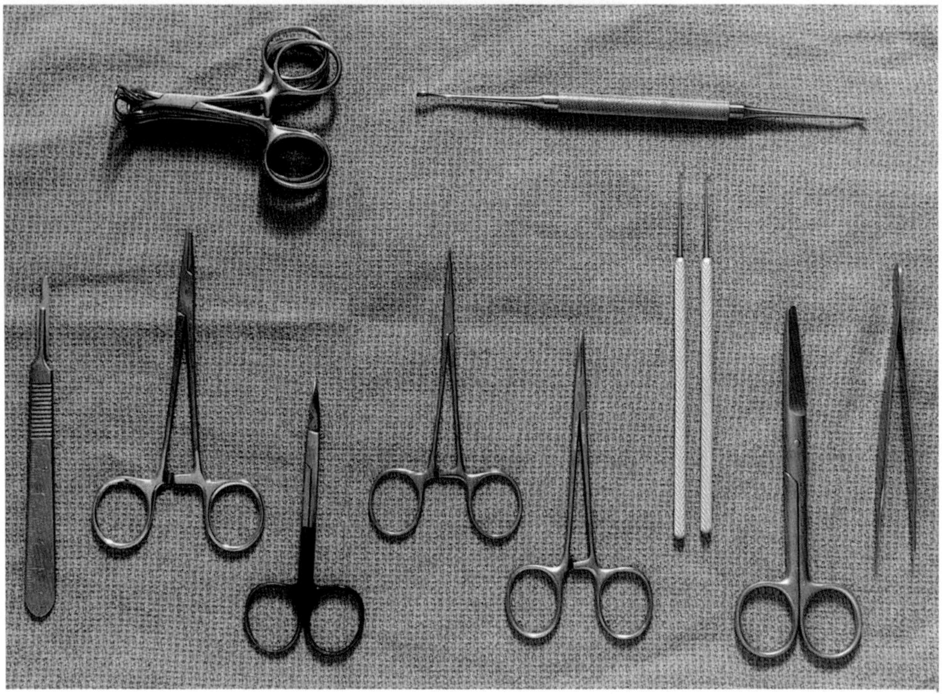

FIGURE 2.1. An example tray of fine surgical instruments used for nail bed surgery.

Patient Preparation

It is important to evaluate any medications (such as warfarin [Coumadin]) being taken by the patient and adjust the dosages as necessary. For diabetic patients undergoing toe operation, it is important that their blood sugar levels be well regulated by their insulin dosage. All aspirin-containing products should be discontinued at least 1 week before surgery. I recommend that patients who have jobs in which dirt gets under the nails soak their nails in soapy water and clean them with a scrub brush for several days before surgery. After surgery, the hand or foot is elevated for several days to decrease swelling and promote healing. Patients should be advised ahead of time so they can make arrangements for accommodations in their daily lives.

Instrumentation

When dealing with small fine structures such as the nail bed, it is important to use appropriate instruments. These instruments should be delicate and function well (Fig. 2.1). I prefer fine Iris scissors for dissection, no. 15 surgical blades for cutting and scraping, fine-tooth Addison pick-ups for handling tissue, and a Kleinert-Kutz elevator for nail removal. In addition, fine mosquito clamps and a Webster needle holder should be available.

Suture Material

It is essential to use fine surgical sutures. I prefer 5-0 or 6-0 nylon suture on a P-1 needle for closure of incisions in the eponychium, paronychium, or fingertip. Absorbable sutures (chromic or pain gut) are best for suturing the germinal and/or sterile matrix. Nonabsorbable sutures are difficult to remove and, when located under bandages, stick to serum and blood. Soaking the finger in warm soapy water beginning approximately 1 week postoperatively will promote removal of the dressing and sutures. I prefer a double-arm 7-0 ophthalmic chromic suture (Fig. 2.2) for suturing of the nail bed. The suture is cut in the

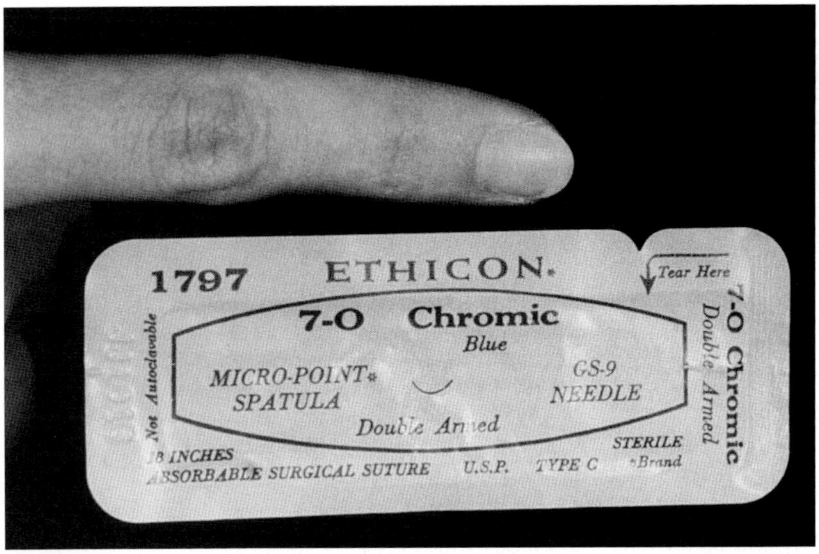

FIGURE 2.2. Chromic suture used for nail bed repair.

middle so an extra needle should be available, if needed. Frequently, one needle is bent when suturing so close to the periosteum and bone. Fine sutures are preferable so that excess tension on the nail bed that could be created by a heavier suture is prevented. Also, resorption is quicker and more predictable. If nonabsorbable suture is necessary, a fine (5-0 or 6-0) monofilament nylon is least reactive.

Anesthesia

Almost all perionychial procedures require some type of anesthesia due to the large number of sensory endings in the area. A finger block is done in most patients by injecting 3 mL of lidocaine (Xylocaine) 1% without epinephrine at the base of the finger in the web space in the area of the digital nerve. A bead of local anesthetic is applied across the dorsum of the finger to block the sensory branches of the radial or ulnar nerve. When a paronychia is being drained, a short general anesthetic may be necessary because a local anesthetic is not as effective in areas of inflammation.

I prefer to use a 25-gauge or smaller needle directed toward the base of the proximal phalanx over the digital nerve and inject 1 mL of lidocaine without epinephrine. I then angle the needle 45 degrees dorsally and ventrally, injecting approximately 1 mL in each of the three angles (Fig. 2.3). Attempts to inject local anesthetic in the pad of the finger or along the paronychium is very painful to the patient and should be avoided.

Tourniquets

It is important that blood be extravasated from the finger to avoid oozing blood and allow visualization of the delicate structures. I prefer using a 1-inch Penrose drain that is wrapped flat (Fig. 2.4). The Penrose drain is wrapped around the finger from distal to proximal, extravasating the blood with the distal end free. When the base of the finger is reached, a loop of the drain is caught with the finger and elevated. This loop is clamped to the proximal end, creating a tourniquet. The distal end of the tourniquet is pulled, which unwraps from distal to proximal past the area to be operated on. The Penrose drain must not be allowed to roll and create pressure on the underlying digital nerves. Rubber band pressure is too localized over the nerves, and sometimes the rubber band is not removed with disasterous complications of skin or digit loss.

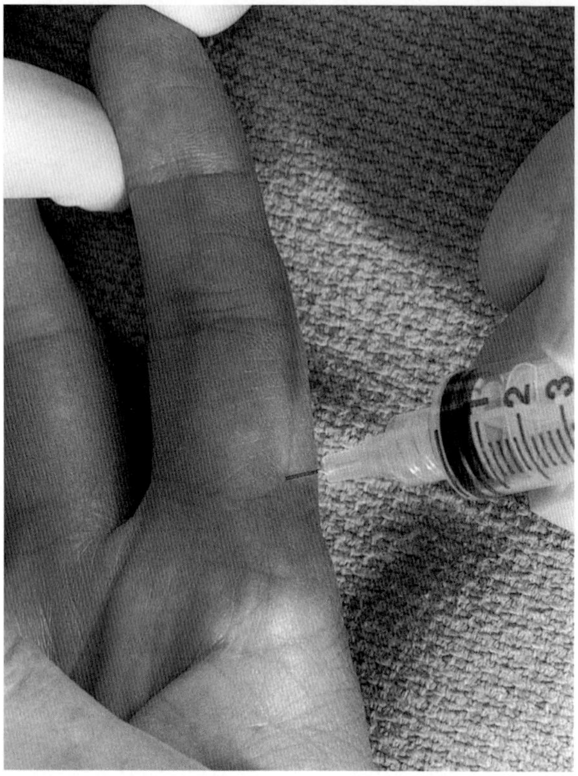

A

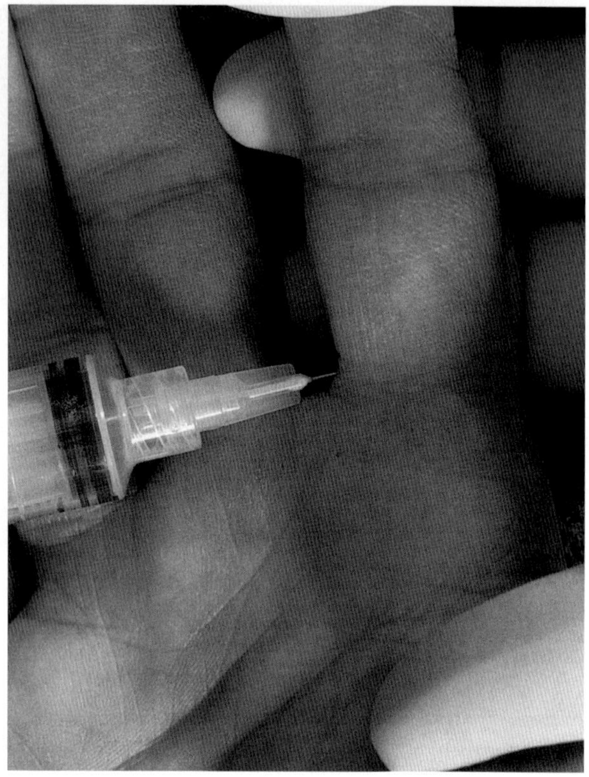

B

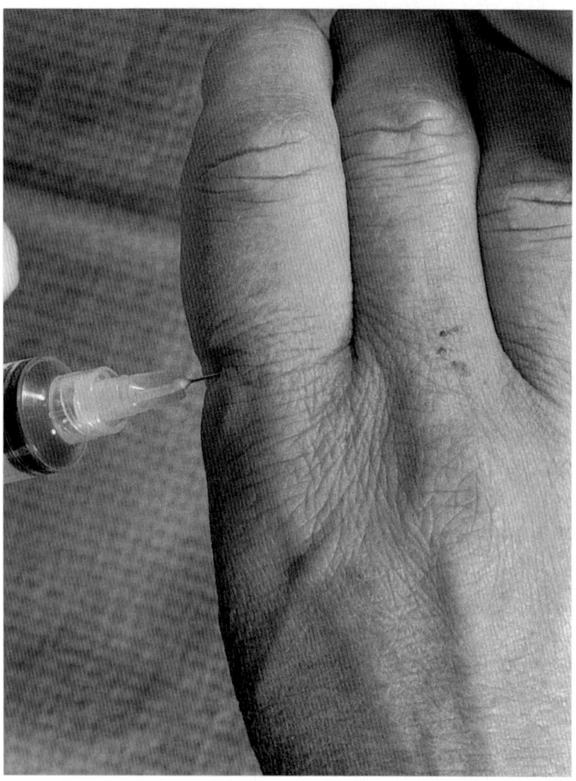

C

FIGURE 2.3. A,B: Lidocaine 1% (no epinephrine) injected bilaterally in the area of the common volar digital nerve. **C:** A bead of anesthetic across the dorsum usually is necessary for nail bed operations.

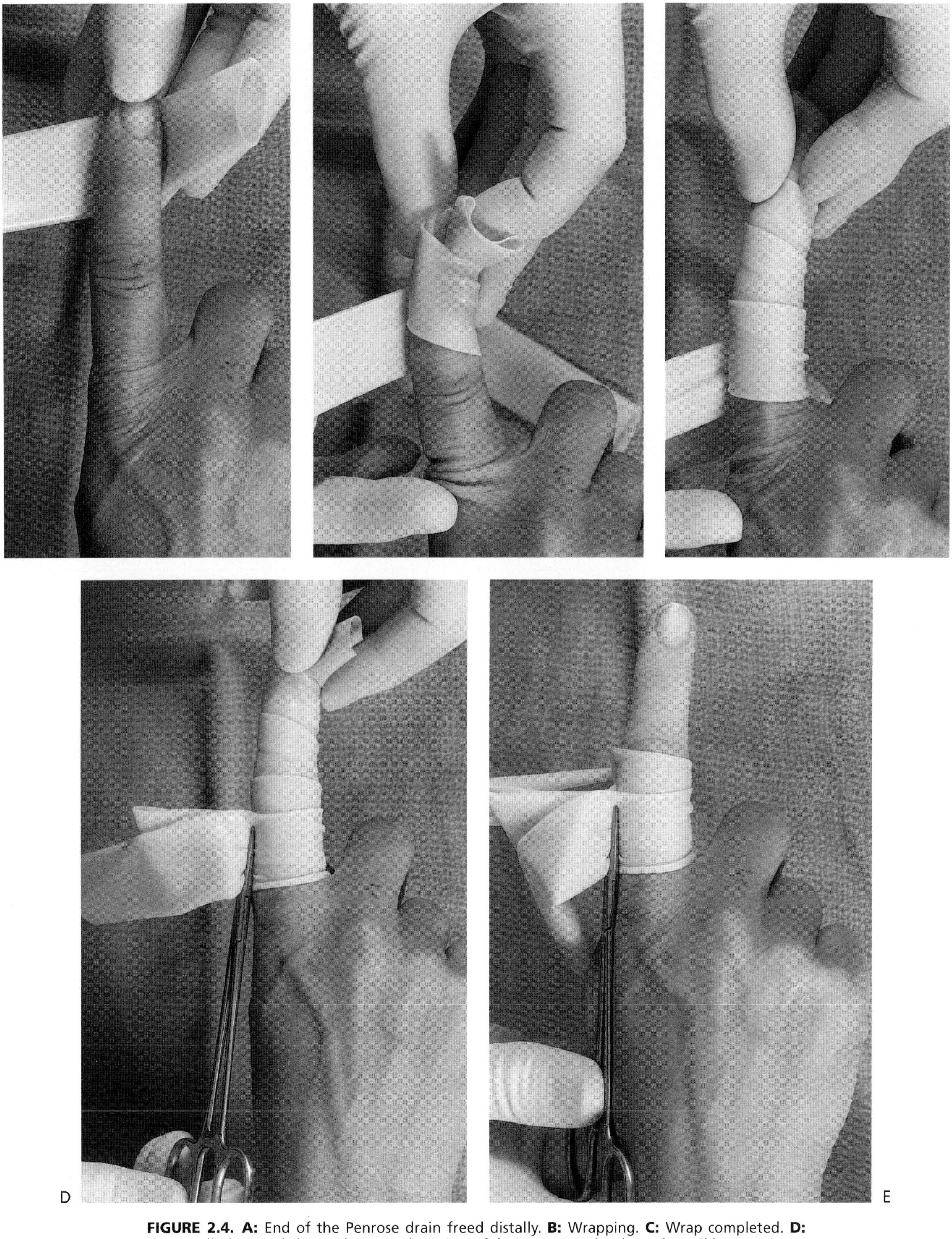

A,B C D E

FIGURE 2.4. A: End of the Penrose drain freed distally. **B:** Wrapping. **C:** Wrap completed. **D:** Hoop pulled up and clamped. **E:** Distal portion of drain unwound to bare the nail but continues to act as a tourniquet. Note the Penrose drain is flat, not rolled.

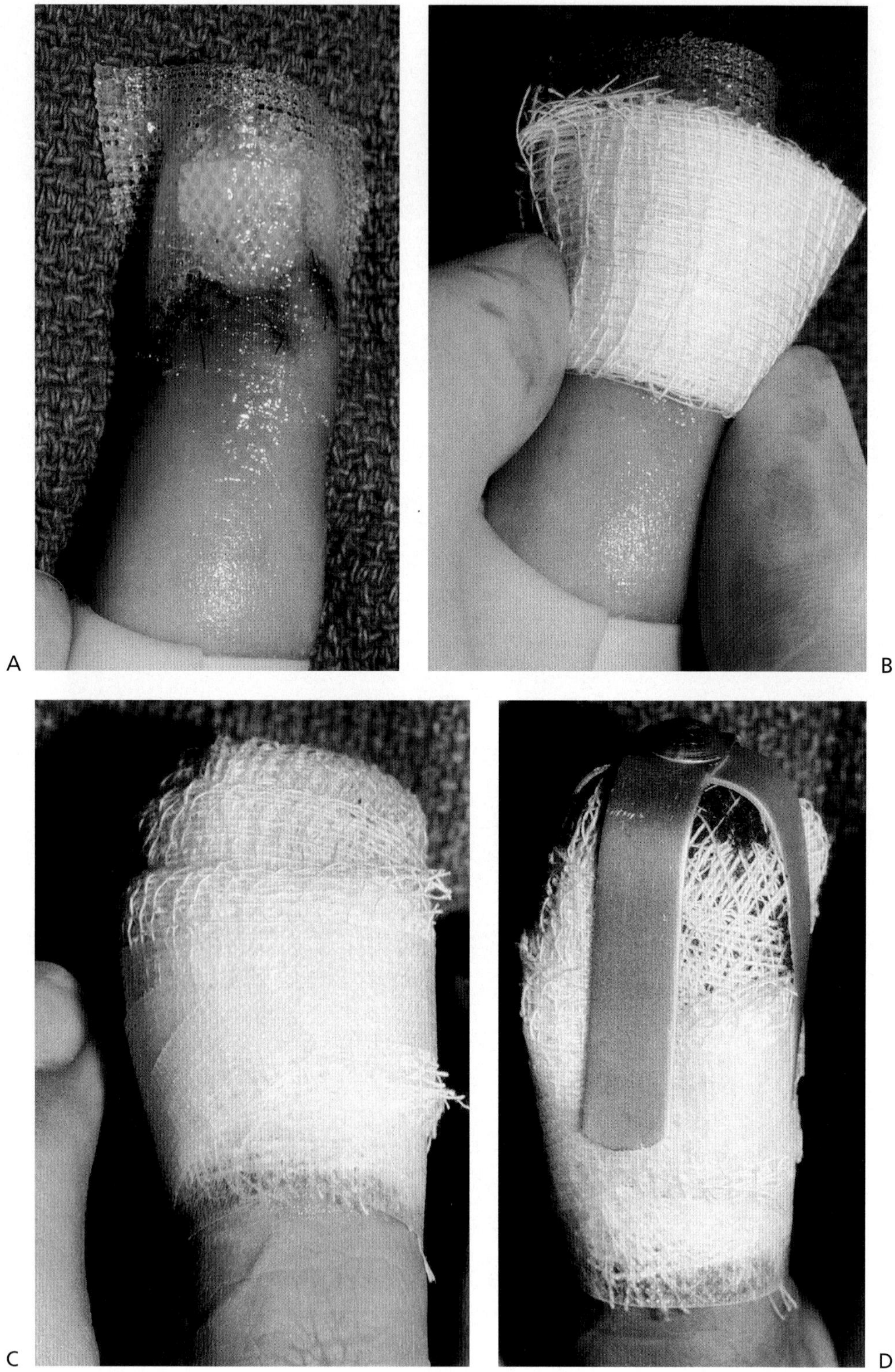

FIGURE 2.5. A: Adaptic placed over nail bed. **B:** Gauze to absorb drainage. **C:** Cling to hold gauze. **D:** Metal or plastic four-prong splint.

Tourniquets should not be left on any longer than necessary. Plan and draw on the perionychium before placing the tourniquet. Tourniquet release every 30 minutes is advisable, but they may be used for as long as 60 minutes, if necessary.

Prepping and Draping

The sequelae of infection after operation on the perionychium are serious, and precautions should be taken. Antibiotics are rarely indicated unless infection is present. Soaking in soapy water and cleaning with a nail brush for several days will help to remove embedded dirt. Before placement of a tourniquet, the finger should be washed with surgical soap (I prefer povidone iodine [Betadine]) as instructed by the soap manufacturer. Towels and/or eye hole drapes are convenient to drape the finger. Drapes should be sterile and kept sterile throughout the procedure. If the nail is appropriate for replacement into the nail fold after the procedure, I prefer to place the nail in a container of povidone iodine solution during the procedure to clean it and reduce bacterial counts. Sterile gloves should be worn to protect the patient and surgeon.

Dressings

Dressings should be as nonadherent as possible so that the nail or nail bed is not disturbed when the dressing is changed. To accomplish minimal nonadherence, use a water-soluble gauze such as Adaptic xeroform (Fig. 2.5A). A bulky dressing is placed over the wound to absorb serum and blood (Fig. 2.5B–C). A plastic or metal four-poster splint can be used to immobilize the finger and protect it from bumping (Fig. 2.5D). The dressing is left in place for 7 to 10 days. Soaking with saline or hydrogen peroxide before removal will loosen clotted blood in the bandage. After dressing removal, the patient is encouraged to soak the finger in warm soapy water three to four times a day for 4 to 10 minutes to loosen extraneous material and clean the finger. After removal of the original bandage, an adhesive bandage (Band-Aid) and the protective splint frequently are all that are needed.

Complications

Infection is a serious complication after nail bed surgery. It will cause destruction of the nail bed by direct involvement of bacteria or by pressure from edema. Any purulence should be drained quickly by removing the bandage or the nail.

Nail deformity secondary to injury of the nail bed during an operation may cause nonadherence, splitting or irregularity of the surface, or lack of growth. Gentle technique and fine instruments are necessary to prevent trauma to the soft tissue.

CHAPTER 3

BASIC TECHNIQUES

3A

NAIL REMOVAL

EDWARD A. KRULL

The principles of dermatologic surgery, including appropriate history, physical assessment, and especially determination of the adequacy of the vascular system, must apply to all nail surgery (Table 3A.1) (1). Doppler studies and consultation with vascular surgeons may be indicated. The nail surgeon must develop a disciplined approach, especially focusing on the vascular system, to determine whether the patient is a suitable candidate for surgery. The absence or presence of pulses (1,2), quality of the skin, presence or absence of hair on the digits, and rate of capillary refill following pressure on the pulp of the digit are some of the important clues to circulation to the digits (Table 3A.2). Assessment of the patient may result in the decision to perform more conservative measures of nail care rather than nail surgery.

NAIL PLATE AVULSION

Avulsion of the nail plate is a basic procedure that is necessary for exploration of the nail bed and matrix. Removal of the entire nail plate permits examination for subungual tumors and other lesions involving the nail matrix and nail bed. It is an essential step in examination to confirm diagnoses and to undertake some surgical procedures (Fig. 3A.1). Partial nail plate removal is indicated for ingrown toenail surgery (Fig. 3A.2), both scalpel and phenolization. Removal of the nail plate helps in management of onycholytic nails by removal of the separated nail plate and curetting away the underlying debris to aid in reattachment of the regrowing nail (3). Removal of the proximal third of the nail plate may clear acute paronychia that is resistant to antibiotics (4).

There are a number of instruments and techniques that are used to remove the nail plate (2,3,5–9). Dental spatulas (5), nail elevators, Freer septum elevators (Figs. 3A.3, 3B), especially for proximal-to-distal nail avulsion, and mosquito hemostats have their advocates (2,3,5,8–11). In the author's experience, a small-nosed mosquito hemostat is especially useful (Fig. 3A.4). The flat surface, rather than the sides of the mosquito hemostat, are inserted under the nail. The uniformity of the low height of the flat part of the hemostat elevates the nail less than the hemostat on edge. This is especially useful when only a partial nail plate removal is desired. Under those circumstances, a greater height of the hemostat on edge or another instrument could separate an unintended part of the nail from its nail bed and matrix. Some authors prefer to open the jaws of the hemostat and insert one of the jaws under the nail plate with the serrations or teeth in contact with the nail plate (2). Similarly, the single jaw of the instrument is used to free the nail plate from the overlying nail folds. I find that the serrations of the jaws of the hemostats cause resistance to insertion of the instrument.

Nail elevators or dental spatulas (5) are satisfactorily used for nail plate avulsion. They have the desirable low height, but are additional instruments to be purchased with little

E. A. Krull: Department of Dermatology, Henry Ford Hospital, Detroit, Michigan 48202.

TABLE 3A.1. GENERAL ASSESSMENT

1. General medical status
   A. Heart disease
   B. Hypertension
   C. Arthritis
2. Intermittent claudication
   A. Nocturnal rest pain in calf and foot
   B. Pain with leg elevation
      I. Suggestive of arterial disease
   C. Pain in calf with walking
3. Heart valvular disease and/or heart valve replacement may require prophylactic antibiotics
4. Joint replacement may require antibiotics
5. Artificial vessel replacement
   A. Stents in cardiac vessels reepithelialize in about 4 to 6 weeks and probably do not require prophylactic antibiotics
6. Diabetes
7. Raynaud's phenomenon
8. Bleeding diathesis
   A. Inherent
   B. Warfarin (Coumadin)
   C. Aspirin

Many of the conditions in the general assessment require discussion with the physician responsible for the patient or other appropriate physicians.

TABLE 3A.2. LOCAL ASSESSMENT OF CIRCULATION TO THE DIGITS

1. Pulses
2. Color of digits
3. Hair on digits
4. Nails: thickness, ridging, and other changes
5. Skin texture
6. Blanch with elevation; rubor with dependency
7. Refill time of pulp of digits after compression

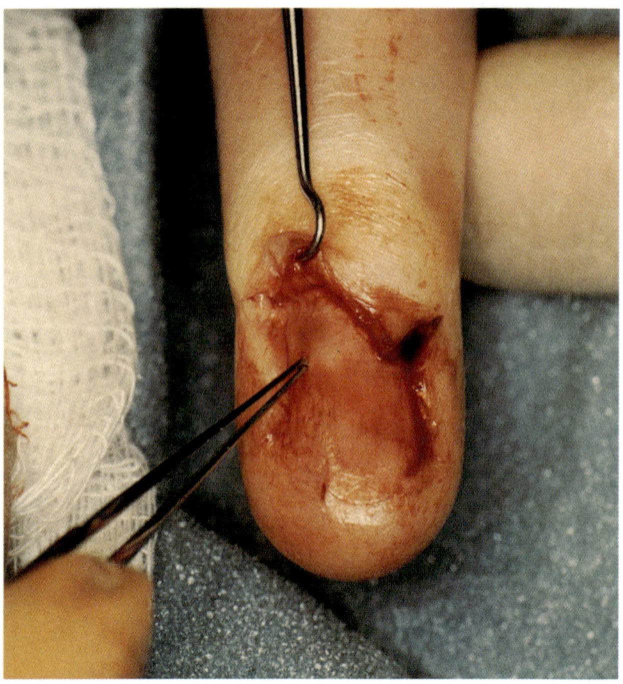

FIGURE 3A.1. Avulsion of the nail plate revealed bluish glomus tumor.

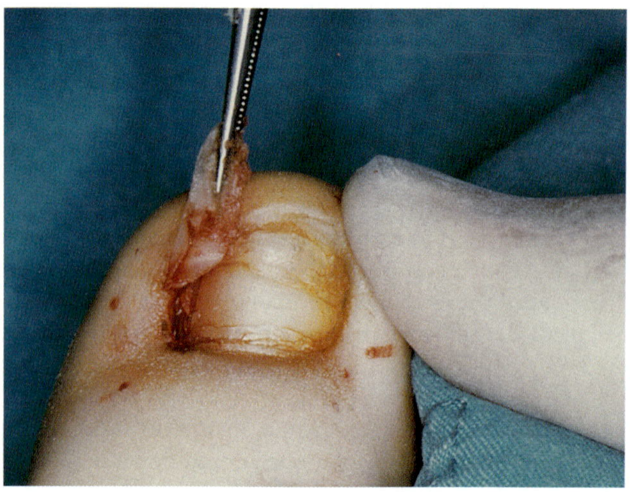

FIGURE 3A.2. Avulsion of the edge of the nail plate to repair ingrown toenail.

A

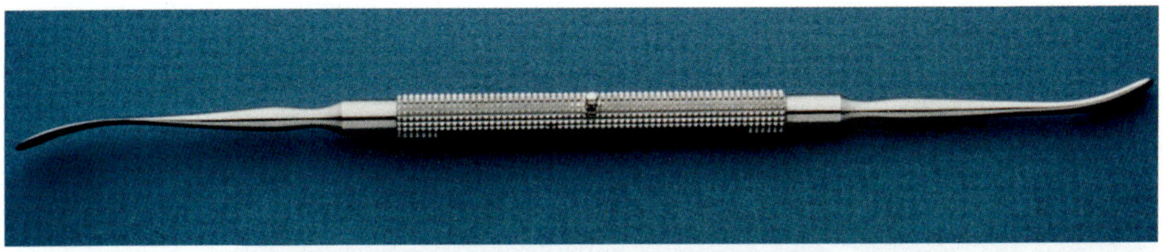

B

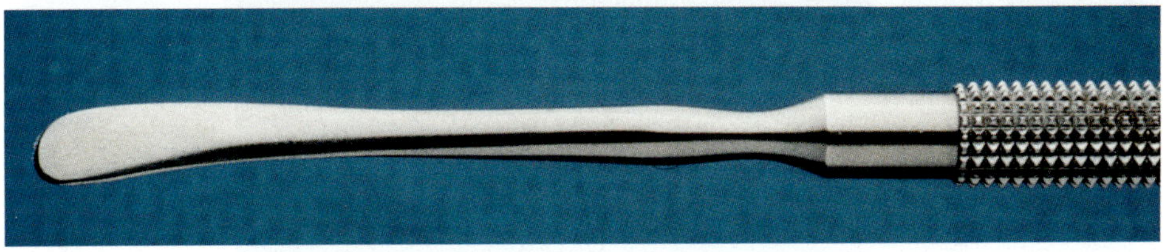

FIGURE 3A.3. A: Freer septum elevator with upturned ends. **B:** Closer view of Freer septum elevator demonstrating the width of the end.

other purpose. The sharper edge of the dental spatula can create false planes when separating tightly adherent nail plates. For these reasons, the author prefers to use a fine-nosed mosquito hemostat for nail avulsion.

If only a part of the nail is to be avulsed, then the English nail splitter or double action nail clipper is very helpful in cutting through the nail plate that is to be removed (Figs. 3A.5 and 3A.6). I usually prefer the English nail splitter because the smooth undersurface of the lower jaw glides along the nail bed and matrix, and the anvil-like upper surface easily slides under the nail plate (Fig. 3A.5). For longitudinal lateral wedges of nail, the mosquito hemostat or other instrument is used to free the nail plate from the underlying matrix and nail bed and the overlying nail folds (Figs. 3A.7A, 7B) (12,13). The anvil of the English nail splitter is inserted beneath the nail plate; the cutting part is worked between the proximal nail fold and the underlying nail plate (Fig. 3A.8). The nail plate

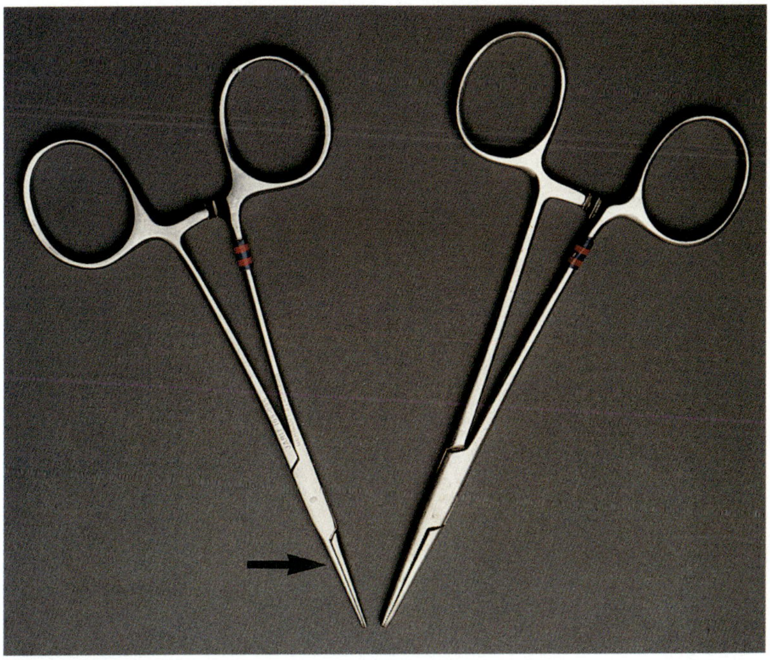

FIGURE 3A.4. Of the two hemostats, the author prefers the smaller-nosed hemostat for avulsion of the nail plate *(arrow)*.

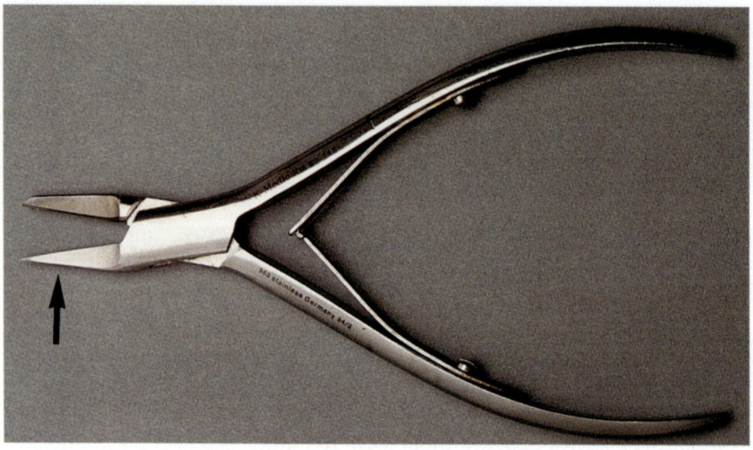

FIGURE 3A.5. English nail splitter with lower anvil-like jaw *(arrow)*.

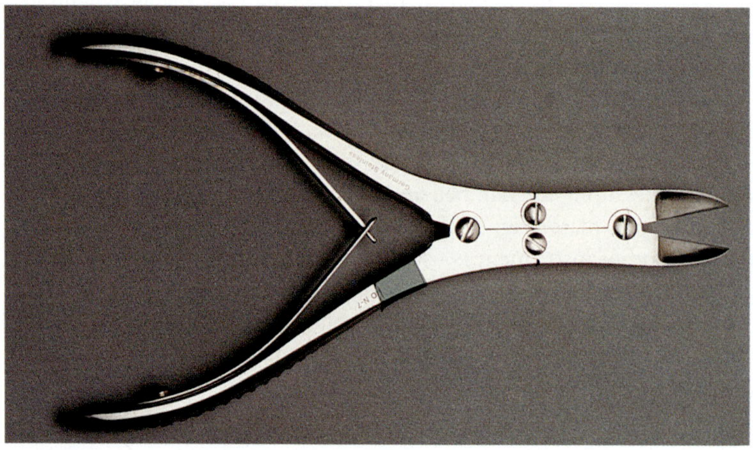

FIGURE 3A.6. Double action nail clipper.

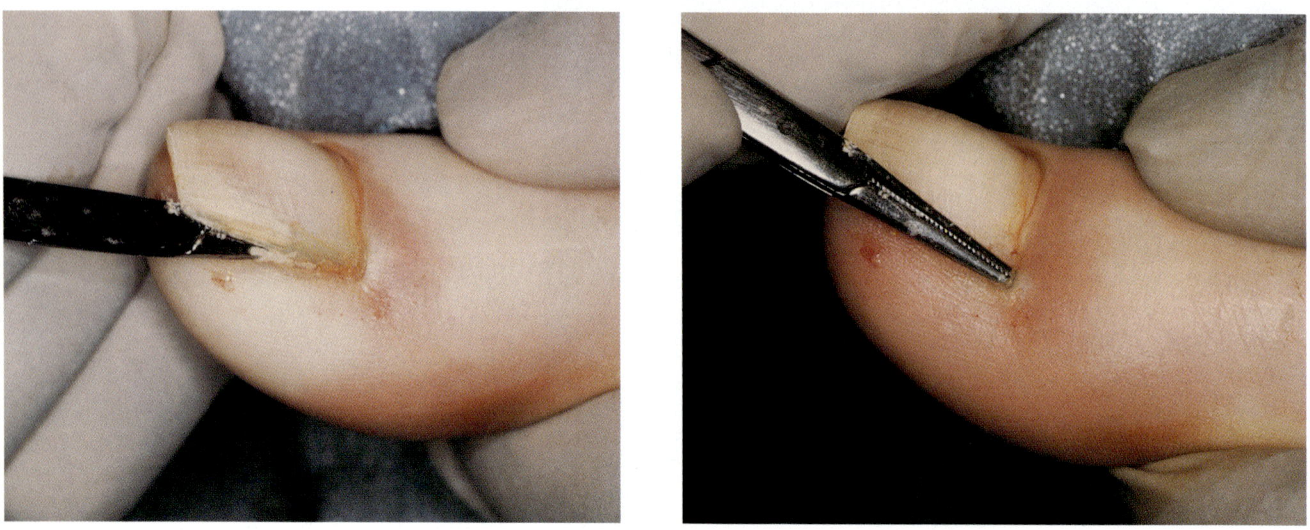

A B

FIGURE 3A.7. A: Mosquito hemostat inserted under the nail plate. **B:** Mosquito hemostat inserted beneath the proximal nail fold and above the nail plate.

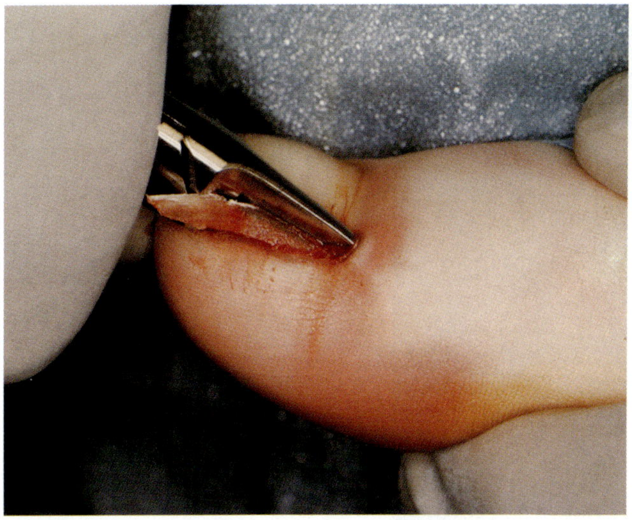

FIGURE 3A.8. English nail splitter inserted beneath and above the nail plate and the nail plate transected. Note the cutting edge of the nail splitter has been slid beneath the proximal nail fold to avoid injuring it.

then is cut through. For longer nails, such as the first toenail, or thicker nails that resist cutting, two steps may be required. If the nail edge is not removed easily, there are usually two causes: (i) the nail plate has not been cut through its entire longitudinal length; or (ii) the nail horn has not been adequately separated from its surrounding tissue.

A number of techniques have been used to free the nail plate from the underlying tissues. One method is to insert the hemostat or other instrument under the distal nail plate, pushing it proximally until there is a "give" (Fig. 3A.9). This sensation of release or "give" is the freeing of the nail plate from the proximal matrix. Until that point, there is a distinct resistance to the insertion of the hemostat under the nail plate. Additionally, the wall of the proximal nail groove blocks further insertion of the hemostat under the nail plate. The author prefers to grasp the handle of the instrument in the fingers and palm of the

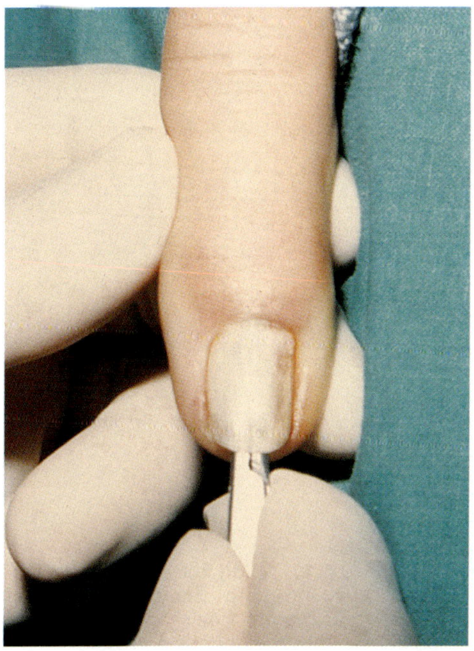

FIGURE 3A.9. Handle of hemostat grasped in fingers and palms and inserted between the nail plate and underlying nail bed and matrix.

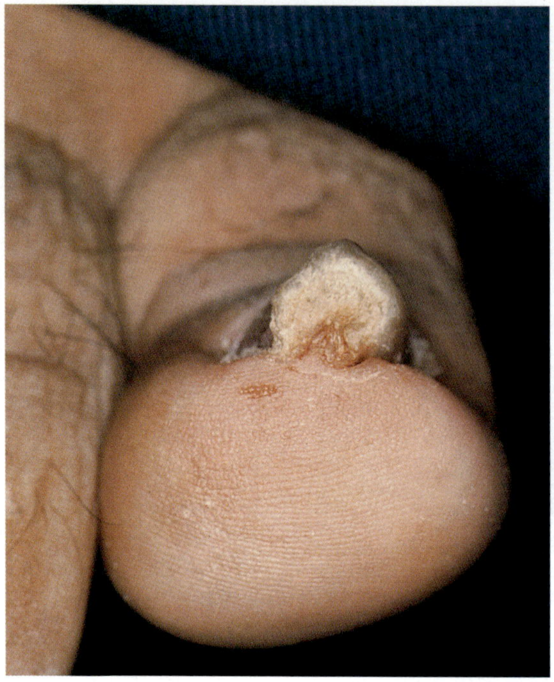

FIGURE 3A.10. Insertion of the mosquito hemostat under this pincer nail would require significant effort.

hand rather than inserting fingers into the two handle holes (Fig. 3A.9). Gripping of the entire handle results in better control and more strength in this position. The greater strength is more useful with thick nails, especially if they are deeply incurvated, such as in pincer nail (Fig. 3A.10). These nail plates offer significant resistance to insertion of the hemostat or other instruments.

Once the hemostat is inserted under the nail plate, a number of methods can be used to free the entire nail plate from the underlying nail bed and matrix. One method is to grasp the mosquito hemostat by the index and other fingers and use the thumb to grip the patient's toe or finger as a fulcrum (Fig. 3A.11). The hemostat is pulled toward the thumb side (Fig. 3A.11). The procedure is reversed, using the index finger as a fulcrum

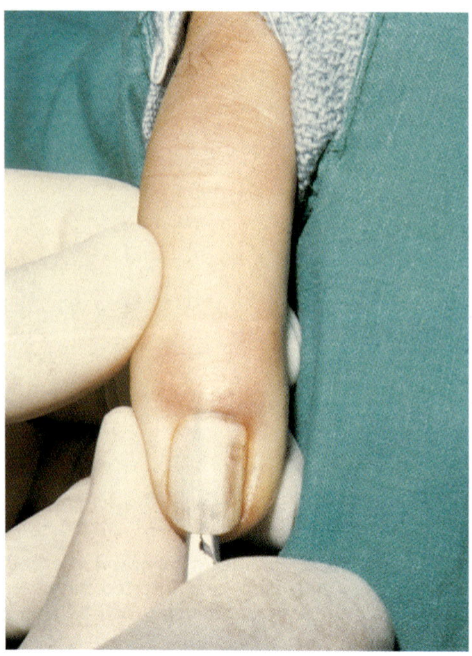

FIGURE 3A.11. Using the thumb as a fulcrum, the mosquito hemostat is pulled toward the thumb, freeing one-half of the nail plate from the underlying tissue.

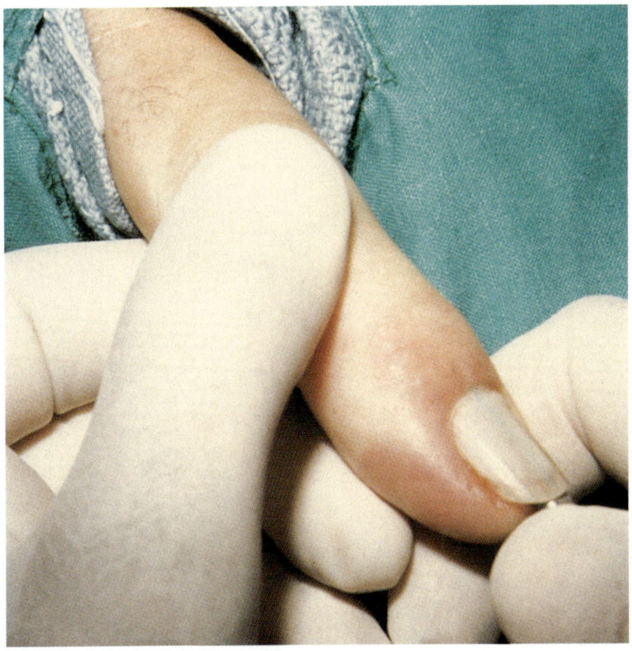

FIGURE 3A.12. Using the index finger as a fulcrum, the hemostat is pulled toward the index finger, freeing the remaining part of the nail plate from the underlying tissue.

and the thumb grasping the hemostat and pulling the hemostat toward the index finger (Fig. 3A.12). Through these two steps, the entire nail plate is separated from the underlying nail bed and matrix.

A similar process is used to free the nail plate from the overlying proximal and, to a lesser extent, lateral nail folds (Fig. 3A.13). The mosquito hemostat is inserted between the nail fold and the underlying nail plate. Then using the thumb and index finger alternately as a fulcrum, the instrument is pulled in both directions to free the underlying nail plate from the overlying lateral and proximal nail folds.

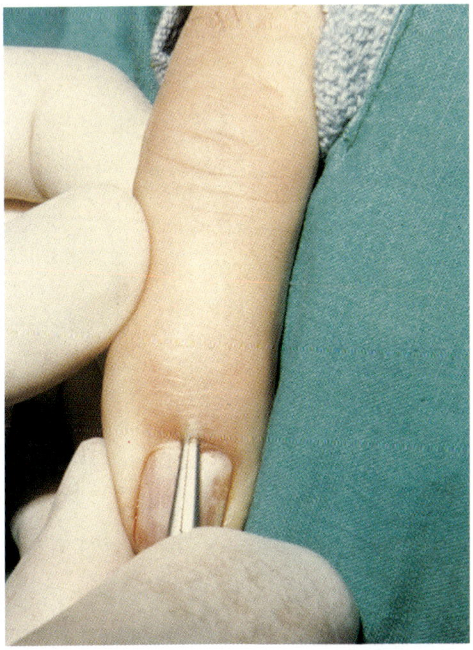

FIGURE 3A.13. In a similar manner as shown in Figs. 3A.11 and 3A.12, the proximal nail fold is freed from the underlying nail plate.

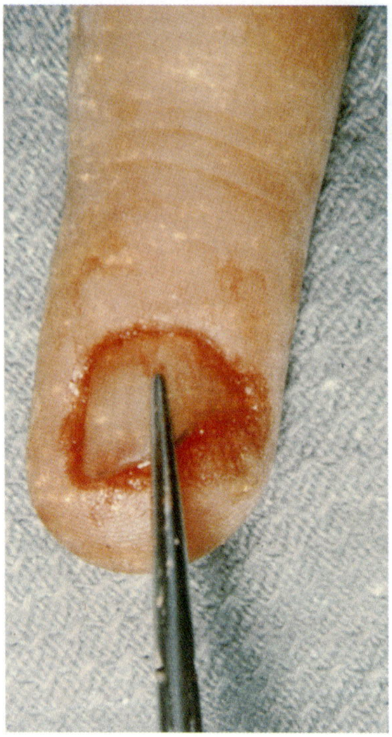

FIGURE 3A.14. A direct pull or rolling action should extract the nail plate.

The nail plate is grasped by the hemostat and pulled distally to remove it (Fig. 3A.14). Some surgeons use a rolling motion to free the nail plate. The author prefers the pull technique. At times, the nail plate resists removal. Usually the problem is the failure to free the corners or horns of the nail plate from the underlying nail matrix and overlying nail folds (Figs. 3A.15, 3A.16). If this occurs, the instrument is reinserted beneath the nail plate to focus the effort on freeing the nail plate from the underlying horns of the matrix. For the same reason, the instrument is inserted between the nail folds and the

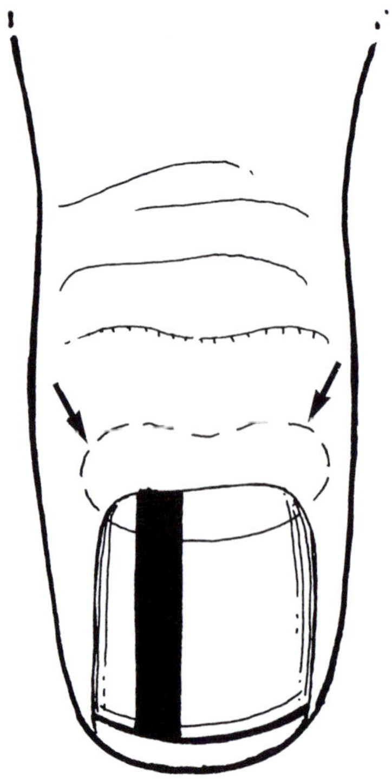

FIGURE 3A.15. Horns of the nail plate and matrix *(arrows)*. Nail plate has a longitudinal pigmented streak.

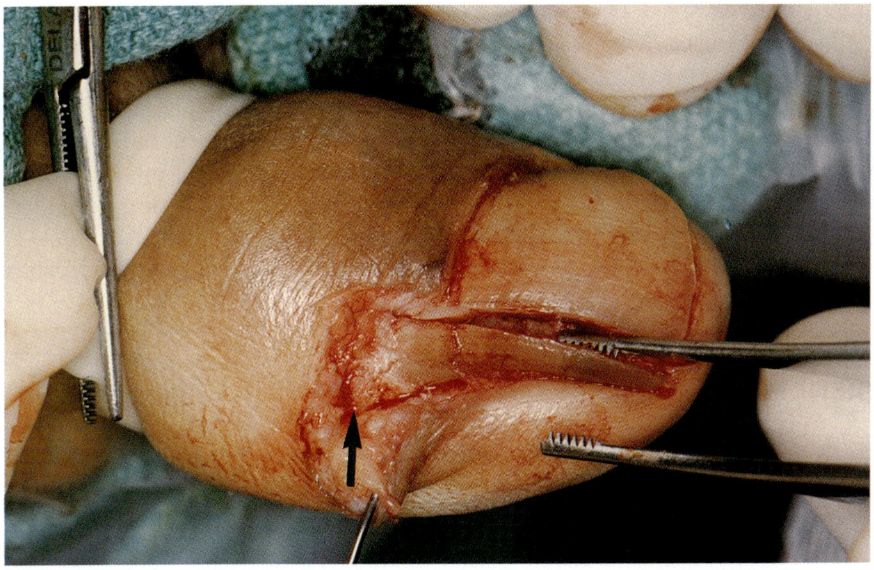

FIGURE 3A.16. Nail horn *(arrow)*.

underlying horns of nail plate. This may have to be repeated a number of times to allow easy removal of the nail plate.

Instead of using the fulcrum technique, multiple insertions of the hemostats may be used. The hemostat is inserted repeatedly across the width of the nail plate. The proximal and lateral nail folds overlying the nail plate are freed in a similar manner. The nail plate is extracted by a direct pull or rolling motion. If the nail plate resists removal, the instrument may need to be reinserted to free the horns of the nail.

Rather than using either of these methods to separate the nail plate from the underlying tissue, once the tip of the hemostat is inserted under the nail plate, the instrument handle can be opened, spreading the two jaws of the hemostat similar to the technique

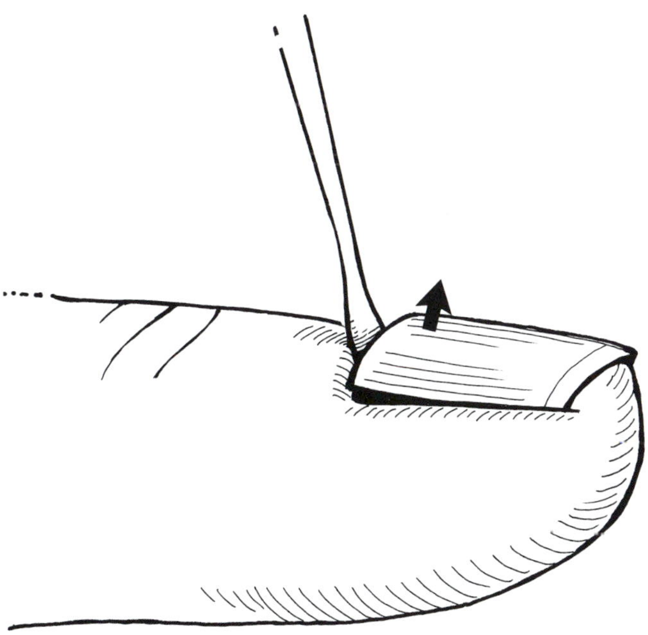

FIGURE 3A.17. Freer septum elevator frees the nail proximally to distally. This is especially useful when subungual material obscures the plane beneath the distal end of the nail plate.

used in undermining tissue. Because of the lesser strength in the function of abducting, or spreading of the fingers, as compared to adduction, or grasping, this method is not preferred by the author, especially in strongly attached nail plates.

At times because of the extensive subungual material and firm adherence of the nail plate, it is difficult to determine the plane between the nail plate and nail bed. This plane usually can be established by careful probing and exploration. However, if this is not possible, then it may be necessary to establish this plane from the proximal end of the nail plate, which is attached to the matrix and is free of the subungual material (Fig. 3A.17) (9–11). Separation of the nail plate begins proximally, first freeing the nail plate from the proximal matrix, and progresses distally.

REFERENCES

1. Balle M, Krull EA. Medical evaluation and universal precautions. In: Roenigk RK, Roenigk HH, eds. *Roenigk and Roenigk's dermatologic surgery: principles and practice*, 2nd ed. New York: Marcel Dekker, 1996:53–64.
2. Haneke E, Baran R. Surgery and traumatic abnormalities. In: Baran R, Dawber RPR, eds. *Diseases of the nail and their management,* 2nd ed. London: Blackwell Science, 1994:345–371.
3. Daniel CR III. Onycholysis: an overview. *Semin Dermatol* 1991;10:34–39.
4. Haneke E, Baran R, Bureau H. Surgery of the nail region. *Z Hautkr* 1957;15:1107–1116.
5. Albom MJ. Avulsion of a nail plate. *J Dermatol Surg Oncol* 1977;3:34–35.
6. Siegle RJ, Swanson NA. Nail surgery: a review. *J Dermatol Surg Oncol* 1982;8:659–666.
7. Baran R, Haneke E. Surgery of the nail. In: Epstein E, Epstein E Jr., eds. *Skin surgery,* 6th ed. Philadelphia: WB Saunders, 1987:534–547.
8. Daniel CR III. Basic nail plate avulsion. *J Dermatol Surg* 1992;18:685–688.
9. Baran R. More on avulsion of nail plate. *J Dermatol Surg Oncol* 1981;7:854.
10. Cordero FA. Albacion ungueal: so uso en la onicomicosis. *Dermatol Int* 1965;14:21–23.
11. Scher RK. Surgical avulsion of nail plate by a proximal distal technique. *J Dermatol Surg Oncol* 1981;7:296–297.
12. Salasche SJ, Peters VJ. Tips on nail surgery. *Cutis* 1985;35:428–438.
13. Rich P. Nail biopsy. Indication and methods. *J Dermatol Surg Oncol* 1992;18:673–682.

3B

EXPLORATION OF NAIL TISSUE

EDWARD A. KRULL

Once the nail plate has been removed, there may be need to explore the entire nail unit. The proximal nail fold obscures the proximal matrix and lateral horns of the nail unit. Incisions through the proximal nail fold can be specifically designed for the size and location of the lesion (1). If the more central part of the matrix is the focus of examination, incisions are made extending from the edge of the proximal nail fold at the junction of the lateral and proximal nail folds (Fig. 3B.1) (2–7). Retraction of the proximal nail fold can be achieved by skin hooks, tissue forceps, or sutures placed in the folds (Fig. 3B.1) (7). A more laterally directed incision will expose more of the horns of the matrix (6). If the lateral nail folds are hypertrophied, as is commonly seen in ingrown nails, or if there is marked transverse curvature of the nail plate, as in pincer nails, a straight incision through the proximal nail fold will not be adequate to visualize the most lateral part of the matrix (Fig. 3B.2). This visualization of the most lateral part of the matrix, or the horns of the matrix, is critical in procedures to narrow the nail plate for treatment of ingrown nails or pincer nails, to perform lateral nail biopsies, and to undertake total matricectomies. Retention of a small piece of matrix may lead to regrowth of a spicule of

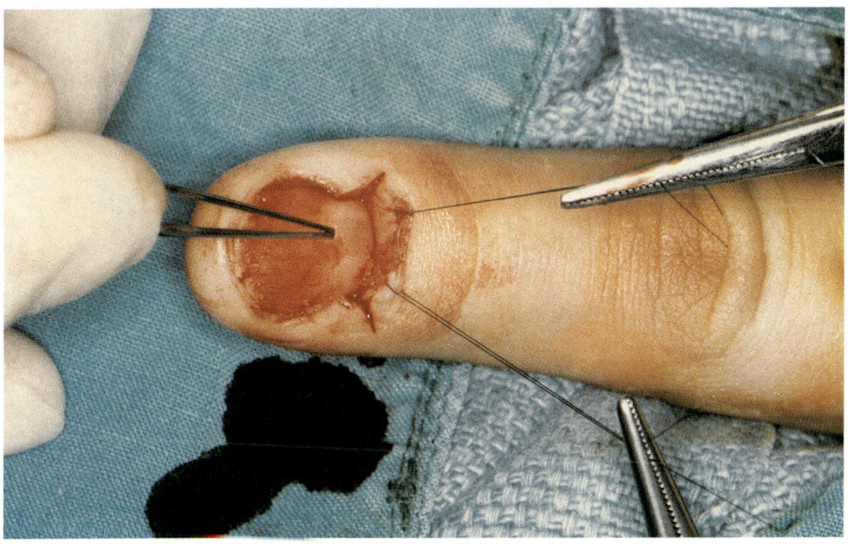

FIGURE 3B.1. Laterally directed incisions through the proximal nail fold to expose the origin of a pigmented nail streak. Sutures placed through the proximal nail fold are used as retractors.

E. A. Krull: Department of Dermatology, Henry Ford Hospital, Detroit, Michigan 48202.

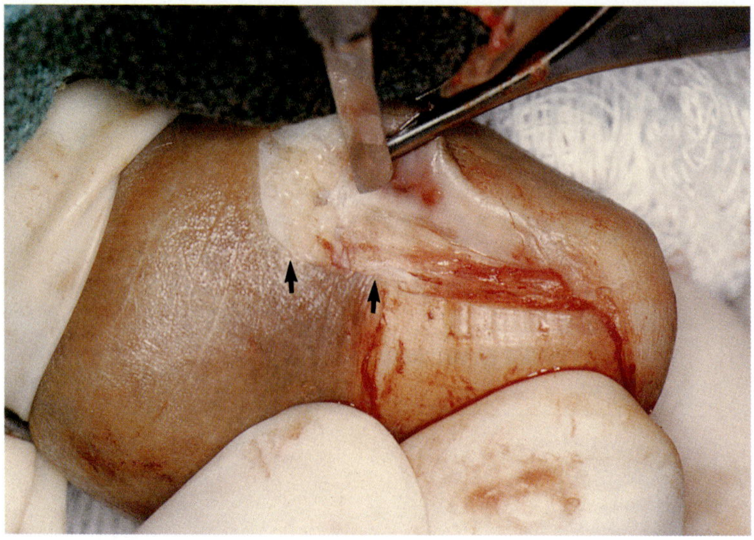

FIGURE 3B.2. A laterally curved incision through the proximal nail fold may be necessary to visualize the lateral matrix horns. The straight part of the incision *(arrows)* corresponds to the length of the proximal nail grove.

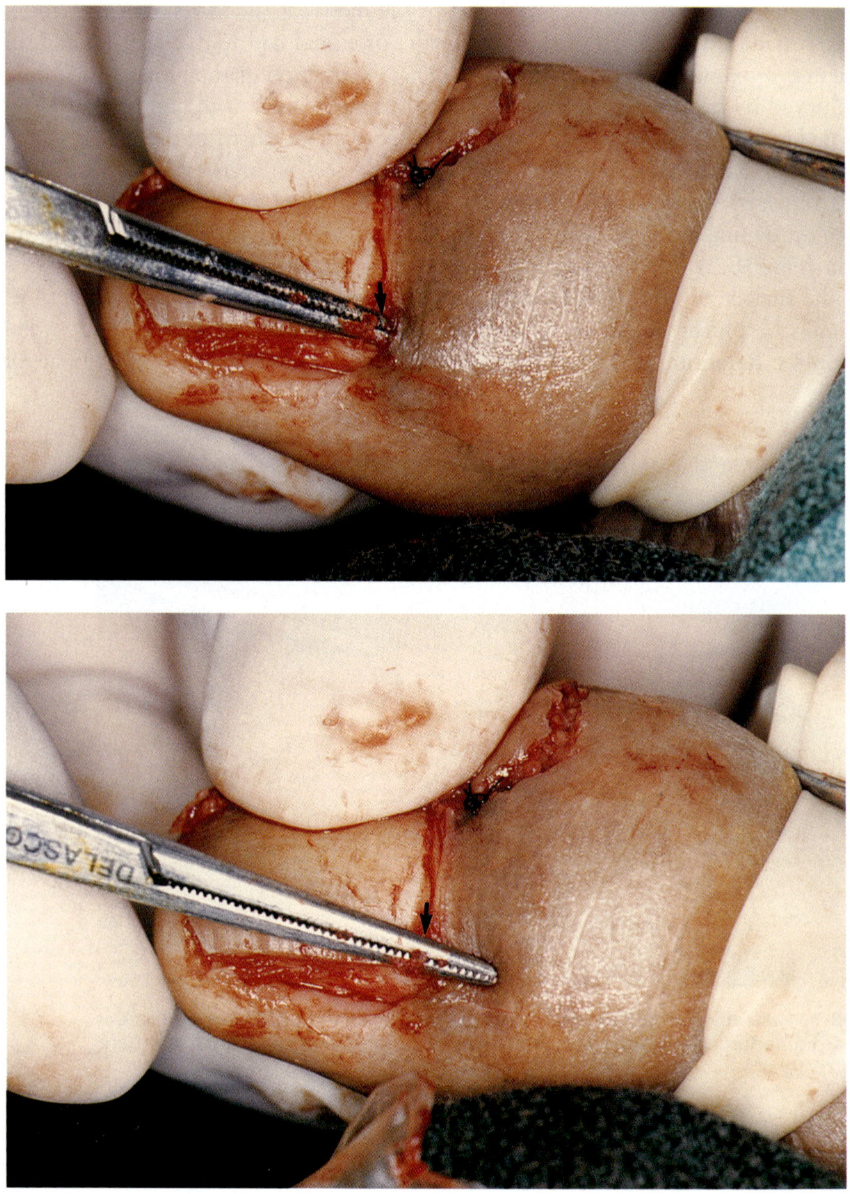

A

B

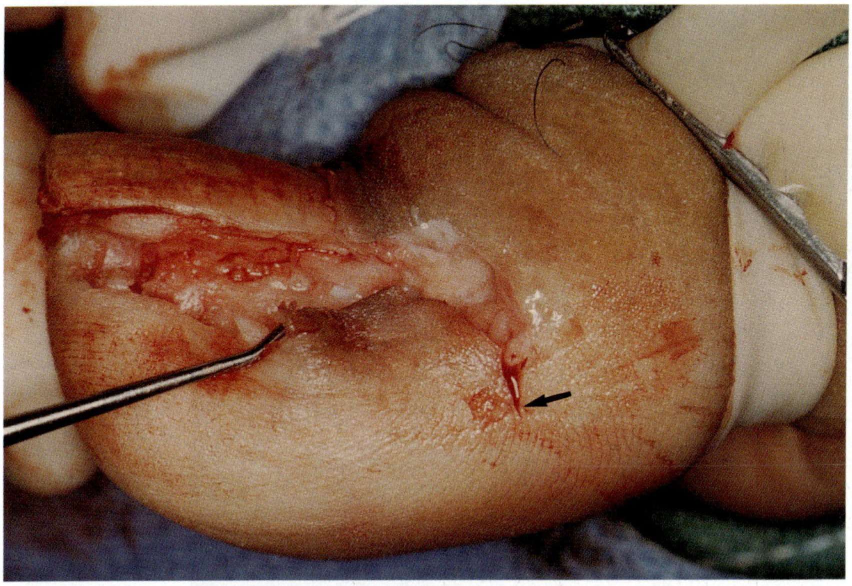

FIGURE 3B.4. The laterally directed curved incision may extend to about the junction of the dorsal and ventral surfaces of the toe *(arrow)*.

nail or the formation of an epidermal cyst. A curved incision through the proximal nail fold will achieve better visualization of the lateral horns of the matrix. The length of the straight part of the incision to the beginning of the curve is closely related to the length of the proximal groove, which corresponds to the invagination of the nail plate (Figs. 3B.2 and 3B.3). The laterally directed part of the curved incision can be extended to whatever distance is needed to visualize the most lateral parts of the matrix. In markedly incurvated nails, the incision may extend half way around the digit to the junction of the dorsal and ventral surfaces (Fig. 3B.4).

For more specific or localized lesions, retractions of part of or the total proximal nail fold may be all that is needed (7). The incisions of the proximal nail fold should be appropriately placed to allow this retraction (Fig. 3B.5) (7).

Care should be taken to avoid injury to the underlying matrix by inadvertently extending the incision through the entire thickness of the proximal nail fold and cutting into the matrix. The empty space of the proximal nail groove created by removal of the nail plate usually protects from this injury. If the nail plate is retained, it will act as a barrier to protect the matrix (Fig. 3B.6). An instrument, such as a Freer septum elevator, may be inserted between the proximal nail fold and the underlying matrix to protect the latter from injury. If, however, the matrix is inadvertently cut, closure by absorbable sutures usually is not necessary because of the narrow width of incision. If the injury does not involve the proximal 20% of the matrix, which forms the dorsal nail plate, there is even less need for suture closure.

FIGURE 3B.3. Insertion of a small-nosed mosquito hemostat between the nail plate and the overlying nail fold determines the length of the straight part of the incision in the proximal nail fold (*arrows* in Fig. 3B.2).

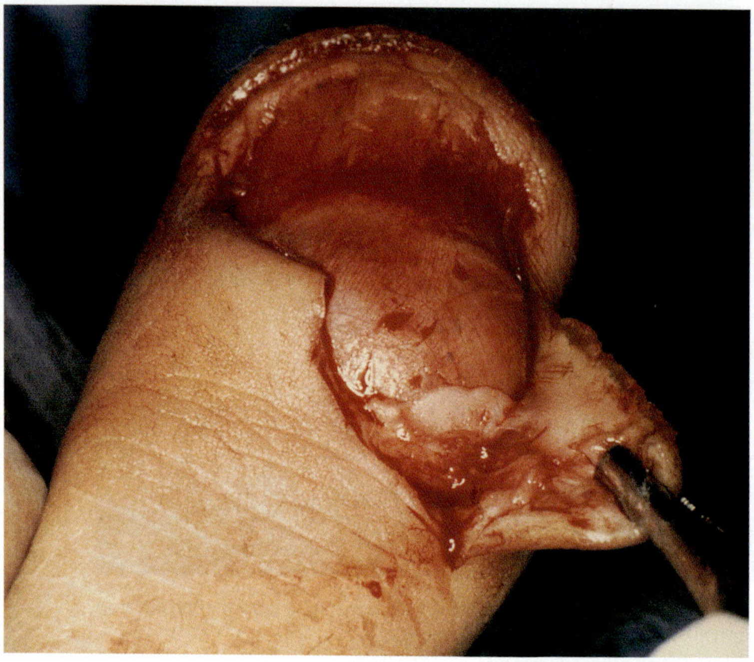

A

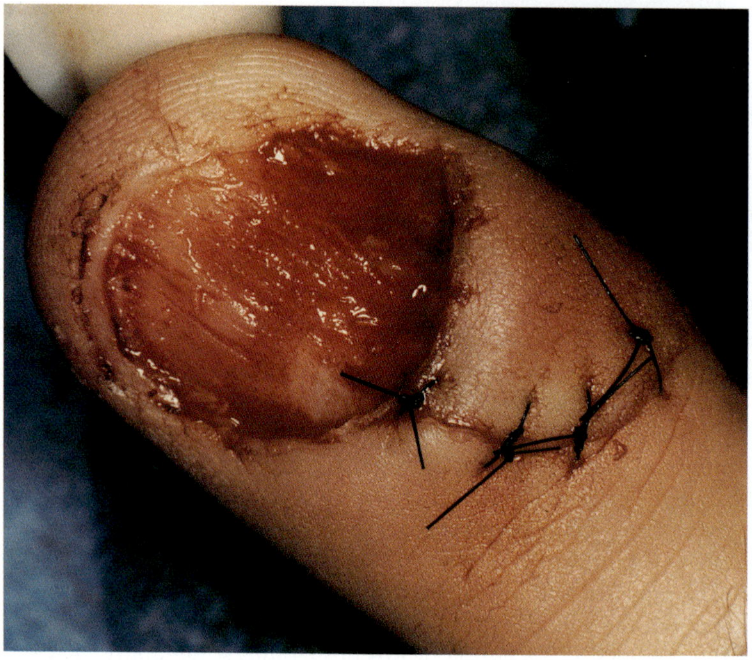

B

FIGURE 3B.5. A: Incision for enchondroma started in the midline of the proximal nail fold and extended laterally for adequate exposure of the tumor. **B:** Incision of the proximal nail fold sutured.

3C

BIOPSY TECHNIQUES

EDWARD A. KRULL

There are a number of ways to perform biopsies of nail tissue. The methods chosen depend on a number of factors, some of which include the following:

A. Location of the lesion under investigation
B. Necessity for nail plate removal or retention
C. Tissue that is being studied
D. Need for exploration beneath the nail folds
E. Minimization of potential nail dystrophy, scarring, and disfigurement, yet obtaining adequate and appropriate tissue for diagnosis (1–4).

There are some basic principles to nail biopsy (Table 3C.1). If the epithelium of the matrix and/or the nail bed is the focus of histopathologic examination, the nail plate preferably should not be removed. The epithelium of the matrix and, especially the nail bed, is tightly adherent to the overlying nail plate, and it can be damaged by nail plate avulsion (5,6). If the lesion is suspected of being dermal or involving the bone, removal of the nail plate has little, if any, effect on the tissue under study.

If an adequate biopsy can be taken while avoiding the proximal 10% to 20% of the matrix, significant nail plate deformity usually can be avoided (4). Because the dorsal nail plate is generated by the proximal matrix, injury to the proximal matrix may lead to a split or dystrophic nail (Fig. 3C.1) (1–5). If the proximal matrix must be included in the surgery, suture closure should be considered. If a transverse rather than a longitudinal incision of the matrix is feasible, for example, when removing a large glomus tumor, the transverse incisions or biopsy that avoids the proximal (2–4,7–9) matrix will be less likely

TABLE 3C.1. BIOPSY OF NAIL

Biopsy methods
1. Avulsion, partial or total, of the nail plate
   A. Tumors of the nail bed and matrix, such as
      I. Glomus tumors
      II. Epidermoid carcinoma
      III. Keratoacanthoma
      IV. Enchondroma
      V. Exostoses
      VI. Split nail
2. Retention of the nail plate
   A. Inflammatory disease or other dermatoses
      I. Psoriasis
      II. Lichen planus
      III. Other
   B. Longitudinal melanonychia

E. A. Krull: Department of Dermatology, Henry Ford Hospital, Detroit, Michigan 48202.

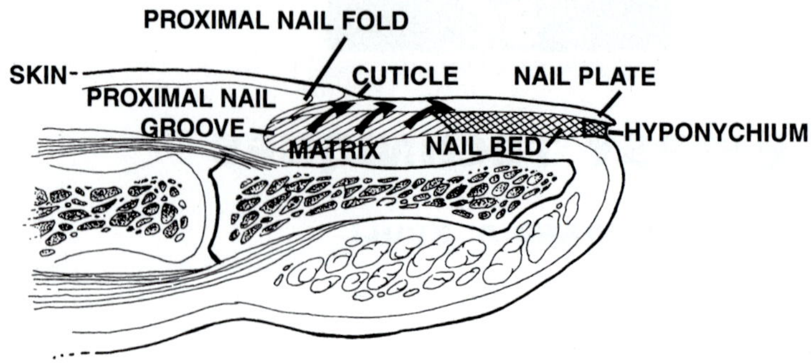

FIGURE 3C.1. Longitudinal cross-section of the nail unit. Proximal matrix produces the dorsal nail, middle matrix gives rise to the intermediate nail, and distal matrix is the origin of the ventral nail *(arrows)*.

to affect the nail plate (Figs. 3C.2 and 3C.3). If possible, it is important to retain the curved distal configuration of the lunula (2,4,7,9). However, Zaias (10) determined that longitudinal elliptical biopsies less than 3 mm wide through the proximal matrix usually will not produce split nails, but a nail dystrophy may occur. Others tend to find this rule unpredictable and prefer to perform longitudinal biopsies on the lateral edges (1). A 2- to 3-mm punch biopsy through matrix and nail bed that avoids the most proximal matrix does not require surgical repair.

In some conditions, such as psoriasis, lichen planus, and other inflammatory diseases, the process may involve the entire width of the matrix and possibly the nail bed. In such cases, instead of a paracentral biopsy with its possible nail plate effect, an excisional biopsy of the lateral matrix and nail bed is preferable (1–4,8). This surgery results in a narrowing of the nail plate, but it avoids a possible split or dystrophic nail plate. The narrower nail plate may be difficult to recognize as having been altered.

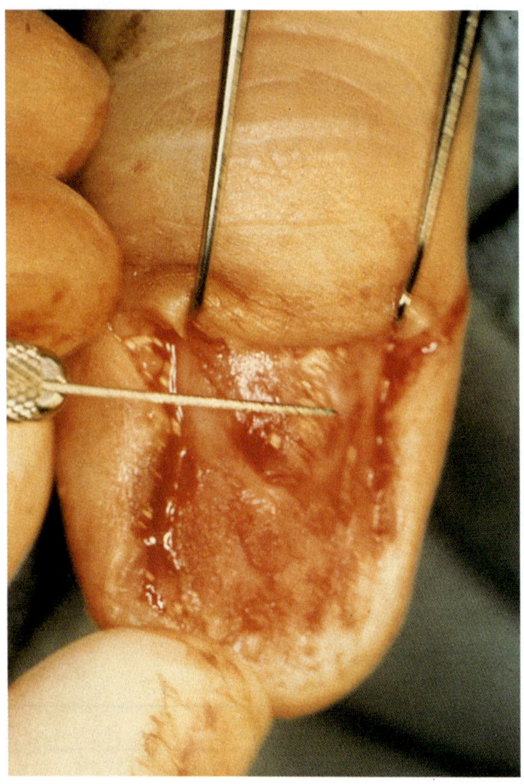

FIGURE 3C.2. Transverse incision is preferable, if possible, to excise this glomus tumor.

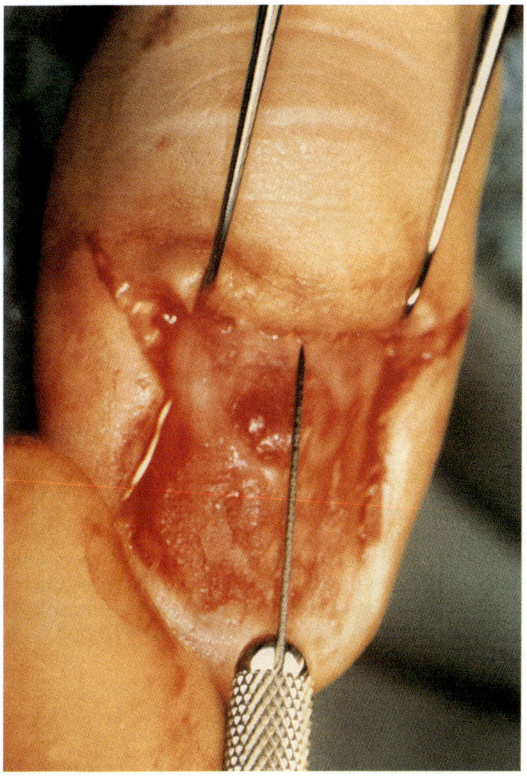

FIGURE 3C.3. Longitudinal incision for excision of the glomus tumor is less desirable because of possible injury to the proximal matrix.

PUNCH BIOPSY

Nail Plate Attached

A 2- to 3-mm punch can be made through the nail plate to include the underlying tissue (Figs. 3C.4 and 3C.5). A sharp disposable punch facilitates the procedures (5). Reflection of the proximal nail folds may be necessary to visualize the lesion, such as the proximal part of a pigmented streak (Fig. 3C.5A) The depth of the punch should extend to the underlying bone (8,9,11). The nail plate overlying the more proximal matrix is softer and more easily incised than the distal nail plate. Soaking the digit in water for 10 to 15 minutes will soften the hard nail plate without distorting the underlying tissue. The nail plate

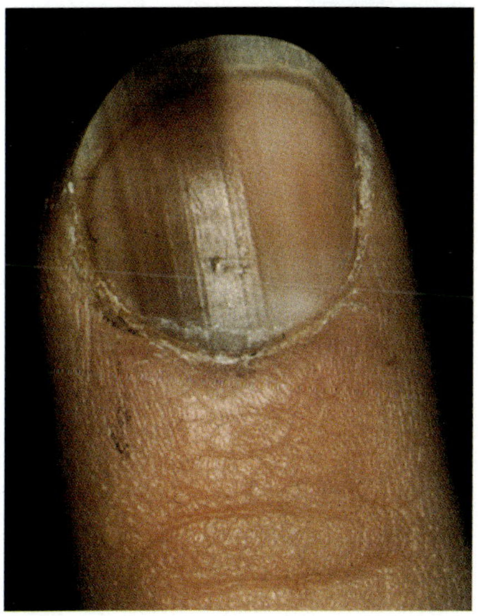

FIGURE 3C.4. Wide, dark, pigmented nail streak (longitudinal melanonychia).

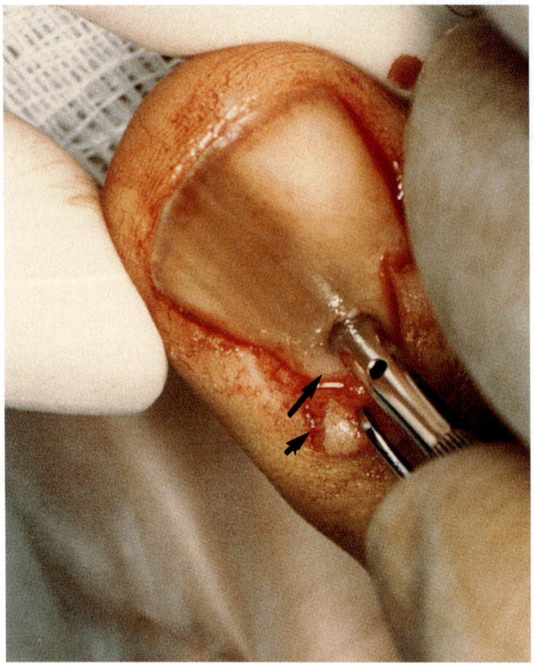

A

FIGURE 3C.5. A: Punch biopsy through nail down to bone. Pigmented streak appears to begin in distal matrix (*large arrow* points to clear proximal nail plate; *smaller proximal arrow* points to incision of proximal nail fold that was necessary to explore matrix and the source of pigmentation). **B:** Pigmented streak. Exploratory incisions in proximal nail fold. **C:** Punch biopsy completed.

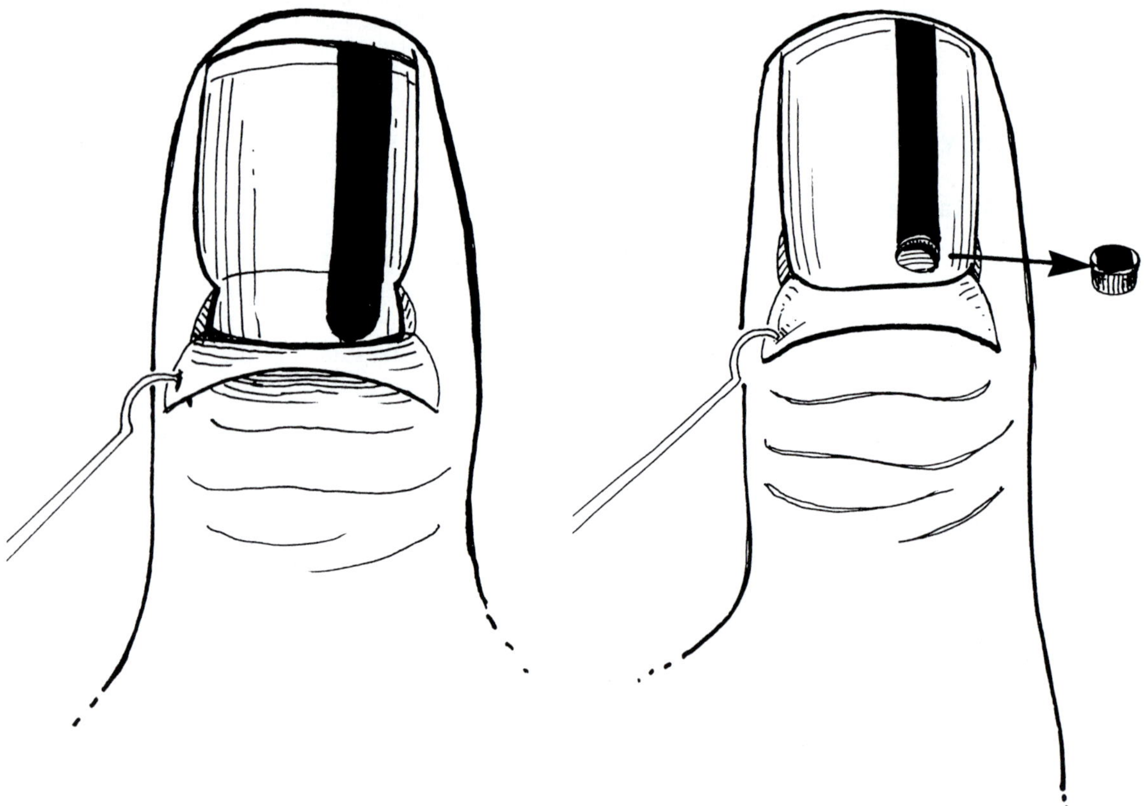

B

C

may be thinned using a hand-held rotary instrument to which a grinding burr is attached to facilitate passage of the punch through the thinned nail plate (6). In removing the tissue, care is needed to avoid separating the nail plate from the underlying tissue. Beaver blades, Bishop Harmon forceps, sharp pointed scissors, and skin hooks are helpful for removal of the tissue. Skewering the specimen with a 30-gauge needle and then bending the distal tip with a hemostat to keep the specimen from slipping off may facilitate

Nail Plate Avulsed

Removal of the nail plate may have a deleterious effect on histopathologic examination of the tissue (5), especially the epithelium of the matrix and nail bed. At times the nail plate must be removed to identify the lesion that needs to be biopsied. Incisions in the proximal nail fold may be necessary to expose the tissue under investigation.

The technique is similar to that described for the adherent nail plate, except that the procedure is facilitated by the absence of the nail (Figs. 3C.9–3C.12). The biopsy specimen is removed more easily than when the nail is retained. The avulsion technique of the total or partial nail plate has already been described. Bleeding is more readily controlled. Electrosurgery can be used more precisely for hemostasis, but usually is unnecessary. If the proximal 10% to 20% of the matrix is involved in the biopsy, the author repairs the defect at the time of surgery.

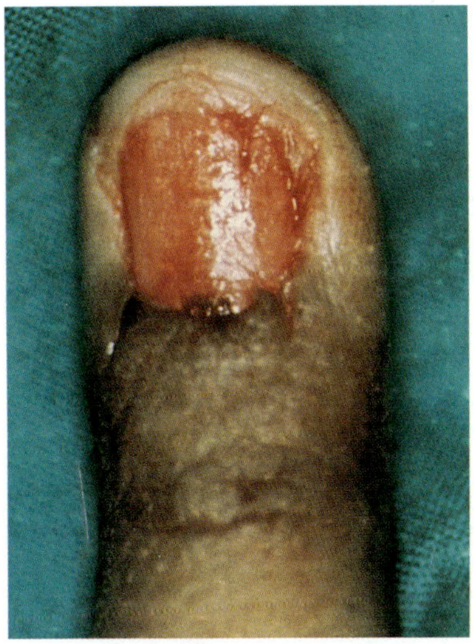

FIGURE 3C.10. Sources of pigmented streak obscured by the proximal nail fold.

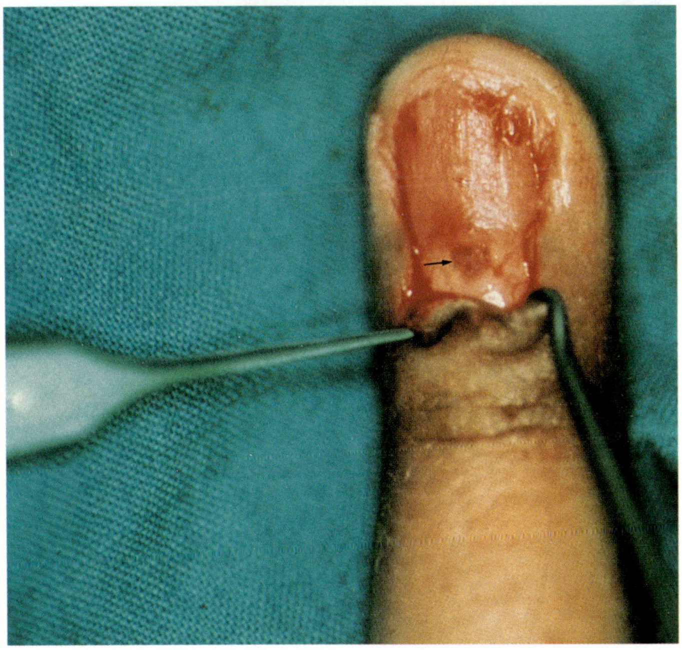

FIGURE 3C.11. Source of pigmented streak identified after incision of proximal nail fold. It appears reddish-brown *(arrow)*. Histopathology revealed melanocytic hyperplasia.

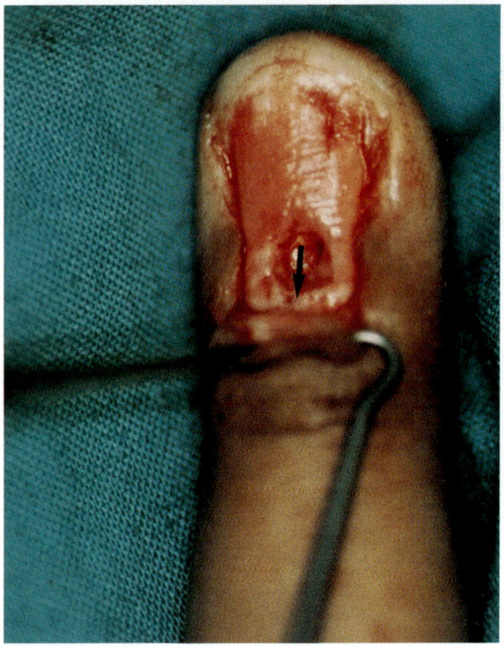

FIGURE 3C.12. Defect after punch biopsy. Note proximal matrix not involved in biopsy *(arrow)*.

Siegle and Swanson (7) described a technique that facilitates removal of biopsy tissue with only limited avulsion of the nail plate (Figs. 3C.13 and 3C.14). A 6-mm punch is made through the nail plate only; the circle of nail plate is removed and a smaller 3-mm punch is made through the underlying soft tissues to the underlying bone. The enlarged hole in the nail plate allows easier access to the biopsy material. The piece of removed nail plate can be replaced as part of the dressing.

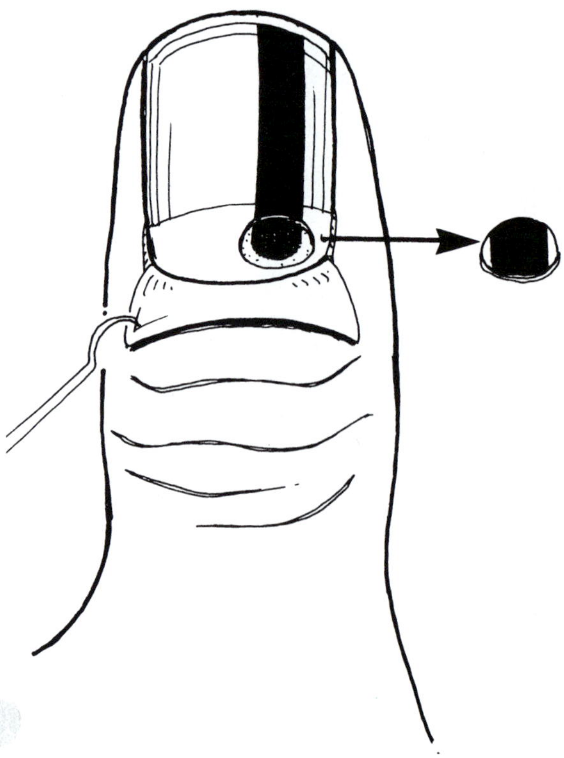

FIGURE 3C.13. Larger punch (6 mm) cuts through only the nail plate. This circle of nail is avulsed to facilitate removal of smaller punch biopsy of matrix.

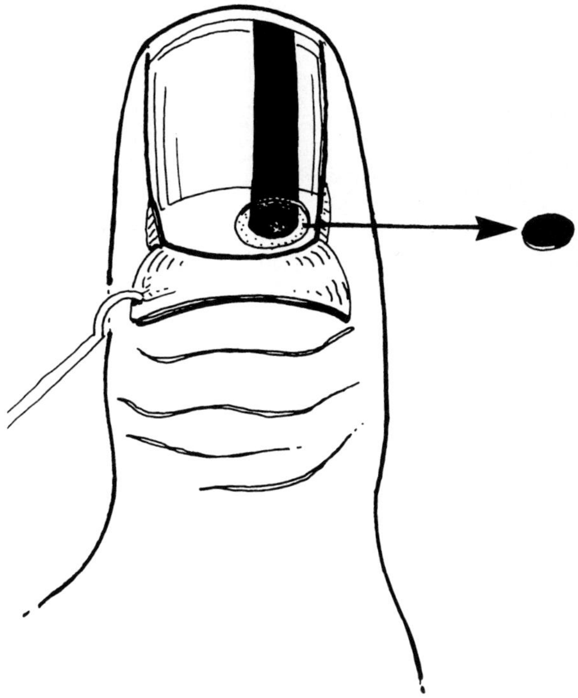

FIGURE 3C.14. Smaller punch biopsy of matrix down to bone.

Dressings include an antibiotic ointment, petrolatum gauze, a fine-mesh gauze (Sof-Wick) cut to size, a self-adherent wrap (Kling), and tubular elastic net (Se Pro Net) for a final dressing. The dressing should afford some compression, but it should not be so tight as to compromise the circulation. In 48 hours the dressing is changed, the wound is washed daily with soap and water, dried, and redressed with the antibiotic ointment. If incisions were made in the proximal nail fold and repaired by sutures, the sutures are removed in 5 to 7 days and Steri-Strips applied for another 2 to 3 days. Less complicated dressings are used for small areas of avulsed nail.

ELLIPTICAL BIOPSIES

Elliptical biopsies may be indicated when an entire linear lesion of the nail bed and matrix is needed for examination. It is desirable to leave the nail plate attached to the underlying tissues if the epithelial tissues are the goal of the surgery. Removal of the nail plate may tear or avulse the epithelium, which can result in a less ideal specimen for histopathologic examination, an especially important factor in the management of pigmented streaks. In some instances, retention of the nail plate is not feasible because exploration of the nail tissue must first be accomplished to identify the lesion to be biopsied or removed.

If the entire matrix and nail bed are involved in the disease process, longitudinal biopsy should be performed on the lateral margins of the nail tissues to avoid a split nail or nail dystrophy. This results in a narrower nail plate but avoids a split nail or nail dystrophy (1).

Elliptical Biopsies of Nail Bed

Nail Plate Attached

Solely nail bed biopsies are indicated less frequently than those involving the matrix. If the elliptical biopsy involves only the nail bed, an incision is made through the nail plate

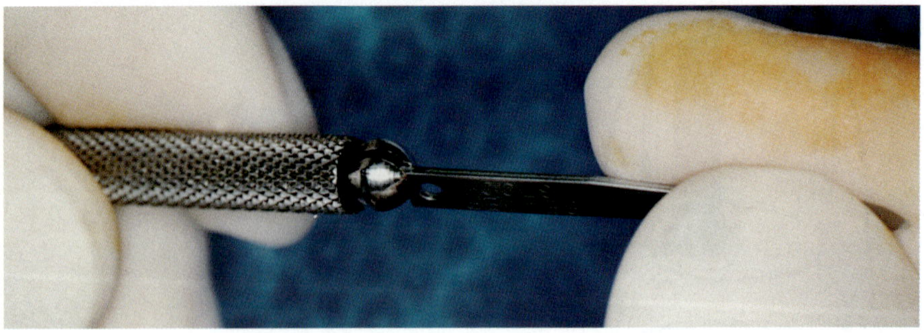

FIGURE 3C.15. Beaver blade inserted into a chuck device in the handle.

and soft tissues to underlying bone. The tissue and attached overlying nail plate are removed intact from the underlying bone. A Beaver blade is useful in the confined area (Figs. 3C.15 and 3C.16) Removal of narrow biopsies does not require nail bed closures. Onycholytic streaks may form from unsutured wider excisions of the nail bed. If the decision is made to repair the excision at the time of surgery, the remaining nail plate is removed and the nail bed is undermined. The undermining may have to extend to the fibrofatty tissue in the lateral nail grooves to adequately mobilize the nail bed for closure. Frequently the undermining to fibrofatty tissue is necessary only on one side; even under those circumstances, a small amount of undermining of the other side of incision is necessary to free the edge of the nail bed to obtain a better closure. If a closure still cannot be achieved, undermining of both sides of the nail bed to the lateral fibrofatty tissue may be necessary. Closure is made with absorbable sutures, such as polyglactin, or with nonabsorbable materials, such as nylon. Because the rete ridges are formed in a longitudinal line, some of the sutures are placed at different distances from the incision line, so that all are not placed along the same rete ridge. This may reduce the tendency for tearing in nail bed closures. The nylon sutures are removed in 5 to 7 days; the absorbable sutures are not removed.

If the nail bed defect is narrow or if there is uncertainty as to whether the defect needs to be closed, the nail plate can be allowed to regrow to ascertain whether a significant abnormality occurs. If an undesirable change occurs, a delayed repair using the same steps can be undertaken. This author tends to close these defects at the time of surgery.

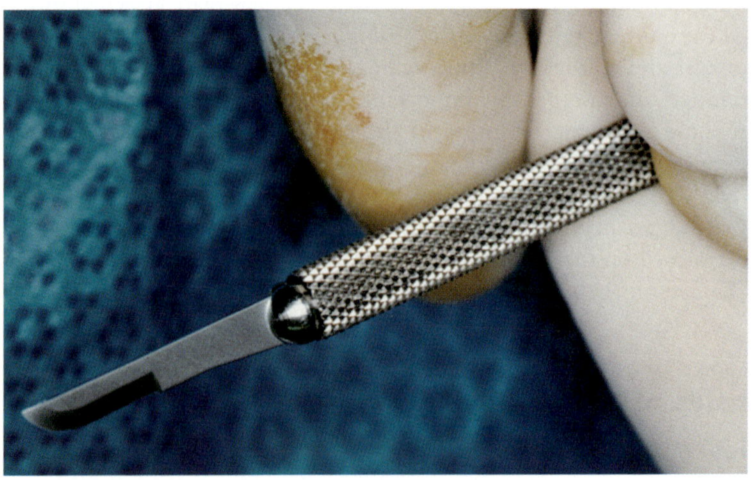

FIGURE 3C.16. Beaver blade and handle is small enough to facilitate nail surgery.

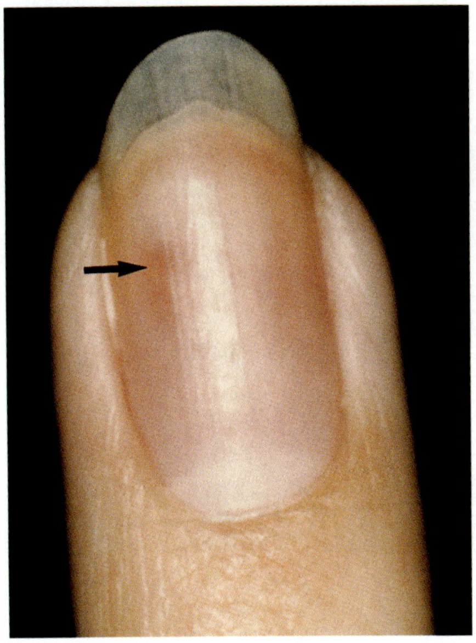

FIGURE 3C.17. Blush-red painful tumor in nail bed *(arrow)*.

Nail Bed with Nail Plate Avulsed

If the nail bed must be examined to determine the site of biopsy, the nail plate is removed, and the biopsy and closure, if necessary, are carried out as previously described (Figs. 3C.17–3C.20).

Elliptical Biopsies of the Matrix

Elliptical biopsies of the matrix or of the matrix and nail bed are more commonly necessary than those of the nail bed alone. They are more difficult to perform with the nail plate in place. Initial removal of the nail plate may destroy some of the underlying epithelium of the matrix and nail bed, possibly adversely affecting the ability for accurate histopathologic interpretation.

Although the author tends to close most of the defects through the proximal matrix, from the work of Zaias (10), longitudinal surgical defects less than 3 mm wide through

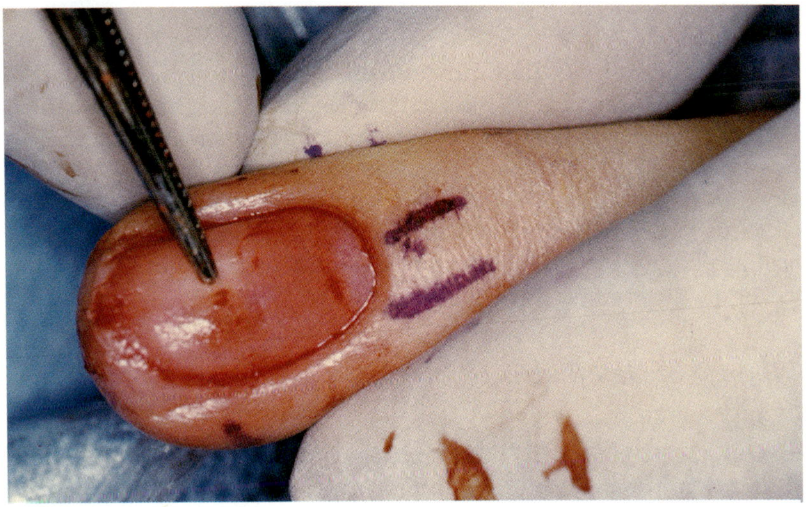

FIGURE 3C.18. Nail plate avulsed. Glomus tumor is identified. Skin marks were used to delineate tumor through nail plate in case its location was difficult to locate after nail plate was avulsed.

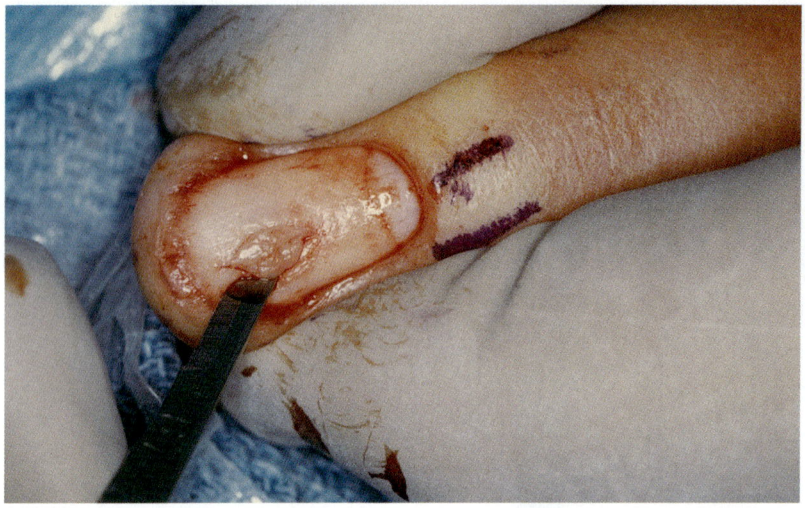

FIGURE 3C.19. Glomus tumor excised.

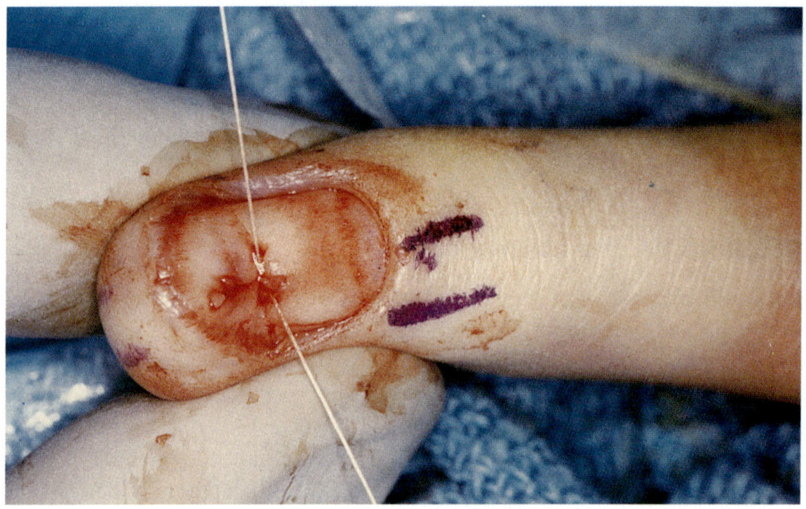

FIGURE 3C.20. Wound closed with 5-0 polyglactin.

the matrix may heal by second intention without resulting in split nails, although nail dystrophies may occur. Therefore, the decision to close defects of the proximal matrix is a function of the width of the excision, the desire to reduce nail dystrophies, and physician preference.

Additionally, the author includes the nail bed with paracentral and lateral matrix biopsies. This is partly related to the difficulty in determining precisely where the matrix ends and the nail bed begins, the possible presence of helpful pathology of the nail bed, and the neatness of the surgery.

Paracentral Biopsy of Nail Matrix

Nail Plate Attached

Incisions are made in the proximal nail fold as described in Chapter 3B: "Exploration of Nail Tissue." If a transverse biopsy that spares the proximal matrix can be used to gain the necessary information, it would be preferable to a longitudinal biopsy that involves the proximal matrix because of the possible resultant split nail or nail dystrophy (7). The transverse biopsy does not need to be sutured because it leaves the proximal matrix intact.

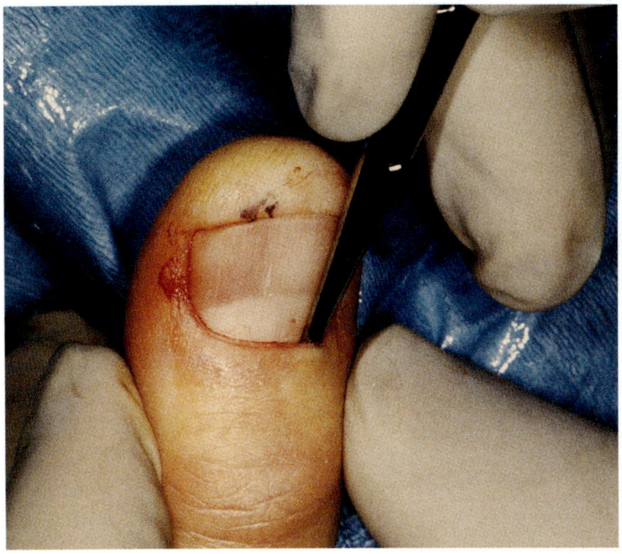

FIGURE 3C.21. Recent onset of a lightly colored pigmented streak. Histopathology revealed only melanocytic hyperplasia. Interestingly, the patient developed primary melanoma of the thigh 1 year later. *Purple lines* demarcate pigmented streak, which may be very light colored in matrix and difficult to identify.

However, if the proximal 10% to 20% of the matrix is included in the biopsy, consideration should be given to closure of the defect.

With the nail plate attached to the matrix, an elliptical biopsy is somewhat more difficult to perform because of the hardness of the nail plate and the difficulty in removing the specimen. However, this better retains the histopathology of the nail matrix (Fig. 3C.21). Soaking the digit in water for 10 to 15 minutes will soften the nail and allow easier cutting through the nail plate with the scalpel. The no. 15 scalpel blades cut through the nail plate more easily than Beaver blades (Fig. 3C.22). A rocking motion of the scalpel

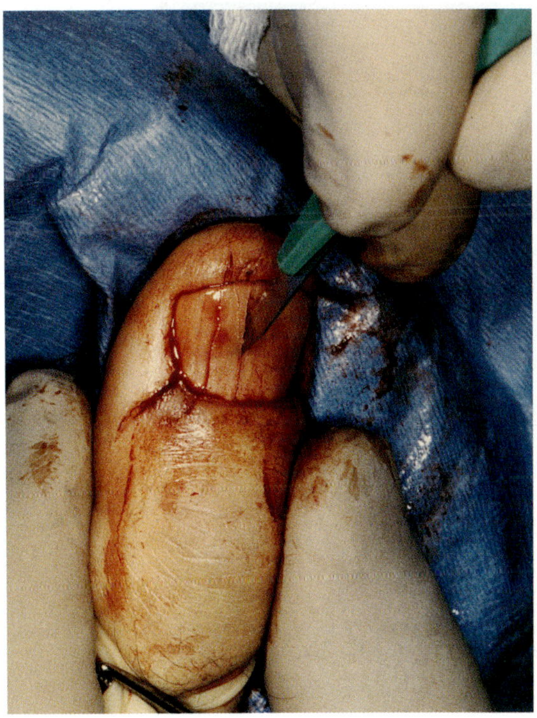

FIGURE 3C.22. Incision through nail plate to underlying bone. Proximal nail folds are already incised to allow exploration for origin of the pigmentation.

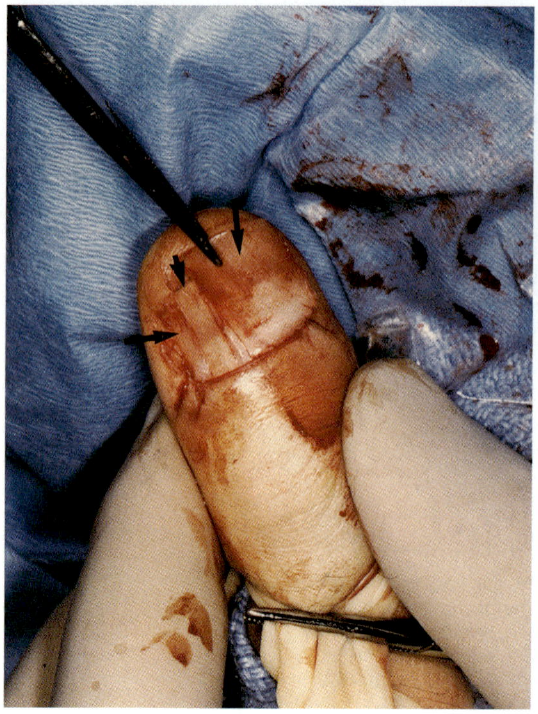

FIGURE 3C.23. Nail plate on both sides of the biopsy specimen are avulsed. *Longer arrows* identify parts of nail plate that will be avulsed. *Short arrow* indicates biopsy specimen.

facilitates cutting through the hard nail plate. The incision is carried to the depth of the underlying bone. After the incision through the nail is completed, avulsion of remaining nail plate, excluding that attached to the biopsy tissue, is completed (Fig. 3C.23). Thus, the only retained nail plate is attached to the biopsy specimen. This facilitates excision of the elliptical biopsy as well as the final repair. This tissue with the nail plate attached is dissected off the underlying bone (Fig. 3C.24). The author prefers to use Beaver blades

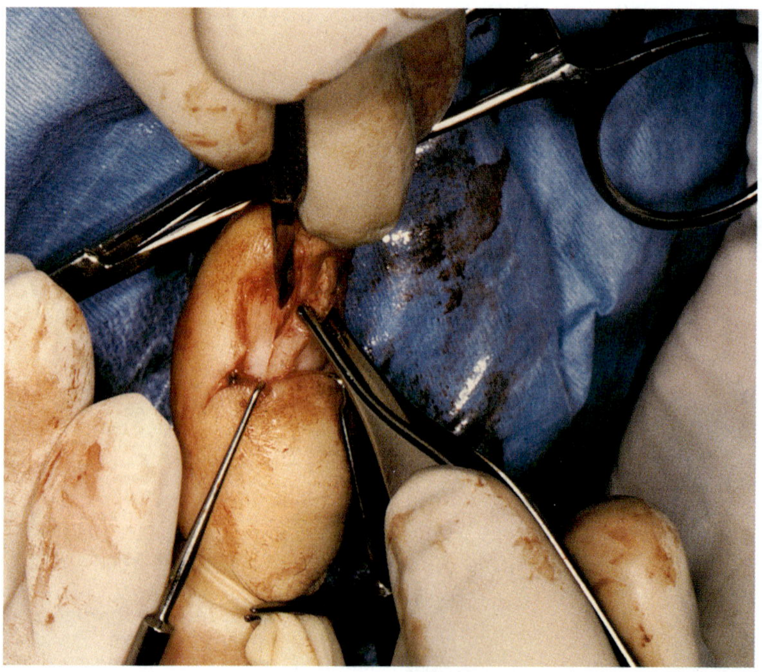

FIGURE 3C.24. Biopsy specimen with attached nail being excised.

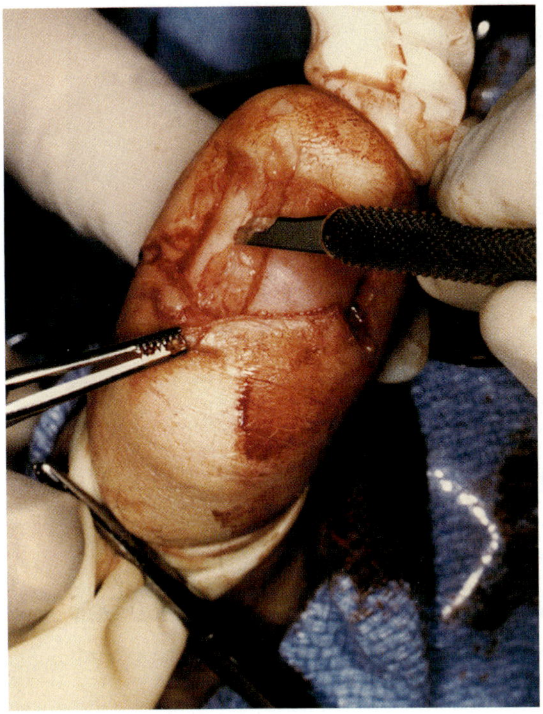

FIGURE 3C.25. To close the defect in the matrix, the sides of the incision, including matrix and nail bed, undermining may have to be extended to the fibrofatty tissue of the lateral nail folds.

for this dissection because of their small size, which makes removal of the specimen in the confined space easier to accomplish (Figs. 3C.15 and 3C.16). This author tends to close all of the defects involving the matrix, especially those extending through the proximal matrix. If the remaining nail plate was not previously avulsed to excise the biopsy specimen, it usually is removed now to facilitate the closure.

The nail matrix and nail bed are tightly adherent to the underlying bone. To repair the defect resulting from a longitudinal biopsy through the matrix, it may be necessary to undermine both sides of the matrix for a short distance to permit approximation of the sides of the excision. If the mobilization still is inadequate, the matrix is further freed from the underlying bone, extending the undermining to the fibrofatty tissue of the lateral nail fold (Figs. 3C.25 and 3C.26). Caution must be taken not to "buttonhole" the matrix in this dissection; if this should occur, a simple absorbable suture will repair the defect if it seems warranted. By extending the dissection to the fibrofatty tissue of the lateral nail fold, the entire matrix on the one side is freed from the underlying bone and the extensibility of the tissues of the lateral nail fold allows the matrix defect to be closed.

Instead of starting from the edges of the excision and undermining toward the lateral nail folds, incisions first can be made through the lateral nail folds and carried beneath the matrix toward the incision line. However, the author prefers to start from the edges of the incision because the plane of the undermining between the bone and matrix is more clearly defined.

Usually the narrowest of the two sides of the incision is undermined first. If this does not allow closure of the wound, the matrix on the other side of the defect is undermined in the same manner. The undermining of both sides usually is adequate to close significant defects. The matrix is fairly strong tissue and allows significant tension on the wound. The author prefers using 5-0 or occasionally 4-0 polyglactin sutures with smaller needles (P-3) for closure of the defect because of the inherent strength of the suture and its braided nature, which has less tendency to tear through tissue than monofilament sutures such as nylon (Figs. 3C.27 and 3C.28).

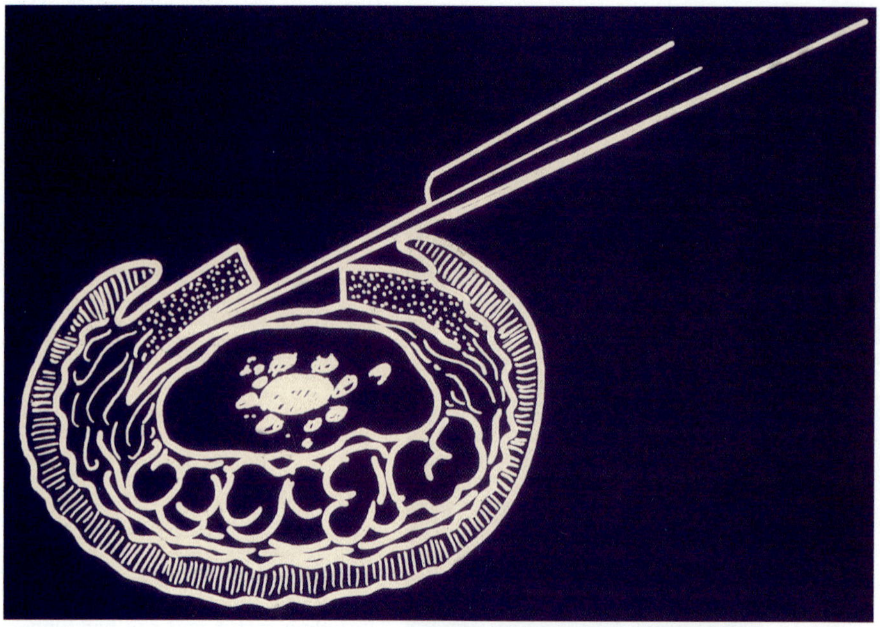

FIGURE 3C.26. Cross-section showing undermining at the level of the bone and extending to fibrofatty tissue of the nail folds.

Nail Plate Avulsed

The procedure and surgical considerations are similar to those for biopsies with the nail plate attached. With the absence of the nail plate, the incision through the matrix and nail bed and the removal of the specimen are less difficult. Closure is identical to that for biopsies with the nail plate attached.

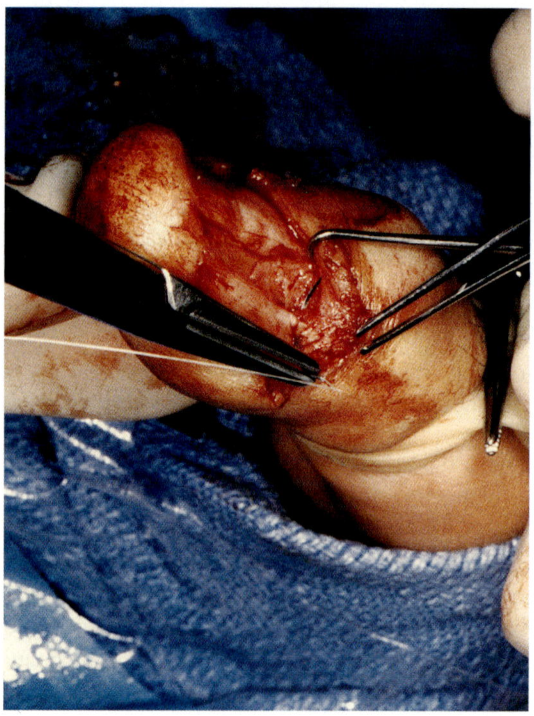

FIGURE 3C.27. First suture usually approximates proximal matrix defect. Suture is 5-0 polyglactin; needle is P-3.

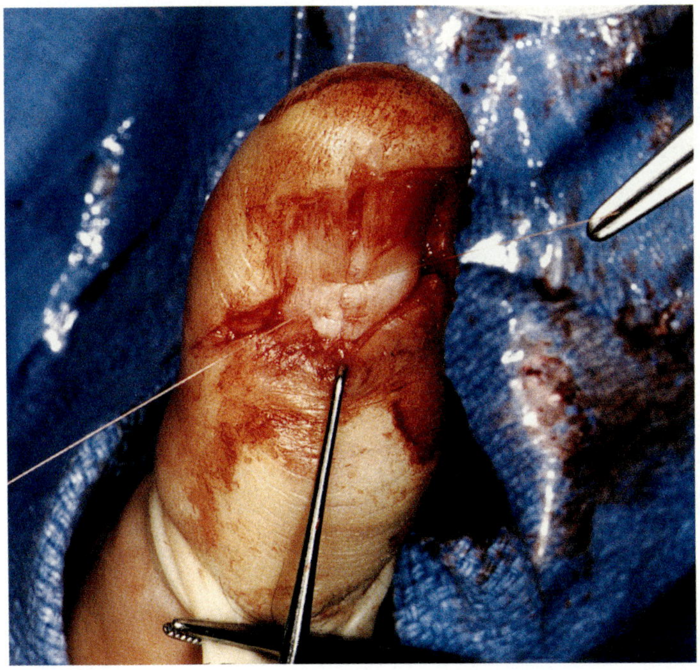

FIGURE 3C.28. Continued closure. Suture material is 5-0 polyglactin.

Lateral Longitudinal Nail Biopsy

If the entire matrix and nail bed are involved in the disease process, which may occur in psoriasis or lichen planus, there is no advantage to making biopsies paracentrally (Fig. 3C.29). This may result in split nails or other nail dystrophies. Instead, the lateral nail bed and matrix biopsies are preferred. These result in only a narrowing of the nail plate. The biopsies can be performed with or without the nail plate attached. For most inflammatory diseases, such as psoriasis and lichen planus, it is desirable to leave the nail plate attached.

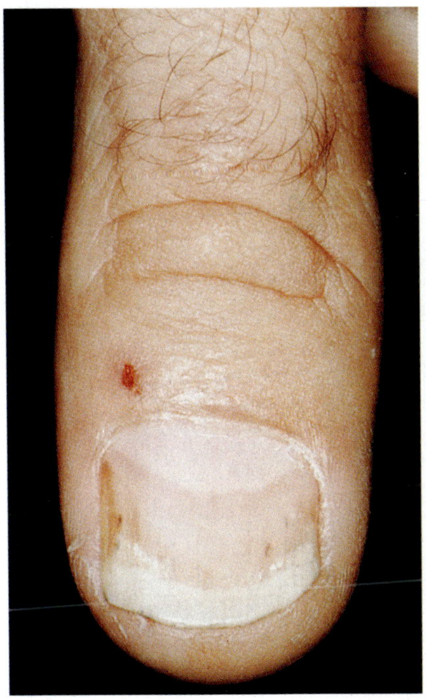

FIGURE 3C.29. All fingernails involved with what appears to be psoriasis.

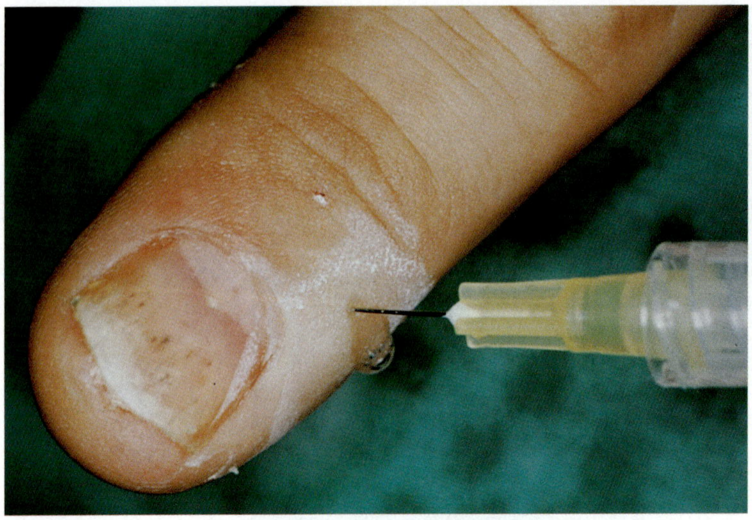

FIGURE 3C.30. Light spray with Frigiderm *(white area)* greatly reduces pain of initial local injections (needle size 30 gauge; 1% plain lidocaine anesthetic). Slow progression with anesthetic injections allow the patient to control pain, i.e., stop when it hurts, and markedly decreases apprehension and pain. Digital and, less frequently, regional anesthetic blocks also are used.

Nail Plate Attached

The procedure is similar to that used for permanent repair of ingrown toenails and biopsies of laterally located pigmented streaks. The same procedure may be applied to an asymmetric split nail in which one part of the split nail is narrow. Soaking the digit in water to soften the nail, use of no. 15 blades, and rocking motion of the scalpel aids the incision through the nail plate. Anesthesia may be local, digital, or, less frequently, regional (Fig. 3C.30).

The surgery is performed in five steps and is aimed at direct visualization of the matrix, especially the lateral horns, and preserving the lateral nail folds for aesthetic reasons. Initially the author and others included the lateral nail fold in the excision (14). However, the author now retains the lateral nail fold for cosmetic appearance.

Step 1: Incision through the nail plate, nail bed, and visible matrix (lunula). An incision is made through the nail plate, nail bed, and visible matrix. The incision stops at the cuticle (Figs. 3C.31 and 3C.32).

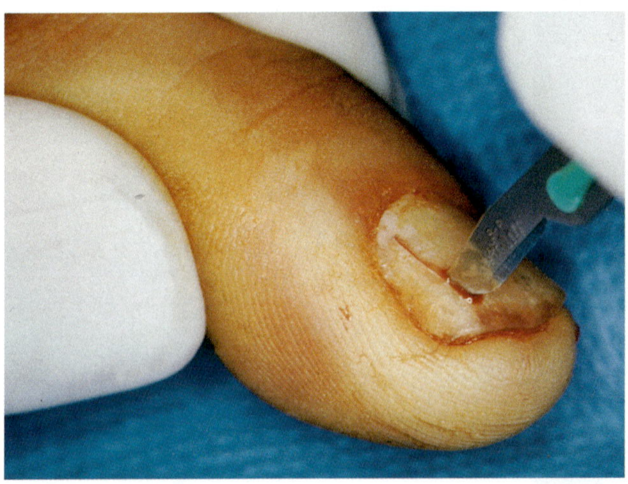

FIGURE 3C.31. A no. 15 disposable scalpel blade is used to cut through the nail plate. A rocking motion is helpful. Rocking motion is better visualized in Figs. 3C.31 and 3C.32.

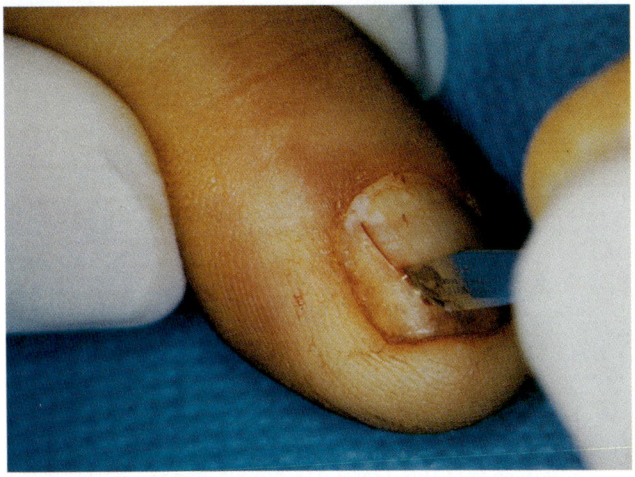

FIGURE 3C.32. In step 1, an incision through the nail plate, visible matrix (lunula), and nail bed. The incision ends at the cuticle.

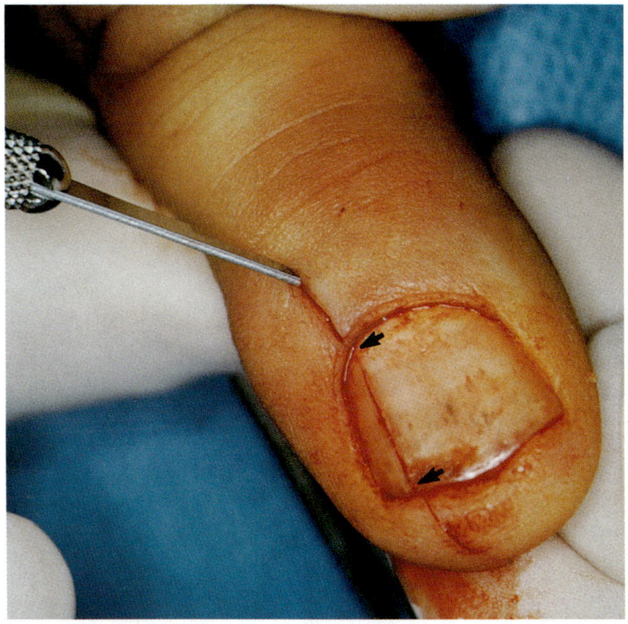

FIGURE 3C.33. In step 2, a straight incision is made through the proximal nail fold to the depth of the nail plate only. The length of this incision corresponds to the nail invagination or proximal nail groove. *Arrows* demarcate incision of nail plate to the cuticle in step 1.

Step 2: Incision and reflection of the proximal nail fold to identify the most lateral part of the horns of the matrix. A straight incision is made through the proximal nail fold to the depth of the underlying nail plate (Fig. 3C.33). The length of the straight part of the incision is equivalent to the length of the nail invagination in the proximal nail groove (Figs. 3C.34 and 3C.35), which is about 5 to 7 mm long. The straight part of the incision of the proximal nail fold will allow a direct extension of the incision of the nail bed and matrix made in step 1. The incision of the proximal nail fold then takes on a laterally curved direction (Fig. 3C.36). It is extended until the most lateral part of the nail plate is visualized (Fig.

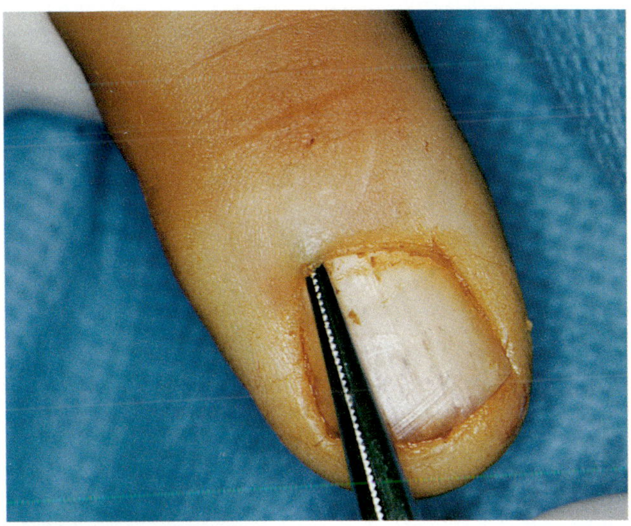

FIGURE 3C.34. The length of the straight incision can be determined by inserting the tip of a small-nosed mosquito hemostat between the nail plate and the overlying proximal nail fold.

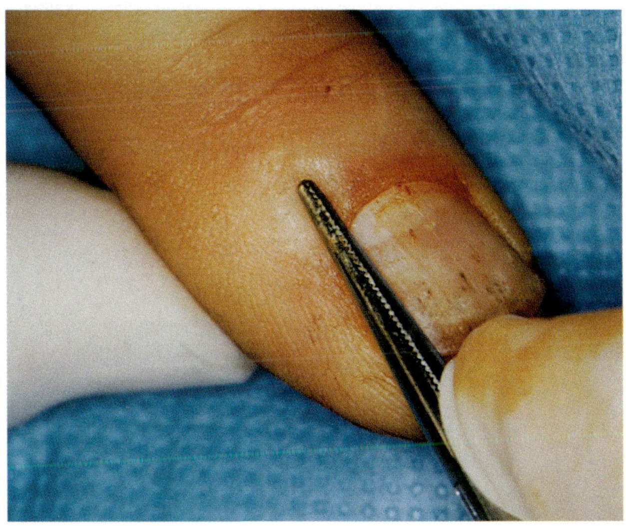

FIGURE 3C.35. The length of the straight incision is the distance the tip of the mosquito hemostat was inserted between the nail plate and overlying proximal nail fold.

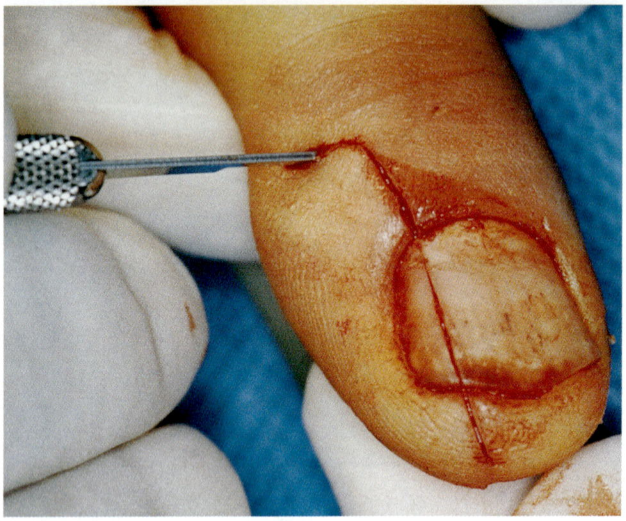

FIGURE 3C.36. After the straight part, the incision of the proximal nail fold takes on a curved lateral direction.

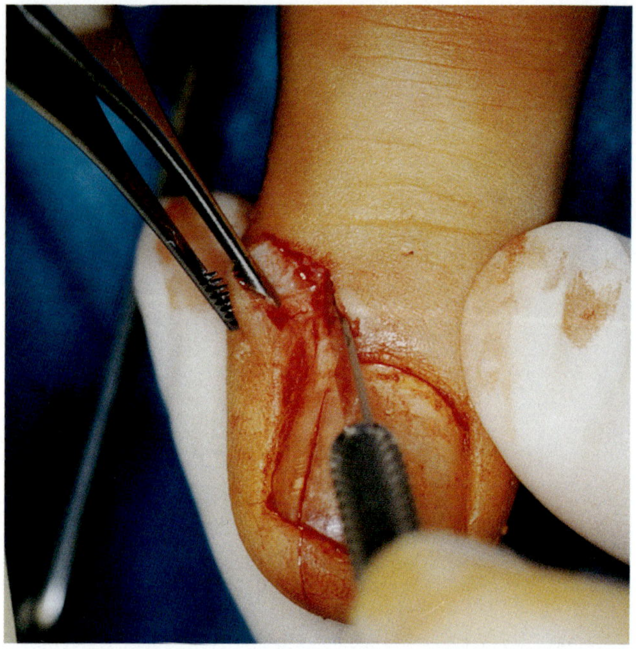

FIGURE 3C.37. The lateral part of the incision must be extended far enough so that the lateral nail groove and the horns of the nail plate and matrix are visualized.

3C.37). This corresponds to the lateral horn of the nail and matrix. In marked hypertrophy of the lateral nail fold, which may result from a chronically ingrown toe nail or the hidden part of the pincer nail that wraps around the phalanx, the curved incision may have to be extended to the junction of the dorsal and ventral surfaces of the digit.

Step 3: Straight extension of the incision through the nail bed and matrix of step 1. The incision of the nail bed and matrix made in step 1 ended at the cuticle. This incision now is continued through the entire longitudinal length of the matrix (Fig. 3C.38).

Step 4: Incision of the lateral nail fold in the lateral nail groove dorsal to matrix. An incision of the lateral nail fold is made in the lateral nail groove 1 to 3 mm dorsal to (or

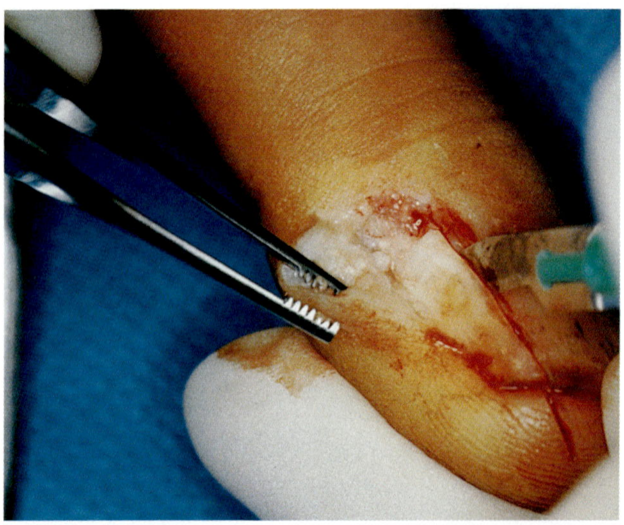

FIGURE 3C.38. In step 1, the initial incision stopped at the cuticle. In step 3, the incision is extended directly through the entire length of the matrix.

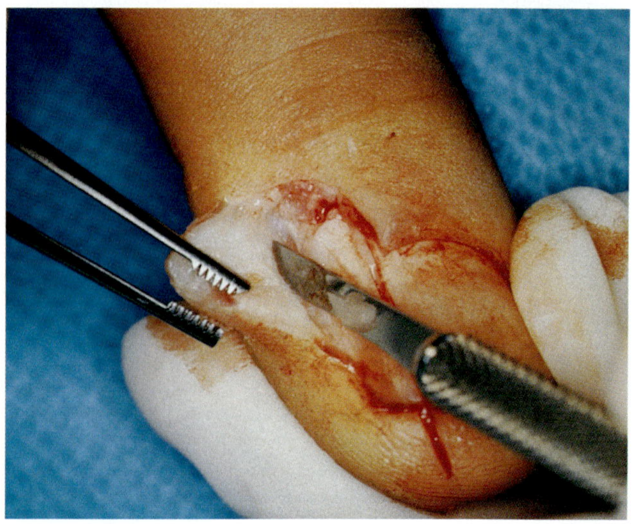

FIGURE 3C.39. In step 4, an incision is made into the lateral nail groove dorsal to the matrix.

above) the most lateral part of the matrix (Fig. 3C.39). It is carried out through the entire length of the lateral nail groove. At the completion of this stage, the lateral matrix and nail bed lie between (a) the medial incision of the nail bed and matrix made in steps 1 and 3, and (b) the incision in the lateral nail groove.

Step 5: Resection of the lateral edge of the matrix and nail bed. Beginning distally, the tissue is removed from the underlying bone. It is preferable to proceed from the medial to the lateral part of the incision, because the bone of the phalanx defines the depth of the excision (Fig. 3C.40). After the specimen is removed, the surgical site is wiped cleaned with gauze wrapped around the hemostat, gently curetted with a small curette such as an ear curette, and rewiped with gauze (Fig. 3C.41). The wound is examined carefully for residual matrix. Two specific anatomic sites are identified. First to be identified is the most proximal part of the nail plate, which invaginates the proximal nail groove and its associated matrix (Fig. 3C.42). Both are closely aligned. This determines whether the excision has been extended through the entire matrix, i.e., that it is proximal enough. Second, the lateral nail groove should reveal only fibrofatty tissue (Fig. 3C.43). If the excision is proximal and lateral enough, there is little likelihood of residual matrix. If some matrix is retained, either a spicule of nail or an epidermal cyst may form, depending on the intactness of the residual matrix in the surgical site.

In such a case, a small area of swelling of an epidermal cyst or protrusion of a spicule of nail, which may be tender or painful, can appear within a few months. Tenderness or pain resulting from a persistent epidermal cyst or spicule of nail should be distinguished from a hypersensitive scar. Differentiation is made by careful examination for a small site of swelling or other evidence of a cyst or spicule of nail or by localization of tenderness to the hypersensitive scar by applying pressure with the wooden end of a cotton application. If the conclusion is a hypersensitive scar, a small amount of triamcinolone (0.2 to 0.4 mL at a concentration of 5 to 10 mg/mL) injected into the scar usually will eliminate the symptoms in a few days. If either of the complications from retained matrix occurs, the sliver of nail or cyst with the corresponding residual matrix can be excised. The incidence in the occurrence of nail spicules or cysts (3) can be greatly reduced by exposure of the matrix, especially its lateral horn, and careful examination of the two landmarks for completeness of removal of the matrix during surgery.

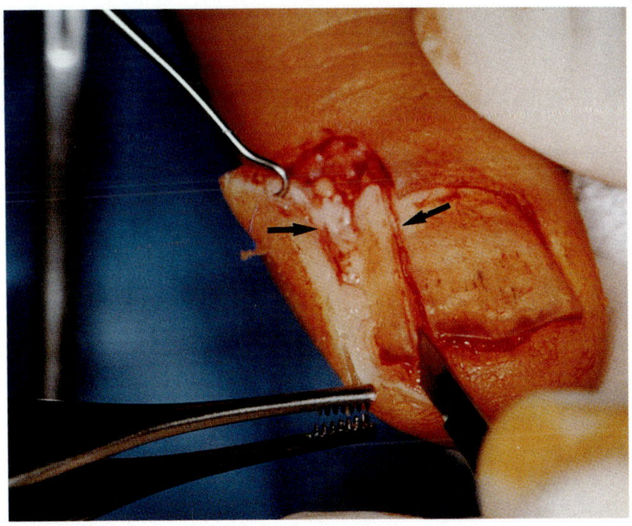

FIGURE 3C.40. In step 5, the tissue between the incision of the nail plate, matrix, and nail bed (steps 1 and 3) and the incision in the lateral nail groove (step 4) is excised *(arrows)*. The dissection begins distally and medially, using the underlying bone to define the depth of the excision.

FIGURE 3C.41. Double layer of gauze wrapped around the hemostat is useful to clean the excision site.

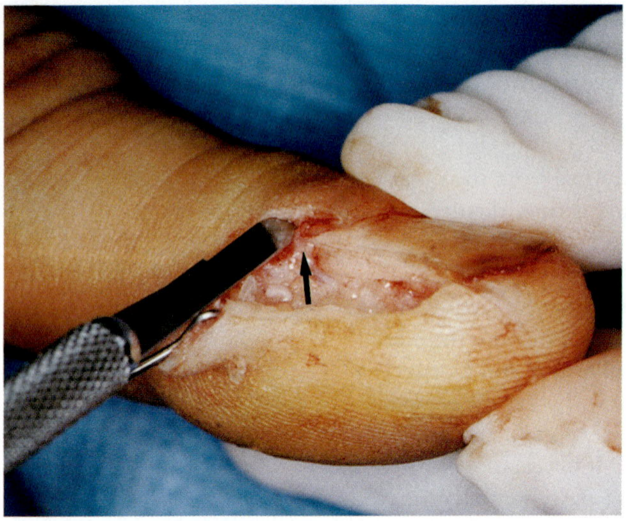

FIGURE 3C.42. Two anatomic points are identified to ensure total removal of involved matrix. The first area is the proximal invagination of the nail plate, the apex of the nail groove, and the proximal matrix, which are essentially the same site *(arrow)*.

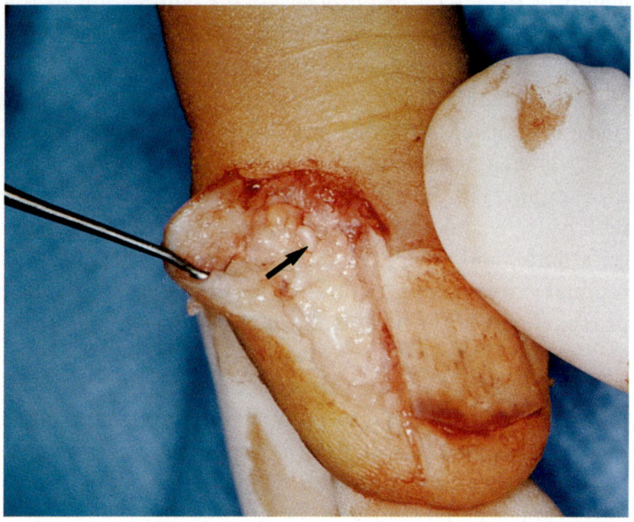

FIGURE 3C.43. The second anatomic landmark is to see nothing but fibrofatty tissue in the lateral nail groove *(arrow)*.

A Penrose drain is applied as a tourniquet. Folded gauze may be applied under the tourniquet to possibly lessen soft tissue injury (3,15). Our use of the tourniquet is very conservative (4,9,12). The tourniquet is applied for 15 minutes, sometimes a few minutes longer. If additional surgery is needed, the tourniquet is removed for 5 minutes and then reapplied.

Nylon sutures are used to close the incisions of the proximal nail fold. The first suture is placed near the cuticle to restore its normal appearance (Fig. 3C.44). If the nail fold is pliable and easily approximated, the author uses the pressure of the dressing to close the sides of the defect of the excision. However, if there is resistance to the approximation of wound edges, a suture is placed through the lateral nail fold and nail plate. A hole is drilled through the nail plate with a no. 11 scalpel blade (Fig. 3C.45), then a suture is

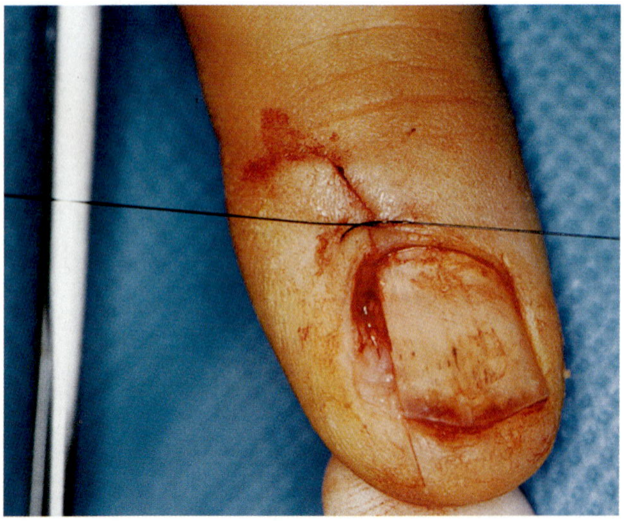

FIGURE 3C.44. The first suture approximates the cuticle.

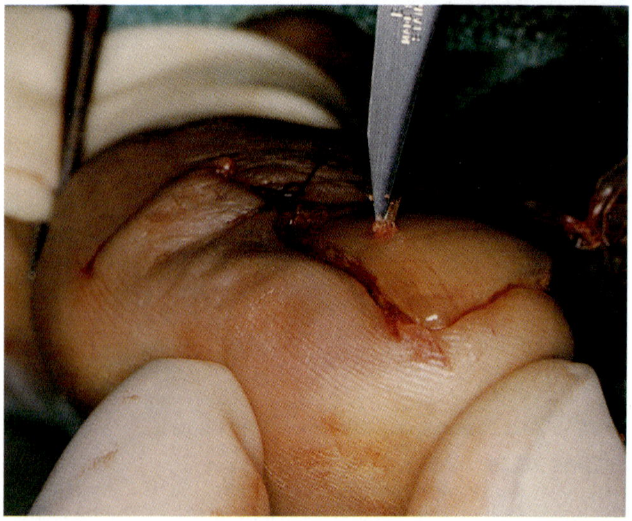

FIGURE 3C.45. A no. 11 blade is used to drill a hole in the nail plate (not same nail as shown in other figures).

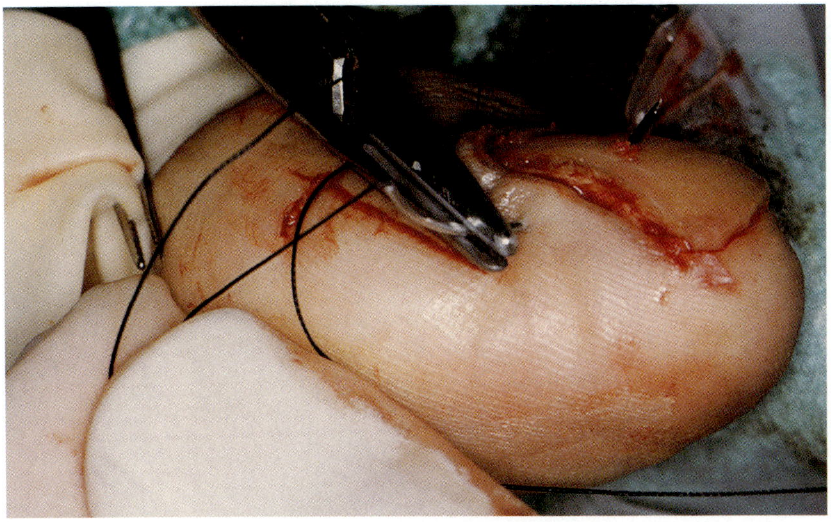

FIGURE 3C.46. Suture (3-0 silk, needle P-2) passed from the soft tissue through the hole in the nail plate (same nail as shown in Fig. 3C.45).

passed from the soft tissue of the lateral nail fold through the subungual tissue and out the hole previously made in the nail plate. It is preferable to proceed from the soft tissue through the nail plate so as not to implant keratin in the soft tissue (Figs. 3C.46, 3C.47).

The dressing the author uses is applied with the tourniquet in place. Topical antibiotic ointment is followed by a layer of petrolatum gauze. A finely woven gauze (Sof-Wick) cut to size is wrapped with a self-adherent material, such as Kling, and finished with tubular elastic net (Se Pro Net). The Kling must be applied carefully so as to produce some pressure on the wound without constriction of blood flow. It is better to err on the side of being too loose rather than too tight.

An open-toed shoe should be brought by the patient or furnished by the surgeon. Elevation of the hand or foot will help reduce the throbbing pain. A long-acting anesthetic, such as bupivacaine (Marcaine), can be used as the initial anesthetic or added at the end of the procedure. About 50% of the patients need analgesics. Usually only for the

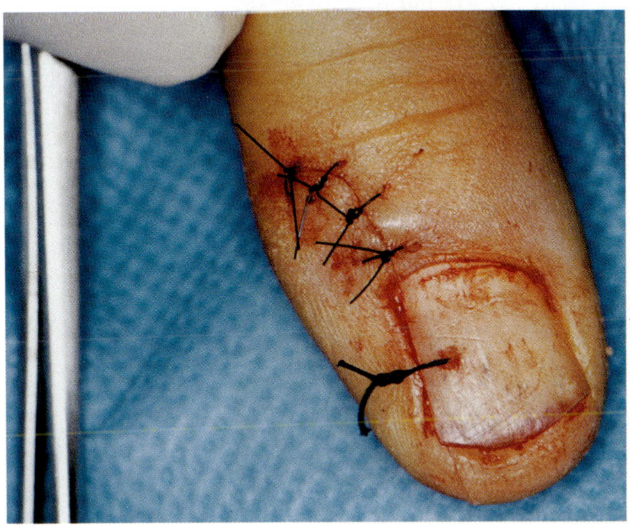

FIGURE 3C.47. Final closure of longitudinal lateral nail biopsy.

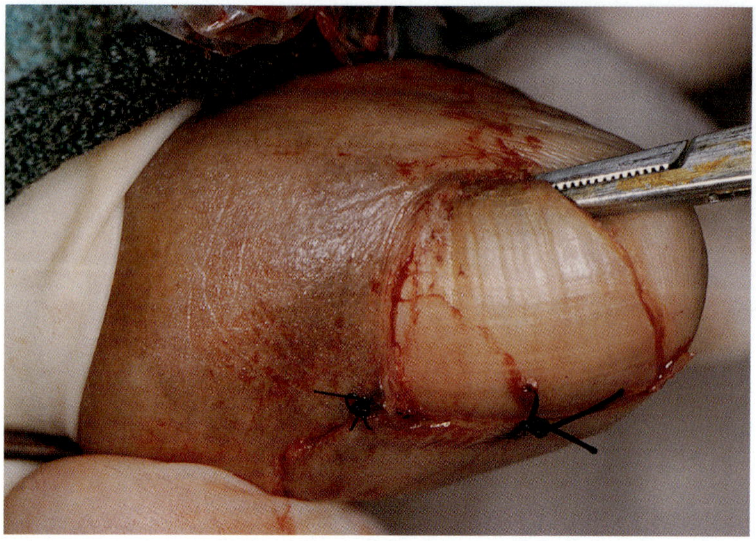

FIGURE 3C.48. Hemostat inserted under the edge of the nail plate.

first 24 hours acetaminophen alone or with codeine is adequate for pain. The first dose is given at the end of surgery. If severe, intense pain persists on the day of surgery or seems disproportionate to that expected from the surgery and the pain is poorly responsive to the analgesics, a dressing that is too tight should be suspected. The bandage should be cut along one side and loosely held in place with tape. Administration of aspirin to reduce the possibility of vascular occlusion should be considered under these circumstances.

Nail Plate Avulsed

The lateral edge of the nail plate is freed from the underlying matrix and nail bed and the overlying nail folds using a fine-nosed mosquito hemostat (Fig. 3C.48). Using the nail splitter, a small wedge of the nail plate is cut through longitudinally (Fig. 3C.49). It then is extracted with the hemostat (Fig. 3C.50). The five steps of the procedure described in

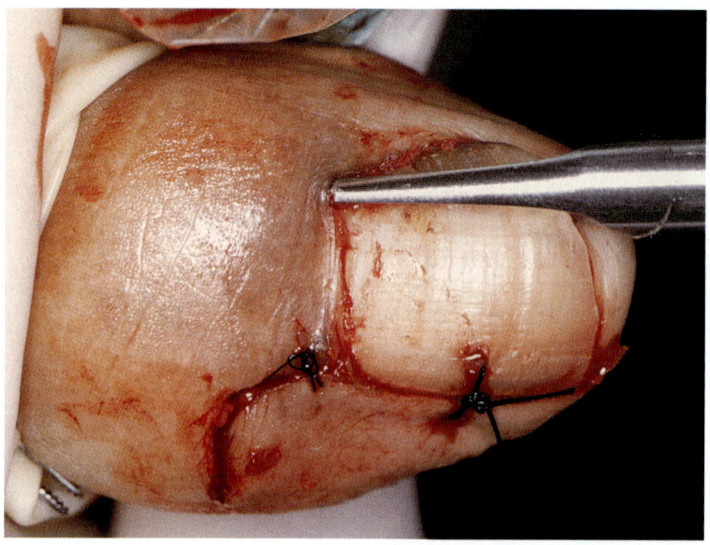

FIGURE 3C.49. English nail splitter cutting through the longitudinal length of the nail plate.

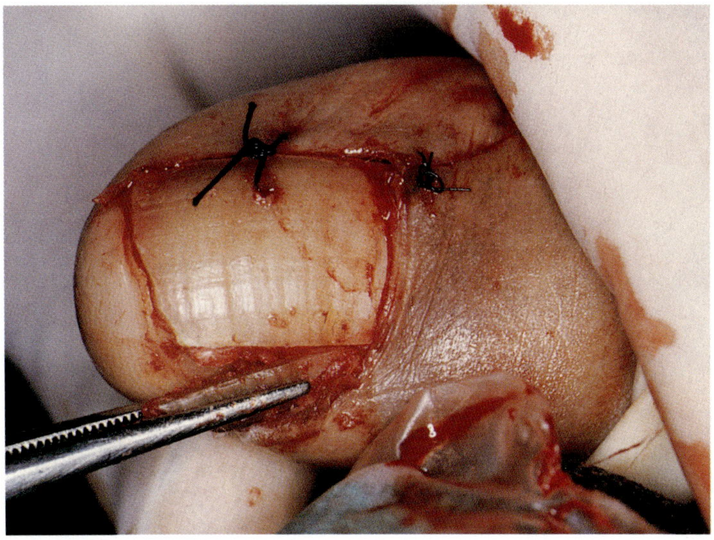

FIGURE 3C.50. Nail edge avulsed.

detail in the section on "Lateral Longitudinal Nail Biopsy, Nail Plate Attached" are followed. Closure, dressing, and postoperative care are the same.

BIOPSY OF LATERAL AND PROXIMAL NAIL FOLDS

Punch biopsies of the nail folds are similar to those in other parts of the body (6); however, the punch biopsies frequently are smaller in diameter (2 to 3 mm) than in other sites. Punch biopsies of the proximal nail fold extend only to the depth of the nail plate. Because the proximal nail plate is softer than the visible nail plate, it is possible to perforate the nail and enter the matrix. Insertion of a Freer septum elevator, mosquito hemostat, or other instrument between the proximal nail fold and nail plate at the site of the biopsy will protect against such an event. If possible, punch biopsies of the proximal nail fold should avoid the most distal cuticular edge of the proximal fold (2,4,6). Closure of punch biopsies with 5-0 fast gut will accelerate healing without the need to remove sutures.

Elliptical biopsies of the lateral nail folds are similar to elliptical biopsies on other skin. The orientation of the biopsies is longitudinal. Closure can be accomplished with nonabsorbable sutures, such as fast gut. If there is not too much tension, bridging tapes such as Steri-Strips can be applied. Elliptical biopsies of the proximal nail fold commonly are crescentic, following the curve of the cuticle (Fig. 3C.51). The width of the tissue removed is 3 to 4 mm. Injury to the extensor tendon, which inserts centrally at the base of the distal phalangeal bone, should be avoided. The tendon insertion is about 12 mm from the cuticle (4). Although most nail biopsies are unlikely to come in contact with the insertion of the tendon, wide deep biopsies of the proximal nail fold may injure the tendon. Such biopsies should be performed carefully under direct visualization and identification of the tendon. The length of the crescentic excision of the proximal nail fold preferably is limited to the transverse width of the nail plate. Closure is not necessary. The excision heals with second intention exposing more of the matrix, i.e., enlarging the lunula, but otherwise it is normal in appearance. The change is not easily noticed. A similar excision may be very useful in the management of mucous cysts of the nail and recalcitrant chronic paronychia (2,4,16). Biopsies of the proximal nail fold, punch, and ellipse may contribute to diagnosis of collagen diseases, especially Raynaud's disease (17,18).

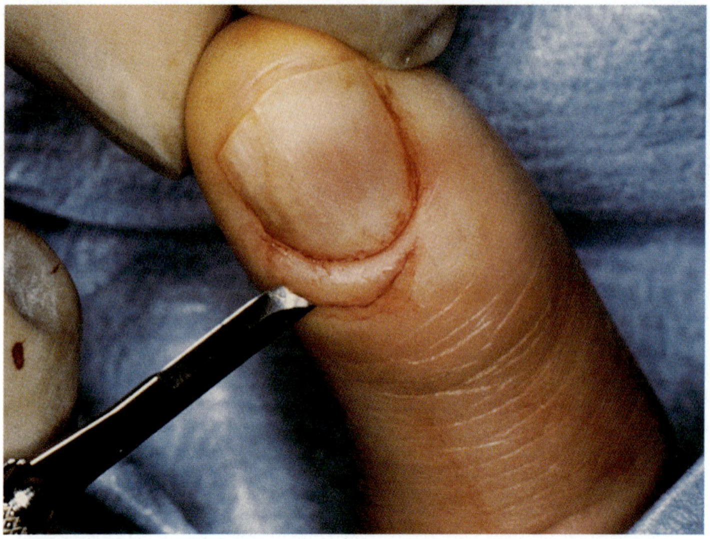

FIGURE 3C.51. Crescentic excision of proximal nail fold for chronic paronychia.

BIOPSY OF NAIL PLATE

Longitudinal nail unit biopsy techniques for diagnosis of psoriasis have been used because of the quality of the tissue specimens (10,11,14). However, some physicians have been reluctant to perform biopsies because of the perceived complexity and the potential for nail dystrophy. Instead, after repeated KOH (potassium hydroxide) preparation examinations and culture studies are negative and a trial of systemic antifungal therapy, which is frequently used, fails, it has been considered most practical and reasonable to clinically settle for the diagnosis of psoriasis without other corroboration (4).

Cytologic examination of nail scrapings (19,20) and shave excision of cornified cells (21) for clumps of parakeratotic cells with neutrophils and absence of fungus with periodic acid-Schiff (PAS) stains has been considered diagnostic for psoriasis. But, some find the interpretation of these tissue specimens difficult.

A technique involving simple partial nail plate avulsion, usually the lateral one-third, with specially processed nail plate, has been helpful in the diagnosis of psoriasis of the nails (22). The avulsed portion of the nail is immersed in cedar oil for 3 days, which softens the nail plate, rinsed with water, and deposited in a cassette with formalin for 2 hours. Hematoxylin and eosin as well as PAS stains for fungus are performed. Diagnostic features of psoriasis often were observed in the epithelium adherent to the nail plate. This method may also be useful in the diagnosis of fungal infections of the nail in the face of repeated negative cultures and KOH examinations. However, punch biopsies of nail bed demonstrated viable fungus more often in the cornified cells of the nail bed than in the nail plate. When identified in the nail plate, the fungi are located in the deepest portion of the nail plate (23).

REFERENCES

1. DeBerker DAR, Dahl MGC, Comash JS, et al. Nail surgery: an assessment of indications and outcome. *Acta Dermatol Venereol* 1996;76:484–487.
2. Haneke E, Baran R. Nail surgery and traumatic abnormalities. In: Baran R, Dawber RPR, eds. *Diseases of the nails and their management,* 2nd ed. London: Blackwell Science, 1994:345–364.
3. Clark RE, Madani S, Betttencourt MS. Nail surgery. *Dermatol Clin* 1998;16:145–164.
4. Rich P. Nail biopsy. Indications and methods. *J Dermatol Surg Oncol* 1992;18:673–682.
5. Scher RK. Punch biopsies of nails: a simple, valuable procedure. *J Dermatol Surg Oncol* 1978;4: 528–530.

6. Stone OJ, Barr RJ, Herten RJ. Biopsy of the nail area. *Cutis* 1978;21:257–260.
7. Siegle RJ, Swanson NA. Nail surgery: a review. *J Dermatol Surg Oncol* 1982;8:659–666.
8. Baran R, Sayag J. Nail biopsy—why, when, where, how. *J Dermatol Surg* 1976;2:322–324.
9. Miller PK, Roenigk RK. Diagnostic and therapeutic nail surgery. *J Dermatol Surg Oncol* 1991;17: 674–680.
10. Zaias N. The longitudinal nail biopsy. *J Invest Dermatol* 1967;49:406–408.
11. Bennett RG. Technique of biopsy of nails. *J Dermatol Surg* 1976;2:325–326.
12. Salasche SJ, Peters VJ. Tips on nail surgery. *Cutis* 1985;35:428–438.
13. Baran R, Kechijian P. Longitudinal melanonychia (melanonychia striata): diagnosis and management. *J Am Acad Dermatol* 1989;21:1165–1175.
14. Hanno R, Mathes BM, Krull EA. Longitudinal nail biopsy in evaluation of acquired nail dystrophies. *J Am Acad Dermatol* 1986;14:803–809.
15. Kechijian P. Nail biopsy vignettes. *Cutis* 1987;40:331–335.
16. Salasche SJ. Myxoid cysts of the proximal nail fold: a surgical approach. *J Dermatol Surg Oncol* 1985; 10:34–39.
17. Schnitzler L, Baran R, Civatte J, et al. Biopsy of proximal nail fold in collagen diseases. *J Dermatol Surg* 1976;2:313–315.
18. Scher RK, Tom DWK, Lalley EV. The clinical significance of periodic acid-Schiff positive deposits in cuticle-proximal nail fold biopsy specimens. *Arch Dermatol* 1985;121:1406–1409.
19. Leyden, Decherd JW, Goldschmidt H. Exfoliative cytology in the diagnosis of psoriasis of the nail. *Cutis* 1972;10:701–704.
20. Koutselinis H, Aronis K, Stratigos J, et al. Cytology as an aid in the diagnosis of psoriatic nails. *Acta Cytol* 1982;26:422–424.
21. Scher RK, Ackerman AB. Subtle clues to diagnosis from nail biopsies: histologic differential diagnosis of onychomycosis and psoriasis of the nail unit from cornified cells of the nail bed alone. *Am J Dermatopathol* 1980;2:255–257.
22. Grammar-West NY, Corvette DM, Diandonis MB, et al. Clinical pearl: nail plate biopsy for the diagnosis of psoriatic nails. *J Am Acad Dermatol* 1998;38:260–262.
23. Scher R, Ackerman AB. Subtle clues to diagnosis from biopsies of nails. *Am J Dermatopathol* 1980; 2:55–56.

3D

NAIL ABLATION AND MATRICECTOMY

ECKART HANEKE
ROBERT BARAN

Nail ablation is the definitive removal of the entire nail organ, and matricectomy is the complete extirpation of the nail matrix resulting in permanent nail loss.

INDICATIONS

Except for malignant tumors of the nail apparatus, nail ablation is rarely indicated. It may be necessary for an excessively painful nail treated several times in vain, but this should be an exception.

PRINCIPLES

The principle of nail ablation is complete removal of the nail unit with hyponychium, nail bed, matrix, and lateral and proximal nail folds. The nail plate may be left in place as a guiding structure (Fig. 3D.1), or it may be avulsed before nail ablation. A sterile staining solution such as methylene blue may be painted into the matrix space and then dried. The matrix surface will be stained and will remain clearly visible throughout the surgical procedure, thus helping the surgeon to dissect the matrix horns and the lateral and proximal nail grooves, which often are not completely extirpated and then give rise to nail spicules.

METHODS

There are several methods for nail ablation.

Scalpel Excision

Scalpel excision is strongly advocated whenever the surgical specimen requires histopathologic examination.

The digit is anesthetized using a ring block or similar technique. If the nail plate is avulsed, the matrix should be stained as described earlier, because it may be difficult to remove all of the lateral matrix horns. An incision is carried down to the bone around the tip of the digit 3 to 5 mm below the hyponychium, along both lateral aspects of the terminal phalanx until the distal dorsal crease of the distal interphalangeal joint and then across the dorsal aspect of the distal phalanx at least 3 mm proximal to the beginning of the matrix (the latter can be made visible by pressing on the free border of the nail plate, which makes a light transverse furrow appear at the base of the proximal nail fold) or along the distal joint crease. The nail unit is dissected from the bone. This is easier using blunt-tipped scissors than the scalpel, especially at the beginning when the nail bed must

E. Haneke: Klinikk Bunæs, Løkkeåsveien 3, 1337 Sandvika, Norway.
R. Baran: Nail Disease Centre, 42, Rue des Serbes, 06400 Cannes, France.

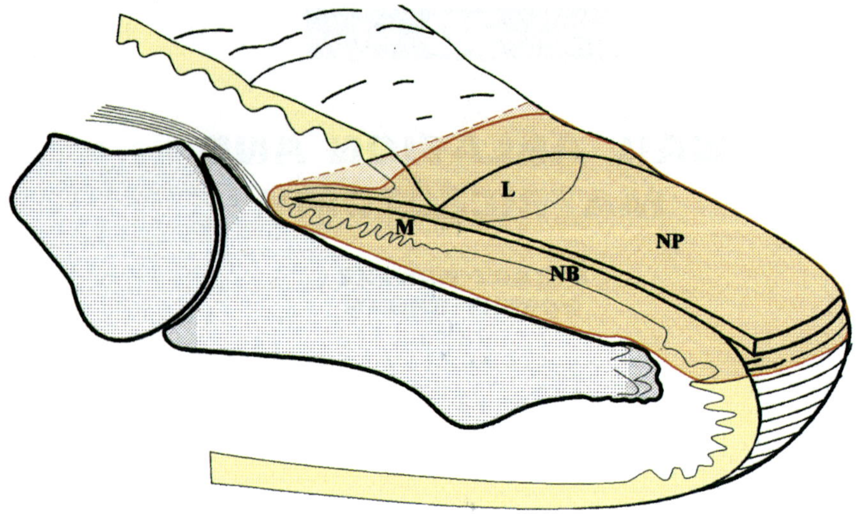

FIGURE 3D.1. Nail ablation.

be separated from the dorsal tuft of the tip of the terminal phalanx bone. The nail bed is removed with the periosteum to the base of the phalanx. Particular attention is given to the extensor tendon insertion, which can be severed when a pointed scalpel is used. This risk is minimized when blunt-tipped curved scissors are used. The dissection is directed to the lateral matrix horns. This procedure is performed easily in fingers and most lesser toes, but it may require dissection of the entire dorsal half of a big toe, particularly in patients with pincer nails or ingrown nails.

If the matrix space has not been stained before dissection, the matrix region must be checked carefully for remnants of the lateral matrix horns. This is particularly important when the patient has been operated on before with insufficient wedge excisions.

Nail ablation cuts the arteries feeding the nail organ. These arteries should be ligated because they are visible even under regional ischemia.

The defect from nail ablation may be covered with a free graft (split thickness, full thickness, reversed dermal graft), which usually takes on the bone in this particular location (Fig. 3D.2). A cross-finger flap is a very useful alternative for a free graft. We prefer

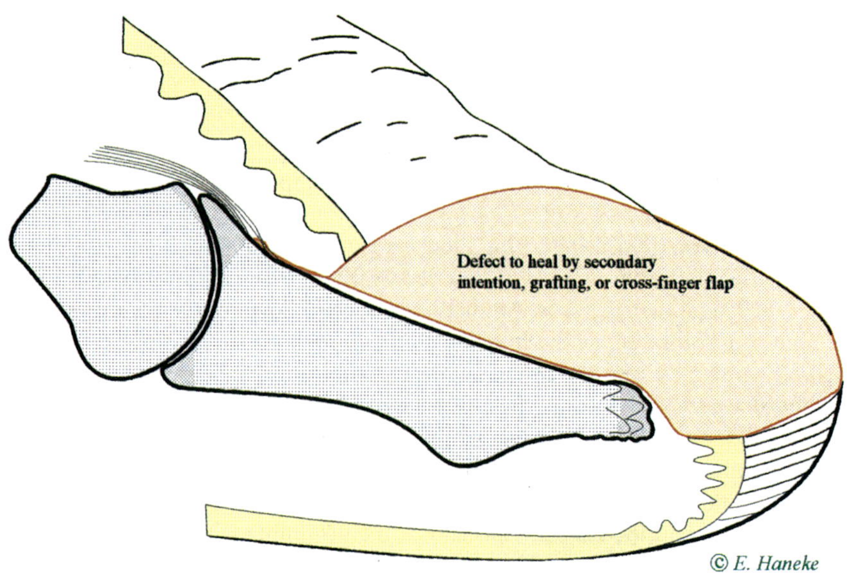

FIGURE 3D.2. Nail ablation. Defect to heal by secondary intention, grafting, or cross-finger flap.

to use skin from the intermediate phalanx of a neighboring finger, which is more convenient for the patient than skin from the thenar area of the palm.

If only permanent nail plate removal is necessary, a matricectomy may be performed (Fig. 3D.3). The proximal third of the nail is avulsed (see Chapter 4A, figure 3b) and the proximal nail fold is incised at both sides to allow the proximal nail fold to be reflected. An incision is made in the nail bed distal and parallel to the lunula border to the lateral aspects of both lateral nail grooves and carried along the lateral nail folds to meet the incisions of the proximal nail fold. Another incision is made along the underside of the proximal nail fold to remove the eponychium and proximal matrix. The entire matrix is dissected from the bone. Hemostasis is obtained by ligation or electrocoagulation after releasing the tourniquet. The remaining portion of the proximal nail fold may be sutured onto the matrix defect so that only a small defect will be left, which usually is even smaller than the lunula was before operation (Fig. 3D.4).

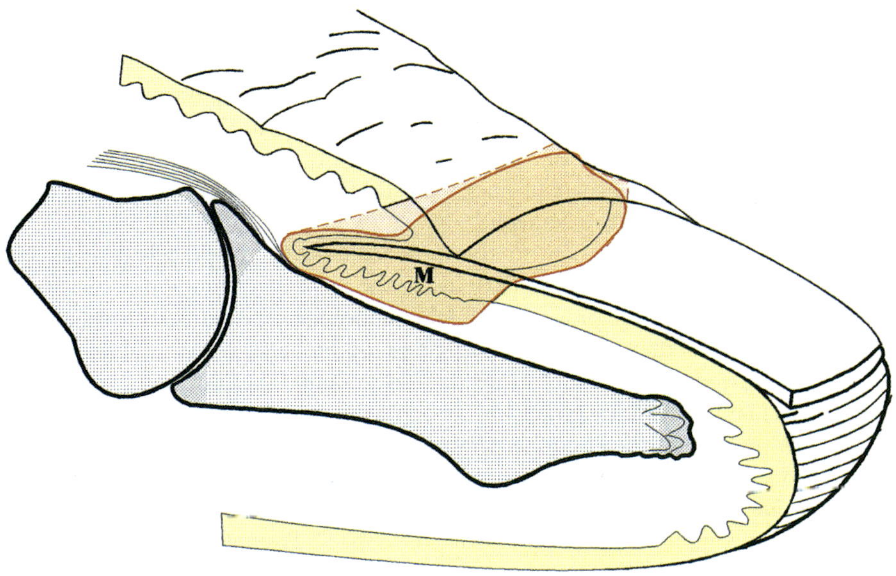

FIGURE 3D.3. Matricectomy.

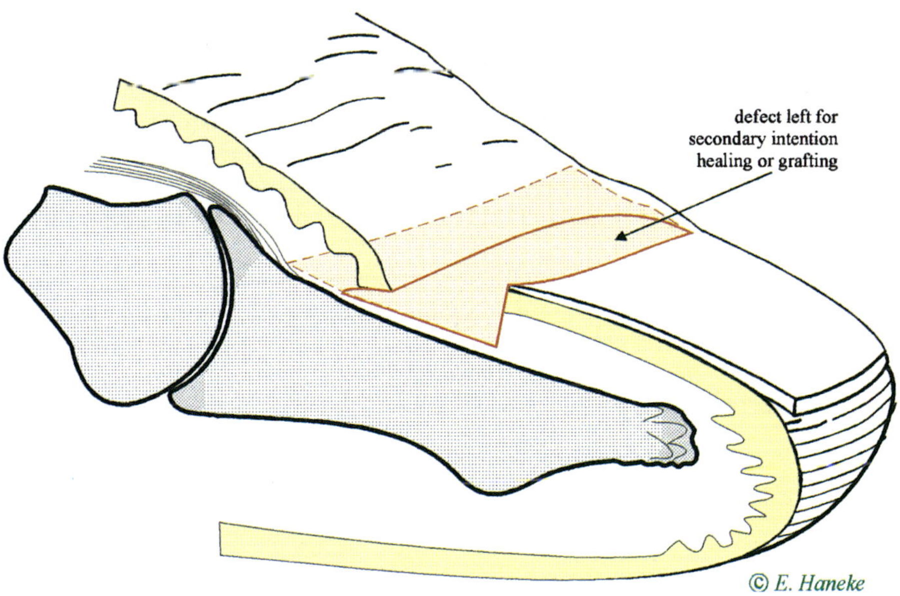

FIGURE 3D.4. Matricectomy. Defect left for secondary intention healing or grafting.

When this approach is used for treatment of ingrown toenails, it is called Zadik's technique (1). Although we believe that adequate primary surgical treatment of ingrown toenails will never make an operation such as the Zadik technique necessary, it is surprising that the relapse rate is very high, with nail spicules emerging from the lateral matrix horns on both side of the toes or even regrowth of dystrophic nail plates (2).

Electrosurgery and Curettage

Electrosurgery may be an alternative to scalpel excision when histopathologic examination of the surgical margins is not important. An electroexcision of the entire nail apparatus, analogous to the scalpel excision described earlier, may be performed. The electric scalpel cuts around the finger or toe tip 3 to 5 mm volar to the hyponychium up to the distal dorsal interphalangeal joint crease. The nail plate and hyponychium are grasped with a sturdy hemostat or pincer and pulled up. The nail unit is easily removed from the bone using the cutting current. Attention must be paid to the extensor tendon insertion and to the lateral matrix horns. Finally, an electroincision is carried out along the distal dorsal interphalangeal crease to remove the pedicle to the nail organ.

A disadvantage of electroresection of the nail unit is heat damage to the underlying structures that unduly delays wound healing and may cause severe postoperative pain.

We do not recommend simple electrodestruction of the entire nail organ. This would leave a large painful burn wound due to heat injury to the periosteum and bone.

Curettage is not suitable for nail matrix ablation because the matrix is too tightly bound to the underlying connective tissue.

Electro-radiosurgery is an easy method for matricectomy (3). A specially designed electrode may be used. The matricectomy electrode has the shape of a spade and is insulated on its upper surface to protect the proximal nail fold's undersurface. After partial nail avulsion of the proximal thirds of the nail plate and applying a tourniquet to achieve a bloodless operative field, the electrode is placed under the proximal nail fold. The electrosurgery unit is set for partially rectified current at a power level of approximately 5. Upon light contact of the electrode with the matrix, the electrode is swept from one side of the matrix to the other about four to six times, with additional electrocautery given to the lateral matrix horns. Because there are no incisions, there is no postoperative bleeding. After releasing the tourniquet, an antibiotic ointment and gauze dressing are applied. Postoperative pain is minimal. Healing usually is uneventful and postoperative oozing is very limited.

Matricectomy using negative galvanic current does not appear to offer any advantages compared to other matrix ablation techniques (4).

Laser Matricectomy

A carbon dioxide laser can be used instead of an electrode (5,6). The proximal nail fold is reflected, the proximal portion of the nail plate removed, and the matrix vaporized using a slightly defocused beam. The laser easily reaches the depth of the recessus of the matrix horns, thereby theoretically reducing the risk of recurrences, particularly of nail spicules. The laser is no better than any other destructive method if the procedure is performed by an experienced surgeon.

There are several new experimental lasers, such as excimer lasers, that allow precise removal of tissue layer by layer and cause less thermal damage to underlying tissue than the carbon dioxide laser.

Phenol Matricectomy

Phenol is a strong protein coagulant. Liquefied phenol, which is a solution of 10% water in crystalline phenol, is a powerful chemical cauterant that has been used successfully for half a century to treat ingrown toenails and onychogryphosis (7,8). The digit is anesthetized and a tourniquet is applied. The nail plate is avulsed—preferably using the proximal approach if it is for onychogryphosis—and the operative field is dried from blood because blood and other proteins inactivate the phenol. The 90% phenol is rubbed into the matrix for about 2 to 3 minutes, with at least three to four changes of the cotton-tipped applicators. The phenol must be rubbed particularly vigorously into the lateral matrix horns. Tapered antibiotic tablets are placed into the matrix horn spaces, the tourniquet is released, and a gauze-antibiotic ointment dressing is applied. This dressing is changed after 24 hours in a luke-warm foot bath to avoid any pain to the patient. Any coagulated blood is removed using hydrogen peroxide or chloramine, and dressings are changed daily with foot or hand baths.

Having used this technique for more than 15 years, we do not find it necessary to apply the classic Vaseline protection on the periungual skin. We do not use isopropyl alcohol flushes to remove or neutralize the phenol at the end of the procedure, and we have seen no cases of excessive cautery necrosis.

We do not curette the lateral matrix horns but have not seen recurrences, probably because the phenol is not inactivated. Application of hemostatic agents is not necessary with our technique. Of all the techniques described in this chapter, phenol matricectomy is the simplest and has the lowest recurrence rate. It can be used even when there is abundant granulation tissue and infection because phenol itself is an antiseptic agent. Because phenol coagulates proteins and coagulated proteins form a barrier against phenol, there is virtually no risk of slow long-term phenol penetration.

Phenol has a local anesthetic effect, so postoperative pain is minimal. However, chemical cautery by phenol causes an epidermal slough, with discharge occurring for about 2 to 3 weeks. Patients do not experience pain during this time, and they usually return to work and physical activities after a few days.

The matrix field is so small that there is no risk of phenol cardiotoxicity from this technique.

Other Methods

Sodium hydroxide has been advocated as an alternative to liquefied phenol for matricectomy. The procedure itself is essentially similar to phenol matrix destruction (9). The nail is avulsed under digital block anesthesia. The surrounding skin is protected with Vaseline. Ten percent sodium hydroxide solution is applied to the matrix with cotton-tipped applicators until a blanching frost develops. This may take 3 seconds to 3 minutes. The alkaline solution is neutralized with 5% acetic acid solution. The authors of the sodium hydroxide technique reported difficulties with phenol cautery, such as a relatively high recurrence rate and an unpredictable period of oozing. However, sodium hydroxide must be applied until blanching of the matrix capillaries occurs. The application time interval of 3 seconds to 3 minutes indicates that the penetration of this chemical is even less predictable than that of phenol. Prolonged application of a strong alkali can cause extreme destruction due to slowly progressive colliquative necrosis with all its untoward sequelae.

The terminal Syme operation is the most radical procedure for nail matrix ablation (Figs. 3D.5 and 3D.6) (10). In addition to surgical excision of the matrix and undersurface of the proximal nail fold, the nail bed and the underlying terminal half of the phalangeal bone are resected. A flap is created from the digit tip tissue, pulled dorsally, and sutured to the defect. It leaves a shortened bulbous toe that is very unsightly. We believe there is virtually no indication for this technique, except for malignant tumors invading the distal portion of the terminal phalanx.

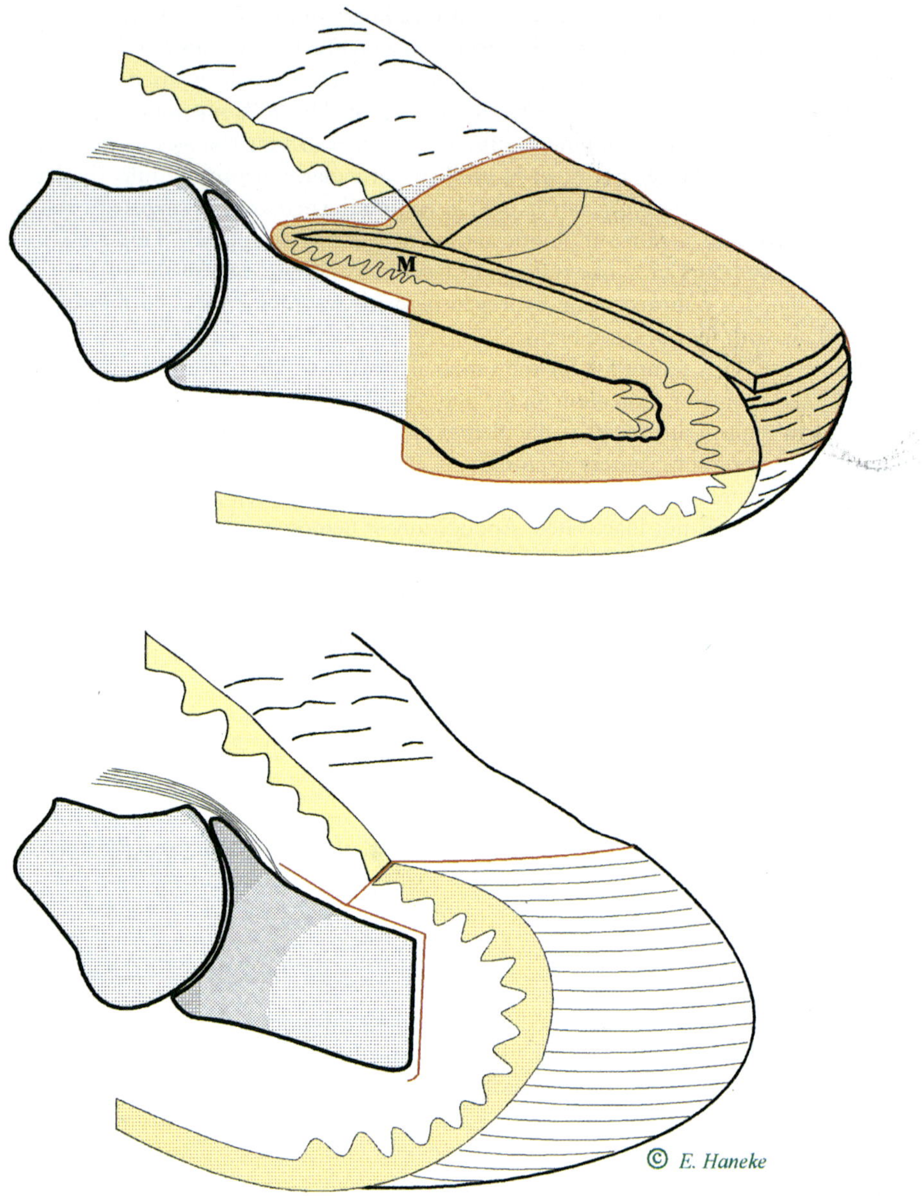

FIGURES 3D.5 and 3D.6. Terminal Syme operation.

REFERENCES

1. Zadik FR. Obliteration of the nail bed of the great toe without shortening the terminal phalanx. *J Bone J Surg* 1950;32B:66.
2. Murray WR, Bedi BS. The surgical management of ingrowing toenail. *Br J Surg* 1975;62:409.
3. Abbot WW, Getto EH. Partial matricectomy via galvanic current. *J Am Podiatr Assoc* 1980;70:239–243.
4. Hettinger DF, Valinsky MS, Nucci G, et al. Nail matrixectomies using radiowave technique. *JAPMA* 1991;81:317–321.
5. Apfelberg D, Rothermel E, Widtfed A, et al. Preliminary report on the use of carbon dioxide laser in podiatry. *J Am Podiatr Assoc* 1984;74:509–513.
6. Leshin B, Whitaker D. Carbon dioxide laser matricectomy. *J Dermatol Surg Oncol* 1988;14:608–611.
7. Siegle RJ, Harkness J, Swanson N. Phenol alcohol technique for permanent matricectomy. *Arch Dermatol* 1984;120:348–350.
8. Haneke E. Surgical treatment of ingrowing toenails. *Cutis* 1986;27:251–256.
9. Travers GR, Ammon RG. The sodium hydoxide chemical matricectomy procedure. *J Am Podiatr Assoc* 1980;70:476–478.
10. Thompson C, Terwilliger C. The terminal Syme operation for ingrown toenail. *Surg Clin North Am* 1951;31:575.

CHAPTER 4

FINGERNAILS

4A

ACUTE INJURY OF THE PERIONYCHIUM

ELVIN G. ZOOK

The perionychium consists of the nail bed and the paronychium surrounding it (Fig. 4A.1) (1). Many authors have defined and redefined the perionychial anatomy (2–8). The nail plate lies on the nail bed and inserts into a proximal fold called the nail fold. The skin over this is the nail wall. The distal border of the dorsal roof of the nail fold is the eponychium (Fig. 4A.2).

The nail fold is divided by the nail into the ventral floor and the dorsal roof. The nail bed is the soft tissue between the nail and the bone and is responsible for the generation and/or migration of the nail. The hyponychium is a plug of keratinaceous material at the junction of the nail bed and the skin of the fingertip.

The nail is formed by production of nail cells in three areas of the nail bed (Fig. 4A.3). The vast majority of the nail (approximately 90%) is produced in the "germinal matrix" area, which is on the ventral floor of the nail fold. The nail from the germinal matrix is produced by duplication of several layers of basilar cells that become flattened and elongated as they are pressed from beneath by newly forming cells. The cell nuclei remain as the cells flow upward and become part of the nail. On the dorsal roof and in

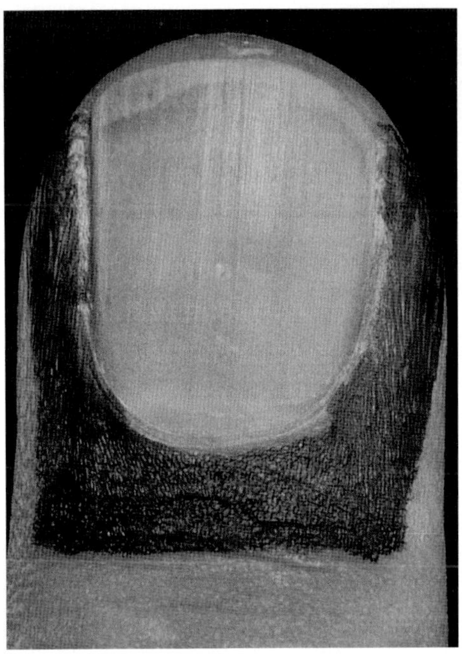

FIGURE 4A.1. The perionychium is composed of the paronychium in dark and the nail bed.

E. G. Zook: Division of Plastic Surgery, Southern Illinois, Springfield, Illinois 62708.

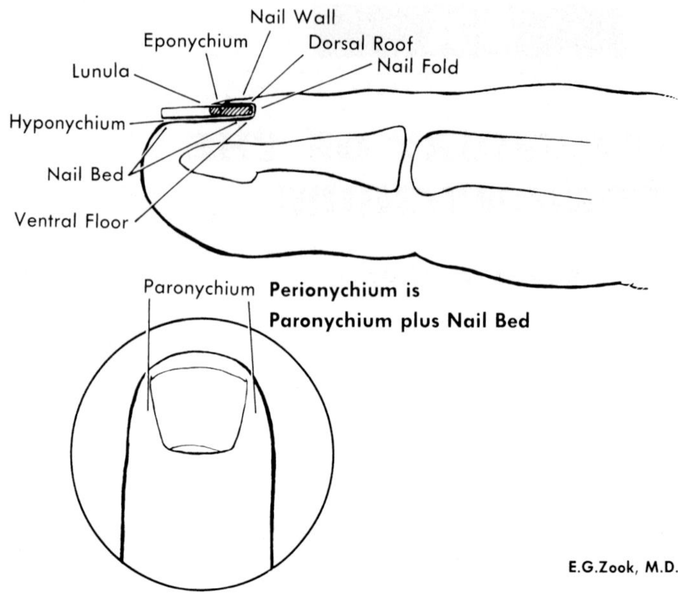

FIGURE 4A.2. Anatomy of the perionychium.

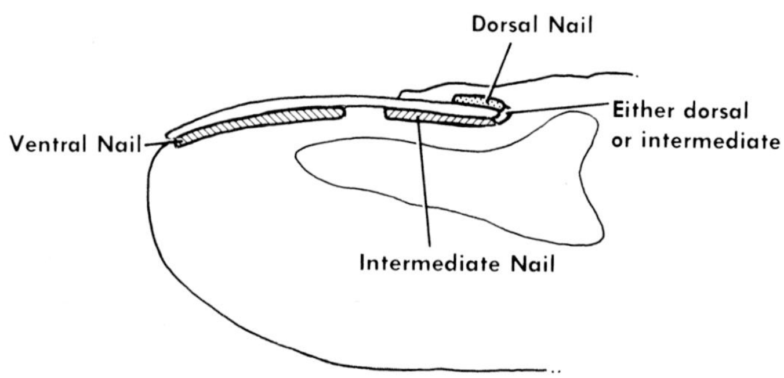

FIGURE 4A.3. Areas of nail production.

the "sterile" matrix portion of the ventral floor, nail is produced by one or two layers of cells that again are compressed and elongated by reproducing cells. Their nuclei enlarge and then disintegrate prior to the cells being incorporated into the nail (9–11).

Two terminal branches of the volar digital arteries, one on each side, communicate with its opposite vessel and supply blood to the nail bed. The perionychium is drained by coalescing veins in the proximal portion of the nail bed and the skin proximal to the nail fold. These veins course in a random fashion over the dorsum of the finger and enlarge as they proceed proximally.

PHYSIOLOGY

Complete longitudinal growth of a nail occurs every 3 to 4 months. After an injury, the appearance of the nail will improve for approximately three to four growths or for about 1 year. Therefore, it is best to wait a year to allow maximum improvement before any reconstructive attempts. Variations of nail growth after injury cause a lump in the normally regrowing nail, which will grow off with the first nail production. The cells formed by the dorsal roof of the nail fold as described previously produce the shine on the nail and those produced by the "sterile" matrix flow into the distally moving nail attaching it to the nail bed.

TREATMENT OF THE INJURED PERIONYCHIUM

The fingertip is the most commonly injured portion of the hand and the most commonly seen in the emergency room. Of these fingertip injuries, a significant proportion involves the perionychium. Injuries of the paronychium plus the nail bed are six times as common as injuries of the nail bed alone. Nowhere is the old adage "an ounce of prevention is worth a pound of cure" truer than in nail bed injuries. "There is never time to do it right, but there is always time to do it over" (12). Correct care administered acutely gives a much better result than attempts at revision.

A closing door is the most common source of injury of the perionychium, followed by crushing injuries, saws, and lawn mowers. The most commonly injured age groups are children and young adults. The injury occurs equally in right and left hands. Being the most exposed, the long fingertip and perionychium are the most commonly injured. This is followed by the ring finger, index finger, small finger, and thumb. The most common type of injury is classified as a simple laceration (really a crushing laceration), followed by stellate lacerations, crush, and avulsions of nail bed. The distal portion of the nail bed is the most commonly injured, with the frequency decreasing as the injuries progress proximally. This is because the more distal portion of the nail bed is the last to be withdrawn from harm's way. Fifty percent of perionychial injuries involve fracture of the distal phalanx or tuft (13).

The vast majority of nail bed injuries are caused by an object striking the nail and compressing the nail bed between the nail and underlying bone. This causes a bursting or crush-type injury to the nail bed in contradistinction to a sharp laceration. If the point of trauma is narrow (Fig. 4A.4A), a crush laceration usually results (Fig. 4A.4B); if it is wide, such as with the ball of a hammer, a stellate laceration (Fig. 4A.4C) or a burst into multiple pieces usually results (Fig. 4A.4D) There is rarely a sharp injury of the nail bed, because any sharp object that will divide the nail usually will pass on through and amputate the tip.

Subungual Hematoma

The nail bed is a highly vascular structure, and injury causes disruption and bleeding. If the nail is not broken or pulled loose by the trauma, blood collects beneath the nail as an expanding hematoma, the pressure of which may cause throbbing pain. Drainage of the hematoma is indicated to relieve pressure and alleviate the throbbing pain. This is best done by a paper clip heated by an alcohol burner (Fig. 4A.5A) or battery-powered ophthalmic cautery (Fig. 4A.5B) When the heated object contacts the hematoma, it is cooled and does not injure the nail bed. Either tool burns a round hole completely through the nail, which stays open to drain. The use of a nail drill or other sharp object to perforate the nail is painful and often leaves only a small opening that quickly clots and allows repeated expansion of the hematoma.

If more than 50% of the nail bed is undermined by hematoma, the nail should be removed and the nail bed explored. Lacerations should be carefully sutured and the nail replaced.

"Simple" Lacerations

Simple lacerations (36%) are the most common injury to the perionychium (Fig. 4A.6). If unrepaired or poorly repaired, excessive scarring of the nail bed results in splitting and nonadherence. A full surgical preparation of the finger and hand is essential before repair to prevent infection. The use of a finger tourniquet to allow bloodless visualization of the nail bed is necessary for the most accurate approximation of the nail bed.

The proximal nail outside of the eponychium is pathognomonic of serious nail bed injury (Fig. 4A.7).

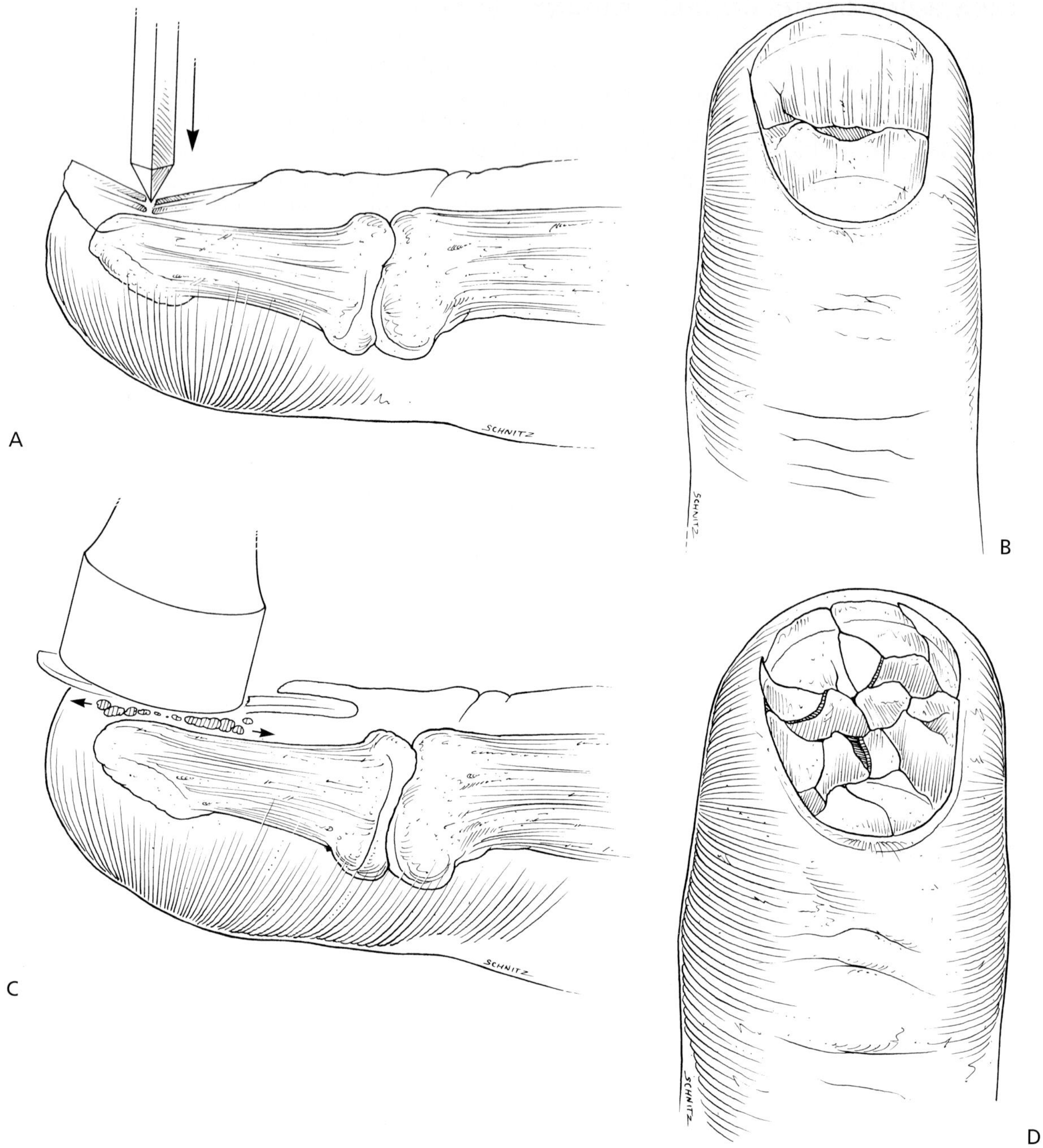

FIGURE 4A.4. A: A small area of compression to the nail causes a crushing laceration of the nail bed as it is pinched between the deformed nail and underlying bone. **B:** The result is a relatively straight laceration. **C:** Wider area of impact, such as from a hammer blow. **D:** Exploded nail bed as a result of the wide trauma.

The nail is removed from the nail bed by gentle elevation with a small periosteal elevator (Kutz elevator) (Fig. 4A.8A) or a fine pair of Iris scissors gently opened and closed while advancing proximally with the points against the nail. Care should be taken to leave all possible nail bed fragments attached to the bone rather than to the nail. After the nail is removed and loose soft tissue is scraped from the nail, the nail is washed and placed in povidone iodine (Betadine) solution while the nail bed is repaired (Fig. 4A.8B–C).

The nail bed is best explored with loupe magnification to aid in evaluation and care of the nail bed injury. Irregularities of the edges of the nail bed wound may be trimmed into

A 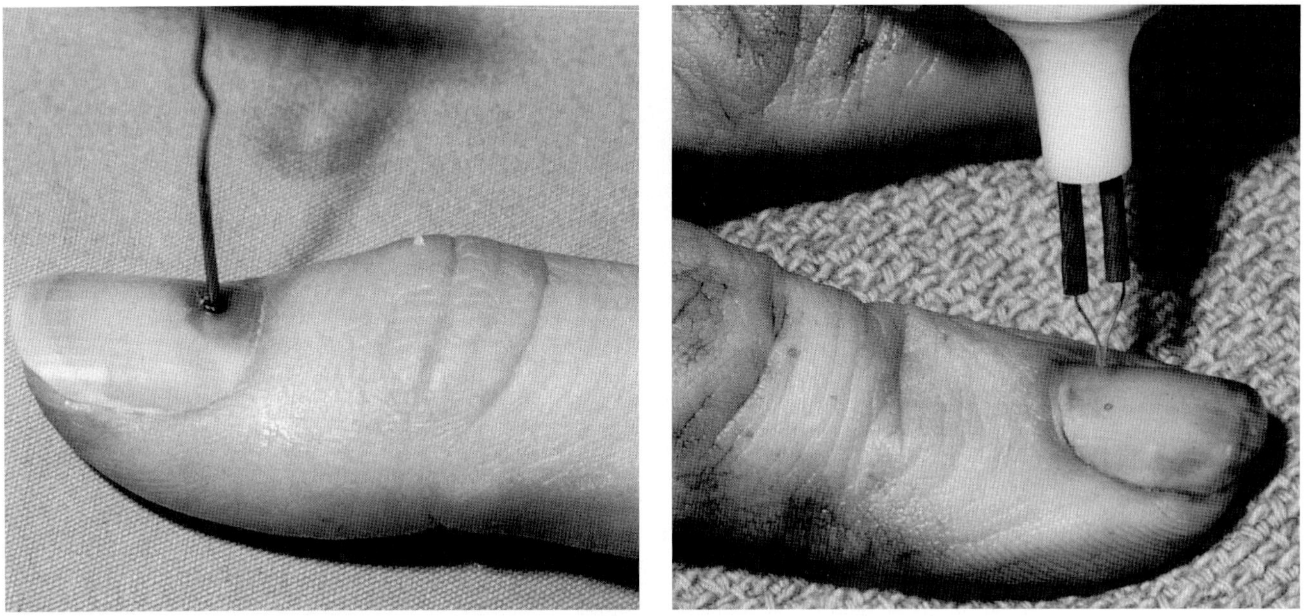 B

FIGURE 4A.5. **A:** Heated paper clip used to penetrate the nail. **B:** Ophthalmic cautery used to perforate the nail.

a straight line if it then can be closed without tension. This is difficult to achieve due to tight attachment of the nail bed to the bone. It is best to approximate the irregular edges as accurately as possible with 7-0 chromic ophthalmic sutures. One millimeter of elevation of the nail bed from the periosteum on each side of the laceration will assist in placing the needles for accurate approximation. I find 7-0 chromic suture on a micropoint spatula, double-armed GS-9 ophthalmic needle (Ethicon) (Fig. 4A.8D) to be ideal for repair. It is a double-armed suture that can be cut in the middle, giving an extra needle because needles are frequently bent while suturing close to the bone. After the nail bed is approximated as

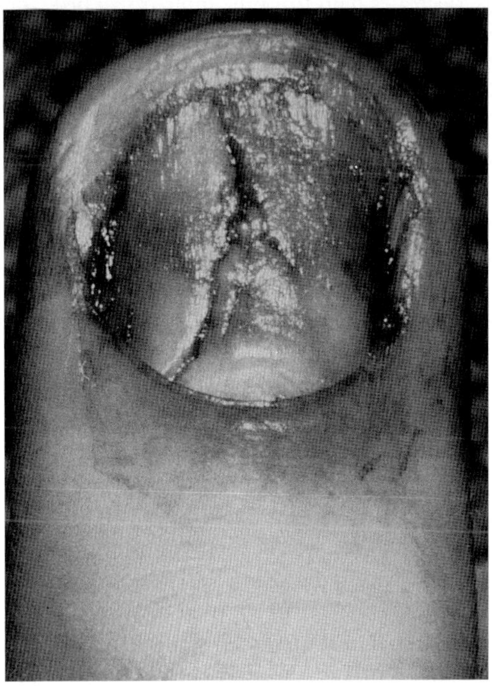

FIGURE 4A.6. Thirty-six percent of nail bed injuries are classified as simple lacerations.

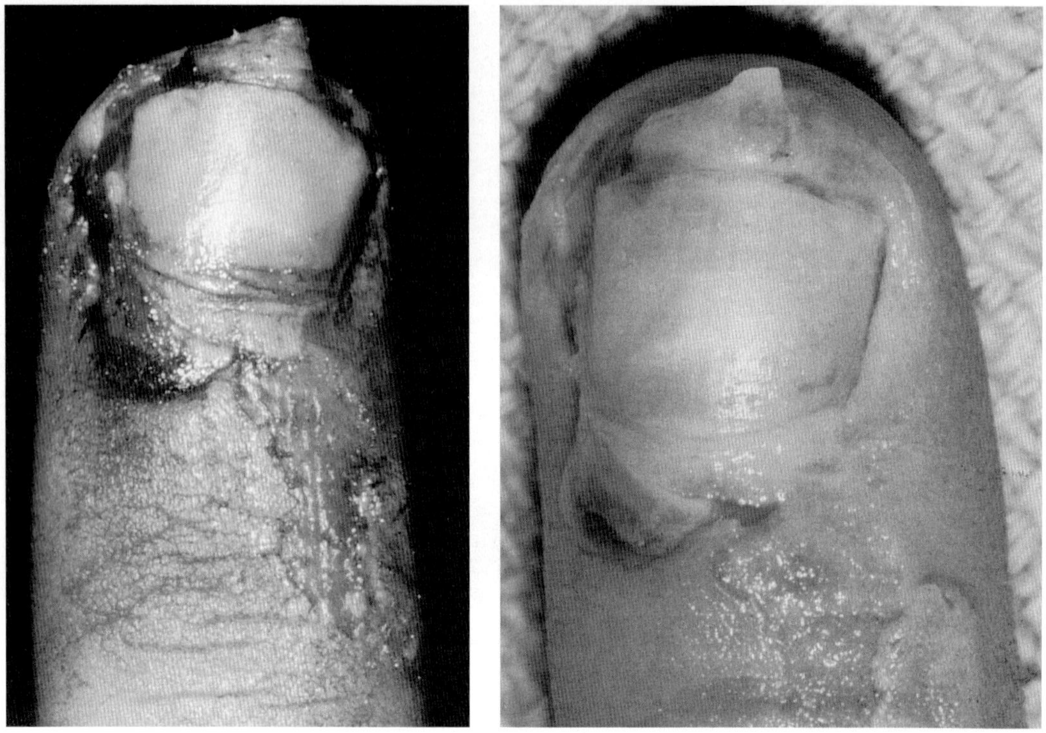

A B

FIGURE 4A.7. A: Acute trauma to the perionychium frequently avulses the proximal portion of the nail out of the nail fold. **B:** After surgical scrub, the nail could be seen on the dorsum of the nail fold.

accurately as possible, the nail is retrieved from the povidone iodine solution. It is rinsed and a hole drilled in it to allow drainage of blood or serum (Fig. 4A.9). The hole is made other than over the sutured laceration. The nail is replaced into the nail fold and held with a 5-0 nylon suture through the nail and hyponychium (14). Replacement of the nail helps mold the edges of the nail bed between the stitches in accurate alignment, maintains fracture alignment of the distal phalanx in most cases, decreases pain from the injured nail bed, and increases sensation in the fingertip. The fingertip is dressed with a nonadherent gauze, an absorbable sponge dressing, and a splint to immobilize and protect the tip. The dressing is changed in approximately 7 days and the nail checked for subungual seroma, hematoma, and infection. If present, the perforation in the nail is reopened, and the nail is elevated along the side or removed to allow drainage. The suture is left in the tip of the nail and finger for 2 to 3 weeks. After removal, the nail usually will stay in place for a few weeks until the newly growing nail pushes it off.

If the nail is unavailable or too badly damaged to be replaced, a nail-shaped piece of 0.020-inch reinforced silicone sheet is used as a substitute. A 6-0 nonfilament nylon suture is passed through the nail wall into the proximal fold, through the silicone sheath, back into the nail fold, and out through the nail wall (Fig. 4A.10). This holds the silicone sheet in the nail fold, as the silicone has no adherent properties of its own. The silicone sheet acts like a nail, molding the repair and maintaining the nail fold open. There are several artificial nails on the market, but most are expensive and many remain too firm to accurately fit and mold the nail bed. If none of these options is available, a single thickness of Adaptic gauze or other nonadherent gauze cut in the shape of a nail is placed in the nail fold.

After the first dressing change, the finger is soaked in warm, soapy water three times a day for 5 to 10 minutes. This helps to clean the wound and allow easier removal of the sutures.

When the nail is avulsed from the nail fold, sometimes it tears the matrix at the proximal portion of the nail fold and the matrix follows the nail out of the nail fold

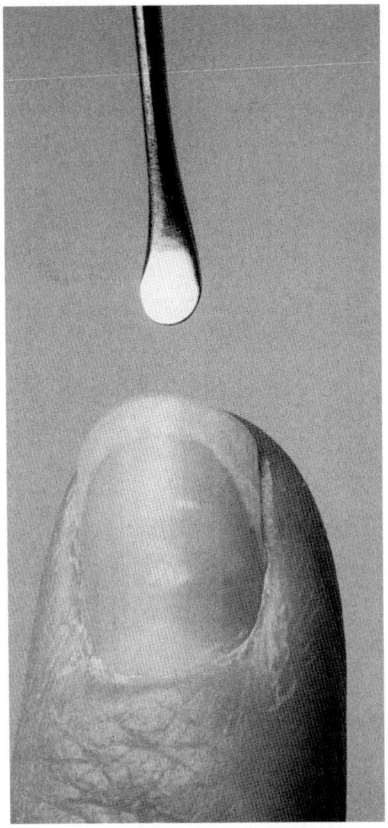

A

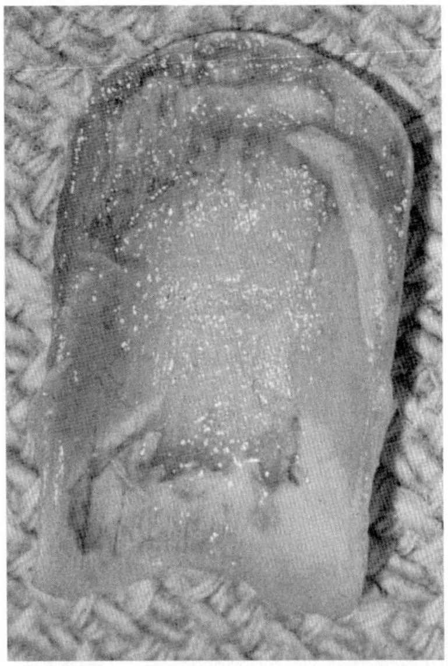

B

C

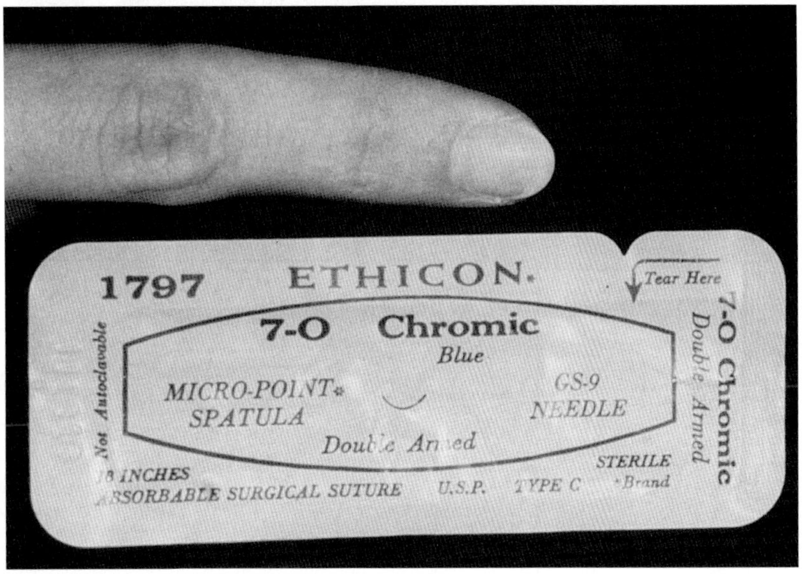

D

FIGURE 4A.8. A: Nail removed with a small periosteal elevator. **B:** Undersurface of the nail with bits of superficial nail bed attached. **C:** Nail scraped clean before placement in povidone iodine solution. **D:** Chromic suture used for repair.

(Fig. 4A.10). When the nail is avulsed from the fold, it is essentially impossible to replace the nail back into the nail fold without detaching it from the sterile matrix. After the nail is separated from the nail bed, sutures are placed through the nail wall into the proximal end of the germinal matrix and back out through the nail wall to accurately reapproximate and stabilize the germinal matrix into the nail fold. This is done with 6-0 monofilament nylon sutures. The nail is replaced into the nail fold to prevent scarring between the dorsal roof and the ventral floor. If more exposure of the nail fold is required, incisions are made at the proximal angles and 90 degrees from the

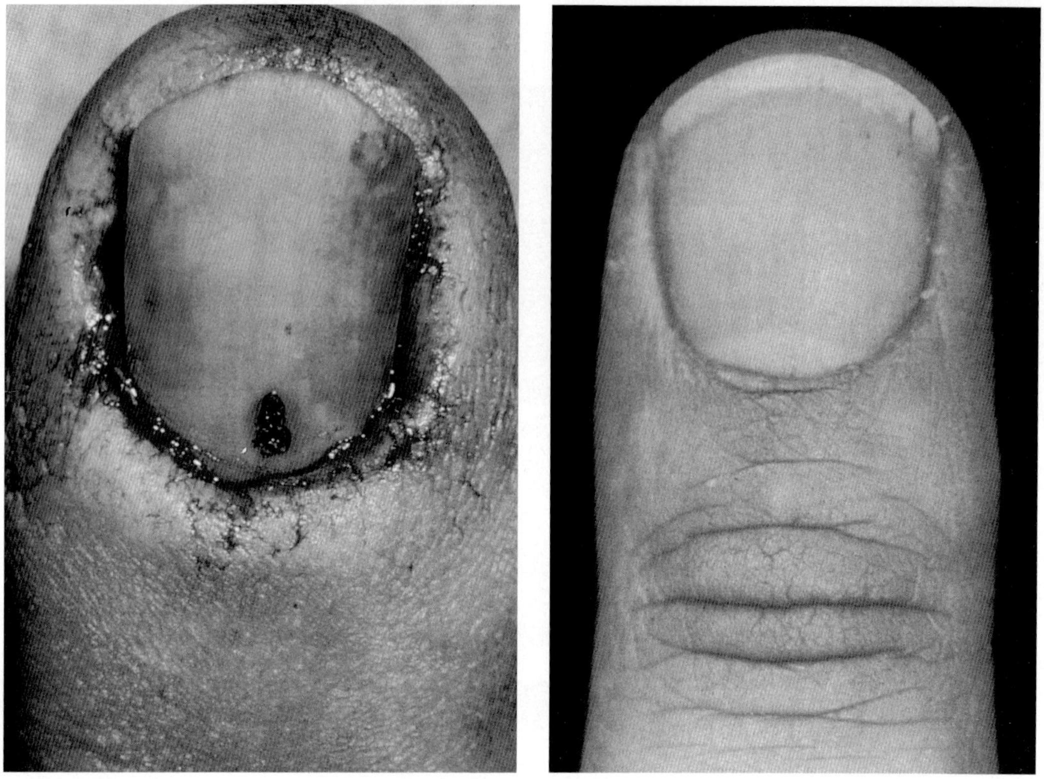

A B

FIGURE 4A.9. **A:** Nail replaced into the nail fold with a hole for drainage. **B:** One year later, the nail is normal.

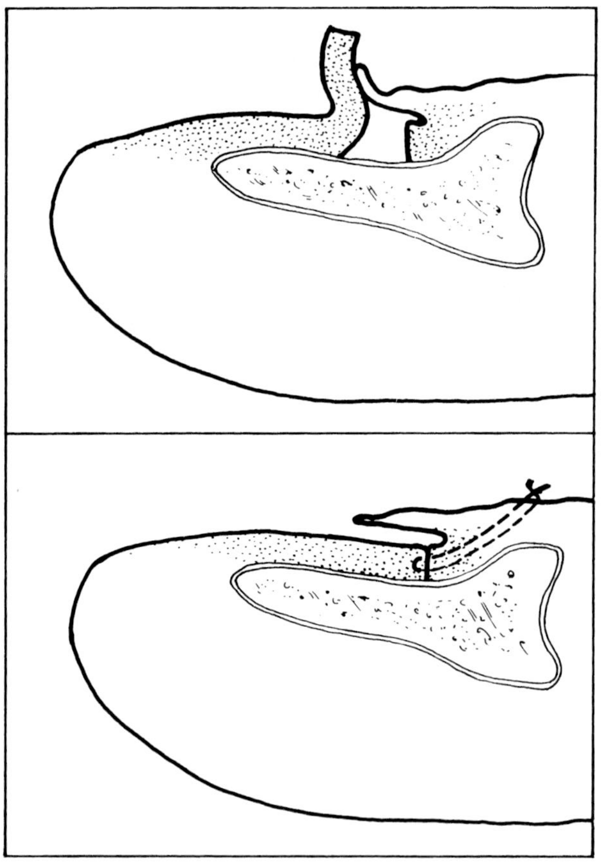

FIGURE 4A.10. Diagram of suture holding the nail bed in the nail fold in a similar fashion as the suture used to hold silicone sheet in the nail fold.

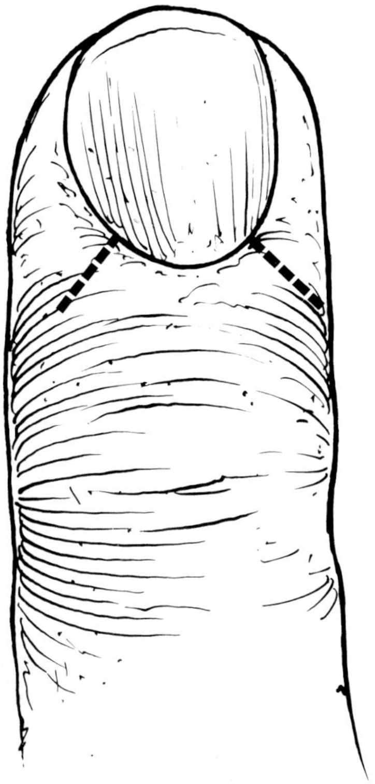

FIGURE 4A.11. Incisions to visualize the nail fold.

eponychium (Fig. 4A.11). This incision is easier to approximate and causes less scarring than an incision made straight proximal.

If a laceration involves the dorsal roof and the ventral floor, it is important that both be repaired (Fig. 4A.12). The knots of the chromic suture should be placed in the nail fold rather than in the tissue, where they will cause inflammation. It is important with this type injury that the nail, silicone sheet, or gauze be placed into the nail fold to prevent scar formation between the dorsal roof and the ventral floor, which almost invariably results in a split nail.

Stellate Lacerations

Stellate lacerations (27%) result from a wider source of injury and have a slightly poorer prognosis than a "simple" laceration (Fig. 4A.13) (1). The same treatment principles apply to this type of laceration as for the "simple" lacerations. The points and irregularities of the nail bed must be approximated as accurately as possible and the nail bed molded by nail, silicone sheet, or gauze.

Severe Crushing Injuries

Severe crushing injuries (22%) result from a wide area of force applied to the nail and causes an explosive type of injury of the nail bed with many fragments and pieces (Fig. 4A.14) (15,16). With this type injury, it is important that all fragments of the nail bed be left attached to the periosteum or, if attached to the nail, removed and returned to the nail bed. No fragments of nail should be debrided and discarded because it is extremely difficult to replace. Again, these fragments are approximated with fine chromic sutures, and the nail or silicone sheet is used to mold the fragments together and prevent scarring.

Fractures of the Tuft and Distal Phalanx

Fifty percent of perionychial injuries have an accompanying fracture of either the distal phalanx or tuft (1). If the fracture is nondisplaced, the nail bed is repaired and the nail

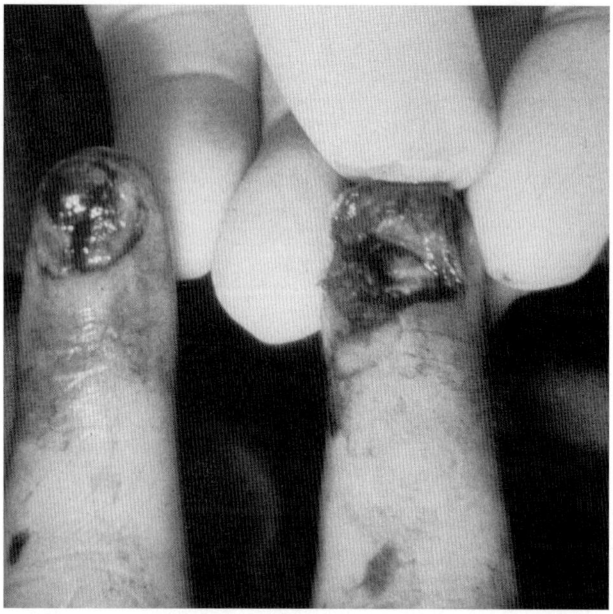

A

FIGURE 4A.12. A: A laceration through the dorsal roof into the nail fold is closed with fine sutures in the matrix of the dorsal roof with knots free in the nail fold. **B:** After skin repair, the nail is reinserted into the nail fold. **C:** One year later, the nail has good shine and appearance.

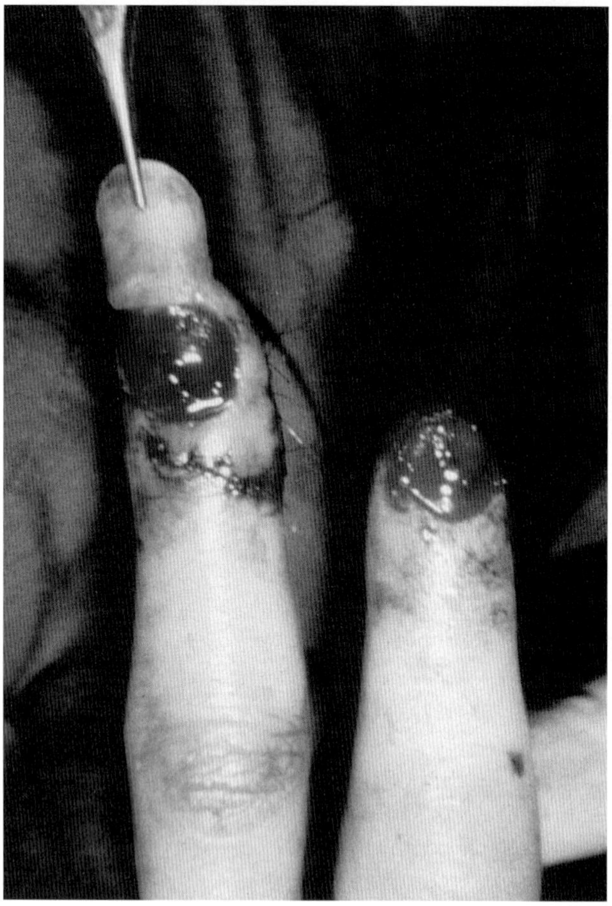

B

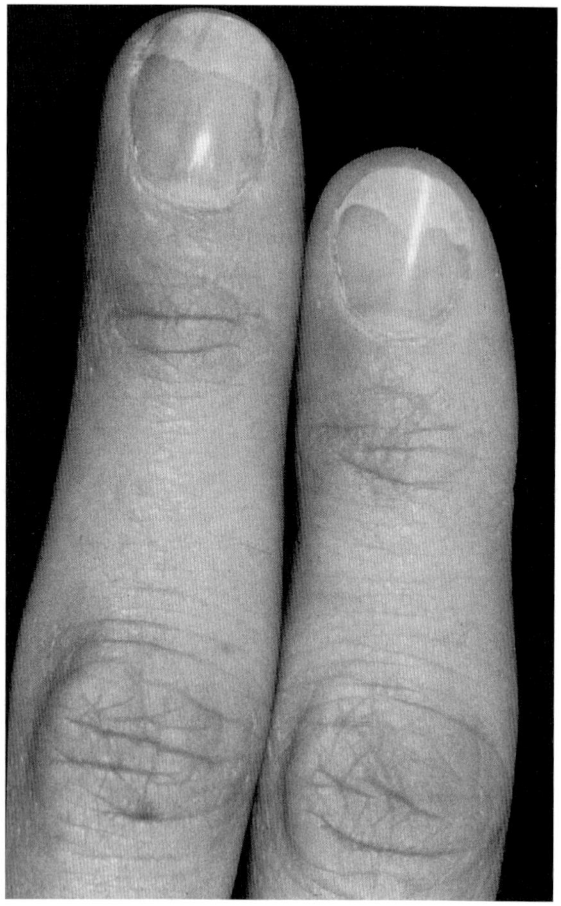

C

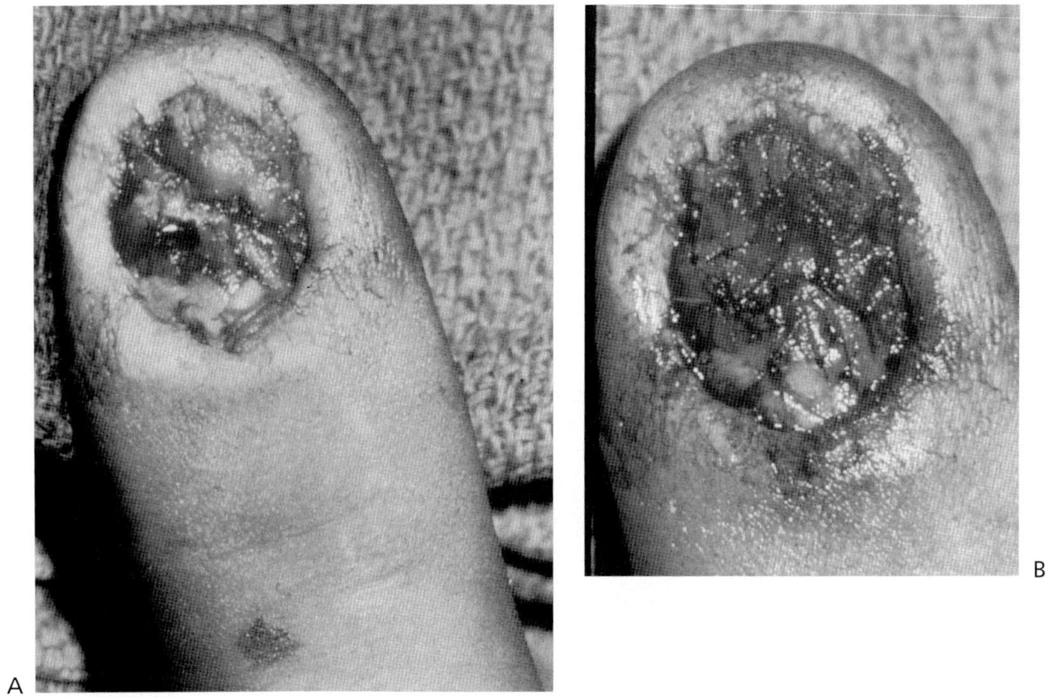

FIGURE 4A.13. A: Stellate lacerations compose 27% of nail bed injuries. **B:** Lacerations are approximated as accurately as possible with fine chromic suture.

used as a splint to maintain reduction. The nail is accurately molded to the bone and serves as an excellent splint. Displaced or unstable fractures are reduced as accurately as possible to achieve accurate alignment of, and a longitudinal pin used to maintain reduction in, unstable fractures. The tuft frequently is exploded into fragments and is remolded as accurately as possible with pressure. The nail is replaced to maintain reduction.

If a patient is seen with a displaced fracture and unrepaired or inadequately repaired nail bed in the first week after injury, the nail bed should be repaired after reduction and pinning of the fracture. Infection, lack of viability, or stiffness and shrinkage of the matrix may contraindicate late repair.

Nail Bed Avulsion

Fifteen percent of injuries are avulsions and for years have been an enigma to surgeons caring for nail bed injuries. Many methods of replacement have been described, including dermal grafts, reversed dermal grafts, gold sheeting, and suture packs. Replacement of the avulsed nail bed gives the best result. The nail should be examined because fragments of nail bed frequently will be attached to the undersurface of the nail (17). Attempts should be made to find the nail if it does not accompany the patient and there is absent nail bed. If the fragment of avulsed nail bed attached to the nail or nail fragment is small, it is replaced as accurately as possible with the nail and held in place with Steri-Strips or sutures (Fig. 4A.15). If the fragment is larger, 1 or 2 mm of the nail is removed from around the edge of the nail bed so that accurate suturing of the nail bed is possible (Fig. 4A.16). In larger fragment avulsions, the nail bed may be removed from the nail, sutured in place, and the nail replaced to mold it (Fig. 4A.17). A free graft 1 cm in diameter usually will live by an osculation and ingrowth of circulation from the periphery, even if it lies on bare cortex of the distal phalanx. Storage of the avulsed fragment in a mild antiseptic solution for a few days before replacement has been recommended by Iselin (18) to determine viability of the bed and maximize the condition of the recipient site. In 1947, Swanker (19) advocated permitting secondary healing of small avulsions while covered

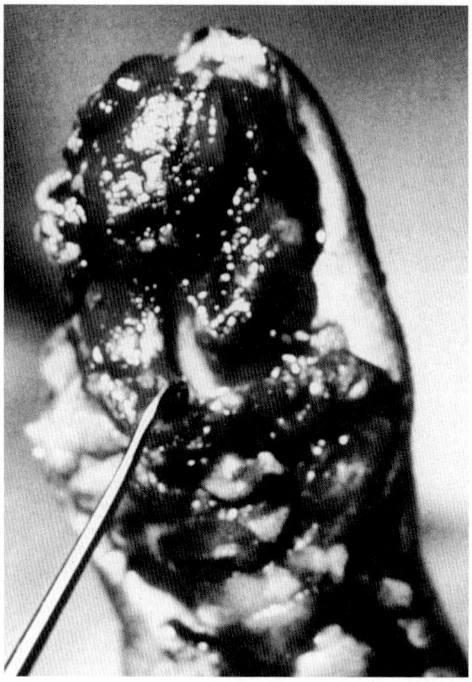

A

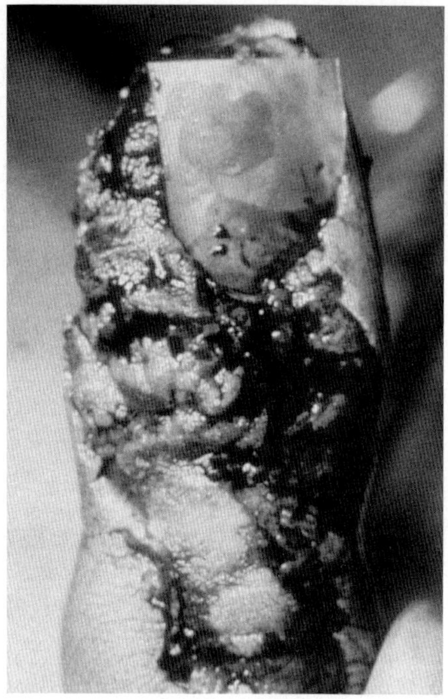

B

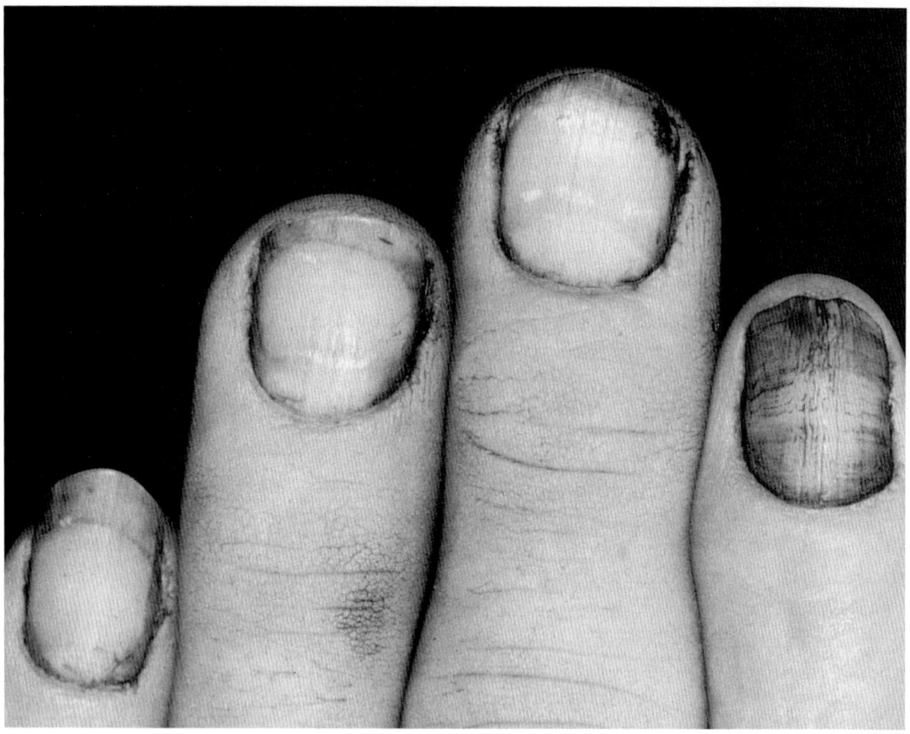

C

FIGURE 4A.14. A: Severe crushing injuries account for 22% of all nail bed injuries. **B:** After accurate approximation of the nail bed and skin of the finger, a piece of silicone sheet is placed into the nail fold to prevent scarring between the dorsal roof and ventral floor. **C:** One year later, the nail has some surface irregularities, narrowing, and curvature. The discoloration is due to the patient's return to work as a printer using ink.

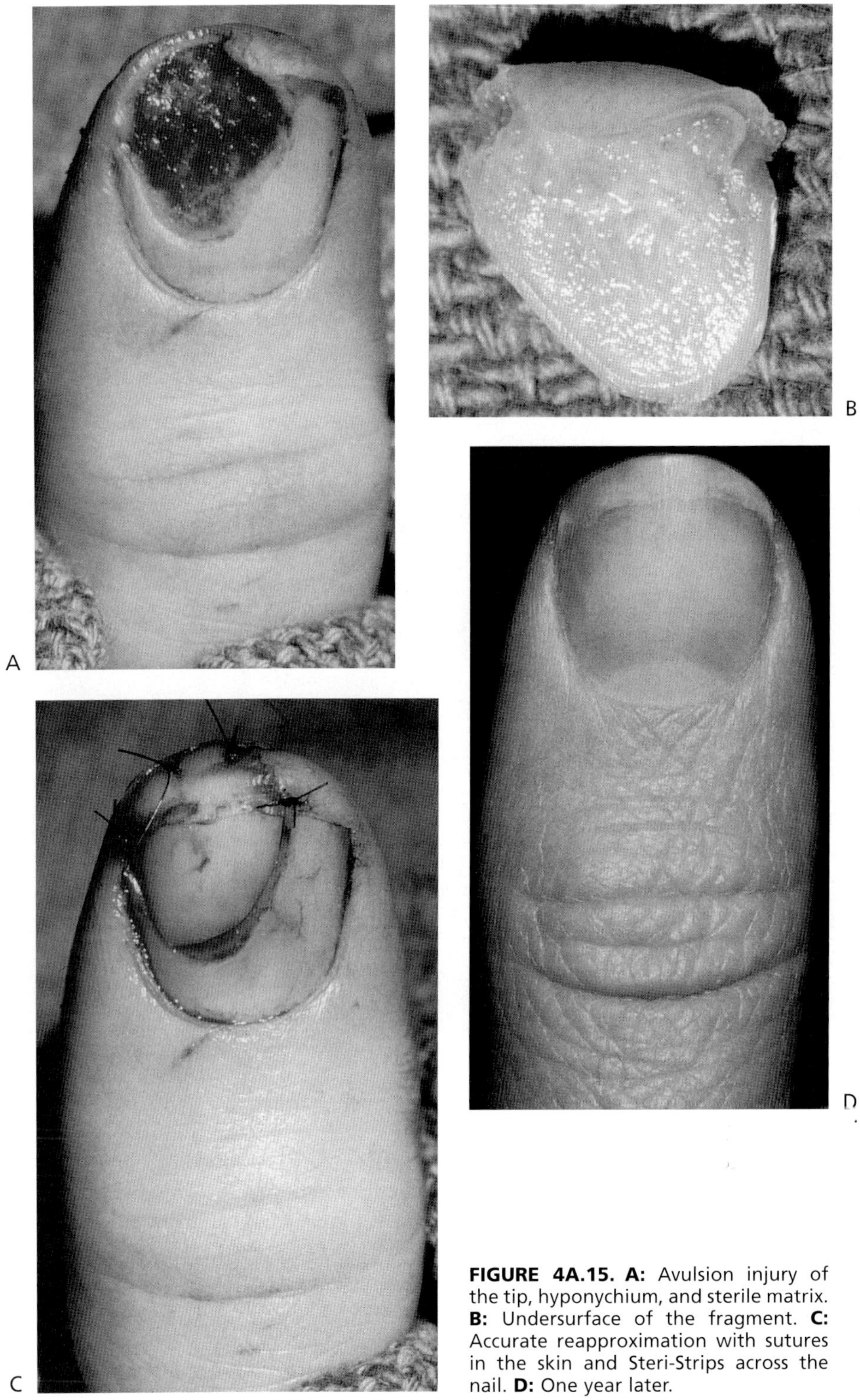

FIGURE 4A.15. A: Avulsion injury of the tip, hyponychium, and sterile matrix. **B:** Undersurface of the fragment. **C:** Accurate reapproximation with sutures in the skin and Steri-Strips across the nail. **D:** One year later.

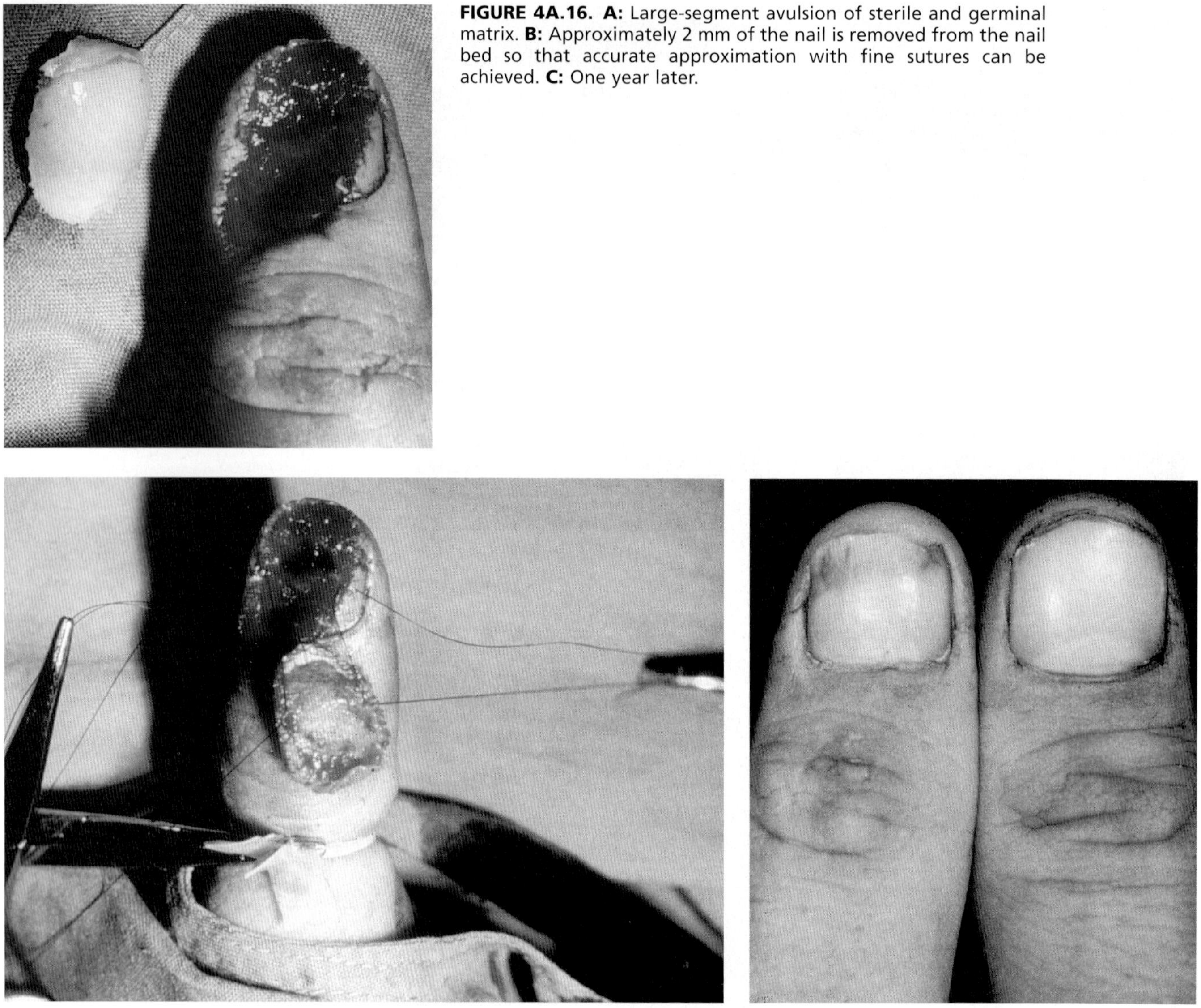

A B C

FIGURE 4A.16. A: Large-segment avulsion of sterile and germinal matrix. **B:** Approximately 2 mm of the nail is removed from the nail bed so that accurate approximation with fine sutures can be achieved. **C:** One year later.

with a tantalum sheet or the use of full-thickness nail bed grafts from an adjacent amputated finger or a 0.020-inch split-thickness nail bed graft from another fingernail bed to cover large avulsion defects. I believe the optimum time for replacement of avulsed nail bed is immediately rather than waiting. Skin graft replacement onto the matrix has been advocated (20–23), but it will not allow adherence of the nail and can cause nonadherence or splitting.

If an adjacent finger has been amputated or so severely crushed that it will be amputated, removal of a full-thickness sterile or germinal nail bed graft or split-thickness sterile matrix nail bed graft to replace the avulsed nail bed is indicated. The use of full-thickness nail bed grafts from the toe has been described to treat avulsion injuries, but this causes deformity of a toe and should be used only with full patient knowledge of the future deformity.

Shepard (24–26) demonstrated excellent results in avulsion injuries of the nail bed by harvesting split-thickness nail bed grafts from an adjacent site on the injured nail bed or taken from a toenail bed. It is my experience that if you cannot harvest enough split-thickness nail bed graft from the remaining sterile matrix of the injured finger, it is necessary to remove the split-thickness sterile matrix graft from the large toenail bed (Fig. 4A.18A–I).

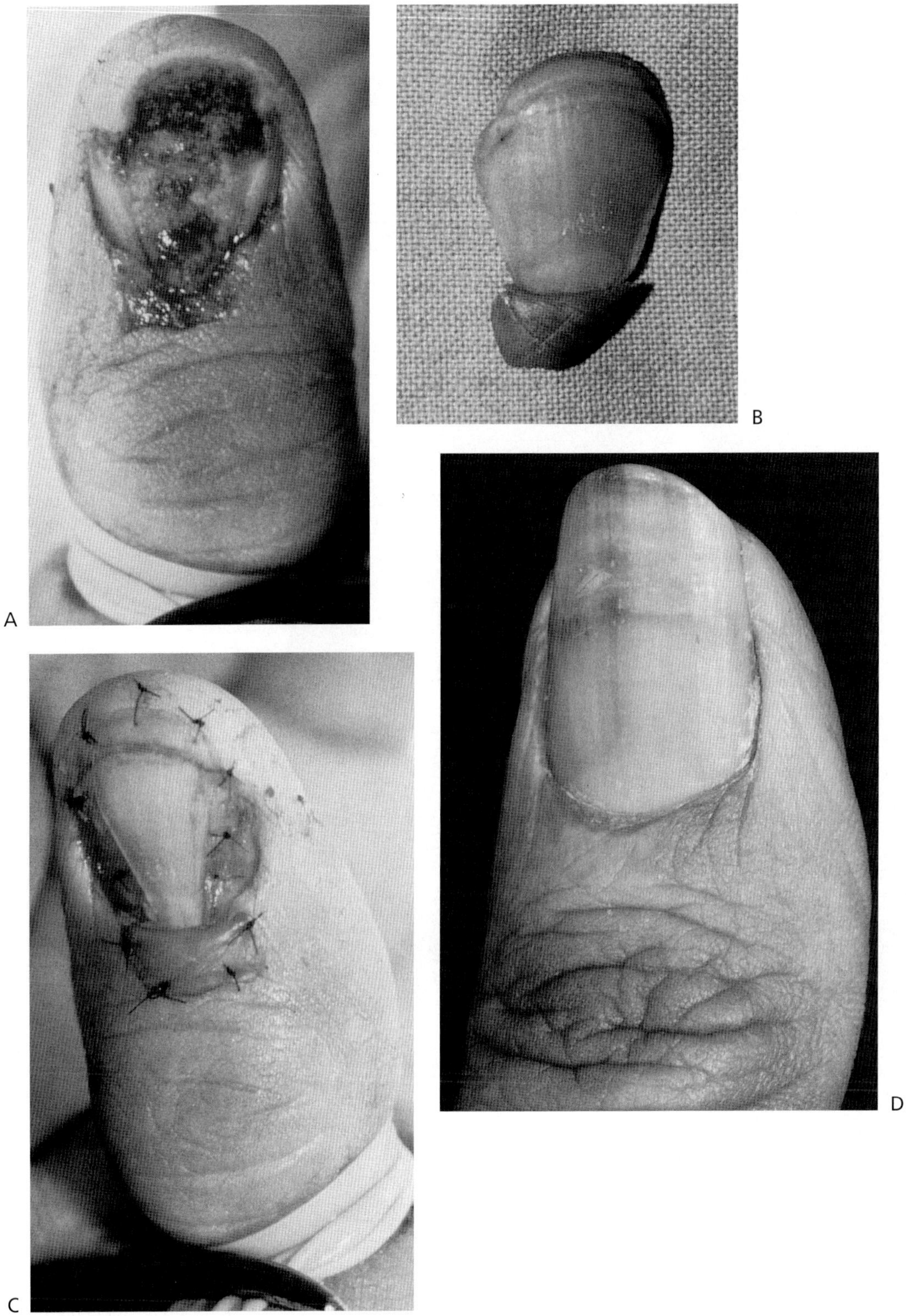

FIGURE 4A.17. A: Avulsion of the sterile matrix, germinal matrix, and eponychium. **B:** Fragment before removal of the nail. **C:** The fragment is sutured into place, nail fold held open, and suture line molded by the 0.020-inch silicone sheet. **D:** One year later.

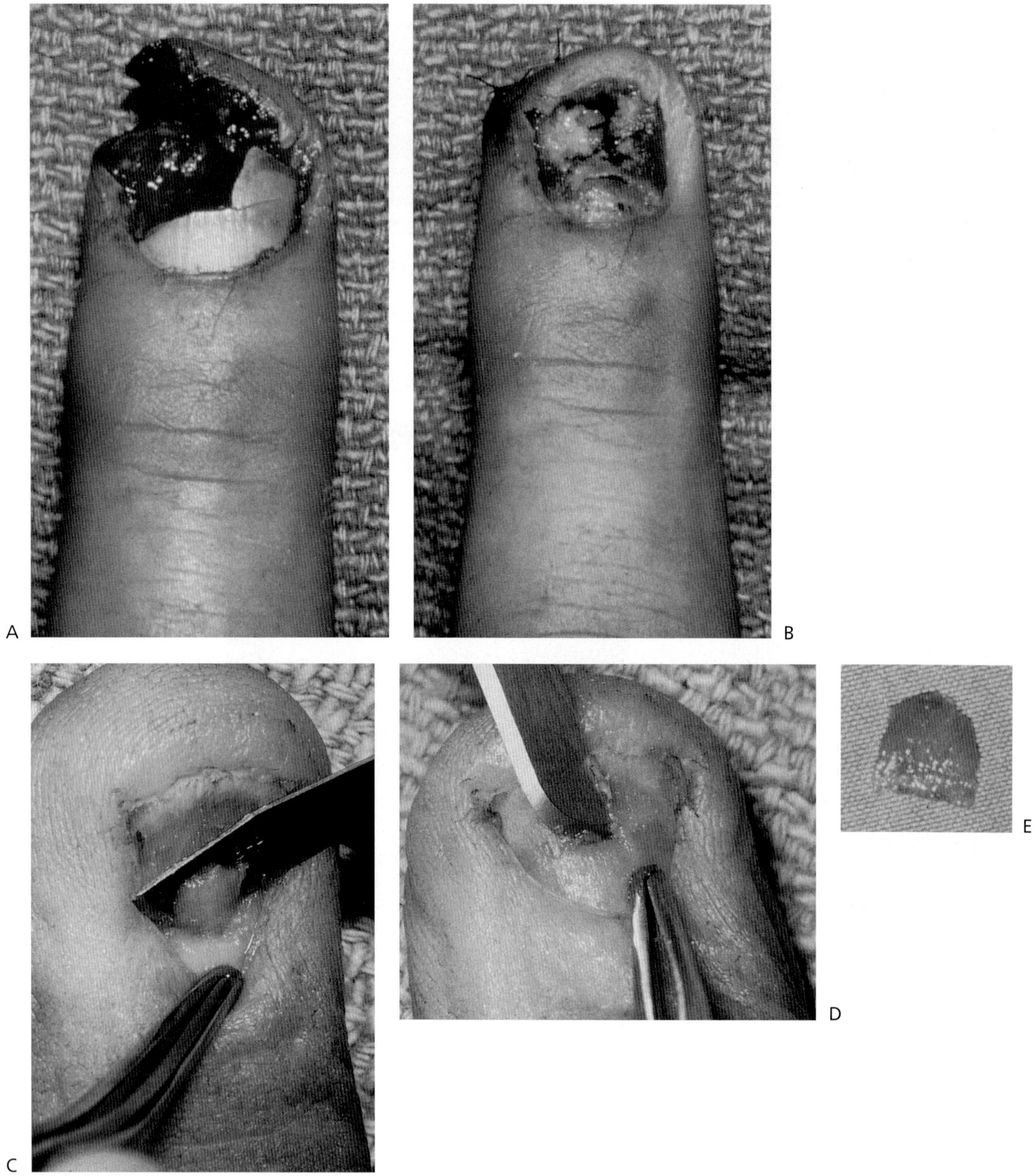

FIGURE 4A.18. A: Avulsion injury of the distal sterile matrix. **B:** After suture of the skin, a defect is seen with bare cortex of the distal phalanx. **C:** A small split-thickness fragment of sterile matrix can be removed by sawing action of a no. 15 blade. Care should be taken to always visualize the edge of the blade through the graft to avoid cutting too deeply. **D:** A larger fragment of split-thickness nail bed graft of the sterile matrix can be taken by using the tip of the blade in a back-and-forth motion. **E:** Fragment of split-thickness nail bed graft. **F:** Split-thickness nail bed graft sutured into place.

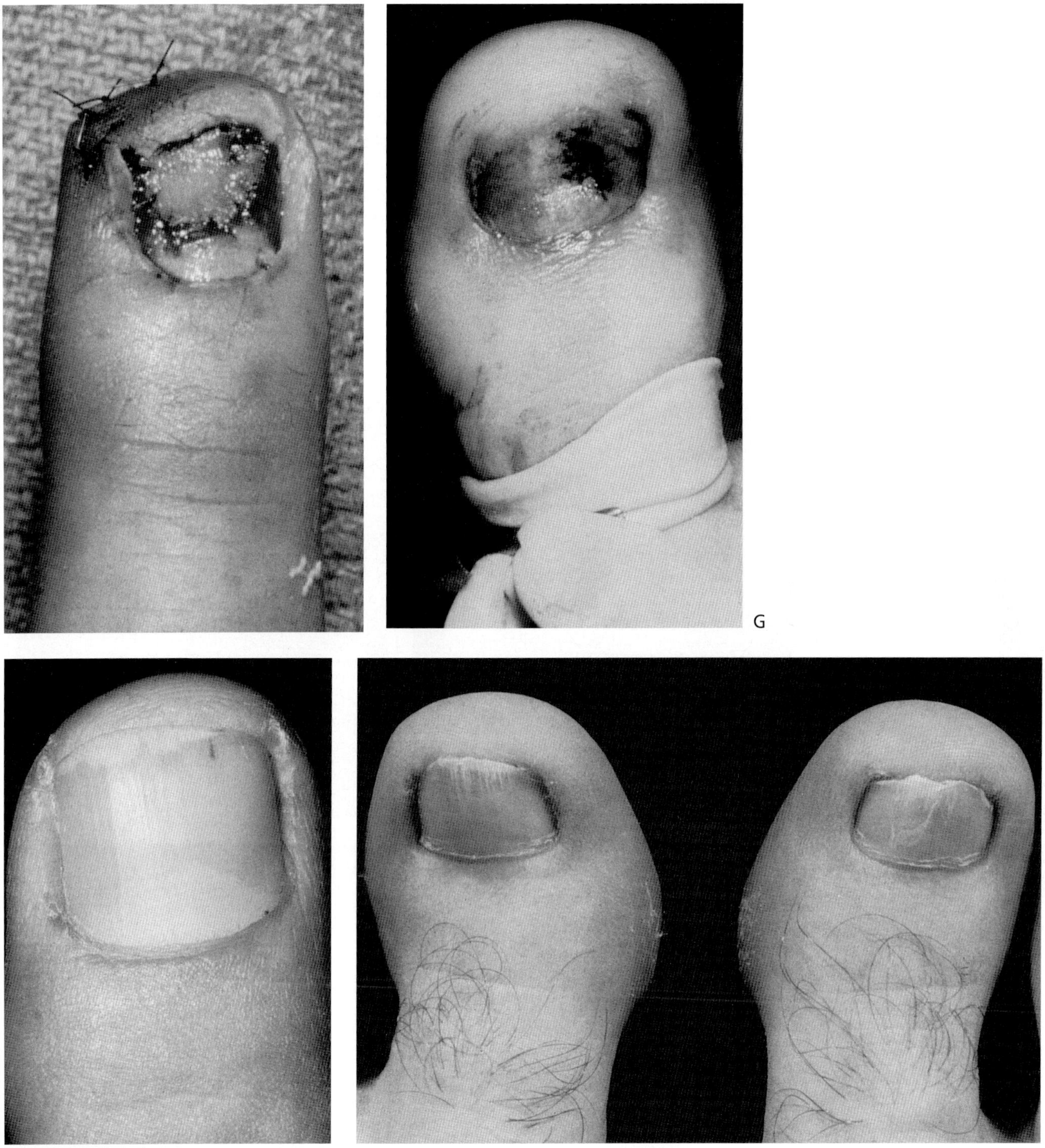

F G H I

FIGURE 4A.18 ***(continued).***

After a toe block, the toenail is removed with the periosteal elevator. A split-thickness graft of the sterile matrix approximately 0.010 inch thick is taken with a surgical blade. The blade can be used in a back-and-forth sawing fashion or the tip used to take a larger area. The curvature of the nail bed makes it impossible to take a large segment with the sawing action. Only split-thickness sterile matrix grafts should be used, because use of a germinal matrix graft requires inclusion of the basilar layer to achieve the desired result. After suturing of the split-thickness matrix graft into the defect, the nail, silicone sheet, or gauze is used to mold it into place. Frequently a portion of the paronychium or hyponychium is avulsed along with the nail, bed and this should be sutured accurately into place.

Avulsion of a composite portion of the nail bed, eponychium, and perionychium may be replaced by a matched composite graft from a finger that is to be, or has been, amputated. Also, a matched piece of tissue may be obtained from a toe to replace that on a finger (Fig. 4A.19A–D). This tissue is carefully matched and sutured as accurately as possible into place.

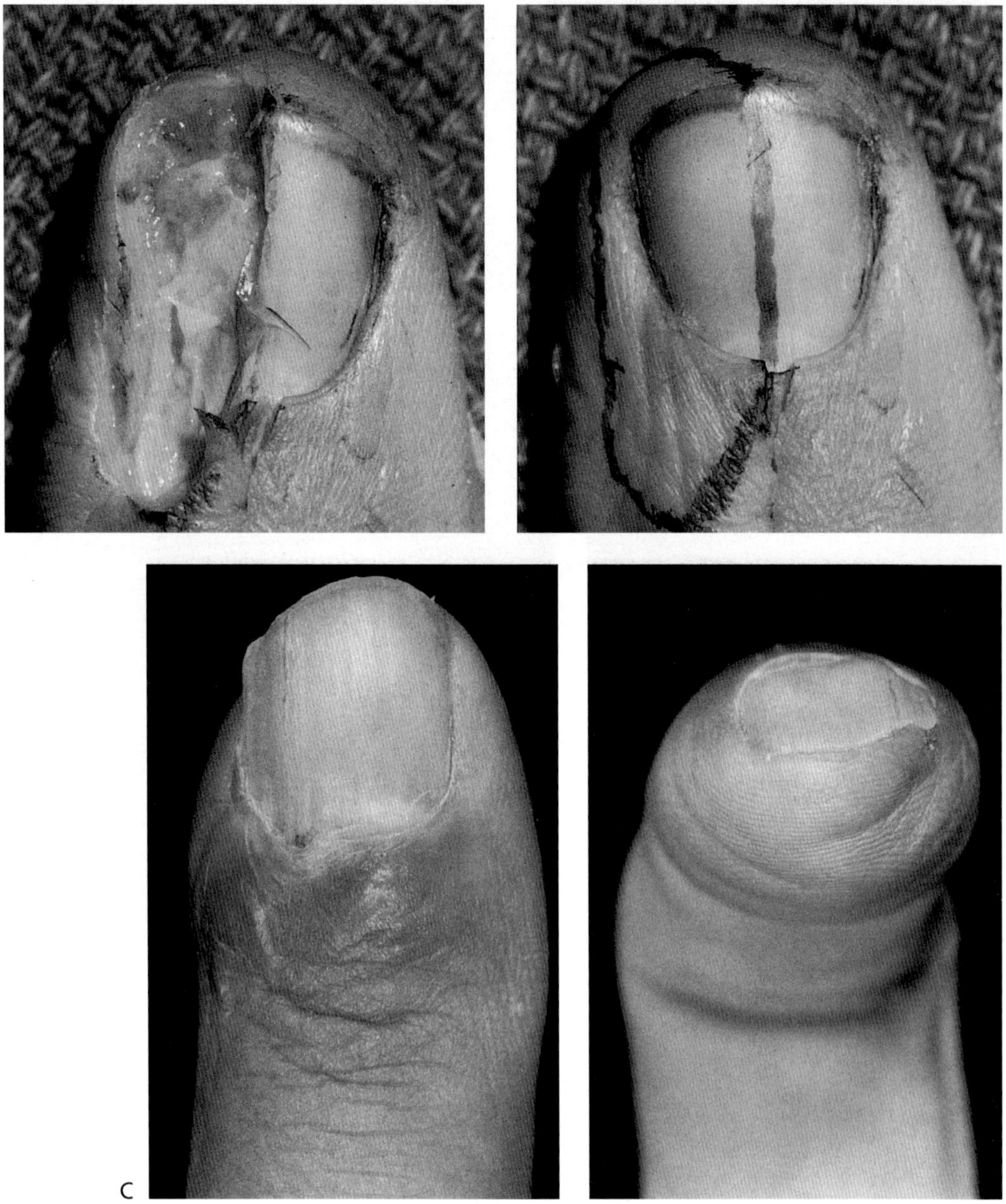

A B C D

FIGURE 4A.19. A: Avulsion of nail plate, germinal and sterile matrix by a saw. **B:** Matched area on non-replantable finger marked prior to transfer as a composite graft. **C:** Dorsal view of perionychium one year later. **D:** Distal view one year later.

REFERENCES

1. Zook EG, Guy RJ, Russell RC. A study of nail bed injuries: causes, treatment and prognosis. *J Hand Surg* 1984;9A:247–252.
2. Barron JN. The structure and function of the skin of the hand. *Hand* 1970;2:93–96.
3. Jones FW. *The principles of anatomy as seen in the hand,* 2nd ed. London: Bailliere, Tindall, and Cox, 1941.
4. Lewin K. The normal finger nail. *Br J Dermatol* 1965;77:421–430.
5. Pardo-Castello V. *Disease of the nail,* 3rd ed. Springfield, IL: Charles C. Thomas, 1960.
6. Samman PD. *The nails and disease,* 3rd ed. Chicago: Year Book Medical Publishers, 1938.
7. Sammon PD. *The nails in disease*, 2nd ed. Springfield, IL: Charles C. Thomas, 1972.
8. Stone OJ, Mullins JF. The distal course of nail matrix hemorrhage. *Arch Dermatol* 1963;88: 186–187.
9. Lewis BL. Microscopic studies of fetal and mature nail and surrounding soft tissue. *AMA Arch Dermatol* 1954;70:732–747.
10. Zook EG, Van Beek AL, Russell RC, et al. Anatomy and physiology of the perionychium: a review of the literature and anatomic study. *J Hand Surg* 1980;5:528–536.
11. Zook EG. The perionychium: anatomy, physiology and care of injures. *Clin Plast Surg* 1981;8: 21–31.
12. Meskinen JK. 117. In: Dickson P, ed. *The official rules.* New York: Delacorte Press, 1978.
13. Zook EG: Injuries of the fingernail. In: Green DP, ed. *Operative hand surgery.* New York: Churchill Livingstone, 1982:1355.
14. Schiller C. Nail replacement in fingertip injuries. *Plast Reconstr Surg* 1957;19:521–530.
15. Zook EG. Fingernail injuries. In: Strickland JW, Steichen JB, eds. *Difficult problems in hand surgery.* St. Louis: CV Mosby, 1982:322.
16. Zook EG. Nail bed injuries. *Hand Clin* 1985;1:701–716.
17. Schiller C. Nail replacement in fingertip injuries. *Plast Reconstr Surg* 1957;19:521–530.
18. Iselin M. Avulsion injuries of the nail. In: Pierre M, ed. *The nail.* Edinburgh: Churchill Livingstone, 1981:83.
19. Swanker WA. Reconstructive surgery of the injured nail. *Am J Surg* 1947;74:341–345.
20. Horner RL, Cohen BI. Injuries to the fingernail. *Rocky Mt Med J* 1966;63:60–62.
21. Flatt AE. Minor hand injuries. *J Bone Joint Surg* 1955;37B:117–125.
22. Flatt AE. Nail-bed injuries. *Br J Hand Surg* 1956;8:34–37.
23. Hanrahan EM. The split thickness skin graft as a covering following removal of a fingernail. *Surgery* 1946;20:398–400.
24. Shepard GH. Treatment of nail bed avulsions with split thickness nail bed grafts. *J Hand Surg* 1983; 8:49–54.
25. Shepard GH. Management of acute nail bed avulsions. *Hand Clin* 1990;6:39–56.
26. Shepard GH. Nail grafts for reconstruction. *Hand Clin* 1990;6:79–102.

4B

RECONSTRUCTION OF THE PERIONYCHIUM

ELVIN G. ZOOK

Reconstruction of the nail bed frequently is not as successful as would have resulted from optimal initial treatment. Lack of improvement has been disappointing to surgeons and has discouraged attempts at reconstruction. In the past several years, there has been a resurgence of interest in nail reconstruction. Many successful techniques have been developed and reported and, in most cases, have improved the appearance and function of the nail. Still, results are rarely perfect and occasionally are detrimental both in appearance and function. Every nail problem is different, and the anatomy and physiology must be used to formulate a treatment plan that will have a chance of success. Reconstruction can be categorized in various ways, such as problems of the "sterile" matrix (nonadherence, splits), the "germinal" matrix (nonproduction of nail, split), or the paronychium (trumpet nail, loss of eponychium, hypertrophy of the hyponychium) (1).

NAIL RIDGES

Nail ridges are caused by irregularity of the nail bed secondary to scar in or under the nail bed or irregular bone surfaces beneath the nail bed that cause irregularity of the bed. The nail follows the shape of the nail bed as it grows distally, and irregularities of the nail bed cause irregularities of the nail. If a ridge in the nail bed is transverse, the nail deformity varies from indentation to nonadherence. Longitudinal ridges are caused by a misshapened nail bed that creates a misshapen nail plate. Flattening of nail bed ridges requires excision of scar that causes the ridge, with preservation of the nail bed if possible or with repair or replacement if necessary (2,3). An irregularity on the dorsal cortex of the distal phalanx must be smoothed to flatten the bone and subsequently the nail bed.

SPLIT NAIL

A split nail may be caused by a scar ridge that prevents the sterile matrix cells from adhering the nail to the nail bed. Nonproduction of either germinal or sterile matrix in a scarred area that either splits the nail or does not produce cells in the center of the nail to hold it together will cause splitting. Many treatments in the past have been advocated for treatment of a split nail with varying success. If the split is due to scar in the "sterile" matrix, the following treatment is advised. The simplest technique, if the scar is narrow, is excision and primary closure with minimal tension (Fig. 4B.1). If the scar is so wide that closure cannot be accomplished without tension, resection of the scar

E. G. Zook: Division of Plastic Surgery, Southern Illinois, Springfield, Illinois 62708.

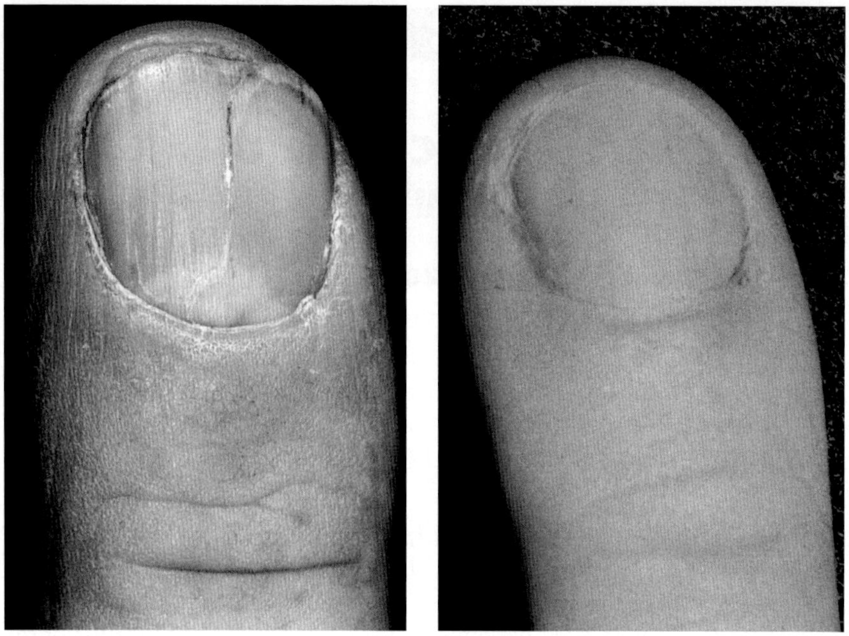

A B

FIGURE 4B.1. A: Relatively fine longitudinal split of the nail bed after trauma. A small scar of the distal portion of the lunula was excised and closed. **B:** One year later, the split is not present.

and a split-thickness graft of sterile matrix placed in the defect is advised (Fig. 4B.2). If more than minimal undermining of the nail bed edges is necessary to obtain tension-free closure, a split- or full-thickness nail bed graft should be used. A split starting in the germinal matrix may be caused by either nonproduction of nail cells in the area of the scar that causes a nail only to be formed on each side of the scar (Fig. 4B.3) or a cicatrix between the dorsal roof and the ventral floor that causes splitting of the nail as it grows distally.

Under finger block, the nail should be removed and the nail fold explored. If there is an obvious cicatrix between the dorsal roof and the ventral floor, this should be divided sharply. A decision must be made whether the split is due to scarring between the dorsal roof and the ventral floor or whether there is scarring of the ventral floor of a magnitude to prevent production of nail. This is not always an easy decision. If it is thought that scarring of the dorsal roof to the ventral floor is the cause of the split, then a split-thickness sterile matrix graft from the distal nail bed should be removed and placed as a graft on the undersurface of the dorsal roof (Fig. 4B.4). A 0.020-inch silicone sheeting is sutured into the nail fold to prevent recurrent scar between the dorsal roof and the ventral floor.

If the scarring of the germinal matrix is thought to be the cause, the scar must be excised and a decision made for primary closure (must be without tension, which is uncommon). Relaxing incisions in the paronychial fold to permit approximation of the germinal matrix has been unsuccessful in our hands. A free nonvascularized full-thickness germinal matrix graft from a toe also may be used (Fig. 4B.5). We prefer the second toe and ablate the remainder of the germinal matrix so that a split or deformed toenail is not left behind. Split-thickness grafts of germinal matrix will not be successful and should not be attempted. When germinal matrix graft is used, it must be full thickness.

A transverse scar in the germinal matrix can cause a horizontal split (double nail) (Fig. 4B.6). If the remaining volar nail is adequate, the dorsal portion of the nail and its germinal matrix can be surgically excised.

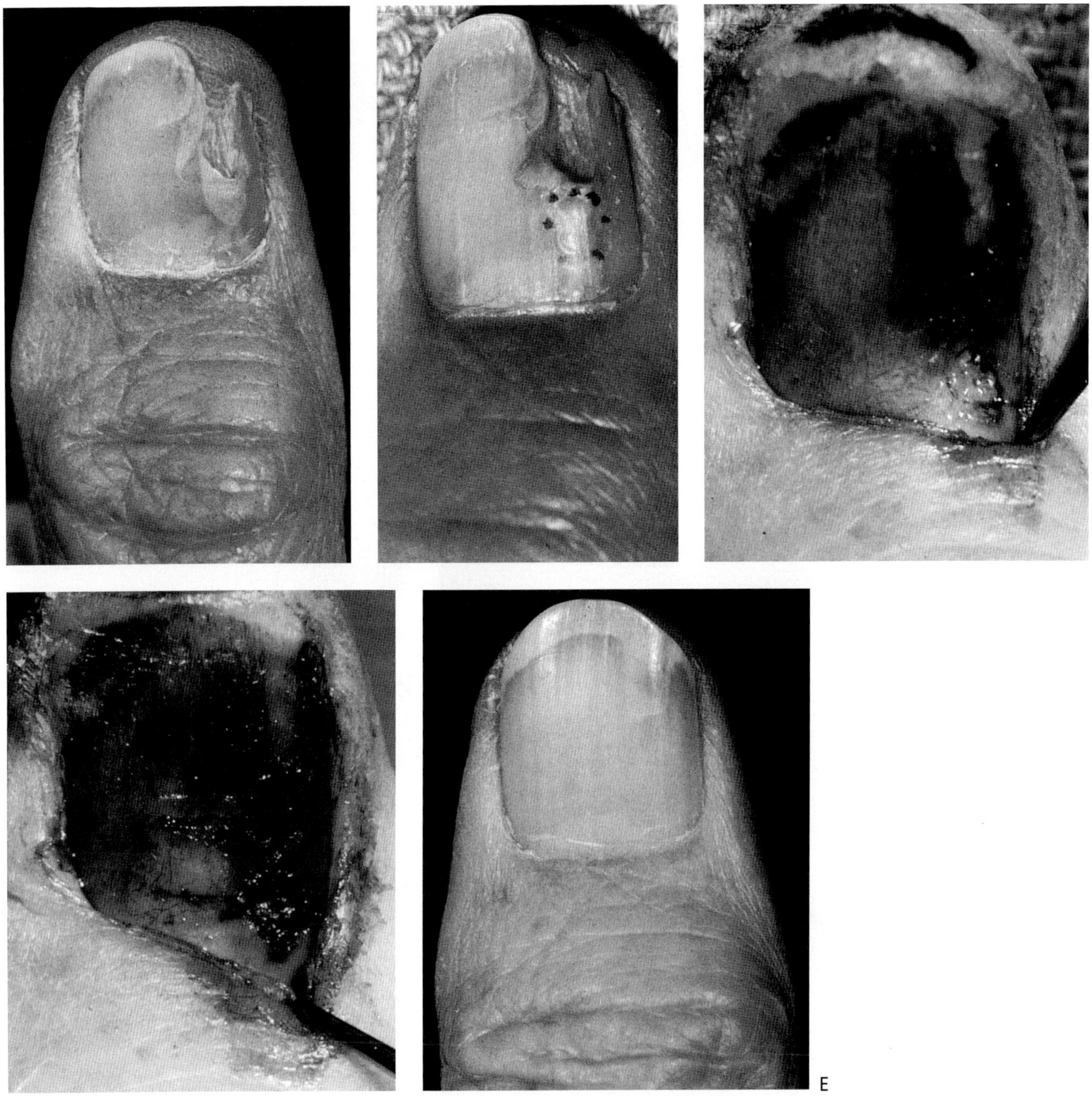

A B,C D E

FIGURE 4B.2. **A:** Split nail secondary to biopsy of the eponychial fold for ganglion. **B:** A 3-mm punch was used for the biopsy, and the defect marked on the nail is exactly 3 mm in diameter. **C:** After the nail is removed, the area of scar can be seen at the distal border of the lunula. **D:** A split-thickness nail bed graft was taken from the more distal portion of the nail bed and sutured into the defect with four 7-0 chromic sutures. **E:** Several months later, the split is not seen.

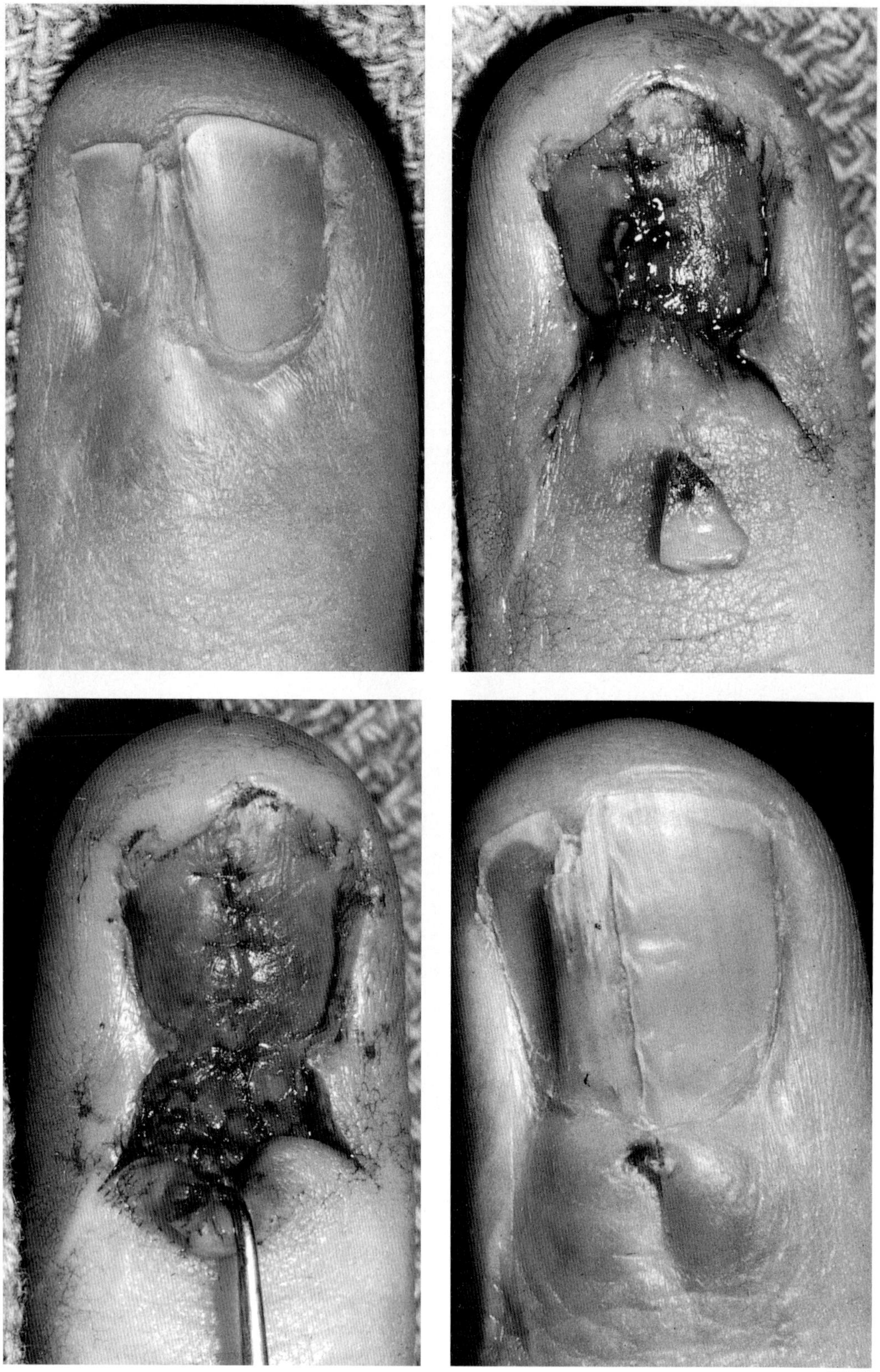

A B C D

FIGURE 4B.3. A: Saw injury that has obliterated a portion of the germinal matrix. **B:** The scar in the sterile matrix was resected and closed. A piece of germinal matrix from the second toe has been harvested to replace resected germinal matrix. **C:** Germinal matrix graft sutured in place. **D:** Four months later, the grafted germinal matrix is growing and incorporating into the nail.

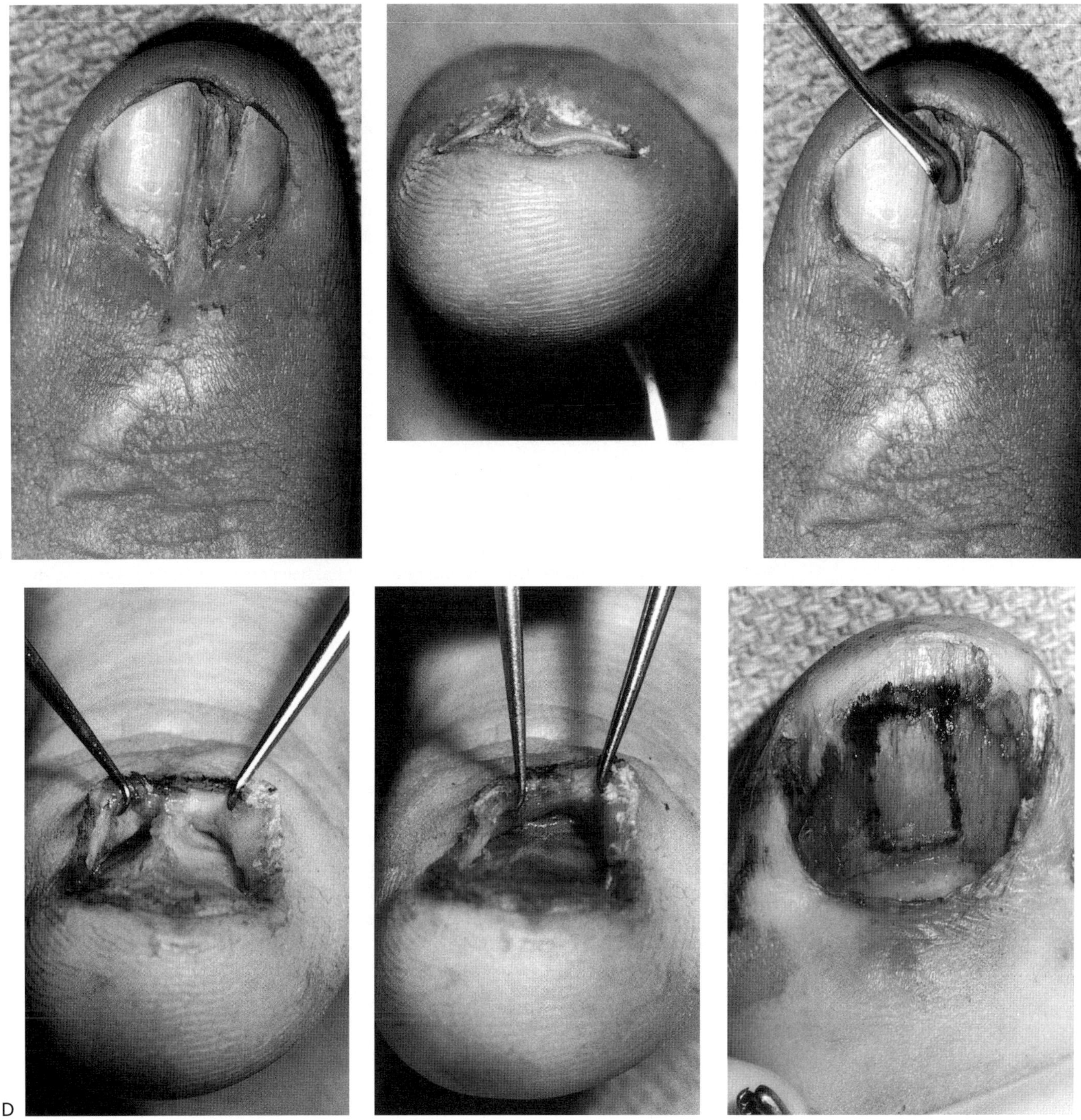

A B,C D E,F

FIGURE 4B.4. A: Unsightly and uncomfortable split of the nail. **B:** End-on view showing the deformity of the nail on each side of the split. **C:** Pressure by an elevator reveals no hard nail between the two sides. **D:** Elevation of the dorsal roof reveals a synechia between the dorsal roof and ventral floor of the nail bed causing splitting of the nail. **E:** The scar is surgically incised, and the germinal matrix was judged to be relatively normal. This left a raw area on the dorsal roof. **F:** Split-thickness sterile matrix graft to be placed on the undersurface of the dorsal roof of the nail fold is outlined.

(continued)

G 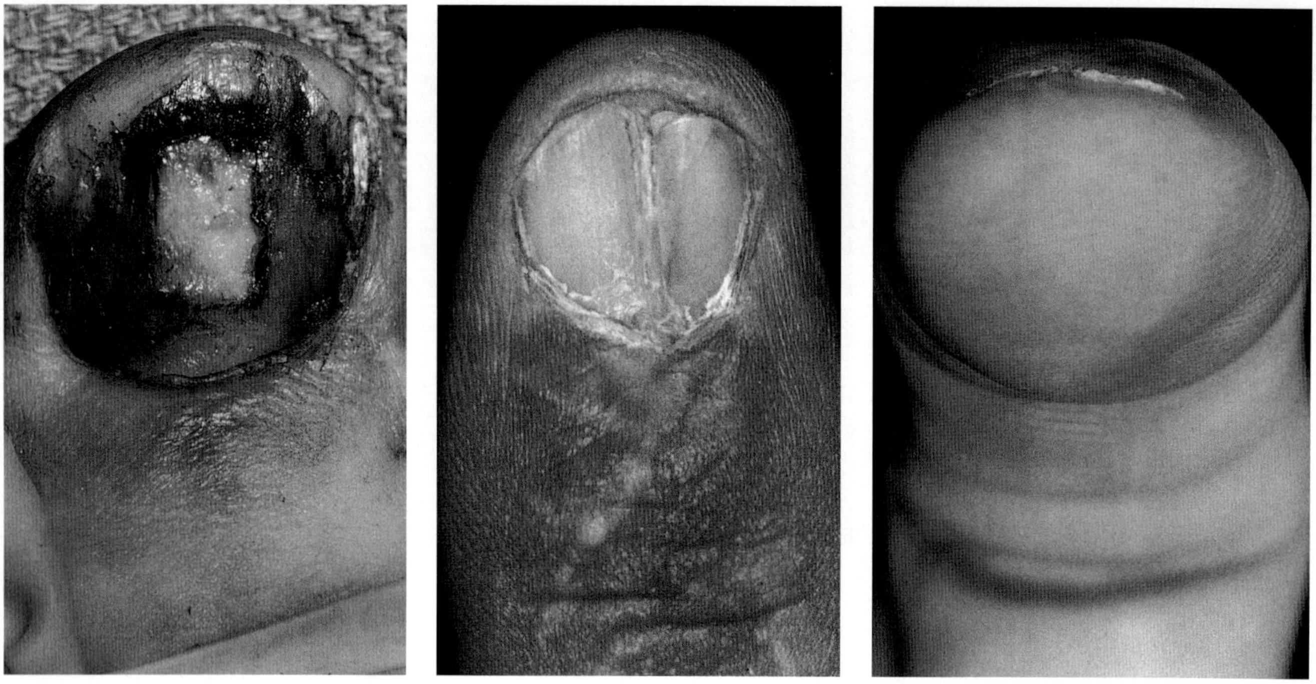 H,I

FIGURE 4B.4 (*continued*). G: The split-thickness graft has been removed. **H:** One year later, the nail is much improved, firm across the middle, and asymptomatic. **I:** End-on view of the nail showing much improvement of the nail.

NONADHERENCE (ONYCHOLYSIS)

Nonadherence of the nail in many patients is caused by transverse scarring of the nail bed so the cells streaming up from the sterile matrix are interrupted by scar and do not reattach distally. If the scar is narrow, the cells seem to jump over it. If the scar is wide, scar excision and split-thickness grafts of sterile matrix must be considered (Fig. 4B.7).

Occasionally, nonadherence will be seen in patients who chronically or acutely pry the nail loose from the hyponychium. When this happens, a buildup of sterile matrix cells forms a keratinous layer that prevents adherence of the nail to the bed (Fig. 4B.8A). In this case, the nail over the nonadherent area is removed (Fig. 4B.8B) and the keratinous buildup is scraped off with the sharp edge of a scalpel blade until punctuate bleeding is seen (Fig. 4B.8C–D). The nail is allowed to grow distally without the impediment of the dead cells, and usually it will readhere (Fig. 4B.8E). A large area of nail bed may be excised and grafted with a split-thickness sterile matrix graft (Fig. 4B.9).

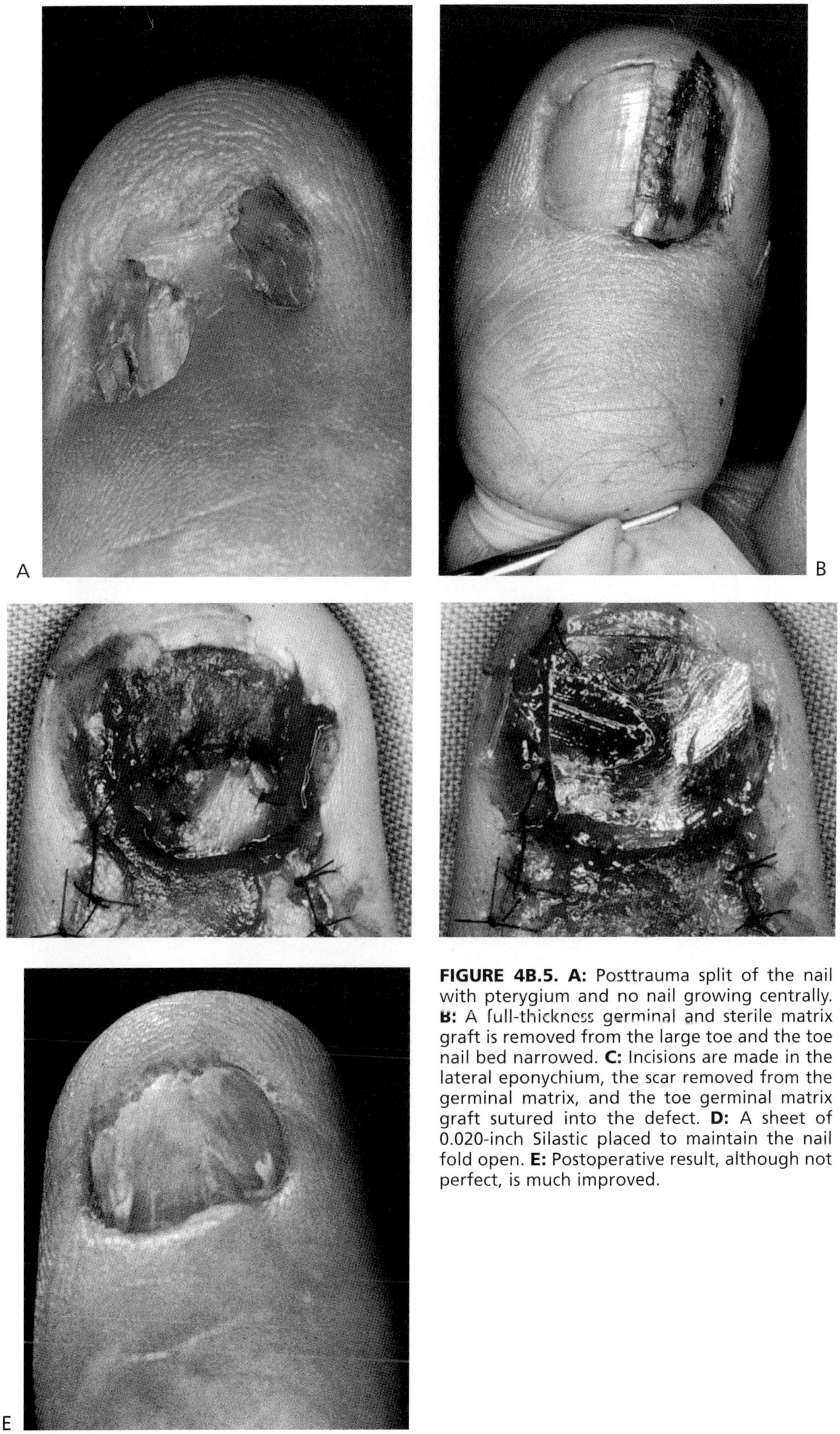

FIGURE 4B.5. A: Posttrauma split of the nail with pterygium and no nail growing centrally. **B:** A full-thickness germinal and sterile matrix graft is removed from the large toe and the toe nail bed narrowed. **C:** Incisions are made in the lateral eponychium, the scar removed from the germinal matrix, and the toe germinal matrix graft sutured into the defect. **D:** A sheet of 0.020-inch Silastic placed to maintain the nail fold open. **E:** Postoperative result, although not perfect, is much improved.

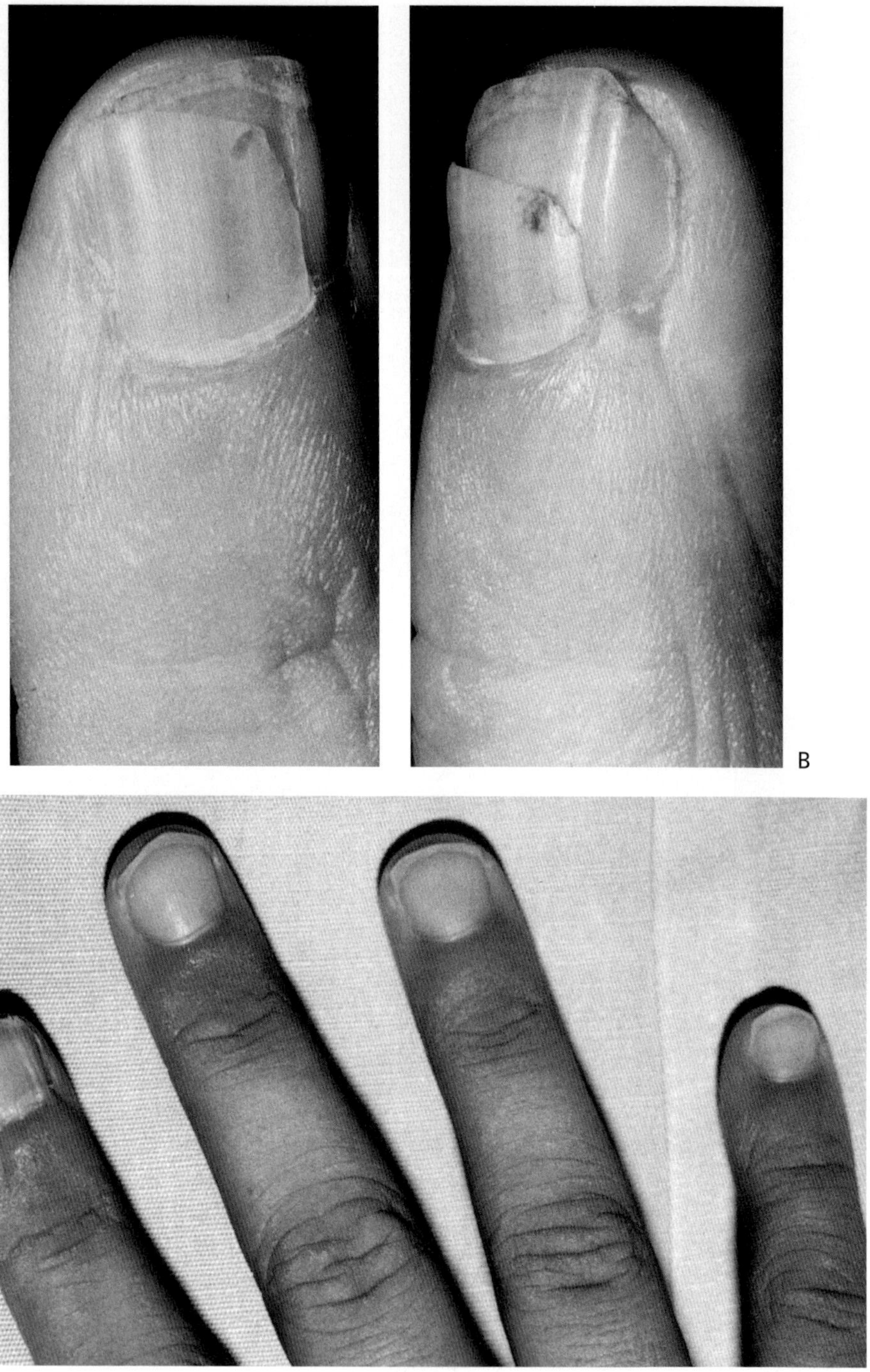

A B C

FIGURE 4B.6. A: Double nail due to a transverse scar across the germinal matrix. **B:** The deep portion of the nail is adequate without the dorsal fragment, so the fragment and its germinal matrix was excised. **C:** One year after the procedure, the nail is asymptomatic and the appearance satisfactory.

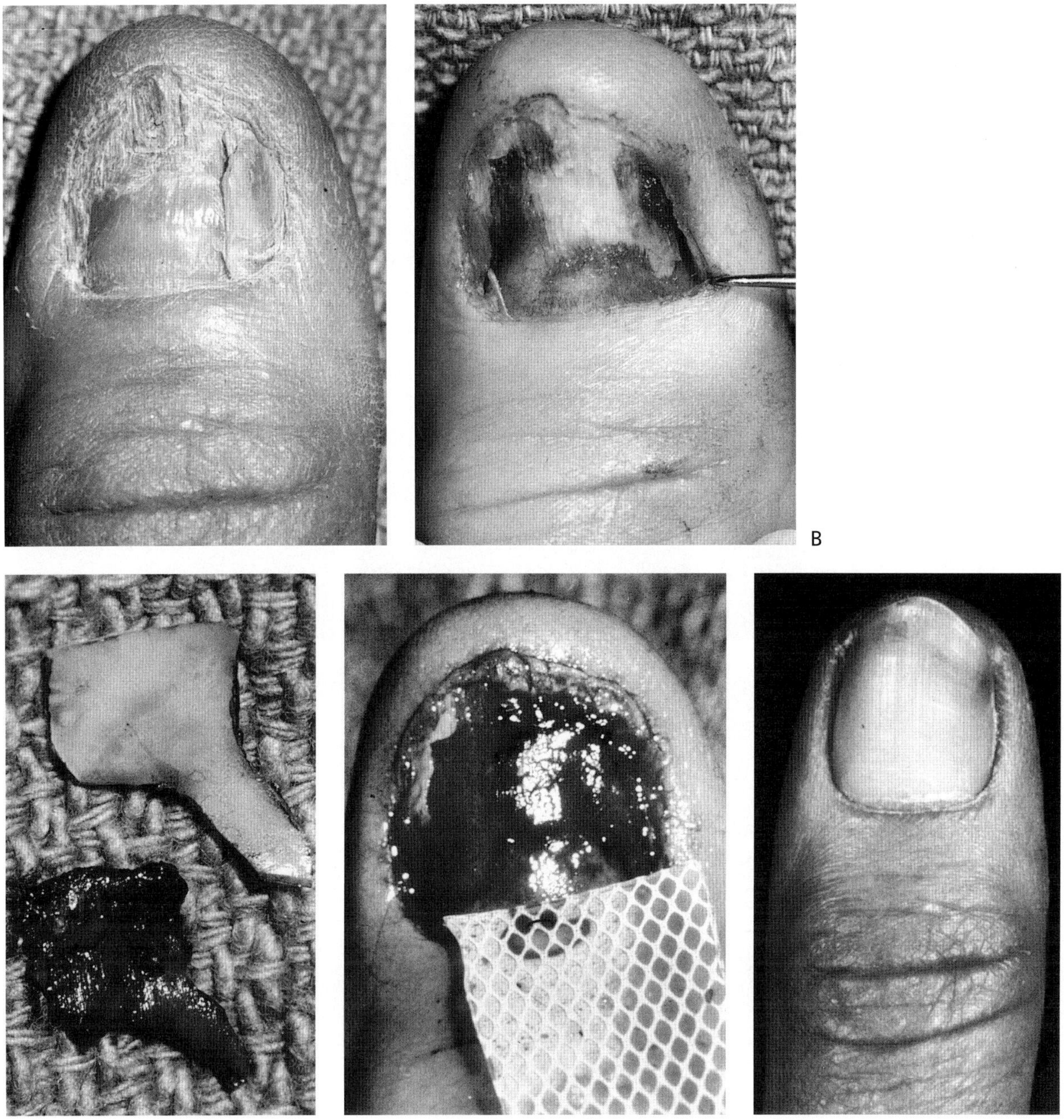

A B C D,E

FIGURE 4B.7. A: Painful, unsightly, nonadherent fragmented nail after nail bed injury caused by a car door. **B:** With the nail removed, the white scar is seen in the sterile matrix. **C:** A pattern is made with a piece of rubber glove **(top)** and a split-thickness sterile matrix graft harvested from the large toe **(bottom). D:** The graft is sutured in place and a piece of 0.020-inch reinforced silicone sheet placed over it and into the nail fold. **E:** One year after surgery, the patient has a near-normal nail.

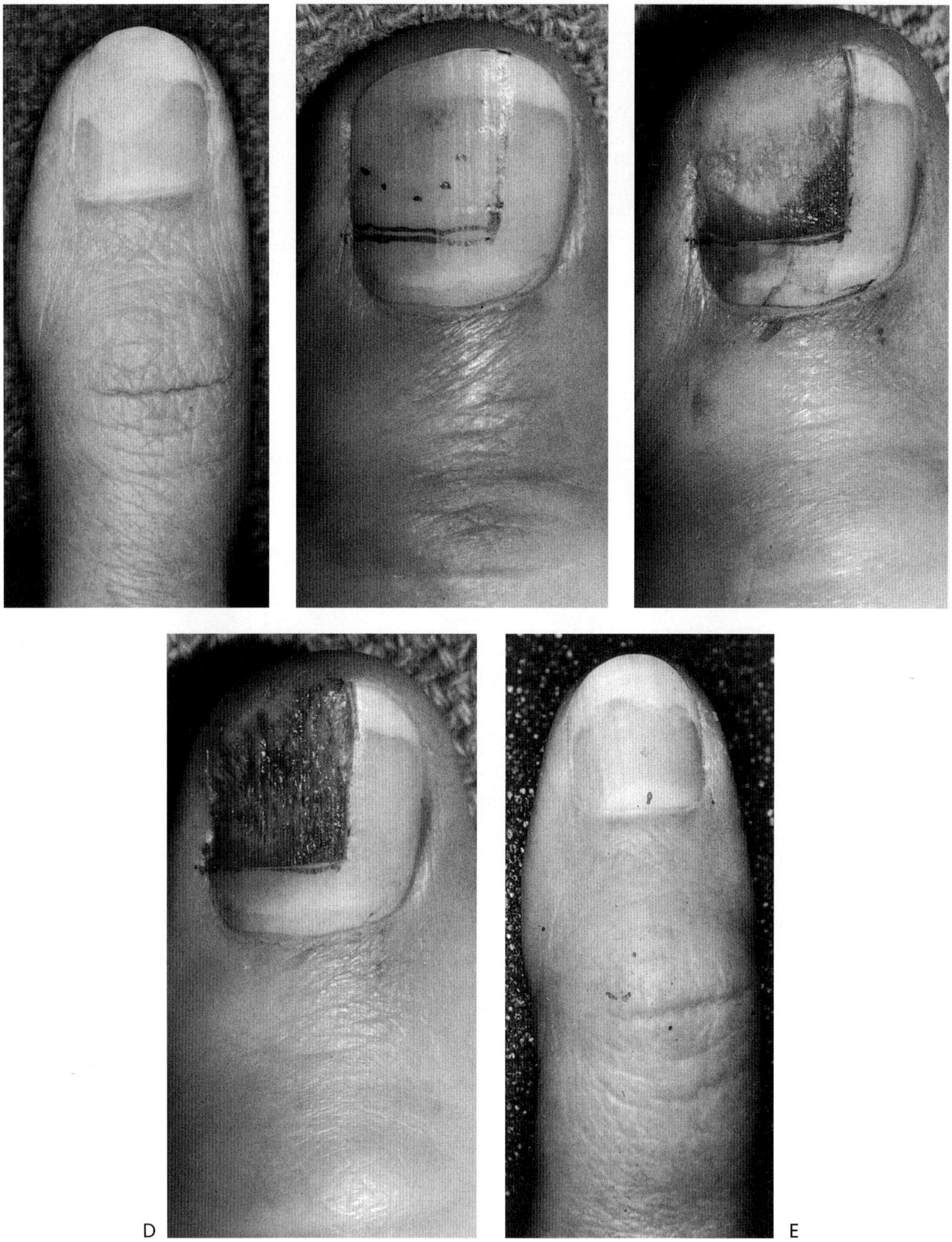

A,B

C

D

E

FIGURE 4B.8. **A:** Nail that was forcibly separated from the nail bed a few years previously and has not reattached. **B:** Nail overlying the nonadherence is removed. **C:** Hypertrophic sterile matrix and hyponychium can be seen. **D:** Hypertrophic keratin scraped from the nail bed. **E:** One year later, the adherence of the nail is much improved.

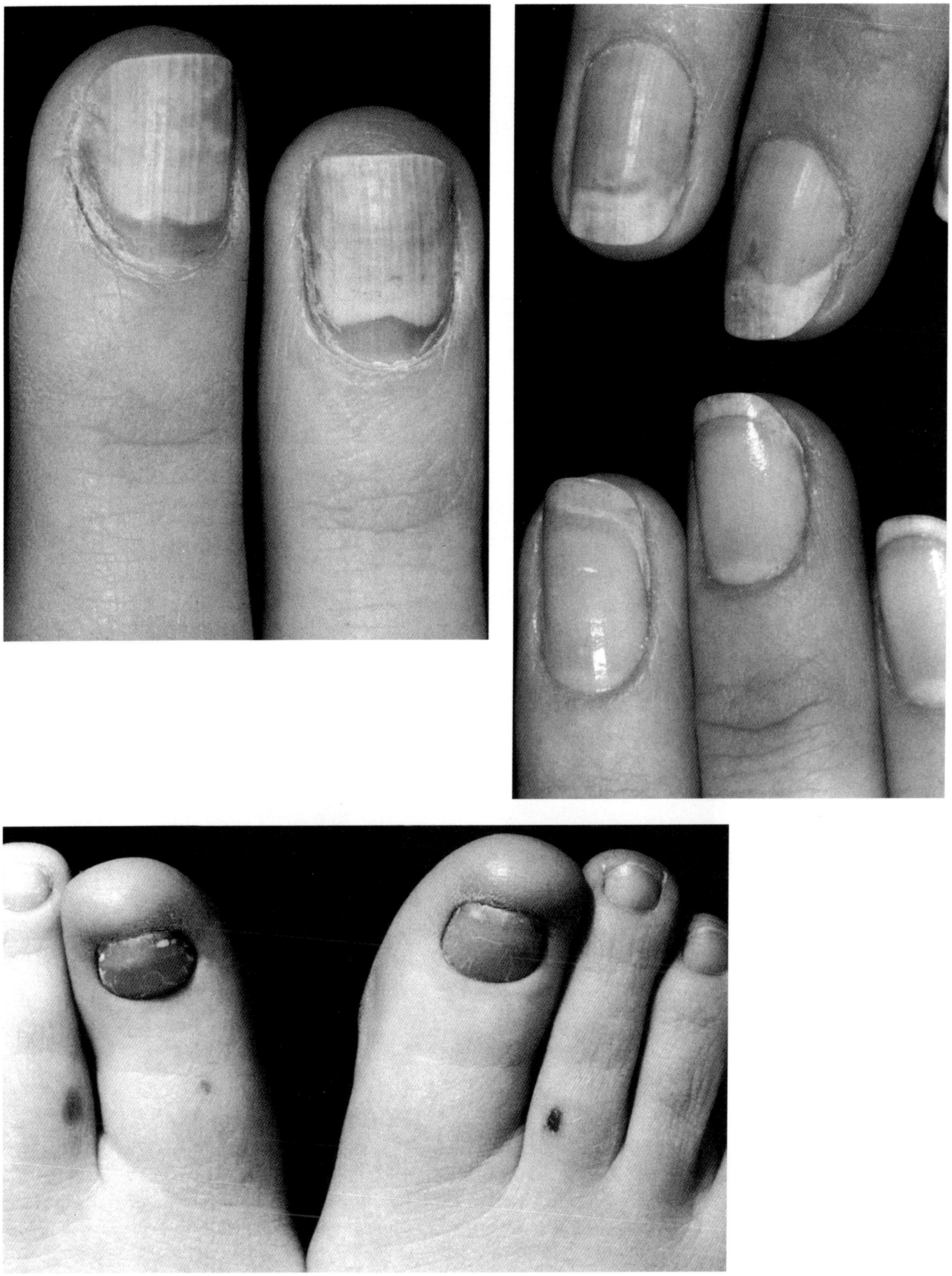

A B C

FIGURE 4B.9. A: Chronic nonadherence can be treated by split-thickness sterile matrix grafts after excision of nonnail-producing sterile matrix. **B:** Improved adherence **(top)** is shown compared with the opposite hand **(bottom)**. **C:** Large toenails 6 months before their full regrowth has occurred.

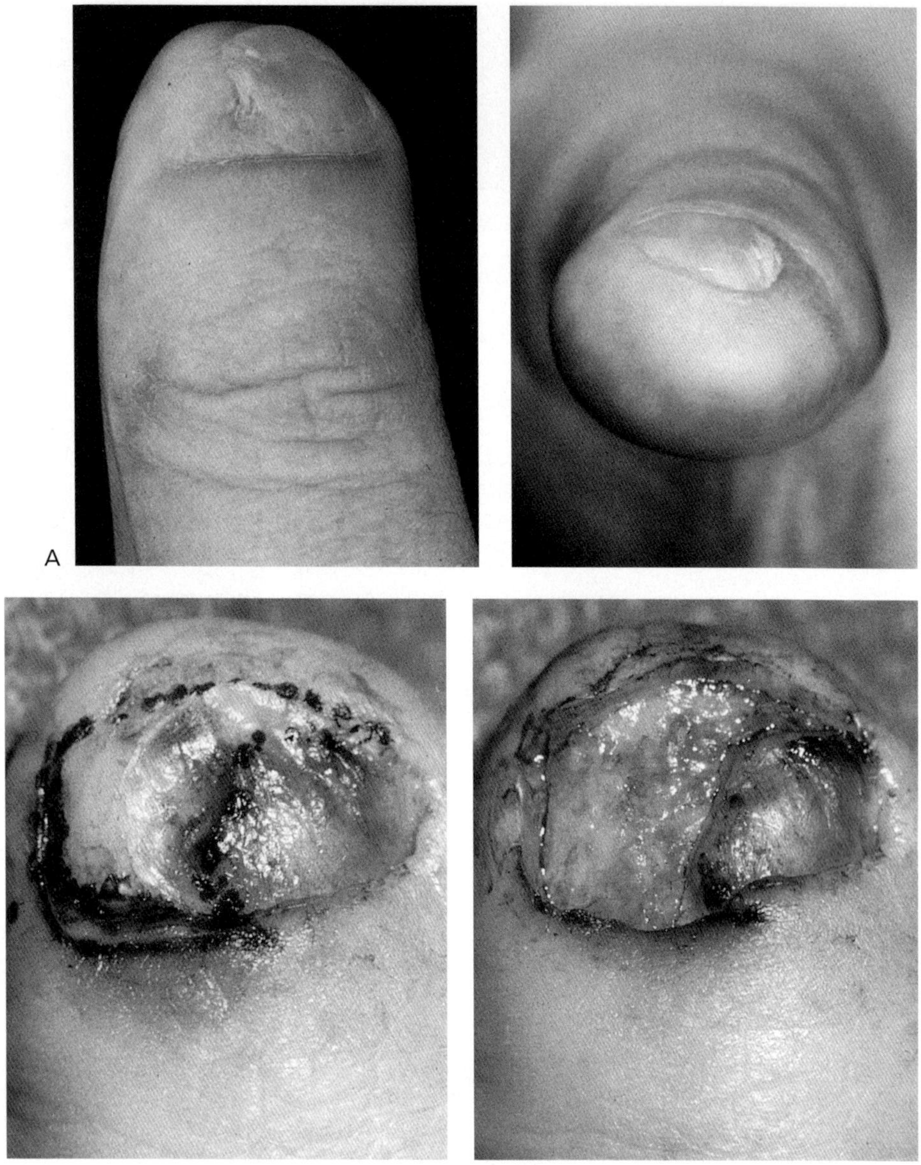

FIGURE 4B.10. **A:** No nail is growing on the ulnar side secondary to traumatic avulsion. **B:** End-on view showing lack of ulnar side nail growth. **C:** Scarred germinal and sterile matrix areas marked for excision after nail removal. **D:** Scarred germinal and sterile matrix areas excised.

ABSENCE

Nail absence may be secondary to trauma or congenital causes. In traumatic absence, portions of the nail or an entire nail bed may be transferred from a toe in a nonvascularized fashion and frequently will be successful. It has been thought that the younger the patient, the better the chances of success. We believe young age is an advantage, but we have had success with older individuals (Fig. 4B.10). Recently, various techniques of vascularized transfer of nail or of nail, bone, and soft tissue from a toe to the finger have been described (Fig. 4B.11). These methods show a more normal nail growth with a more predictable chance of success, but they require the ability to anastomose very small vessels. When a nail bed is transferred, either nonvascularized or vascularized, the nail should be left

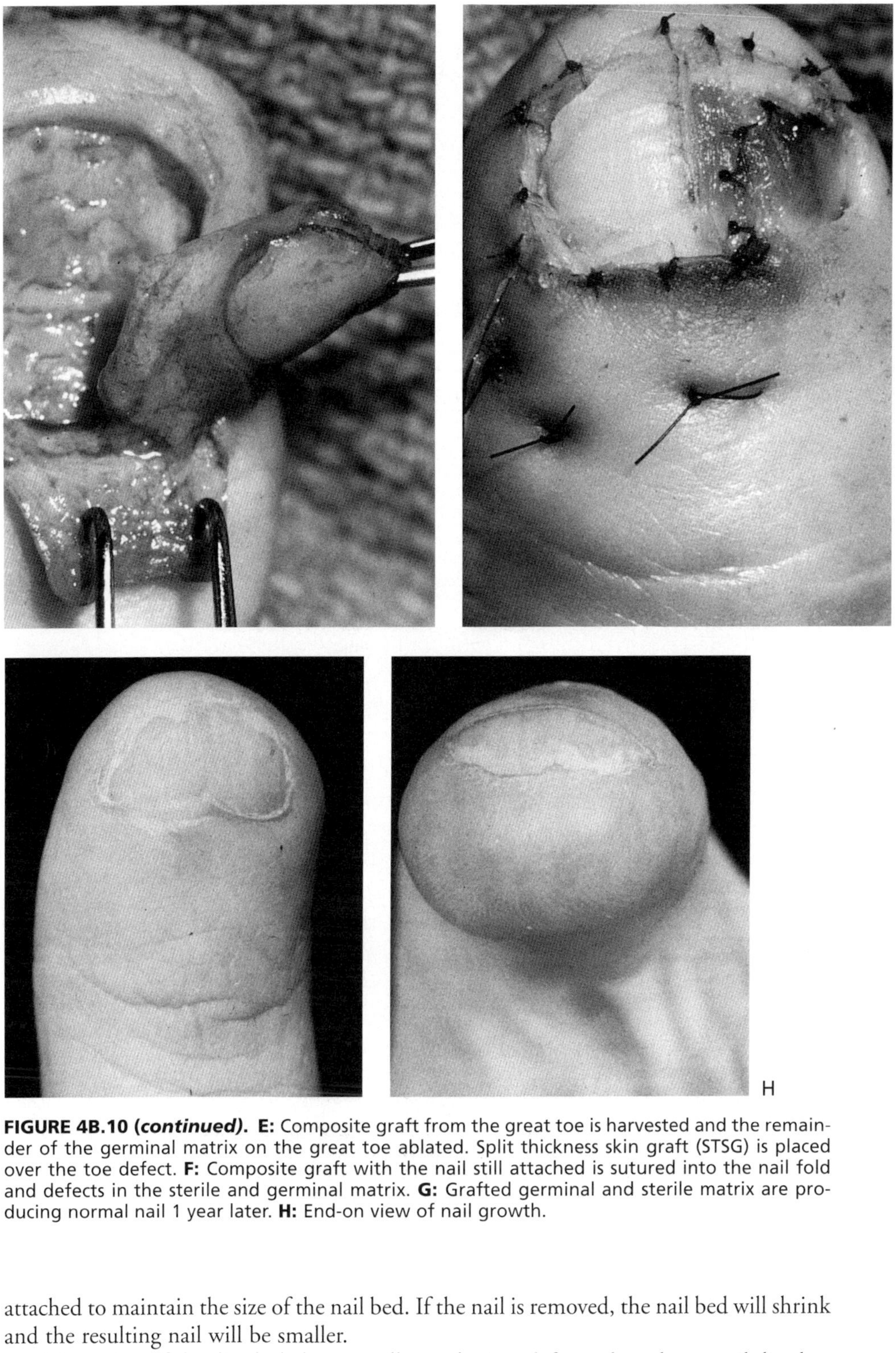

E F G H

FIGURE 4B.10 (*continued*). **E:** Composite graft from the great toe is harvested and the remainder of the germinal matrix on the great toe ablated. Split thickness skin graft (STSG) is placed over the toe defect. **F:** Composite graft with the nail still attached is sutured into the nail fold and defects in the sterile and germinal matrix. **G:** Grafted germinal and sterile matrix are producing normal nail 1 year later. **H:** End-on view of nail growth.

attached to maintain the size of the nail bed. If the nail is removed, the nail bed will shrink and the resulting nail will be smaller.

Nonunion of the distal phalanx usually results in a deformed or absent nail distal to the nonunion and painful tip (Fig. 4B.12). The nonunion must be corrected before revision of the nail bed.

Congenital absence of the nail usually has no other or inadequate numbers of other nails to transfer. A segment of dorsal skin the size and shape of a nail is excised, and a split-thickness skin graft is placed (Fig. 4B.13), which will resemble a nail. Full-thickness grafts can be placed to simulate a white hyponychium and lunulae if desired.

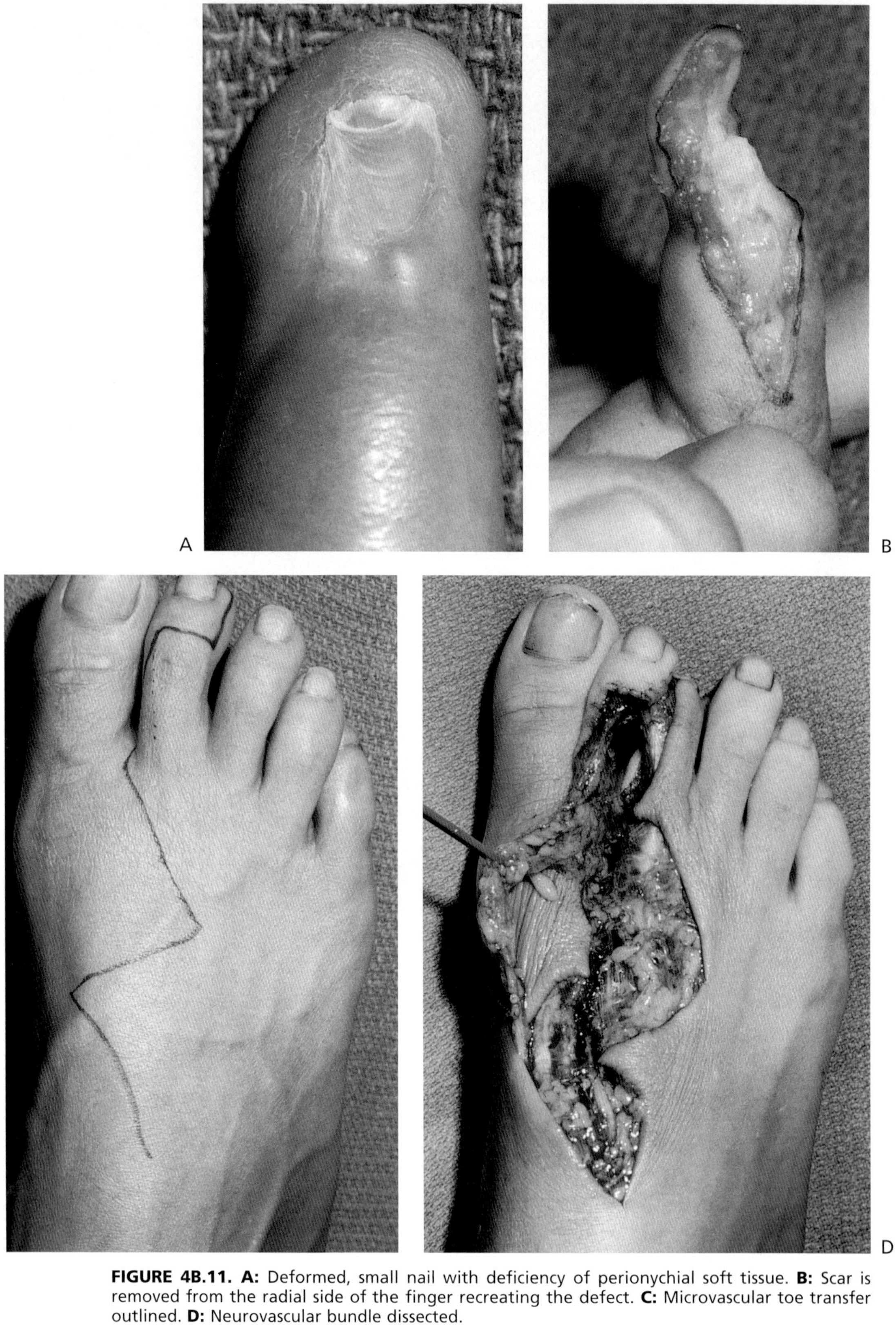

FIGURE 4B.11. A: Deformed, small nail with deficiency of perionychial soft tissue. **B:** Scar is removed from the radial side of the finger recreating the defect. **C:** Microvascular toe transfer outlined. **D:** Neurovascular bundle dissected.

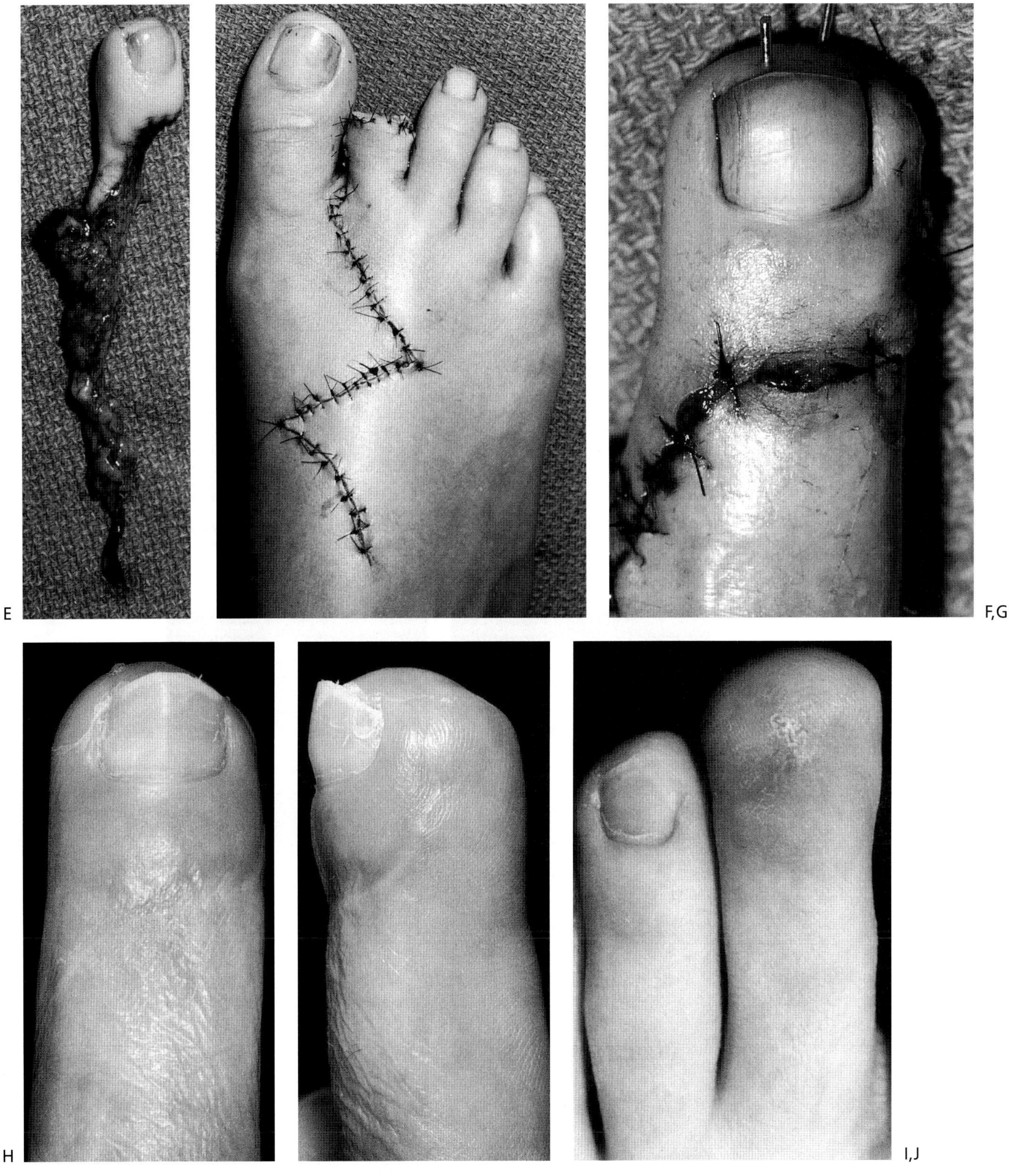

FIGURE 4B.11 *(continued)*. E: Second toe tip and its vascular pedicle. **F:** Donor foot after transfer. **G:** Pins are used to fix the bone. The artery and vein are connected using the operating microscope. **H:** One year later, the finger and nail are growing normally. **I:** Lateral view of the new nail and tip. **J:** Shortened donor toe.

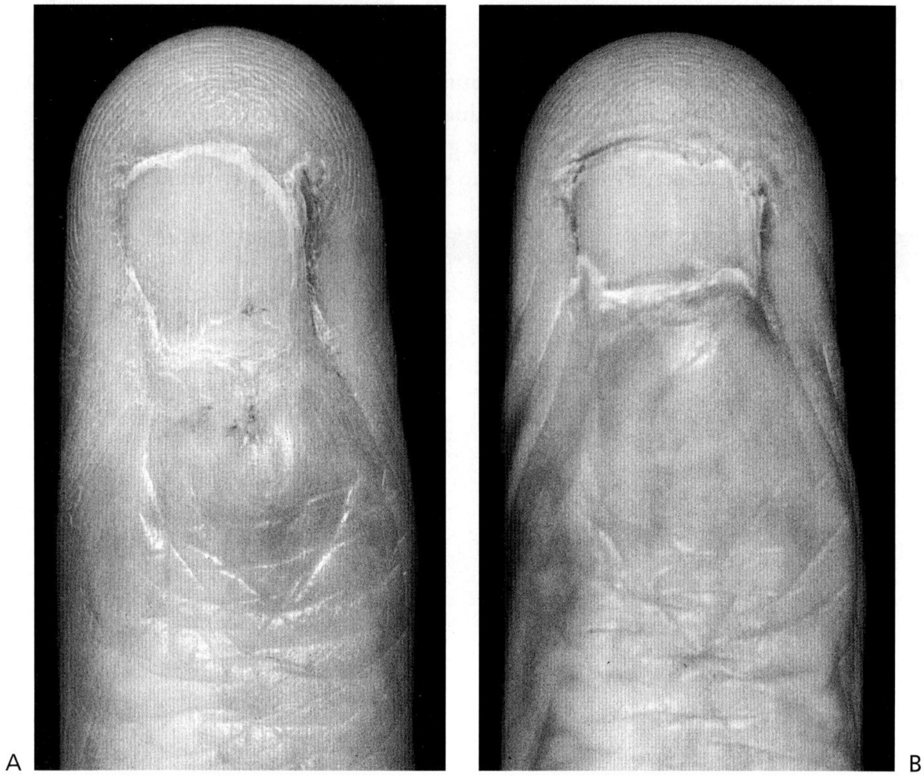

FIGURE 4B.15. A: Overzealous use of the laser for eponychial wart removal has left no dorsal roof of the nail fold and a very tender, painful eponychial area. **B:** Composite eponychial graft from the large toe has replaced the lost eponychium. The appearance is improved, and the discomfort caused by the nail is relieved.

large toe eponychium graft (Fig. 4B.15). The donor site heals secondarily and gives little deformity. Replacement of the eponychium frequently will restore the shine and smoothness to the nail as well as improve the contour and appearance of the eponychium.

TRUMPET OR PINCER NAIL

The progressive tubulization of a nail is not only unsightly but usually causes progressive digital discomfort and pain (Fig. 4B.16) (see also Chapter 5A). The nail progressively tubes itself until it actually pinches off the nail bed (Fig. 4B.17). There are many described methods of correction, but the only one we have found effective other than permanent obliteration is as follows. First, remove the nail. Small incisions are made on the tip of the digit and sharp dissection used to free the paronychial fold or folds from their bony attachments so that they can be elevated flattening the nail bed. The elevation of the paronychium is maintained by dermal grafts placed beneath the elevated paronychium through the tip incisions.

EPONYCHIAL DEFORMITIES

The eponychium may be split secondary to trauma, burns, tumor, nonhealing after division for drainage of paronychia, or absence due to aggressive treatment of paraungual warts. The defect is recreated, and a composite graft of skin and dorsal roof matrix from a toe is used to fill the defect (Fig. 4B.14). The size of the defect may require use of the

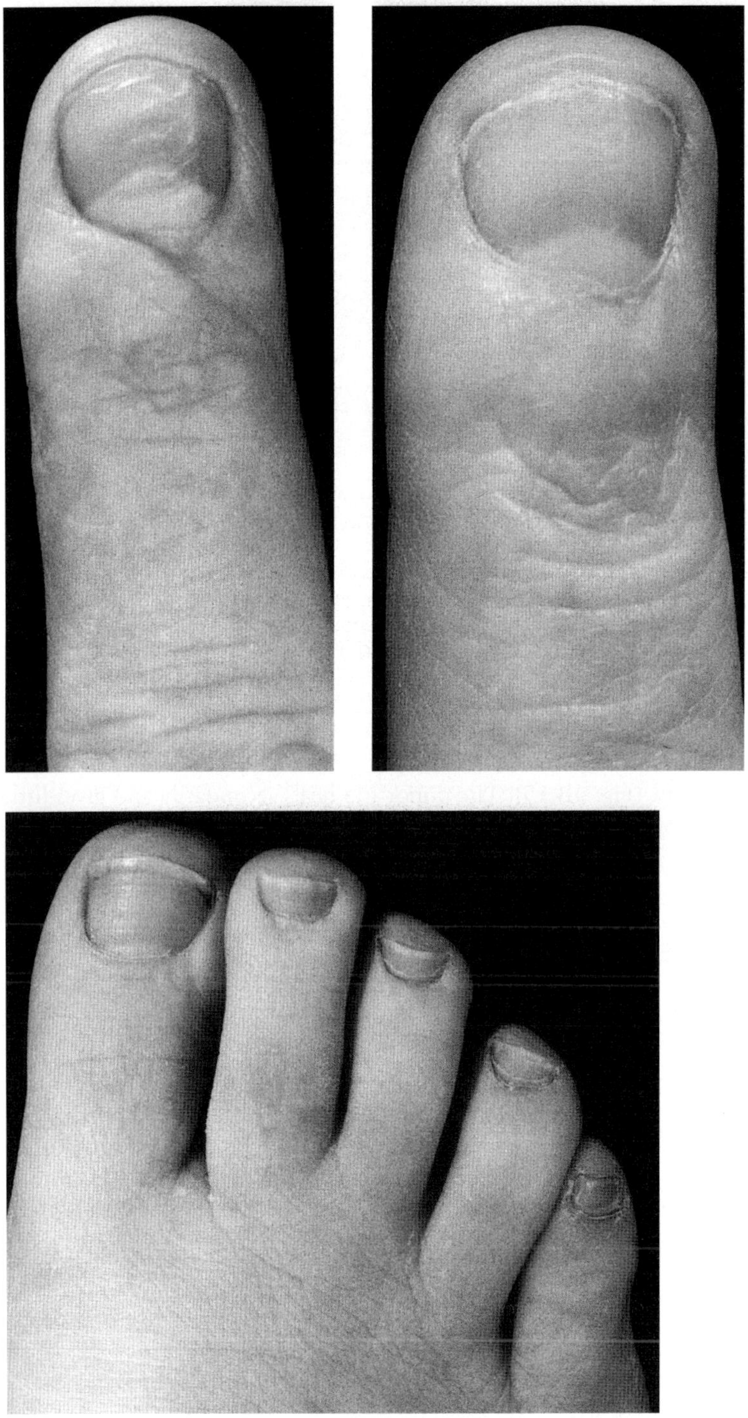

A B C

FIGURE 4B.14. A: Traumatic loss of the dorsal roof of the nail fold and its shine-producing cells leaves a notch and an irregular nonshiny surface on the nail. **B:** Composite graft of eponychium from the second toe. After the graft, the notch and nail surface are much improved. **C:** The eponychial donor site on the toe is allowed to heal spontaneously and shows little deformity.

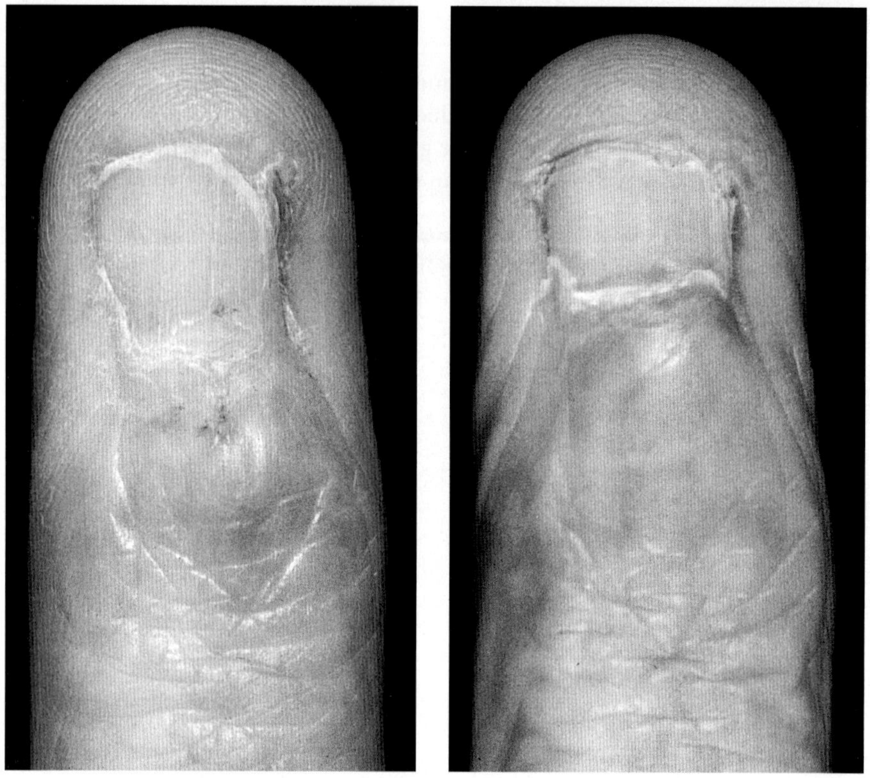

A B

FIGURE 4B.15. A: Overzealous use of the laser for eponychial wart removal has left no dorsal roof of the nail fold and a very tender, painful eponychial area. **B:** Composite eponychial graft from the large toe has replaced the lost eponychium. The appearance is improved, and the discomfort caused by the nail is relieved.

large toe eponychium graft (Fig. 4B.15). The donor site heals secondarily and gives little deformity. Replacement of the eponychium frequently will restore the shine and smoothness to the nail as well as improve the contour and appearance of the eponychium.

TRUMPET OR PINCER NAIL

The progressive tubulization of a nail is not only unsightly but usually causes progressive digital discomfort and pain (Fig. 4B.16) (see also Chapter 5A). The nail progressively tubes itself until it actually pinches off the nail bed (Fig. 4B.17). There are many described methods of correction, but the only one we have found effective other than permanent obliteration is as follows. First, remove the nail. Small incisions are made on the tip of the digit and sharp dissection used to free the paronychial fold or folds from their bony attachments so that they can be elevated flattening the nail bed. The elevation of the paronychium is maintained by dermal grafts placed beneath the elevated paronychium through the tip incisions.

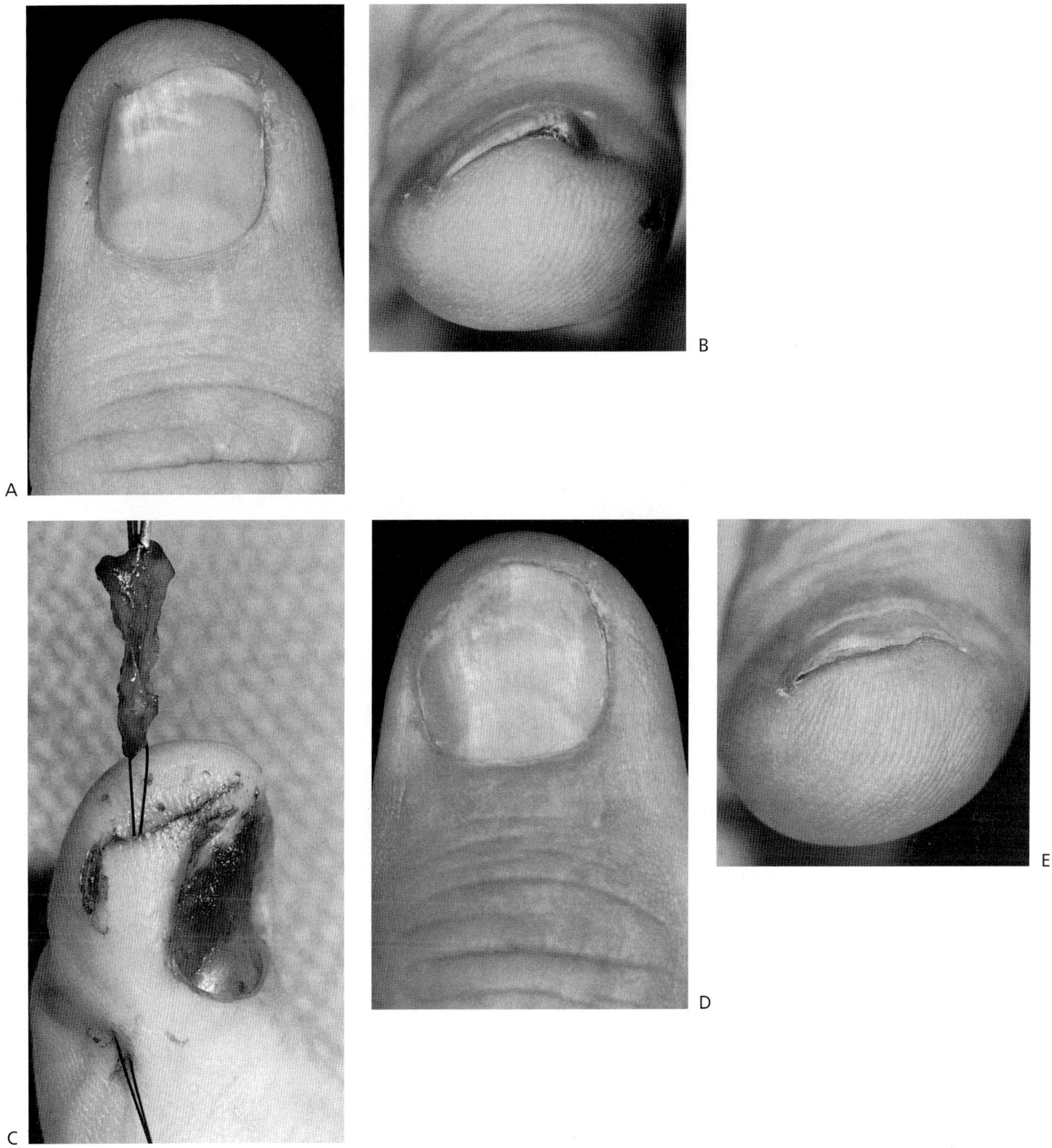

FIGURE 4B.16. A: Unexplained curling of the lateral side of the nail causes ingrowth and pain. This was of recent onset and is progressive. **B:** End-on view of the unilateral curling nail. **C:** The paronychium has been freed from the bone of the distal phalanx, and a dermal graft is placed beneath it to flatten the lateral nail bed. **D:** One year later, the nail is much flatter and nonpainful. **E:** End-on view of the nail.

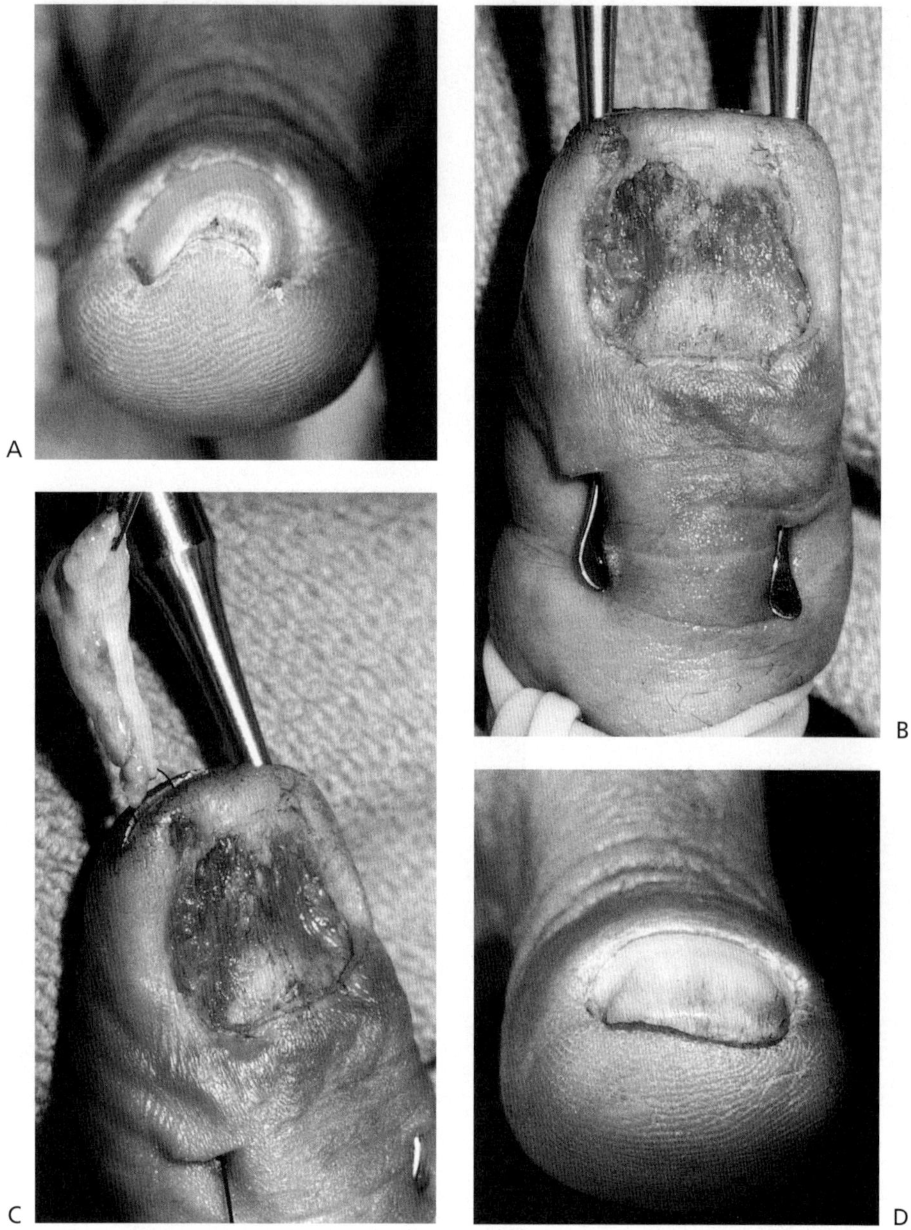

FIGURE 4B.17. **A:** Typical bilateral trumpet (pincer) nail causing severe pain as the nail bed is progressively pinched. **B:** The nail is removed and the lateral paronychial attachments to the bone freed bilaterally. **C:** Dermal grafts are placed beneath the paronychium bilaterally to flatten the nail bed. **D:** A flatter asymptomatic nail seen several months postoperative.

HYPERTROPHIC HYPONYCHIUM

Sometimes the hyponychium becomes hypertrophic and causes discomfort and lack of adherence of the distal nail (Fig. 4B.18). This is treated by removal of the distal portion of the nail, removal of the hypertrophic hyponychium, and placement of a split-thickness nail bed graft from the proximal nail bed.

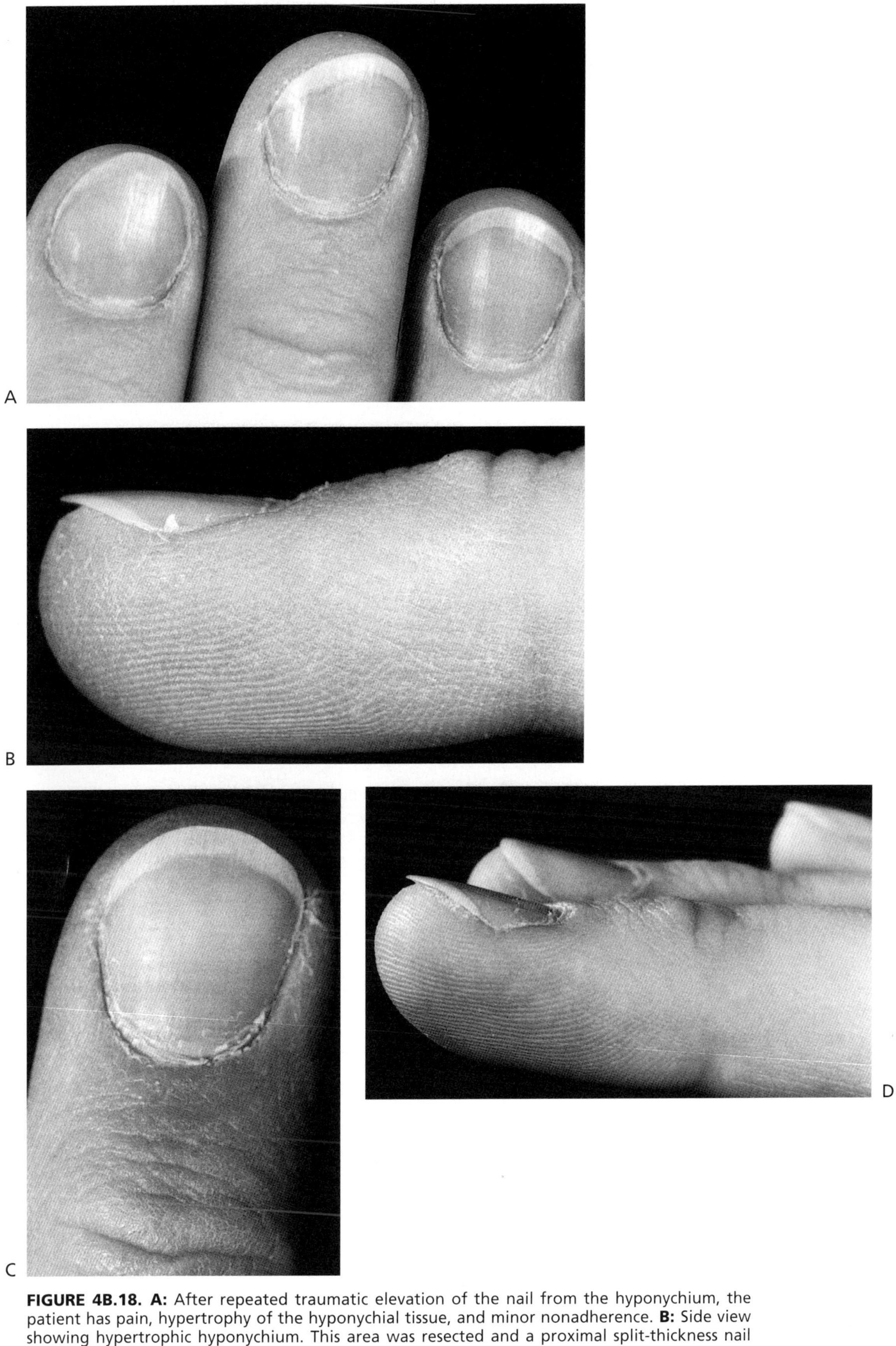

FIGURE 4B.18. A: After repeated traumatic elevation of the nail from the hyponychium, the patient has pain, hypertrophy of the hyponychial tissue, and minor nonadherence. **B:** Side view showing hypertrophic hyponychium. This area was resected and a proximal split-thickness nail bed graft placed. **C:** Normal hyponychium with full adherence several months later. **D:** The hypertrophic hyponychium has been corrected and the pain relieved.

HOOKED NAIL

The nail follows the curve of the nail bed. If the nail bed has been used to close a tip amputation or the bony support to the nail bed of the distal phalanx is lost, the nail will follow the nail bed and hook downward over the end of the finger. This is unsightly, sometimes is uncomfortable, and interferes with function of the tip.

If the nail bed has been pulled over the end of the bone, the defect should be recreated to allow the nail bed to return to the dorsum of the bone. The defect is corrected by one of several means, including V-Y advancement flap, Kutler flaps, and cross-finger flaps. Correction of the hooked nail caused by missing bone has been described, with varying degrees of improvement. The nearest to a normal nail and fingertip is achieved by a free vascularized composite graft of bone, soft tissue, and nail bed from a toe but, of course, causes scarring of the finger and the toe at the donor site (4–6).

PROSTHESIS

When the nail and/or tip are deformed, the patient may choose a prosthesis rather than a reconstruction, which in many instances may be a very satisfactory solution to the problem (7). These prostheses may replace the tip or mask a nail absence or deformity.

REFERENCES

1. Zook, EG, Russell RC. Reconstruction of a functional and aesthetic nail. *Hand Clin* 1990;6:59–68.
2. Ashbell TS, Kleinert HK, Putcha S, et al. The deformed fingernail, a frequent result of failure to repair nail bed injures. *J Trauma* 1967;7:177–190.
3. Tajima T. Treatment of open crushing type of industrial injuries of the hand and forearm: degloving, open circumferential, heat press and nail bed injuries. *J Trauma* 1974;14:995–1011.
4. Koshima I, Soeda S, Takase T, et al. Free vascularized nail grafts. *J Hand Surg* 1988;13A:29–32.
5. Morrison WA. Microvascular nail transfer. *Hand Clin* 1990;6:69–76.
6. Shibata M, Seki T, Yoshizu T, et al. Microsurgical toenail transfer to the hand. *Plast Reconstr Surg* 1990;88:102–109.
7. Beasley RW, deBeze GM. Prosthetic substitution for fingernails. *Hand Clin* 1990;6:105–111.

CHAPTER

5

TOENAILS

5A

TOENAIL SURGERY

EDWARD A. KRULL

The principles of dermatologic surgery, including appropriate history and physical examination, must apply to all nail surgery (Table 5A.1). However, because the potential for arterial disease is greater in the toes than in the hands, there is a special need to assess circulation to the toes. The surgeon performing toenail surgery needs to develop a disciplined approach to evaluating adequacy of blood flow to the toes (Table 5A.2). Loss of hair on the toes, cool temperatures of the feet, thinning and shininess of the skin of the toes, loss of pulp from the digits (the "crinkly skin sign") (1), a mottled duskiness, and a reddish-blue discoloration may indicate arterial disease. Pallor with elevation and rubor with dependency are signs of arterial insufficiency. The slow return of the pink color to the soft tissue of the distal pulp of the toes after blanching with pressure is a sign of impaired circulation; capillary refill should occur within 5 seconds (1). Palpation of peripheral pulses is an important part of the examination before undertaking toenail surgery. Doppler testing may be necessary to assess patients in whom the findings and the pulses are difficult to evaluate. Vascular surgical consultation may be required in borderline cases. More conservative measures, such as grinding and trimming of the nails, may be the preferable treatment because of arterial disease.

TABLE 5A.1. GENERAL ASSESSMENT

1. General medical status
   A. Heart disease
   B. Hypertension
   C. Arthritis
2. Intermittent claudication
   A. Nocturnal rest pain in calf and foot
   B. Pain with leg elevation
      I. Suggestive of arterial disease
3. Heart valvular disease and/or heart valve replacement may require prophylactic antibiotics
4. Joint replacement may require antibiotics
5. Artificial vessel replacement
   A. Stents in cardiac vessels reepithelialize in about 4 to 6 weeks and probably do not require prophylactic antibiotics
6. Diabetes
7. Raynaud's phenomenon
8. Bleeding diathesis
   A. Inherent
   B. Coumadin
   C. Aspirin

Many of the conditions in the general assessment require discussion with the physician responsible for the patient or other appropriate physicians.

E. A. Krull: Department of Dermatology, Henry Ford Hospital, Detroit, Michigan 48202.

TABLE 5A.2. ASSESSMENT OF CIRCULATION TO THE DIGITS

1. Pulses
2. Color of toes or fingers
3. Hair on digits
4. Nails: thickness, ridging, and other changes
5. Skin texture
6. Blanch with elevation; rubor with dependency
7. Refill time of pulp of digits after compression

INGROWN TOENAILS

Ingrowing toenails, especially of the first toe, are common problems. Hereditary or constitutional factors producing a nail plate wider than the nail bed have been implicated in the pathogenesis of ingrowing nails. Other contributing factors, such as improper cutting of the toenails, hyperhidrosis, poorly fitting shoes, poor foot hygiene, faulty foot mechanics, and deformities of the feet, have also been implicated (1–3). The prominence of the lateral nail fold and a broad flat nail may be important factors in the development of ingrown toenails during adolescence, especially for young men ages 14 to 18 years. Increased nail curvature, such as pincer nails, may produce ingrown nails in adults (2,4). Nail thickness (thinner nail plates lacerate the lateral nail groove more easily) and medial rotation of the toe in walking or standing have been designated as risk factors in developing ingrown toenails (5). It is of interest that ingrowing toenails tend to occur more often in adolescents, especially those with increased sweating of the feet. In this age group, conservative measures may prevent ingrowing of the nail and control the inflammatory response. The tendency for ingrowing toenails frequently disappears in adulthood. Pincer nails occurring in older patients who tend not to have hyperhidrosis uncommonly result in an inflammatory ingrowth of nails despite the increased curvature of the toenails (Fig. 5A.1). Ingrowing toenails are uncommon in infants; when this occurs, it may be related to alignment of the big toes, congenital hypertrophic lip of the hallux, sleeping prone, wearing tight clothes, improper nail cutting, and wearing tight fitting shoes. Some of these factors are aggravated at the time the infant starts actively kicking (6,7). Chronic inflammation may produce a fibrosis and inelasticity of the lateral nail folds, perpetuating the ingrown toenail (8). Diabetes, obesity, nail changes in the elderly, bony deformity, and arthritis have been implicated as factors in ingrown toenails (9).

Regardless, the result is the nail lacerates the side of the lateral nail groove. This permits the ingress of bacteria and yeast organisms, and the nail acts as a foreign body. The

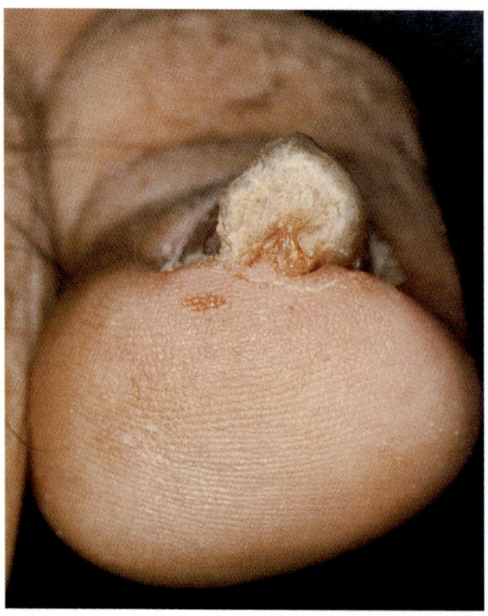

FIGURE 5A.1. Marked pincer nail without evidence of ingrowing or inflammation.

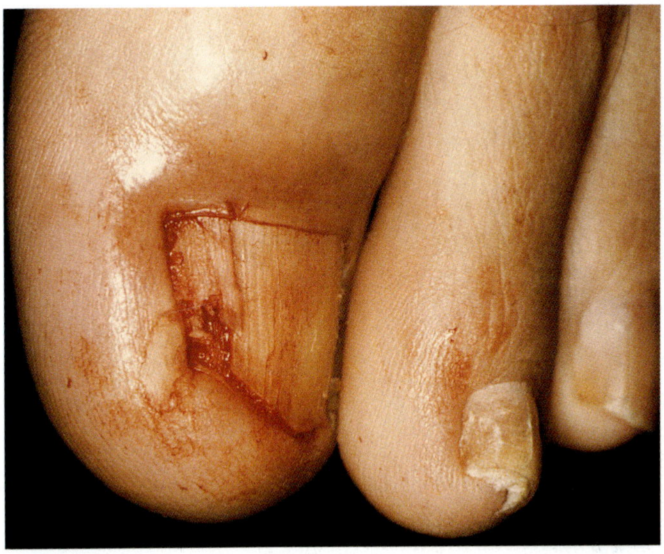

FIGURE 5A.2. Ingrowing nail with mild swelling at the distal end of the nail. Edge of the nail has been freed from underlying and overlying tissues and transected.

presence of the ingrown nail does not allow the fissure in the side of the lateral nail groove to heal. Therefore, antibiotics alone, systemic or topical, often are ineffective in the treatment of ingrown toenails. The goal of treatment is to eliminate or prevent the nail from cutting into the side of the lateral nail groove.

Management of inflammation of the ingrown toenail to some extent depends on the location of the laceration in the side of the lateral nail groove. Because of hypertrophy of the distal nail fold and excessive granulation tissue in some ingrowing toenails, the patient may not be able to visualize the entire width of the nail plate when cutting it. Failure to cut across the entire width of the nail, especially at the edge, may splinter the nail, leaving a pointed remnant at the lateral margin. This is the so-called "fishhook nail" (Figs.

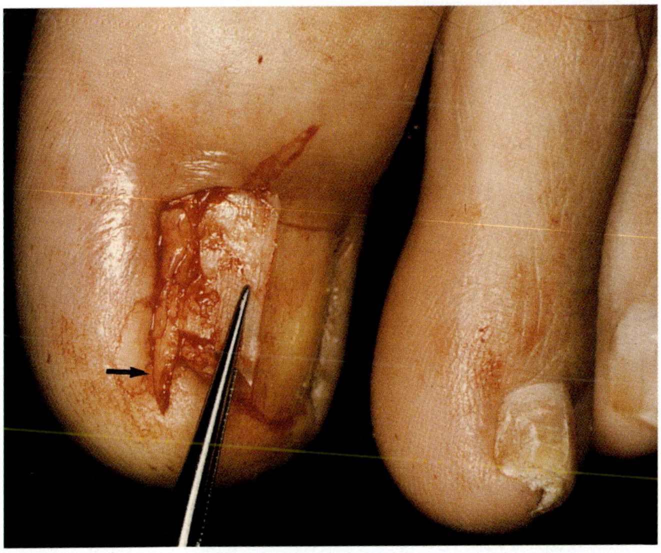

FIGURE 5A.3. Avulsion of nail edges shown in Fig. 5A.2 reveals "fishhook nail." *Arrow* points to part of nail that was incompletely cut repeatedly by patient because the lateral nail was hidden by lateral nail fold.

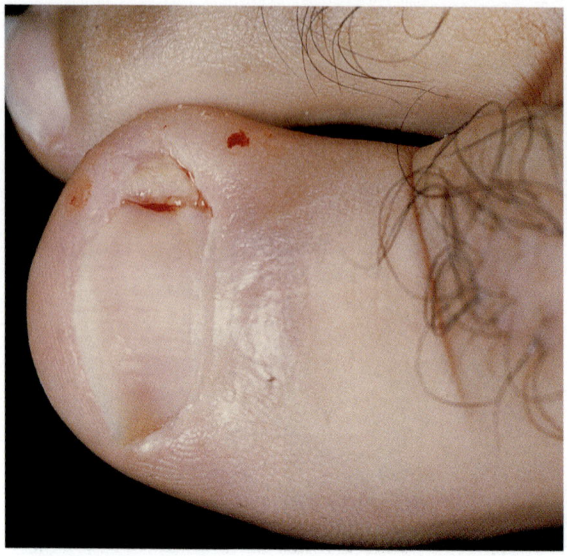

FIGURE 5A.4. Ingrown toenail with bulging of lateral nail fold to be treated by splinting with plastic tubing.

5A.2 and 5A.3). This pointed edge digs deeply into the soft tissue, aggravating the ingrowing nail.

If the break in the integrity of the lateral nail groove is distal, a number of conservative corrective measures can be undertaken. For ingrowing in the distal nail groove, cotton wool or cotton impregnated with collodion or other material may be packed between the edge of the nail plate and the side of the lateral nail groove (10,11). Moistening the cotton facilitates the packing of the cotton, which then acts as a barrier between the nail fold and the edge of the nail plate and allows the laceration to heal. A nail file used as a probe can be used by the patient to replace the cotton (12). Splinting with a flexible tube

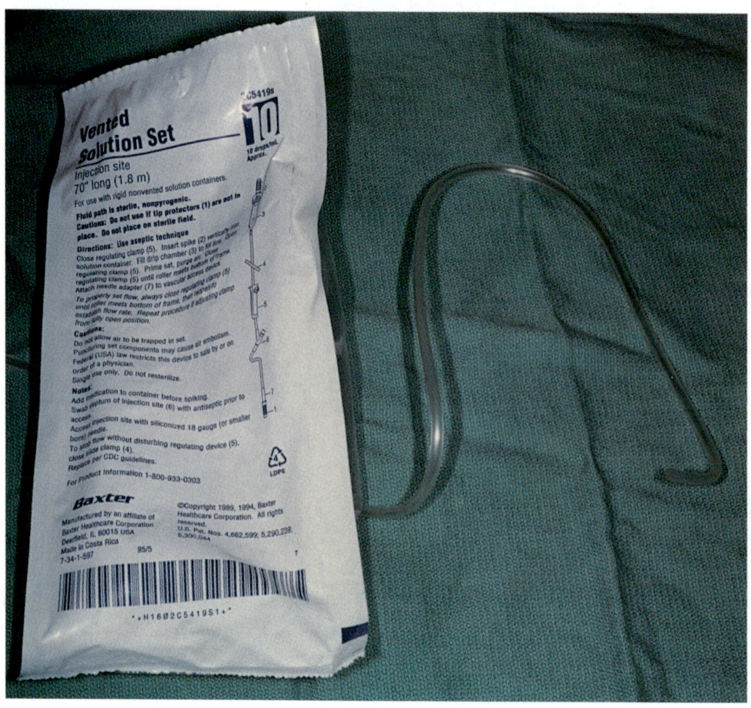

FIGURE 5A.5. Plastic tubing from an intravenous kit.

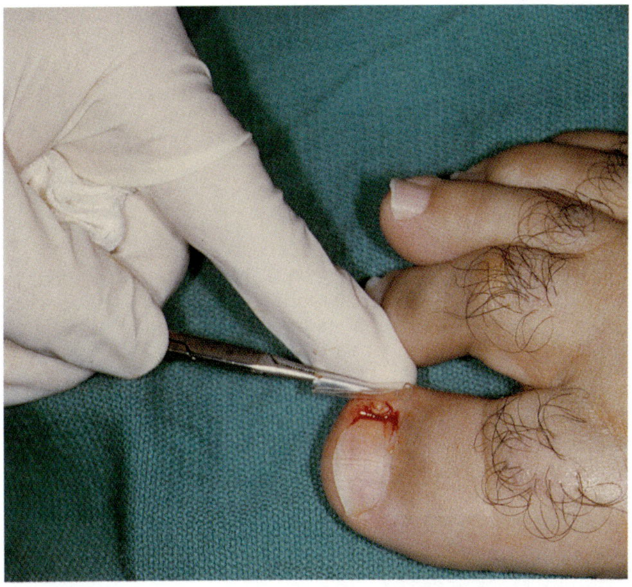

FIGURE 5A.6. Plastic tubing measured for length of the lateral nail edge.

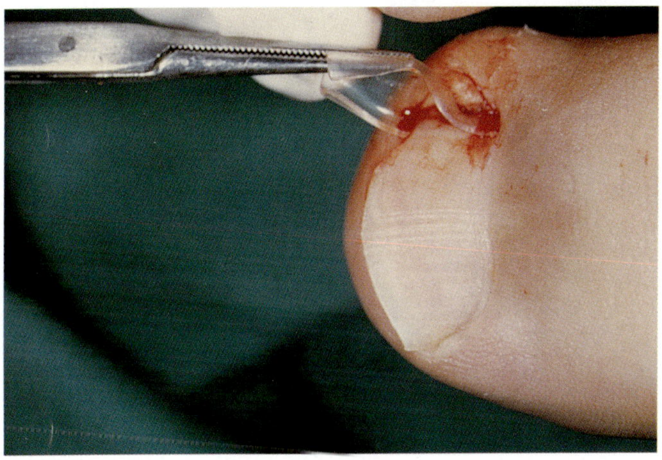

FIGURE 5A.7. Plastic tubing divided lengthwise and beginning to be inserted along the edge of nail plate. This is performed under local anesthesia.

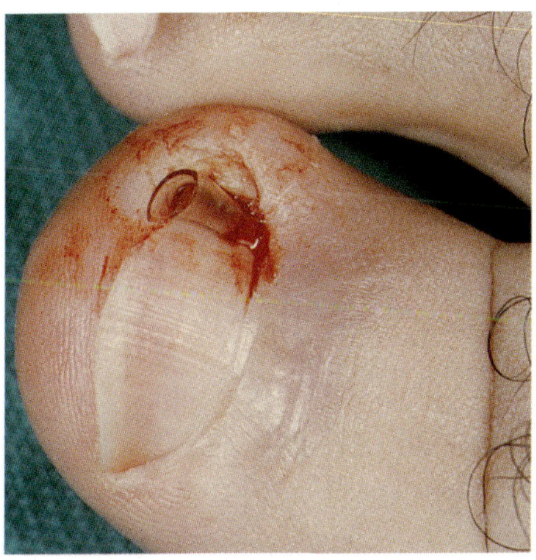

FIGURE 5A.8. Plastic tubing in place, protecting nail wall from edge of nail plate.

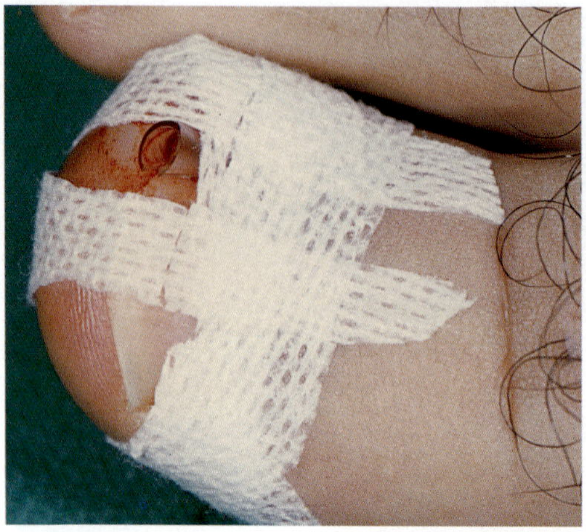

FIGURE 5A.9. Plastic tubing taped in place. Patient had loss of pain soon after the procedure. The treated toe should be washed with an antiseptic solution, such as povidone iodine, for up to 3 or 4 weeks. Our patient lost the tube in about 10 days but had no pain or ingrowing over 4 months of observation.

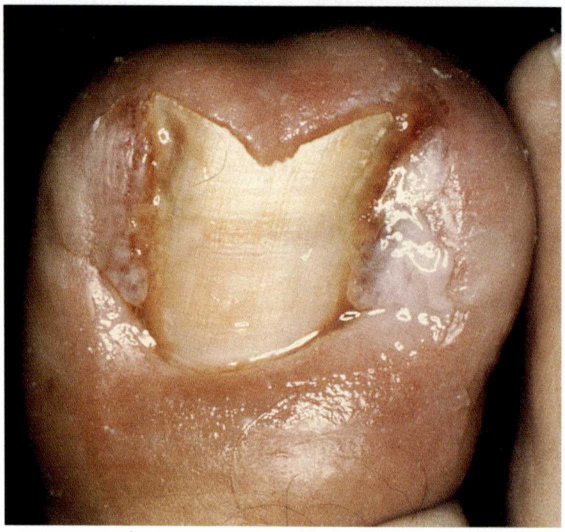

FIGURE 5A.10. Marked ingrowing toenail. Nail folds are hypertrophied, fibrotic, and nonflexible. Marked granulation tissue is evident. Partial or total avulsion is likely to result in recurrence of ingrown toenail. Notch in the nail was an attempt by a physician to reduce pressure on the edges of the nail by splitting the center of the distal edge of the nail plate.

that is maintained for 3 to 4 weeks has been described for noninvasive treatment of ingrown toenails (Figs. 5A.4–5A.9) (13); it is performed under local anesthesia. Abrading the entire nail plate surface except for the margins to make the nail more flexible has been recommended for the conservative management of ingrown toenails (14). Taping the skin of the toes so as to pull the lateral folds from the nail has been successful and does not require surgical intervention. One end of the tape is placed along the lateral fold of the involved side with the rest of the tape wrapped around the toe (15).

Many of the nails ingrow more proximally in the side of the lateral nail groove and may require removal of the nail edge that caused and perpetuated the ingrown nail. Treatment methods, such as phenolization and laser surgery, are presented in other parts of the book. If the ingrowing toenail is to be corrected by scalpel surgery, its management may be divided into two stages: (i) the inflammatory stage (Fig. 5A.10) and (ii) the final repair.

THE INFLAMMATORY STAGE

Appropriate local anesthesia is established with either a local or digital block with plain lidocaine or long-lasting anesthetics such as plain bupivacaine (16). I do not use lidocaine

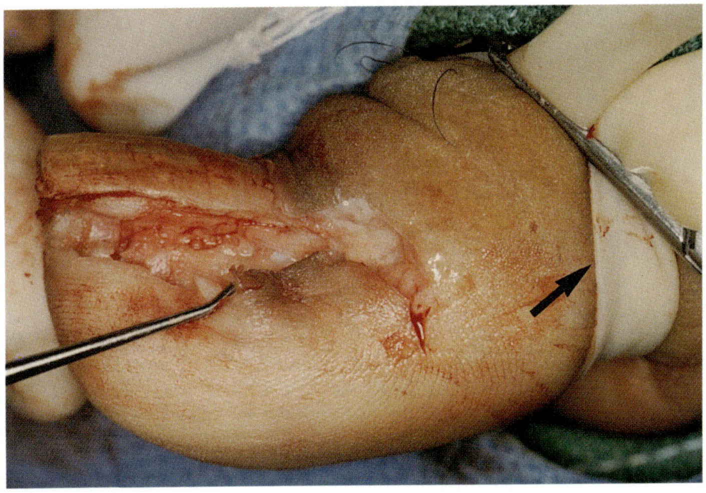

FIGURE 5A.11. Penrose drain used for hemostasis (*arrow*).

with epinephrine in nail surgery because of the risks of untoward vasoconstriction. Instead I use a Penrose drain, clamped with a hemostat, which is left in place for approximately 15 minutes, released for about 5 minutes, and then reapplied if needed for another 15 minutes (Fig. 5A.11). These are very conservative time limits (17). A small-nosed mosquito hemostat is inserted under the offending nail plate to free it from the underlying nail bed and matrix (Fig. 5A.12). Determination of whether the instrument has reached the most proximal part of the nail is the distinct feeling of a sudden "give" as the nail is freed from the proximal matrix. The resistance of the proximal nail fold blocks the instrument from penetrating through the nail fold. The width of the edge of the nail plate may require more than one insertion of the instrument to free the entire nail plate edge from the underlying tissue. Particular attention must be given to the horn of the nail plate, which may be overlooked if it extends far laterally. The proximal nail fold must be freed from the underlying nail plate (Fig. 5A.13), again paying particular attention to the lateral nail horn. The anvil-like jaw of the English nail splitter or one side of the double-action nail clipper is inserted under the nail plate; at the distal edge of the proximal nail fold, the other jaw is inserted under the fold to avoid its laceration. I prefer to use the English nail splitter because the smooth undersurface glides along the nail bed and the anvil-like upper surface easily slides under the nail plate (18). The nail plate is cut in a lon-

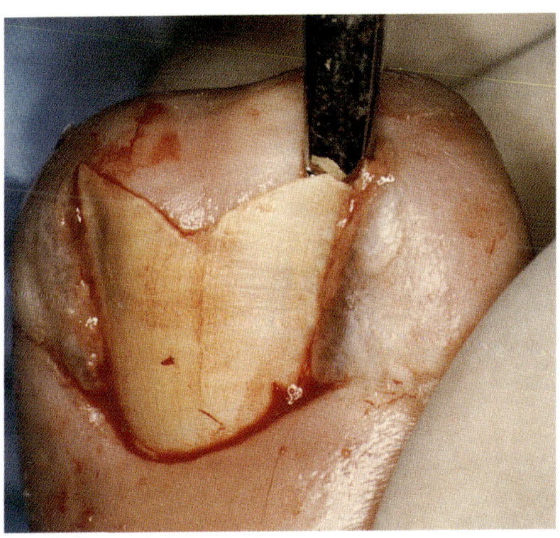

FIGURE 5A.12. Small-nosed mosquito hemostat inserted under edge of ingrown toenail to free it from underlying tissues.

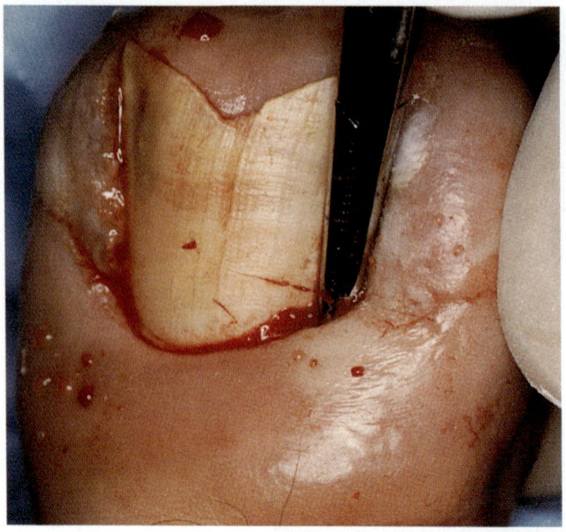

FIGURE 5A.13. Small-nosed mosquito hemostat inserted between the nail plate and proximal nail fold to free the nail plate from the overlying proximal nail fold.

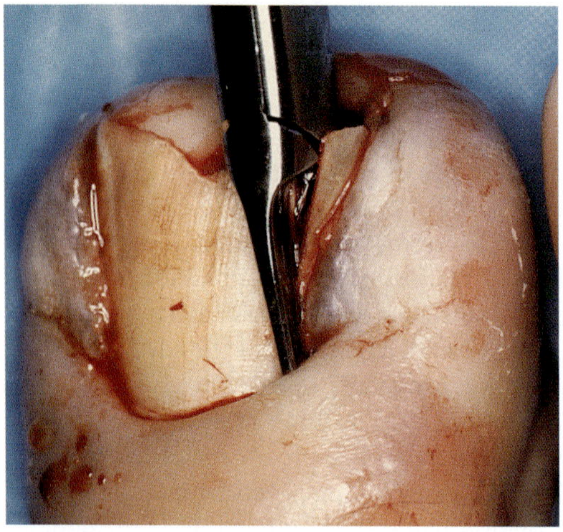

FIGURE 5A.14. English nail splitter inserted beneath the nail plate and proximal nail fold. Nail edge is transected longitudinally.

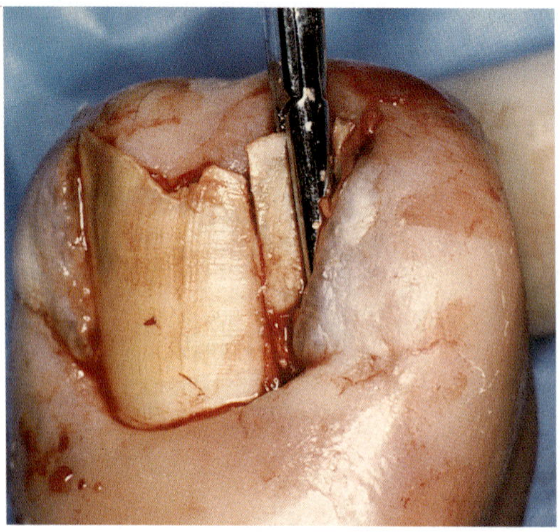

FIGURE 5A.15. Edge of nail is extracted with the hemostat.

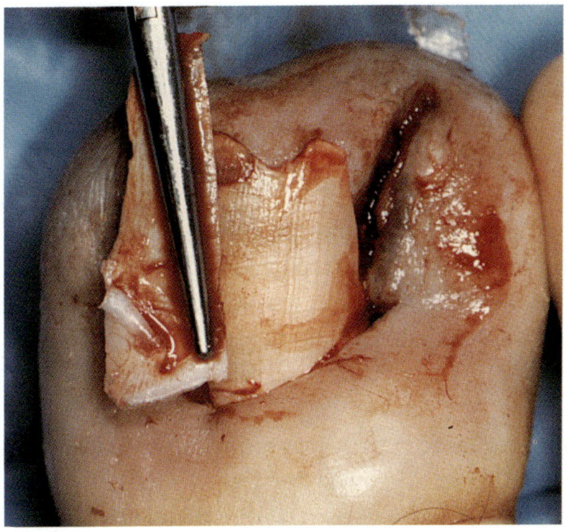

FIGURE 5A.16. The width of the nail edge reflects degree to which lateral nail fold had hypertrophied and overgrown the nail plate.

gitudinal fashion. The edge of the nail plate is grasped with the hemostat and extracted by a pulling motion (Figs. 5A.15,5A.16). If the resistance to the removal of the edge of the nail is significant, it usually is related to one or both of two reasons: (i) The piece of nail plate has not been adequately separated from the nail bed, matrix, or proximal nail fold; the persistently adherent site is usually the lateral horn of the nail plate; (ii) The nail plate has not been cut through its entire longitudinal direction. Repeating the described steps should result in easy removal of the edge of the nail plate.

After the nail edge has been removed, the granulation tissue is curetted away (Fig. 5A.17). At this point, a probe or curette excavator can be carefully run along the lateral nail groove to locate the laceration into the side of the lateral nail groove. However, nothing specifically needs to be done to this laceration. Removal of the edge of the nail allows the wound to heal.

Dressings with antibiotic ointment, petrolatum gauze, fine-mesh gauze (such as Sof-Wick) cut to size, and a self-adhering wrap (such as Kling) are applied. The dressing is finished with tubular elastic net (Se Pro Net). The patient begins soaks the next day, twice daily for 15 to 30 minutes each time for 7 to 10 days.

Systemic antibiotics usually are not required. Because the initiating factor—the laceration of the side of the lateral nail groove by the nail plate—has been corrected, the

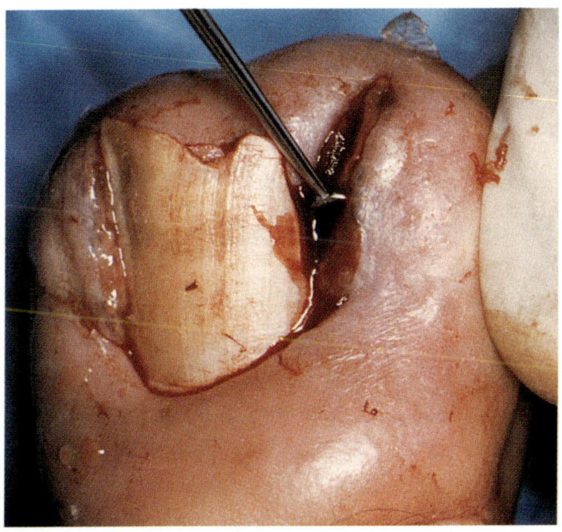

FIGURE 5A.17. Granulation tissue is curetted.

inflammation subsides. However, with diabetics and patients with impaired circulation or other local or systemic conditions, systemic antibiotics may be indicated. Dicloxacillin, erythromycin, or cephalosporins are reasonable initial choices of antibiotics. If the inflammation does not show some signs of subsiding in 2 to 4 days, additional antibiotic therapy should be selected on the basis of cultures and sensitivity results, and there should be an intensification of local wound care.

The final repair usually is scheduled at least 4 weeks after treatment of the inflammatory stage (19). It is not advisable to undertake the final repair and dissection to the level of the underlying bone during the inflammatory stage because of the greater risk of soft tissue infection and osteomyelitis. It is very unusual for some of the inflammatory response to persist beyond the 4 weeks, although shorter time intervals for clearing of the inflammation are not uncommon.

The growth rate of fingernails is approximately 3 mm per month and of toenails is approximately 1 to 2 mm per month (17). The regrowth for first toenails is 12 to 18 months and the ingrowing usually occurs in approximately the same site. Therefore, the window of time in which to perform the final repair is determined to be from 1 month after the initial treatment to allow the subsidence of the inflammatory stage to the time at which the regrowing nail plate reaches the previous site of the laceration. This time interval is usually from 1 to 4 or 5 months for more proximally ingrown toenails.

Do all ingrown toenails need definitive repair? The answer is clearly no (20). Often, especially for adolescents who are experiencing their first episode of ingrown toenails, treatment of the acute inflammatory stage may lead to a permanent cure. As the patients become older, the chance of recurrence is greatly reduced. However, with repeated episodes the lateral nail fold becomes fibrotic, firm, and inflexible (8) and may lead to narrowing of the width of the space for the regrowing nail. Under these circumstances, the likelihood of further ingrowing is great and permanent. In general, avulsion of the edge or total nail plates alone has resulted in high recurrence rates of over 60%, which increases with each episode (21,22).

PERMANENT REPAIR

The object of this method of permanent repair of an ingrown toenail is to narrow the nail plate so that it does not cut into the lateral nail groove. It also preserves the lateral nail folds, which enhances the final aesthetic appearance. In other methods, it may depend on the retraction or destruction of the lateral nail fold or protecting the side of the lateral nail groove from the lateral edge of the nail plate. The goal of narrowing the nail plate is achieved by excision or destruction of part of the lateral nail matrix. Destruction can be accomplished by phenolization, laser surgery, or electrosurgery. These methods are described elsewhere. The method that will be described involves the excision of the lateral matrix and can be adapted for nail biopsy, laterally placed pigmented streaks, and laterally located split nails.

STEPS IN PERMANENT REPAIR OF INGROWING NAIL

Permanent repair of an ingrowing nail is divided into five steps, which also are described in the section on "Lateral Longitudinal Nail Biopsy" in Chapter 3C. If the edge of the nail plate has regrown by the time of permanent repair but is not inflamed, then it must be removed in the manner described under inflammatory ingrowing nail (Figs. 5A.12–16 and 5A.18–22). The amount of nail plate and matrix that should be removed may be significant because of the hypertrophy of the nail fold or the marked curvature of the nail plate, as in pincer nail (Figs. 5A.23 and 5A.24). A general guide to the width of the excision is that the incision of the nail plate, nail bed, and matrix corresponds approximately

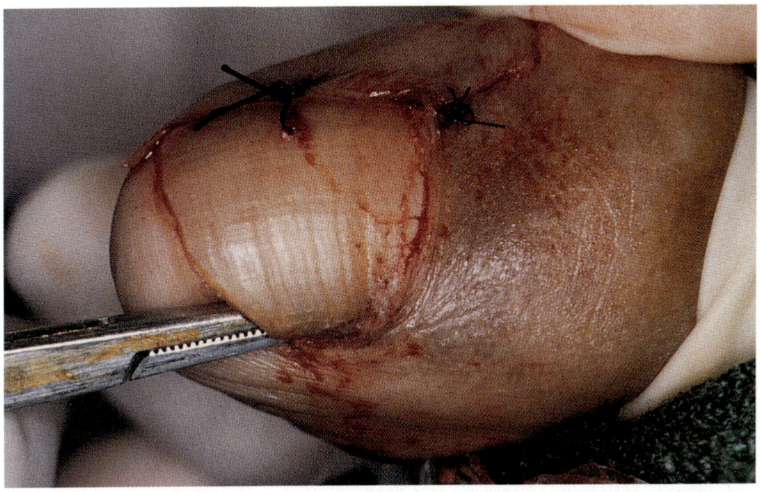

FIGURE 5A.18. Nail plate has regrown after initial partial avulsion months before. The other side of the nail had a biopsy of a pigmented streak on the same day.

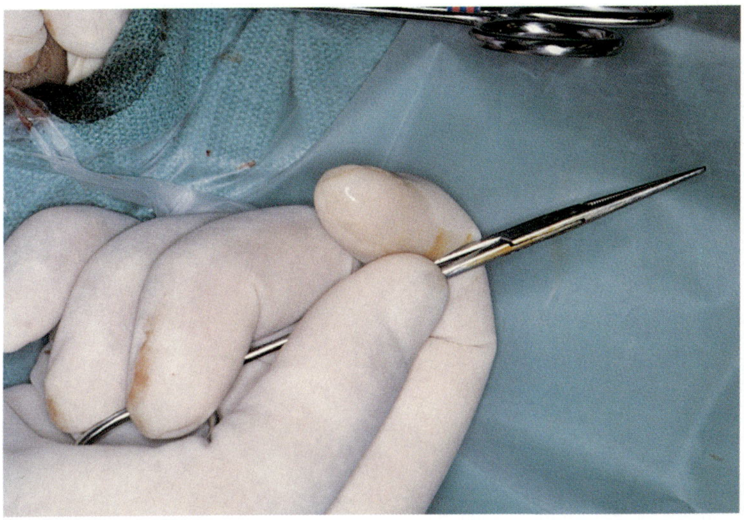

FIGURE 5A.19. Often for thick nails, especially pincer nails, it takes significant effort to insert the hemostat under the nail plate. Therefore, I prefer to hold the handle in the palm of the hand, which affords greater strength and control.

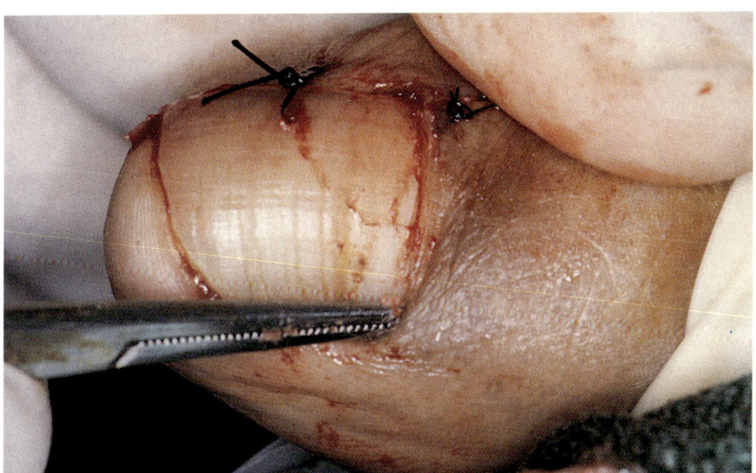

FIGURE 5A.20. Tip of hemostat inserted between nail plate and proximal nail fold.

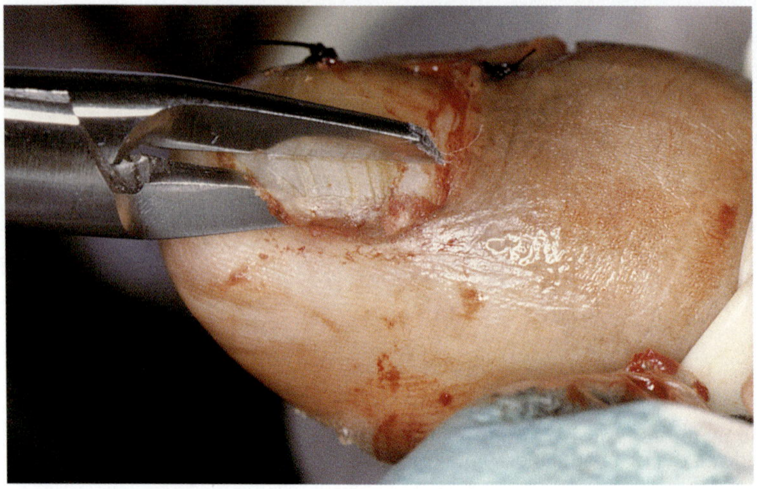

FIGURE 5A.21. English nail splitter inserted under nail plate.

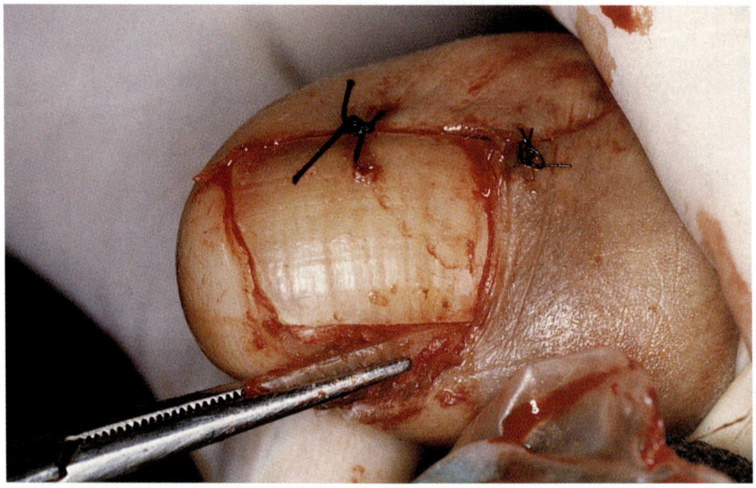

FIGURE 5A.22. Lateral edge of nail plate is extracted.

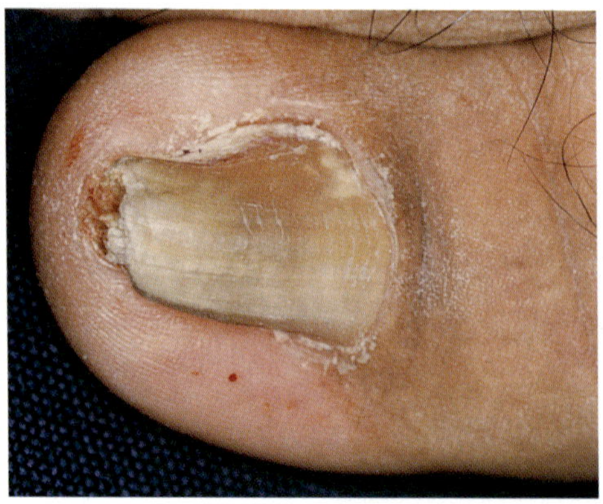

FIGURE 5A.23. Pincer nail.

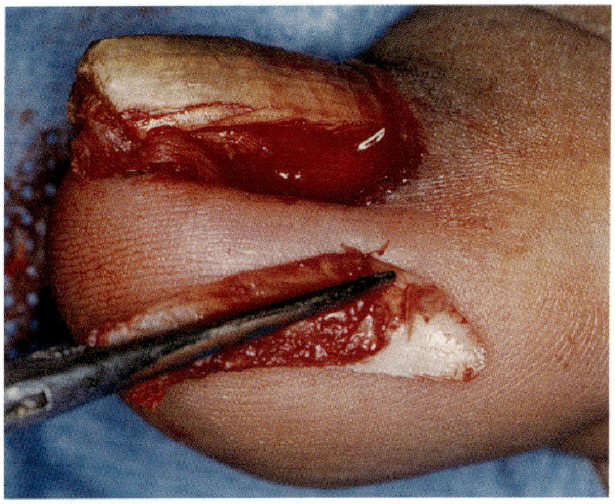

FIGURE 5A.24. The width of the avulsed nail wedge may be significant in pincer nails and hypertrophic nail folds. Same patient as shown Fig. 5A.23.

to the medial edge of the lateral nail fold, or 2 to 3 mm more medialy. This should allow removal of the appropriate amount of matrix and yet results in excellent aesthetics.

The permanent repair is divided into five steps.

Step 1: Incision through the nail bed and visible matrix (lunula) stopping at the cuticle. The depth of the incision is the underlying bone (Figs. 5A.25 and 5A.26).

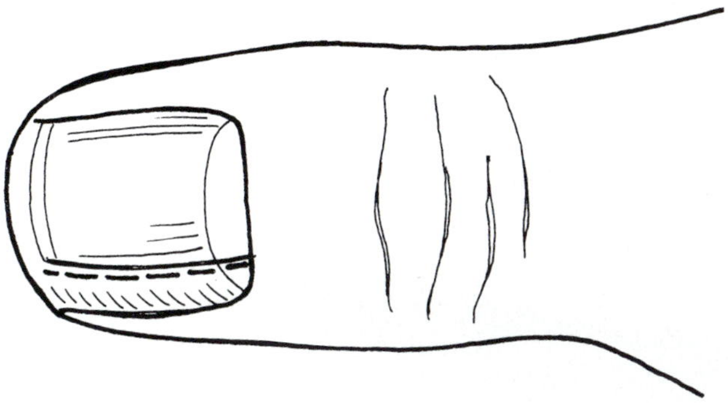

FIGURE 5A.25. Step 1: incision through nail bed and visible matrix down to bone. Proximally the incision stops at the cuticle.

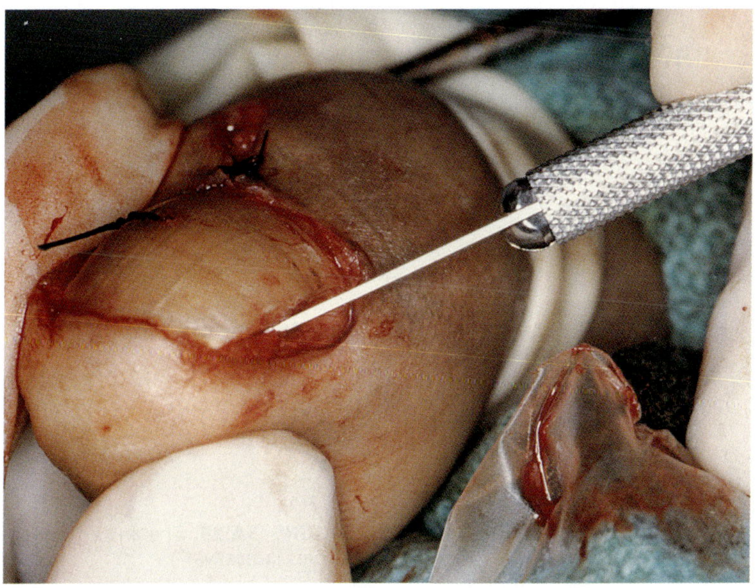

FIGURE 5A.26. Step 1 being performed on a patient. Lateral edge nail plate had been previously avulsed.

Step 2: Incision and reflection of the proximal nail fold to identify the lateral extent of the matrix (Fig. 5A.27). This incision is carried through only the depth of the proximal nail fold in a straight line extension of the incision in step 1. The incision is first straight and then curves laterally to better expose the horn of the matrix (Figs. 5A.28 and 5A.29). The length of the straight part closely approximates the depth of the invagination

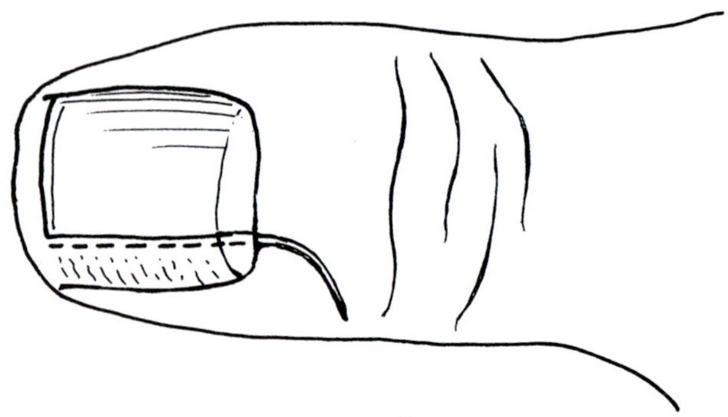

FIGURE 5A.27. Step 2: incision in proximal nail fold to explore lateral nail and matrix horns.

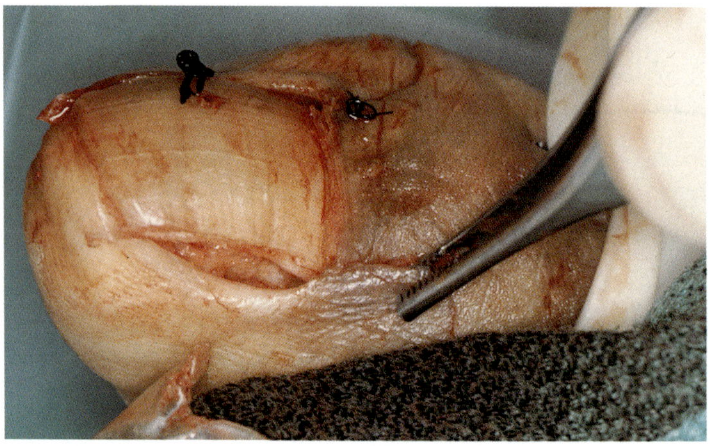

FIGURE 5A.28. The incision in the proximal nail fold initially is a straight line extension of step 1.

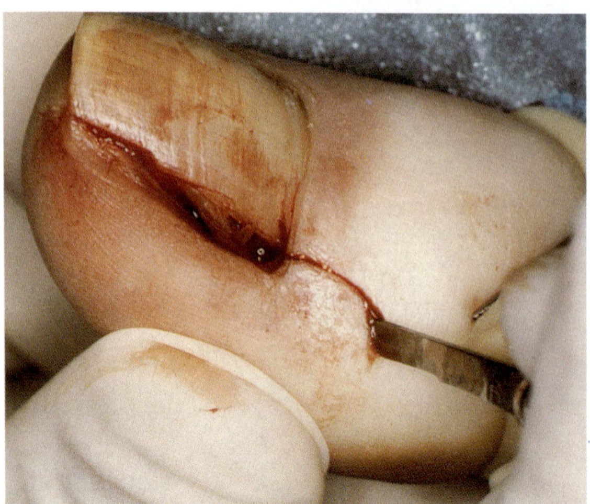

FIGURE 5A.29. The incision then curves laterally.

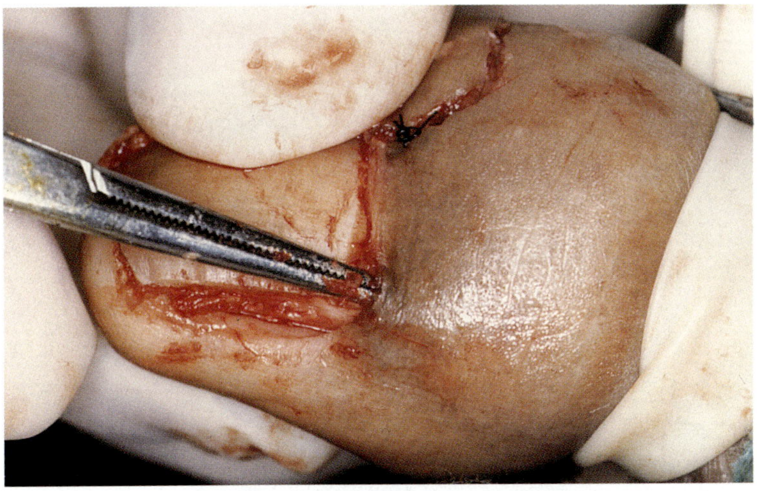

FIGURE 5A.30. The length of the straight part of the incision corresponds to the length of the proximal nail groove, which can be determined by the length of the tip of the hemostat that can be inserted between the nail plate and the proximal nail fold. This is usually 5 to 7 mm.

of the nail plate. The depth of the invagination can be ascertained by the depth of the insertion of the mosquito hemostat under the nail fold during the procedure to remove the lateral edge of the nail plate (Figs. 5A.30 and 5A.31). This usually is approximately 5 to 7 mm. The initial incision should be superficial to essentially mark the skin and not an attempt to cut through all layers of the nail fold. This will allow a more precise deeper dissection of the proximal nail fold. After 5 to 7 mm, the incision is curved laterally, terminating at what is judged to be the lateral extent of the matrix (Figs. 5A.29 and 5A.32). The incision is completed through the nail fold to the space created by removal of the nail plate. This creates a flap of tissue which covers the lateral matrix. This incision, which develops as flap, must be carried laterally enough to visualize the horn of the matrix and the lateral nail groove (Fig. 5A.32) The horn of matrix may extend much further than originally judged, necessitating further elongation of the incision in the proximal nail

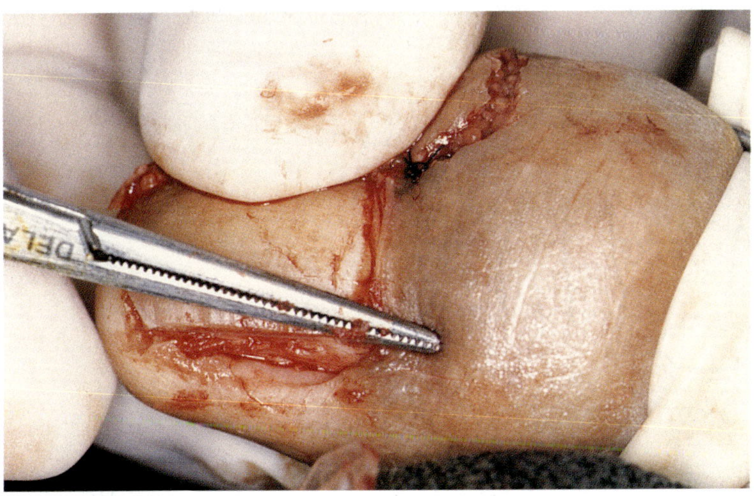

FIGURE 5A.31. The tip of the hemostat lying on the surface of the proximal nail fold reflects the depth of the proximal nail groove which represents the invagination of the nail plate.

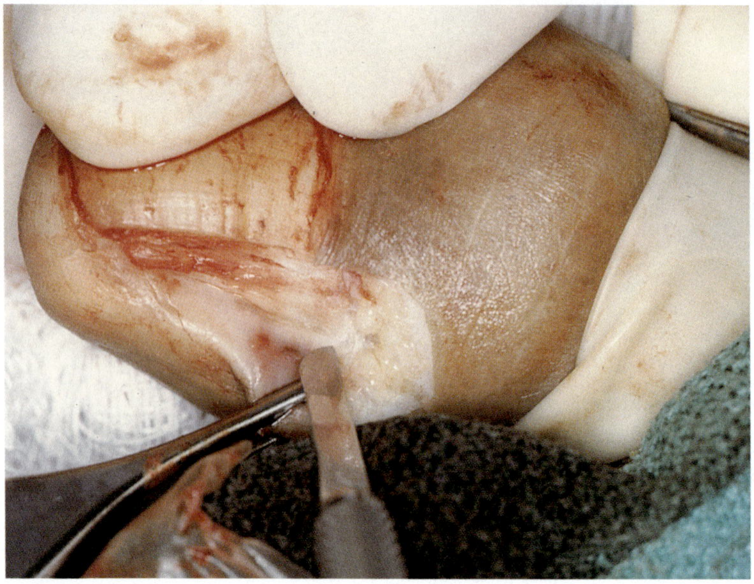

FIGURE 5A.32. The curved part of the incision of the proximal nail fold should be extended laterally until the lateral nail groove and the horn of the matrix can be visualized. This visualization is critical to prevention of recurrent spicules of nail or epidermal cysts resulting from retained matrix. The tip of the nail points to the horn of the matrix.

fold. The key to prevention of recurrences of nail or keratinous cyst formation is visualization of the junction of the horn of the matrix and the beginning of the lateral nail groove.

Step 3: Completion of the incision directly in line with the original incision that stopped at the cuticle (Fig. 5A.33). This incision should be carried through the longitudinal length of matrix (Fig. 5A.34), which has been made visible because of the retraction of the flap of skin in step 2.

Step 4: Incision of the side of the lateral nail groove dorsal to or above the matrix (Figs. 5A.35 and 5A.36). With reflection of the proximal nail fold and visualization of the most lateral part of the matrix, an incision is made in the lateral nail groove 1 to 2 mm dorsal to or above the matrix (Fig. 5A.37). At the completion of this stage of the surgery, the matrix to be removed lies between the medial incision in the nail bed and the matrix of steps 1 and 3 and that of the lateral incision through the nail groove of step 4 (Figs. 5A.38 and 5A.39).

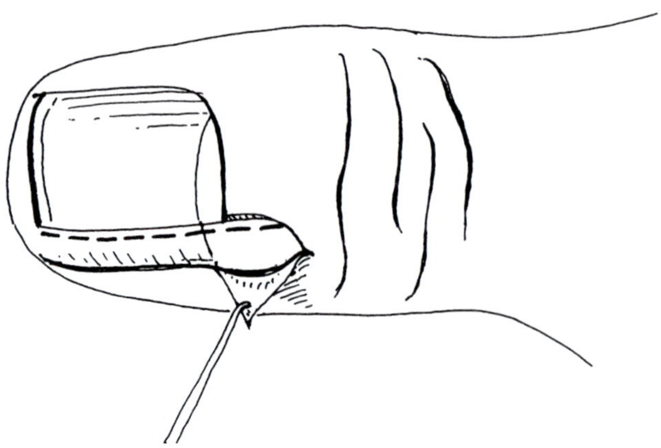

FIGURE 5A.33. Step 3: a straight line extension of the incision in step 1 through the entire matrix.

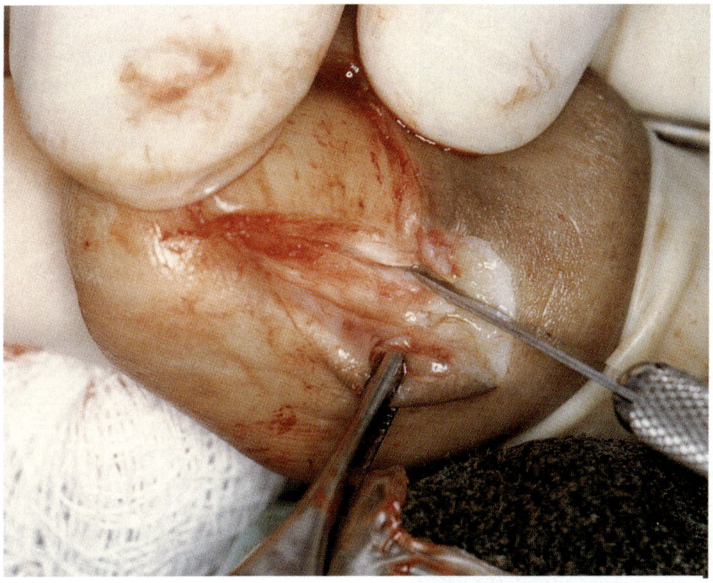

FIGURE 5A.34. Step 3 being completed in the patient.

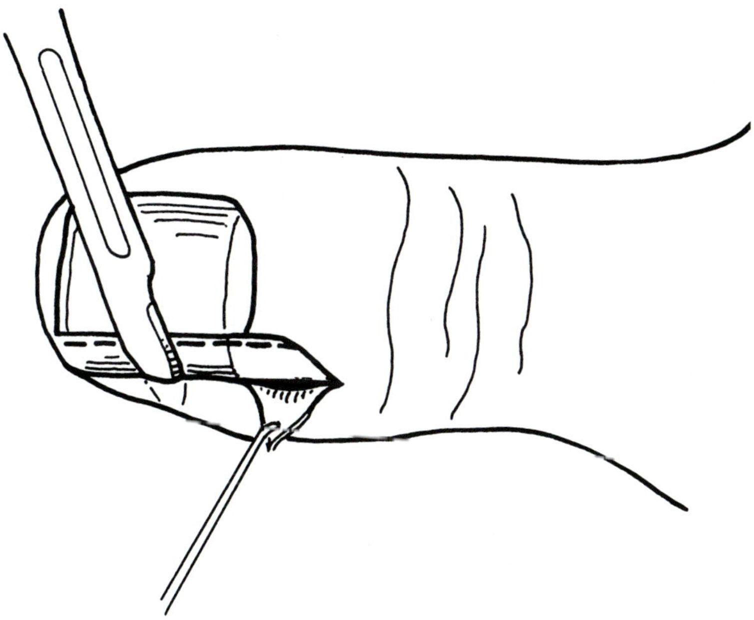

FIGURE 5A.35. Step 4: incision in the lateral nail groove dorsal to the matrix.

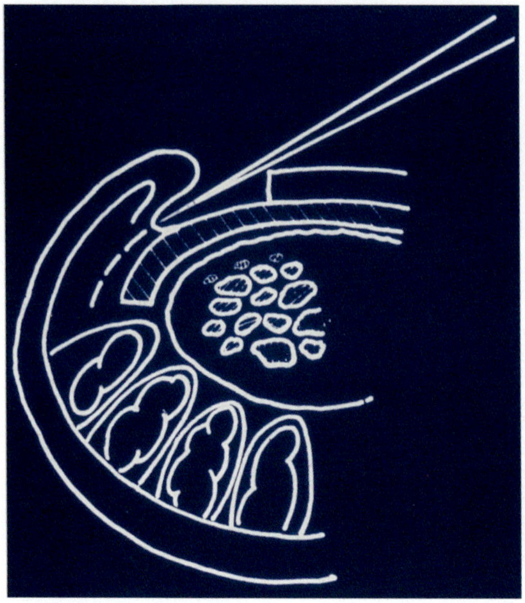

FIGURE 5A.36. Cross section illustrating the incision in the lateral nail groove.

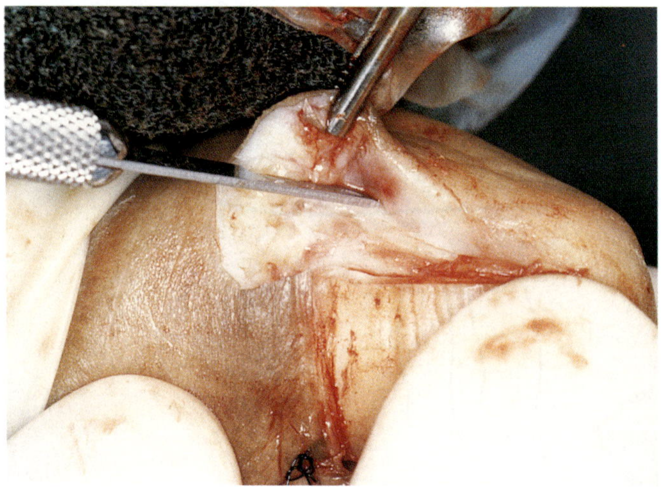

FIGURE 5A.37. Incision in the lateral nail groove 1 to 3 mm dorsal to the horn of the matrix.

Step 5: Removal of the lateral edge of the matrix (Fig. 5A.40). Starting distally, the nail bed and matrix are removed by dissecting the tissue from the underlying bone. It is preferable to carry the excision from medial to lateral because the underlying bone defines the depth of the excision. Initially it may be difficult to grasp the tip of the tissue with forceps or skin hooks; in that case, a free tip is developed by simple dissection. Once the tip is freed, it can be grasped by the forceps and the dissection continued (Fig. 5A.41). The scalpel is used to peel the tissue from the underlying bone, carrying the resection to the incision previously made in the lateral nail groove.

The excision should encompass the entire width and length of the tissue to avoid leaving residual matrix. During the final stages of the excision, the tissue should be pulled with significant tension. When the dissection extends beyond the proximal matrix into the fibrofatty tissue, there usually is a sudden release of the tissue (Fig. 5A.42).

The defect is examined carefully for residual matrix. All tissue resembling matrix should be removed (23). Matrix that remains can lead to a recurrent spicule of nail or a small keratinous cyst that may take a number of weeks to appear (1). The area is gently

FIGURE 5A.38. The part of the matrix to be removed lies between the incision of steps 1 and 3 medially and step 4 laterally. Cross-section view.

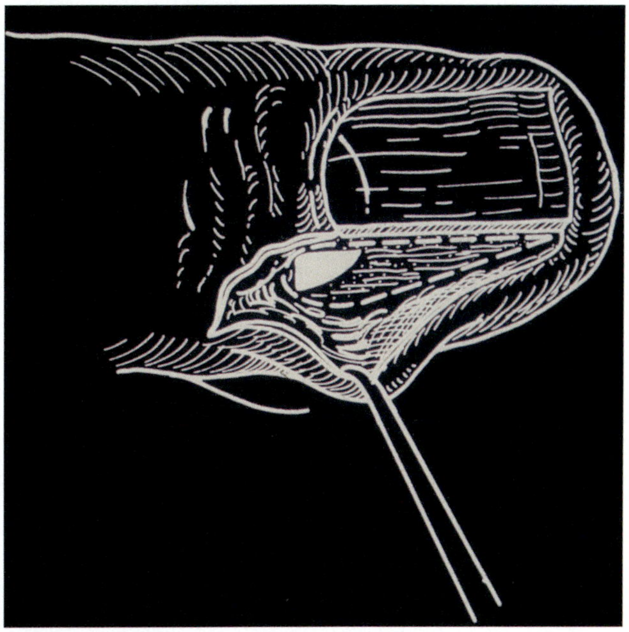

FIGURE 5A.39. The part of the matrix to be removed lies between the incision of steps 1 and 3 medially and step 4 laterally. Longitudinal view.

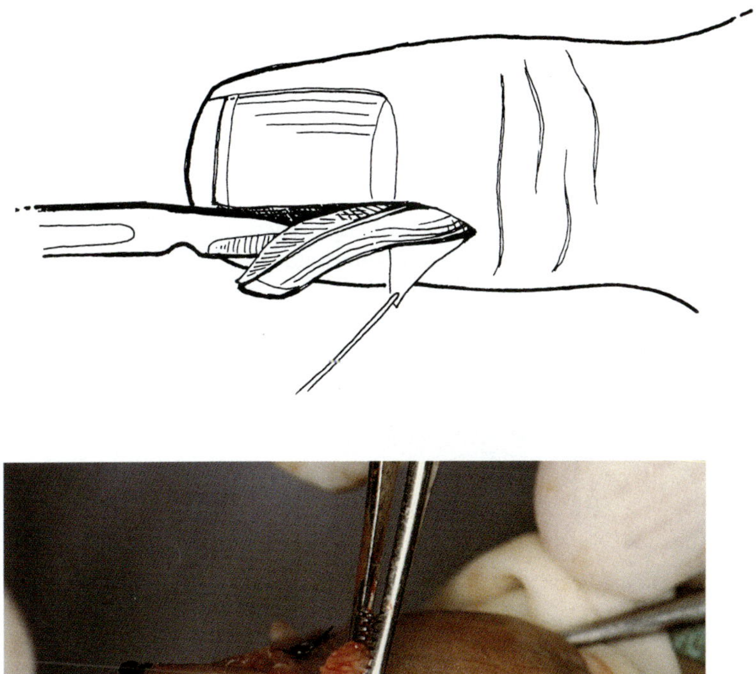

FIGURE 5A.40. Beginning distally and medially, the nail bed and matrix are excised from the underlying bone. By beginning medially the underlying bone defines the depth of the excision.

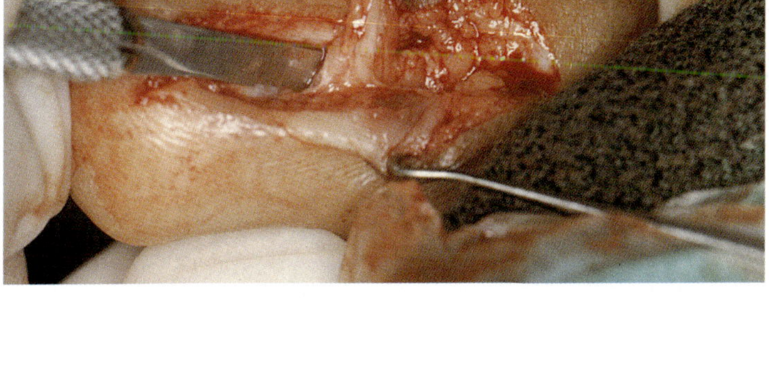

FIGURE 5A.41. Nail bed and matrix being excised, progressing from medial to lateral and distal to proximal.

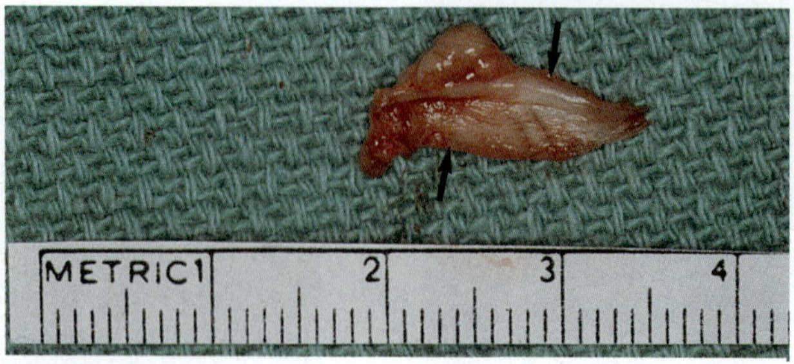

FIGURE 5A.42. The matrix *(lower arrow)*, nail bed, and part of the lateral nail wall *(upper arrow)* are visible in the excised tissue specimen.

curetted to remove fragments of matrix or nail bed and then wiped clean with gauze wrapped around the mosquito hemostat.

At this point of the surgery, two questions are addressed: (i) Is the resection proximal enough?; and (ii) Is the resection lateral enough?

Examining the medial part of the excision should reveal the invagination of the nail into the proximal nail groove (Fig. 5A.43). Therefore, if the excision is proximal to this invagination, then it is proximal enough. If inspection of the lateral nail groove reveals only fibrofatty tissue, then the excision is wide enough (Fig. 5A.44). With good visualization of the entire matrix and examination for these two anatomic landmarks, the recurrence of spicules of nail or development of keratinous cysts should be very infrequent. Prevention of recurrence is technique and surgeon dependent (22–24). Identification of the two landmarks greatly reduces recurrences even for the less experienced surgeon.

The first suture, usually 5-0 nylon, is placed at the distal end of the proximal fold, near the cuticle (Fig. 5A.45). Careful placement of the suture results in a normal-appearing cuticular area and excellent cosmetic results. Other sutures are placed as needed to close the incision of the proximal nail fold.

Approximation of the lateral nail fold to the nail edge is accomplished in a number of ways. After drilling a hole through the nail plate with a no. 11 scalpel blade, a suture is passed from the lateral nail fold through the hole in the nail plate and tied (Fig. 5A.46).

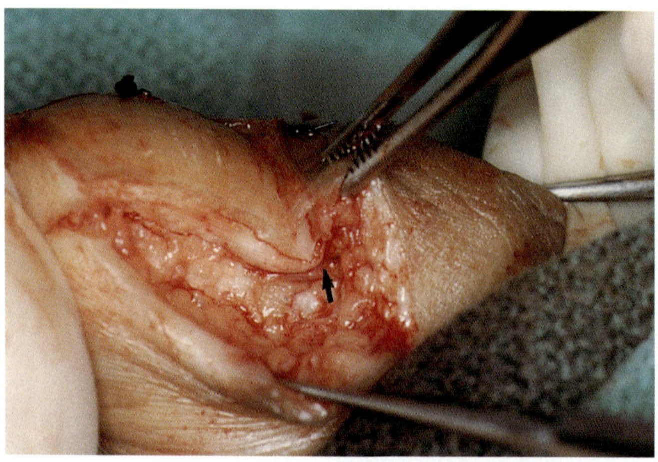

FIGURE 5A.43. *Arrow* points to proximal part of the matrix, which corresponds to the apex of the proximal nail groove. This anatomic landmark determines whether the excision is proximal enough.

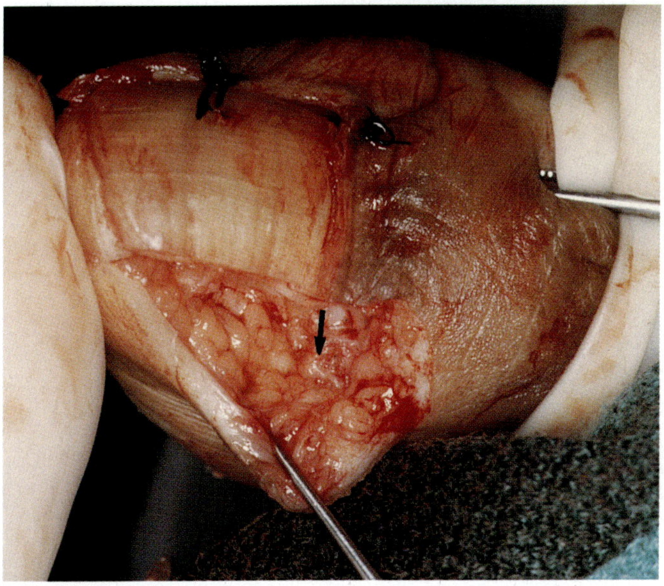

FIGURE 5A.44. Fibrofatty tissue in the lateral nail groove *(arrow)* determines that the excision has been made adequately lateral.

The author prefers larger needles (P-2) and sutures such as 3-0 or 4-0 silk, but 4-0 nylon on a larger needle is adequate. I prefer to proceed from the lateral nail fold through the nail to avoid carrying keratin into the soft tissue. Often it is unnecessary to use sutures to close the defect between the lateral nail fold and the nail. Wrapping of the toe may be sufficient to approximate the wound edges (Fig. 5A.47). At times, Steri-Strips may be used for this approximation.

With the tourniquet in place, an antibiotic ointment is applied, followed by petrolatum gauze and a fine-mesh gauze (Sof-Wick cut to size), wrapping with self-adherent gauze, such as Kling, and finished with tubular elastic net (Se Pro Net) (Figs. 5A.48–5A.52). The wrapping should be firm but not tight. The patient should be driven home by the person accompanying the patient; driving with a painful toe may cause an accident. The patient is told to elevate the foot, advised on limited ambulatory privileges, and given a cutout shoe, if he or she does not have appropriate footwear. The patient is

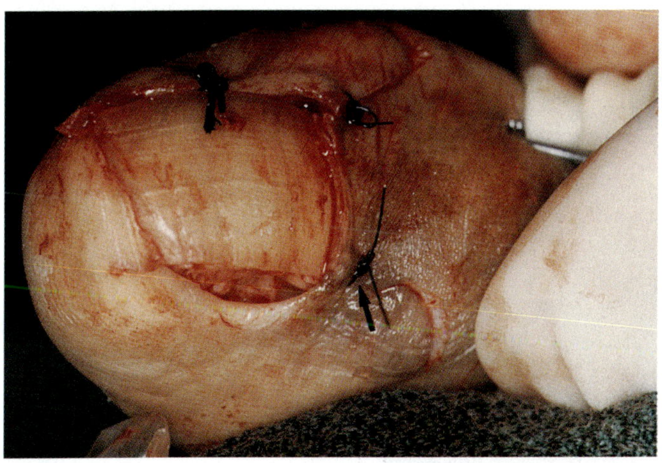

FIGURE 5A.45. The first suture approximates the cuticular area *(arrow)*.

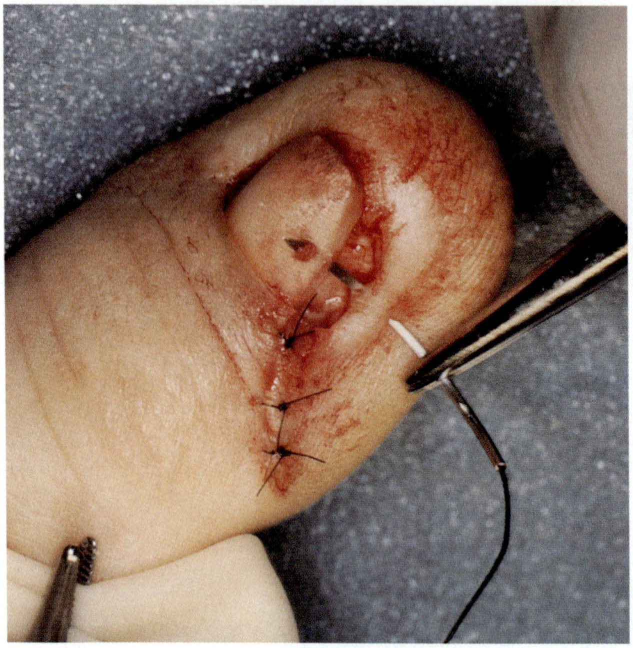

FIGURE 5A.46. If the defect is wide, the suture placed through the lateral nail fold and the nail plate can approximate the sides of the excision.

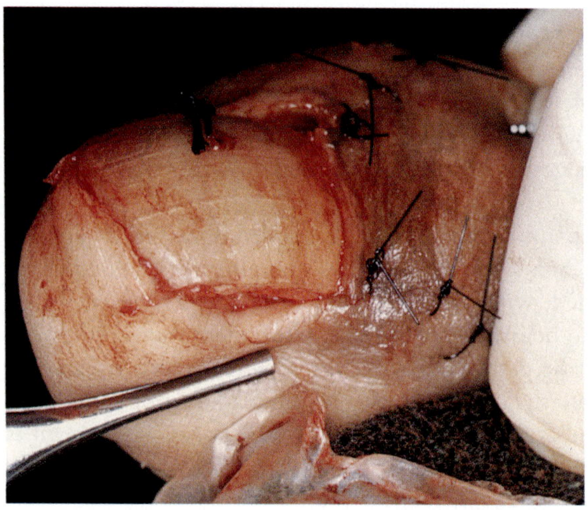

FIGURE 5A.47. Often the space is narrow and the skin movable so that the dressing alone approximates the side of the excision.

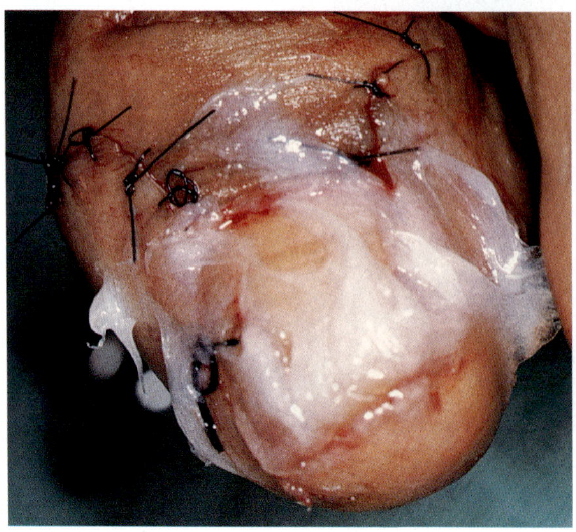

FIGURE 5A.48. Antibiotic ointment is applied.

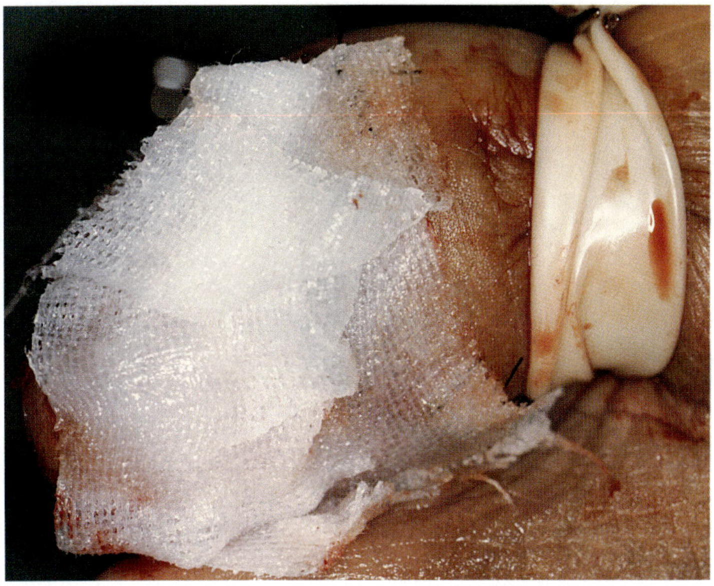

FIGURE 5A.49. A petrolatum gauze is put in place. In the author's experience, application of Telfa at this point of the dressing still resulted in a very adherent dressing.

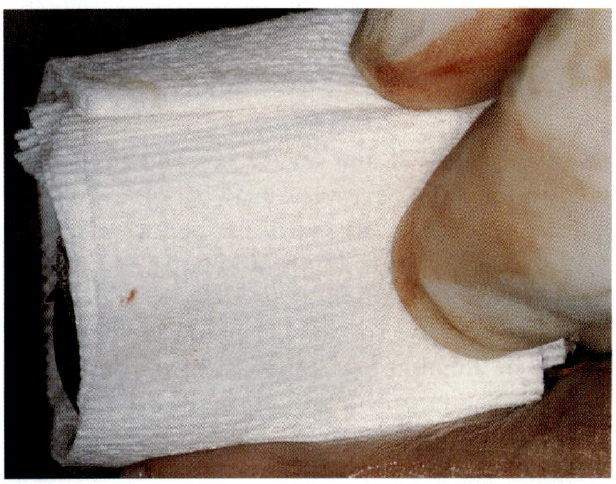

FIGURE 5A.50. Fine-mesh gauze cut to size is placed around the toe.

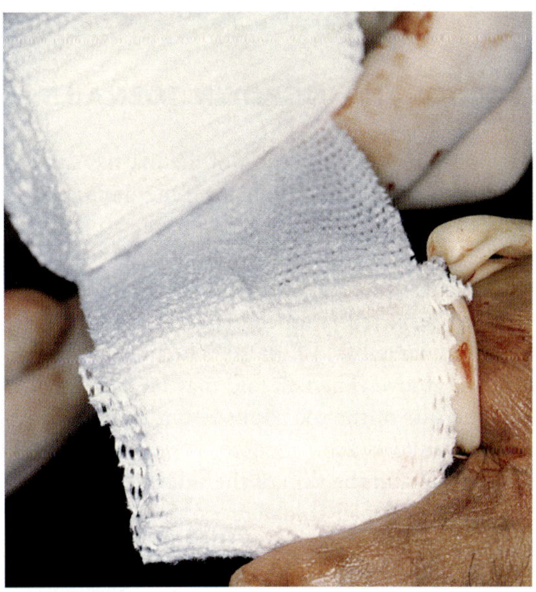

FIGURE 5A.51. Self-adherent dressing, such as Kling, is wrapped around the toe. This should not be applied so tightly that circulation to the toe is adversely affected.

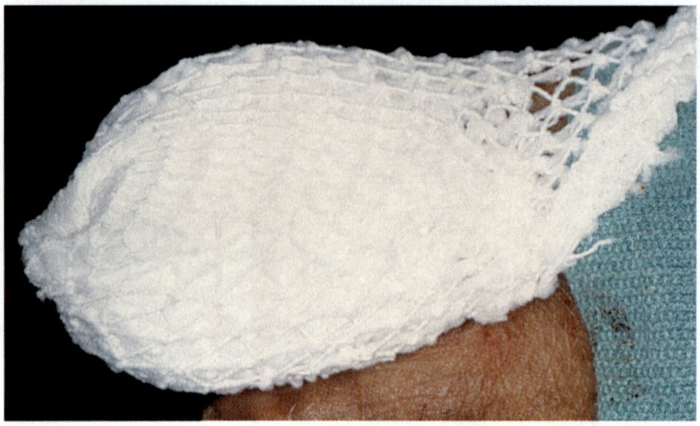

FIGURE 5A.52. The dressing is finished with a tubular elastic net.

asked to return in 2 days for dressing change or is instructed on how to change the dressing. In our experience, waiting longer than 2 days for the initial dressing change tended to increase the wound infection rate. The patient returns in about 6 or 7 days postoperatively for suture removal. After suture removal, the incision in the proximal nail fold can be reinforced with Steri-Strips or similar material.

DISTAL NAIL TISSUE HYPERTROPHY

Occasionally the skin along the distal edge of the nail unit may mound up and form a wall that interferes with nail growth. As a result, the nail plate may be forced to grow through this tissue. This condition may result from an extensive ingrown nail that involves both lateral nail folds and the distal end of the nail unit. Additionally, if the nail plate is avulsed surgically or traumatically or is shed as in onychomadesis, the skin of the distal toe may heap up because of the absence of the counterpressure of the nail plate. Conservative methods of treatment should be attempted initially. If the nail plate can be guided to ride over the elevated distal tissue, the subsequent pressure of the nail plate usually will flatten the mound of skin. The patient can be taught to apply pieces of cotton under the distal nail edge as a wedge to guide the nail over the rim of elevated tissue. The procedure is performed for a number of weeks until nail growth extends beyond the soft tissue barrier. If conservative methods fail, then an elliptical excision of the elevated tissue is performed and the wound closed with nylon sutures.

OTHER SURGERY FOR THE REPAIR OF THE INGROWN TOENAILS

Although used less frequently than procedures directed at ablation of the matrix, soft tissue surgery of the nail folds has been advocated for ingrown toenails. One method is to excise an ellipse of tissue from the sides of toes. Suturing of the excision retracts the lateral nail fold from the edge of the nail plate (Figs. 5A.53 and 5A.54) (9). However, in some repeatedly and chronically ingrown toenails, the resultant hypertrophy and fibrosis of the lateral nail fold prevents it from being retracted from the edge of the nail plate through this procedure. Others resect the entire lateral fold (25,26). The medial incision begins at the edge of the nail in the lateral nail groove. The nail fold may be transfixed by a no. 11 blade exiting in the soft tissues of the side of the toe about 7 mm from the edge of the lateral nail fold. The nail fold is excised *en bloc* (26).

Others make an ellipse at the edge of the nail and the skin of the side of the toe (25). The tissue between the incisions, including lateral nail fold, are excised *en bloc* (Fig. 5A.55). The subcutaneous tissue can be excised down to the periosteum. In both of these

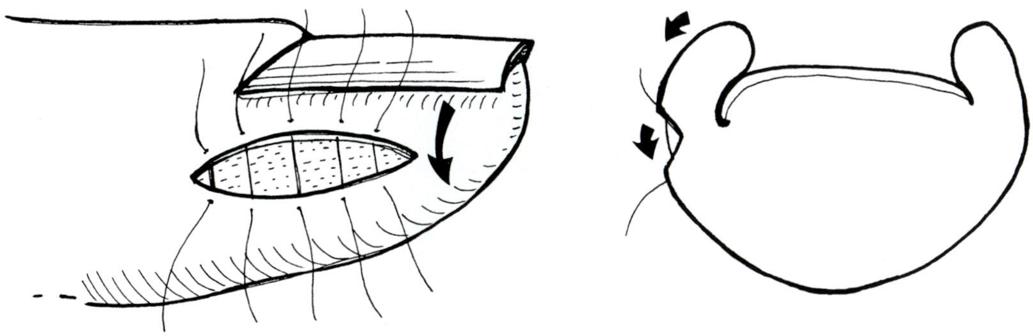

FIGURE 5A.53. An ellipse of tissue is excised from the skin of the lateral nail fold of an ingrown toenail.

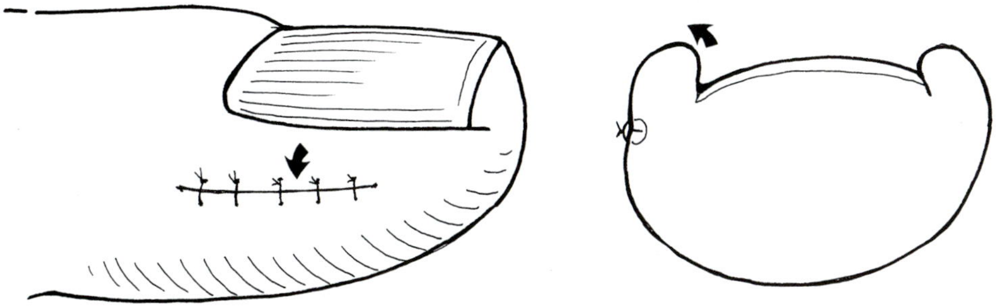

FIGURE 5A.54. Closure of the ellipse retracts the lateral nail fold from the edge of the nail plate.

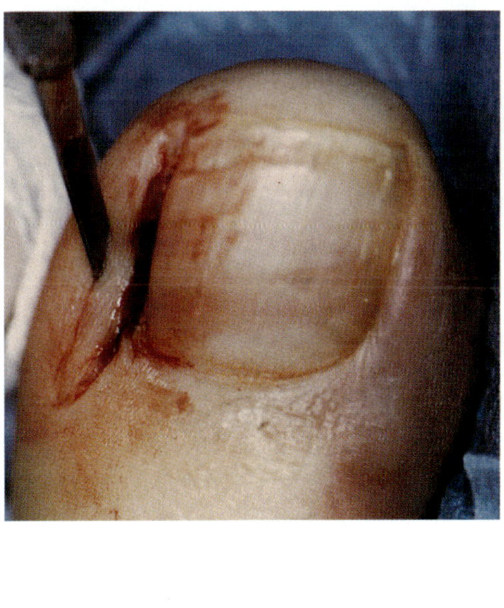

FIGURE 5A.55. Part of the nail fold as well as the lateral matrix and nail bed are excised. This affects the final aesthetic appearance of the surgery.

procedures, healing is by second intention. The aesthetics are less desirable than in the lateral matrix excision with preservation of the lateral nail folds. Second intention healing may be prolonged.

Nail splinting with flexible tubing has been described as a noninvasive treatment for ingrown toenails (13). The procedure has been described in more detail in the section on "The Inflammatory Stage" (Figs. 5A.4–9).

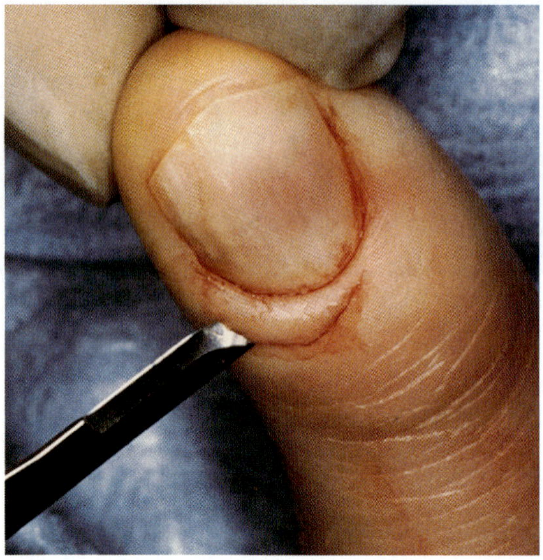

FIGURE 5A.56. Excision of the proximal nail fold.

PROXIMAL NAIL FOLD

Resection of the proximal nail fold has been recommended for chronic paronychia (27,28), myxoid cyst (29), and biopsy for collagen disease (Fig.5A.56) (30).

The excision is a crescent-shaped full thickness of the proximal nail fold extending to the lateral nail folds. Its greatest longitudinal length from the cuticle to mid-point of excision is 5 to 7 mm, approximately the depth of proximal nail groove. Healing is by second intention. More lunula is visible, but this not readily noticeable. The fold reattached to the nail plate prevents the entrance of bacteria, yeast, and foreign materials between the fold and the nail plate, which are some factors important to the development of chronic paronychia.

REFERENCES

1. Ceilly RI, Collison DW. Matricectomy. *J Dermatol Surg Oncol* 1992;18:728–734.
2. Eisele SA. Conditions of the toenails. *Orthoped Clin North Am* 1994;25:183–188.
3. Zuber TJ, Pfenninger JL. Management of ingrown toenails. *Am Fam Physician* 1995;52:181–188.
4. Haneke E. Surgical treatment of ingrowing toenails. *Cutis* 1986;37:251–256.
5. Langford DT, Burke C, Robertson K. Risk factors in onychocryptosis. *Br J Surg* 1989;76:45–48.
6. Baran R, Bureau H. Congenital malalignment of the big toe-nail as a cause of ingrowing toe nail in infancy. Pathology and treatment (a study of thirty cases). *Clin Exp Dermatol* 1983;8:619–623.
7. Baran R. The treatment of ingrowing toe nails in infancy. *J Dermatol Treatment* 1989;1:55–57.
8. Palmer BV, Jones A. Ingrowing toe nails: the results of treatment. *Br J Surg* 1979;66:575–576.
9. Siegle RJ, Stewart R. Recalcitrant ingrowing nails. Surgical approaches. *J Dermatol Surg Oncol* 1992; 18:744–752.
10. Connolly B, Fitzgerald RJ. Pledgets in ingrowing toenails. *Arch Dis Child* 1988;63:71–72.
11. Ileld FW. Ingrown toenail treated with cotton collodion insert. *Foot Ankle* 1991;11:312–313.
12. Murray WR. Onychocryptosis. Principles of non-operative and operative care. *Clin Orthop Rel Res* 1979;142:96–102.
13. Schulte KW, Neumann NJ, Ruzicka T. Surgical pearl: nail splinting by flexible tube—A new non-invasive treatment for ingrown toenails. *J Am Acad Dermatol* 1998;39:629–630.
14. Maeda N, Mizuno N, Ichikawa K. Nail abrasion: a new treatment for ingrown toe-nails. *J Dermatol* 1990;17:746–749.
15. Manca D. Taping toes—effective treatment for ingrown toenails. *Can Fam Physician* 1998;44:275.
16. Connolly AAP, Meyer LC, Tate JJT. Local anesthetic agents in surgery for ingrown toenail. *Br J Surg* 1994;81:425–426.
17. Walters DS, Scher RK. Nail terminology. *Int J Dermatol* 1995;34:607–610.
18. Salasche SJ, Peters VJ. Tips on nail surgery. *Cutis* 1985;35:428–438.

19. Fowler AW. Excision of the germinal matrix: a treatment for embedded toenail and onychocryptosis. *Br J Surg* 1982;45:382–387.
20. Iseli A. The management of ingrown toenails. *Aust Fam Physician* 1990;19:1414–1419.
21. Murray WR, Bedi BS. The surgical management of ingrown toenail. *Br J Surg* 1975;62:409–412.
22. Sykes PA. Ingrowing toenails: time for critical appraisal? *J R Coll Surg Edinb* 1986;31:300–304.
23. Hyder N. Ingrowing toenails: the extent of the germinal matrix. *J Bone Joint Surg* 1993;76-B:501–502.
24. Fulton GJ, O'Donohoe MK, Reynolds JV, et al. Wedge resection alone or combined with segmental phenolization for the treatment of ingrowing toenail. *Br J Surg* 1994;81:1074–1075.
25. Chapeski A. Simple cure for the ingrown toenail. *Aust Fam Physician* 1998;27:299.
26. Bose B. A technique for excision of the nail fold for ingrowing toenail. *Surg Gynecol Obstet* 1971;132:511–512.
27. Baran R. Removal of the proximal nailfold: why, when, how? *J Dermatol Surg Oncol* 1986;12:234–236.
28. Baran R, Bureau H. Surgical treatment of recalcitrant chronic paronychia of the fingers. *J Dermatol Surg Oncol* 1981;7:106–107.
29. Salasche SJ. Myxoid cysts of the proximal nailfold: a surgical approach. *J Dermatol Surg Oncol* 1984;10:35–39.
30. Schnitzler L, Baran R, Civatte J, et al. Biopsy of the proximal nail fold in collagen diseases. *J Dermatol Surg Oncol* 1976;2:313–315.

5B

INGROWN TOENAIL—PHENOL CAUTERY

ECKART HANEKE
ROBERT BARAN

As mentioned in Chapter 3D, phenol is a potent coagulant and cauterizer and has been used for treatment of ingrown toenails. The therapeutic principle is selective cautery of the lateral matrix segments (lateral matrix horns) to obtain a narrower nail that no longer digs into the lateral nail grooves.

A digital block anesthesia is performed as usual. The ingrown lateral strip of the nail plate is freed from the nail groove and nail bed as well as the proximal nail fold and cut longitudinally in its entire length. It then is levered out with an elevatory. The great toe (or any other digit) is exsanguinated using a rubber band, Penrose drain, or other tourniquet. The matrix horn cavities and lateral grooves are cleaned and dried with a cotton swab. A cotton-tipped applicator is dipped into liquefied phenol and excess phenol is allowed to drop off. The phenol is vigorously rubbed into the lateral matrix horns for two to three minutes, first with back-and-forth motions, then with rotation of the cotton applicator. The rotation direction is always away from the nail plate-matrix adhesion. This limits spilling of the phenol under the nail plate in the matrix and nail bed and avoids undue narrowing of the plate and lateral onycholysis (Fig. 5B.1). The cotton is changed three to four times to prevent minute bleeding of the matrix from inactivating the phenol. The matrix horn cavity is disinfected and filled with small antibiotic tablets, which prevents secondary postoperative infection and keeps the lateral horn space open for exudate.

Most patients present with considerable granulation tissue in the distal portion of the lateral nail groove and fold. Antiseptic foot soaks will diminish bacterial infection. Because phenol cautery involves no incision and phenol is a traditional antiseptic agent, no systemic antibiotics are needed. Excessive granulation tissue is gently treated with phenol for about 30 seconds and will disappear within 3 days to 1 week as soon as the offending lateral nail strip is removed. Nail avulsion before phenol cautery is not necessary, neither a few days before phenolization nor during the procedure. Any phenol spilling over the periungual skin is wiped away with a gauze swab soaked with 70% isopropyl alcohol. Flushing of the matrix horn with isopropyl alcohol to stop the phenol action is not necessary.

The procedure is completed with a bulky absorbent dressing with antibiotic tulle gras and gauze pads. The dressing is changed the next day; the outer bandage is removed and the foot is placed into a foot bath containing povidone iodine soap or chlorhexidine gluconate (Hibiclens) to allow the gauze to float off and to remove any exudate. Clotted blood is removed with 3% hydrogen peroxide or 2% tosylchloramide sodium (clorina, chloramine). A new dressing with antiseptic ointment is applied and changed every day until oozing stops.

E. Haneke: Klinikk Bunæs, Løkkeåsveien, 1337 Sandvika, Oslo, Norway.
R. Baran: Nail Disease Centre, 42, Rue des Serbes, 06400 Cannes, France.

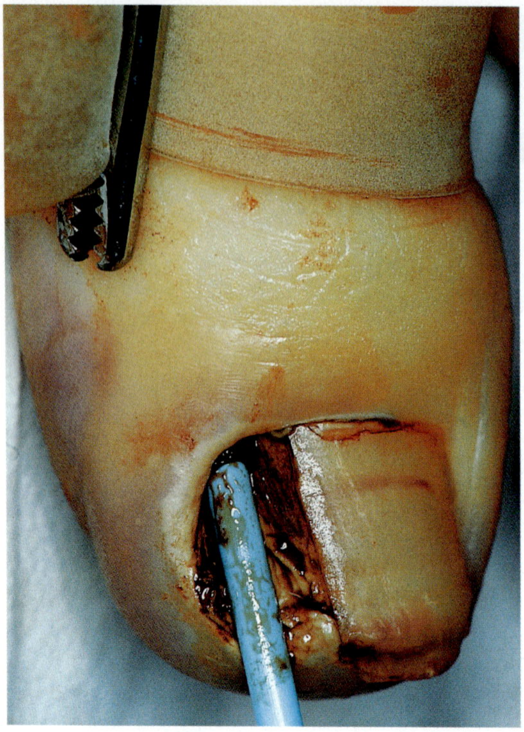

FIGURE 5B.1. Phenol cautery method.

Phenol cautery has several advantages over scalpel surgery:

1. It is technically simple, very easy to perform, and a time-honored procedure.
2. Postoperative pain is minimal or virtually nil.
3. Risk of postoperative infection is extremely low when antiseptic foot soaks are performed regularly every 24 hours.
4. Recurrence rate is very low when 90% phenol is rubbed in vigorously and for a sufficient amount of time (1–6).

When the ingrown toenail has caused considerable hypertrophy of the lateral nail fold this abnormality will regress, at least to a certain degree. Any excess tissue can be removed in a second-stage operation to regain normal proportions.

REFERENCES

1. Andrew T, Wallace WA. Nail bed ablation—excise or cauterize? A controlled study. *Br Med J* 1979;1:1539.
2. Haneke E. Surgical treatment of ingrowing toenails. *Cutis* 1986;27:251–256.
3. Ross WR. Treatment of the ingrown toenail. *Surg Clin North Am* 1969;49:1499.
4. Ramsay G, Caldwell D. Phenol cauterization for ingrown toenails. *Arch Emerg Med* 1986;3:243–246.
5. Young MR, Rutherford WH. Re-operation rate for ingrowing toenail treated by phenolization. 3-year follow-up. *Br J Surg* 1987;74:202–203.
6. Byrne DS, Caldwell D. Phenol cauterization for ingrowing toenails: a review of five years' experience. *Br J Surg* 1989;76:598–599.

5C

ELECTRORADIOSURGERY

ROBERT BARAN
ECKART HANEKE

Electrosurgery is a procedure by which tissue is removed or destroyed by electrical energy. This energy, usually in the form of high-frequency alternating current, is converted to heat by tissue resistance to its passage. The heat is generated in the tissues themselves and, in marked contrast to electrocautery, the treatment electrode in electrosurgery remains "cold" throughout the procedure (1).

Electrosurgery was popular for many years. It was later forgotten but now is back in vogue because of the appearance of radiosurgery and the development of new flexible electrodes with a flat triangular tip for treating ingrowing toenails. The insulated matricectomy electrode is specially coated for protection of the upper tissue while it destroys underlying cells (Fig. 5C.1). Destruction of the lateral horn of the matrix is possible without injuring the ventral aspect of the proximal nail fold by using the uncoated side down on the matrix.

The nail plate is first separated from the proximal nail fold and the nail bed. Using an English nail splitter, the offending lateral border is detached from the nail and the partial segment is removed with a forceps. The current selector is turned to "partially rectified current" and the electrode is applied to the matrix area for 3 seconds. After the power is turned off, the electrode is removed and one waits 10 seconds. This procedure is repeated twice when treating a great toe. A 0.125-mL dose of dexamethasone phosphate is instilled into the wound and a layer of antibiotic ointment is applied, followed by a sterile compressive dressing (2). Complete matricectomy is possible using the same technique after total nail avulsion (3).

Radiosurgery is used by some to treat warts. When the warts are subungual, they first require a partial avulsion of the distal or lateral half of the nail plate, depending on their location. Radiocoagulation is done superficially, followed by careful curettage of the area because of the fragile nature of the tissue in the nail bed. Simple pressure with a dressing made of double-thickness sterile gauze can result in hemostasis. In cases of persistent bleeding, slight pressure with a cotton-tipped swab soaked in 35% aluminum chloride solution usually is sufficient.

Pyogenic granuloma can be excised with a cutting loop electrode, providing a specimen for the histopathologist while coagulating the base (1).

Radiowave procedures should not be performed in patients with cardiac pacemakers even though fixed-rate pacemakers, in contrast to demand pacemakers, are resistant to external electromagnetic interference such as that caused by electrosurgery.

The presence of any flammable or explosive liquids or gases must be avoided.

Under certain conditions, the patient and the therapist may experience minor electric shocks during the course of electrosurgery. These mishaps can be partially prevented if the patient's skin does not contact metal and the therapist wears surgical gloves during the procedure.

R. Baran: Nail Disease Centre, 42, Rue des Serbes, 06400 Cannes, France.
E. Haneke: Klinikk Bunæs, Løkkeåsveien, 1337 Sandvika, Oslo, Norway.

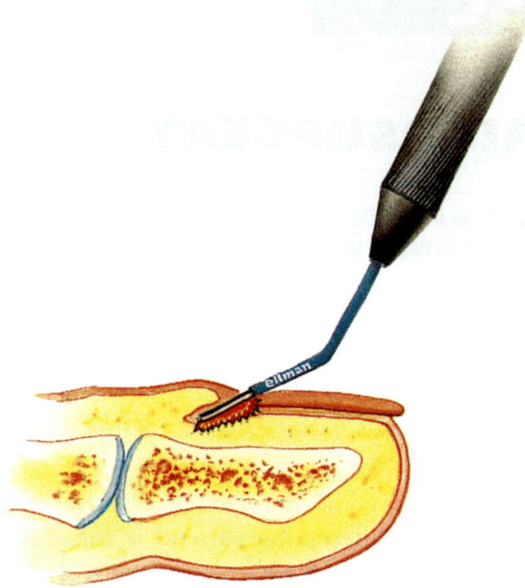

FIGURE 5C.1. Insulated matrix electrode. (Courtesy of W.S. Gurski, DPM.)

REFERENCES

1. Pollack SV. *Electrosurgery of the skin.* New York: Churchill Livingstone, 1991.
2. Hettinger DF, Valinsky MS, Nucci G, et al. Nail matricectomies using radio wave technique. *J Am Ped Med Assoc* 1991;81:317–321.
3. Kendall MV. Radiosurgery—an advanced technique for performing nail matrixectomies. *Podiatry Management.* May 1998;53.

5D

PINCER NAILS

INCURVATED NAIL, UNGUIS CONSTRINGENS, TRANSVERSE OVERCURVATURE, TRUMPET NAIL, CONVOLUTED NAIL, AND OMEGA NAIL

ECKART HANEKE

DEFINITION

Transverse overcurvature pinching the nail bed usually is called pincer nail or unguis constringens. The curvature commonly increases from proximal to distal, giving it a trumpetlike appearance.

Transverse overcurvature of nails occurs frequently on toes, but rarely on fingers.

PATHOPHYSIOLOGY

There are at least four different variants of pincer nails:

1. Hereditary,
2. Acquired due to foot deformation,
3. Acquired due to chronic dermatoses, and
4. Acquired due to degenerative osteoarthritis of distal interphalangeal finger joints.

Hereditary overcurvature almost always is symmetric. Similar nail changes may be seen in other family members. The great toes usually are affected, but the lesser toes also may be involved. The great toe commonly shows a lateral deviation of the long axis of the distal phalanx, but the overcurved nails are even more deviated laterally. When the lesser toes are involved, they exhibit a medial deviation. This anomaly is already seen in adolescents and young adults.

Acquired pincer nails may be due to a number of different dermatoses, of which psoriasis is the most frequent, to repeated trauma and nail avulsions, subungual exostoses, and cysts (1), as well as other processes that cause shrinkage of the nail bed. The most frequent cause of acquired pincer nails is foot deformation with deviation of the phalanges, probably subsequent to wearing ill-fitting shoes.

Pincer nails of the fingers are commonly seen in degenerative osteoarthritis of the distal interphalangeal joints.

Acquired pincer nails are not symmetric, although fingernail involvement may be extensive and appear to be fairly symmetric.

The overcurvature most probably is due to an enlarged base of the distal phalanx to which the matrix is firmly bound by ligamentlike collagen fibers. Because toenails are

E. Haneke: Klinikk Bunæs, Løkkeåsveien, 1337 Sandvika, Oslo, Norway.

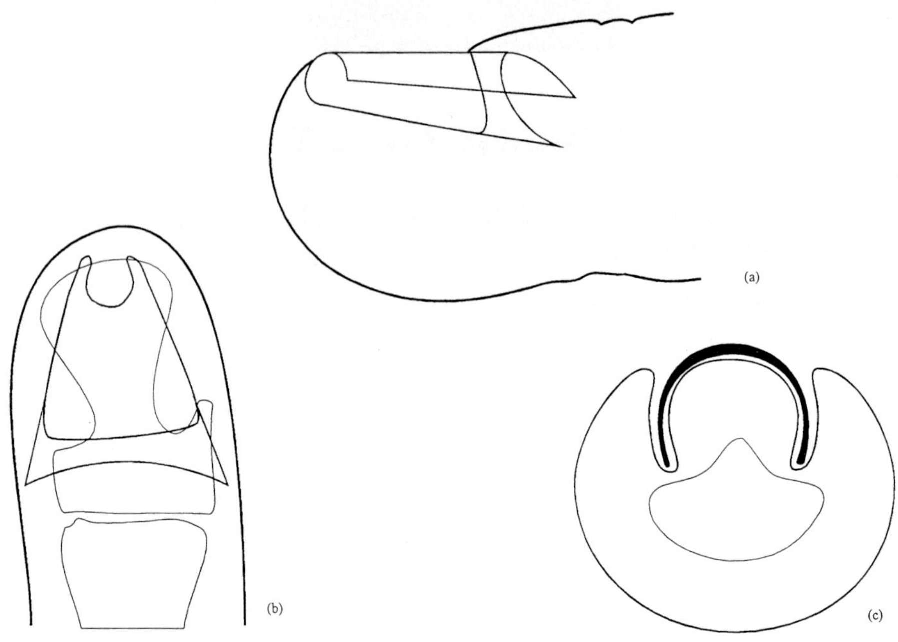

FIGURE 5D.1. A: Lateral view of big toe with overcurved nail. **B:** Enlargement of the base up the distal phalanx. **C:** Transverse section of the distal phalanx.

markedly curved, the curvature of the proximal (matricial) nail plate portion will decrease and consequently the curvature will increase distally (Fig. 5D.1A).

Radiographs of big toes with pincer nails invariably show enlargement of the base of the distal phalanx (Fig. 5D.1B) and often hooklike lateral osteophytes pointing distally. This almost always is more pronounced on the medial aspect, explaining why the nail's longitudinal axis is deviated more laterally than that of the distal phalanx. The distally more pronounced overcurvature exerts traction on the nail bed, which is transduced to the bone by the ligamentlike fibers fixing the solehorn to the tip of the terminal phalanx (Fig. 5D.1C). This eventually results in a traction osteophyte that often is seen on radiographs (2,3).

SIGNS AND SYMPTOMS

The toenails present a physiologic transverse convexity that is more pronounced than that of fingernails. When this transverse curvature becomes exaggerated, whether due to trauma, foot deformation, inflammation, or a hereditary abnormality of the distal phalanx (4), the lateral nail plate edges dig into the lateral grooves and, on further incurving, start to pinch the nail bed.

Morphologically, there are three clinical types:

1. "Common" pincer nail,
2. Tile-shaped nail, and
3. Plicatured nail.

The most frequent type is the trumpet nail deformity, which has an overcurvature increasing along the axis from proximal to distal (4). The lateral plate margins virtually roll in, sometimes forming a tube. The nail bed becomes pinched, shrinks in its transverse diameter, and is lifted up distally; by the latter, continuous traction is exerted to the distal dorsal tuft. The lateral plate margins eventually may break through the epidermis and cause granulation tissue mimicking ingrown toenail. Cutting the nail may become more difficult and painful with increasing overcurvature. This frequently is associated with thickening of the nail plate. To enhance the cosmetic appearance, patients and podiatrists tend to round the distal margin of the overcurved nail plate. This reduces pressure on the most distal portion of the lateral nail grooves; however, there is a great risk of leaving a nail spike that inevitably will pierce into the soft tissue.

Pachyonychia congenita may mimic pincer nails, but it usually is not painful and involves both fingernails and toenails.

Pain is not a consistent symptom of pincer nails. Some extreme cases are completely painless, whereas some mild cases may cause excruciating pain (5).

Tile-shaped nails are characterized by an even transverse overcurvature with the lateral nail edges remaining parallel. This type usually is less severe, does not cause serious symptoms, and frequently is seen in young tall persons with the "unguis incarnatus syndrome" (6) or in fingernail overcurvature.

The plicatured nail presents a moderate convexity with one or both lateral plate edges being sharply bent to form a vertical sheet pressing into the lateral nail groove. Bilateral symmetric involvement of the nails is mainly seen on fingers, but unilateral angling of a nail is common in foot deformities (5).

INDICATIONS FOR TREATMENT

Major indication for treatment are pain and inflammation. Other indications are interference with wearing shoes and cosmetic embarrassment.

Therapeutic approaches vary according to the severity and type of overcurvature, possible risk factors, previous unsuccessful treatment, and a multitude of personal and psychological preferences of both the treating physician and patient.

Conservative Treatment

Hardly any patients consulting a physician have not tried some conservative method. Such patients usually try to clip down the lateral edge of the incurved nail as far proximally as possible, and they only stop when they cut into the skin of the lateral nail groove. Because the nail frequently is thick and hard, it should be softened either using an emollient for some days or in a 10-minute hot foot bath before nail clipping.

In early overcurvature, thinning of the central portion of the nail plate from the lunula to the free margin may alleviate the pain because this technique increases the nail plate's pliability. A single groove or a series of grooves may be cut into the nail plate surface using a burr, or the entire nail plate may be thinned by grinding (7). More recently, the use of 40% urea paste and subsequent removal of the softened nail material, performed regularly over a 1-year period, was shown to normalize hereditary pincer nails in a 38-year-old woman (8). There are several methods to mechanically correct malformed nails, called orthonyx (9–11). The principle is to exert tension on the transverse nail curvature to gradually flatten the plate. After cleaning the lateral nail grooves, a stainless steel brace is inserted on the nail plate and fixed under the lateral edges. An adjustment is made to the brace to exert countertension to the plate. With a series of adjustments, the nail plate will gradually flatten in a period of 6 months or more (10). Baran (unpublished data) successfully treated fingernail overcurvature associated with osteoarthritis (7); however, two of our own cases showed an immediate relapse after orthonyx treatment (E. Haneke, unpublished data).

Most recently, elastic plastic braces glued to the abraded nail surface were used successfully by Effendy et al. (12).

Conservative treatment always appears to be tempting for the patients. However, no publication recommending clipping, grooving, thinning, and orthonyx with steel or plastic braces has mentioned the bone alterations seen on x-ray films in nearly all patients.

The patients we saw all experienced recurrences within a time period usually only about half of that needed to correct the pincer nail. Therefore, we believe that conservative treatment of pincer nails provides only temporary relief.

Surgical Treatment

Repeated nail avulsions were thought to permit the pinched nail bed tissue to spontaneously flatten during the period of nail plate regrowth (13). However, many patients experienced dramatic worsening of their condition after previous nail avulsion, and there is rarely any benefit from this procedure. Furthermore, nail avulsion is known to increase even the physiologic transverse curvature of the normal hallux nails (13–15).

As still often advocated for ingrown toenails, wedge excisions also have been recommended for treatment of pincer nails (7). This is an operation that essentially must remove the entire lateral matrix portions that extend nearly around the dorsal half of the base of the terminal phalanx. Because it is not the nail walls, nail folds, and nail bed that are at fault, this operation, which considerably narrows the whole nail apparatus, is inadequate.

Soft tissue resection also has been recommended (16,17). After avulsion of the nail plate, the tissue around the nail was removed down to the bone, the nail bed dissected free from the phalanx, and the distal tuft rongeured off. The 4-0 sutures used for wound closure were removed after 10 to 14 days. Because of the bone surgery, the procedure took several weeks to heal. The regrowing nail plate should be treated with nail braces (7).

Another technique is aimed at correcting the firm swelling of the distal nail bed. The nail bed is cut by a median longitudinal incision and the soft tissues are dissected from the terminal phalanx. Reversed tie-over sutures are put in the lateral nail folds and tied over a pad on the plantar aspect of the toe to spread the nail bed. The resulting triangular defect is covered with a free skin graft. The stitches are removed after about 3 weeks. An onycholytic area will develop that corresponds to the size of the free graft (18). This technique was slightly modified by primarily suturing the nail bed and inserting a small plastic plate between the curved nail plate and the flattened resutured nail bed (2); however, the result was not permanent.

None of these surgical techniques takes into account the underlying bone alterations that cause the overcurvature. Except for permanent nail eradication by surgical nail ablation or phenolization (5), relapses are frequent.

As judged from the supposed etiopathogenesis of pincer nails, it was believed that either the lateral osteophytes or the matrix horns that are pushed outward and forward by these bony excrescences would have to be removed. Because removal of the lateral osteophytes most probably would result in damage to the lateral ligaments of the distal interphalangeal joint, permanent removal of the lateral matrix horns was considered to be the simplest, least painful, but nevertheless sufficiently effective treatment modality. This must be supplemented by spreading the pinched nail bed and removal of the distal dorsal tuft if this is abnormally prominent (2,3,19,20).

Under regional block anesthesia, lateral nail strips of the entire nail plate are avulsed (Fig. 5D.2). The digit is exsanguinated and the matrix horns are dried and either carefully dissected and removed or phenolized by vigorously rubbing in liquefied 90% phenol for 3 minutes. Small antibiotic tablets are inserted into the wound cavities. A median incision is carried down to the bone in the nail bed from the lunula border to 2 mm beyond the hyponychium. During this incision, the traction osteophyte is felt with the scalpel even when it is not obvious on the radiograph. The pinched nail bed is dissected from the terminal phalanx, the distal dorsal tuft with the osteophyte is rongeured off, and the nail bed is expanded and sutured using 6-0 monofil absorbable sutures (PDS II). Reverse tie-over sutures are put in the lateral nail folds, with small rubber tubes used as a cushion to prevent the sutures from cutting through the nail folds. These sutures keep the nail bed stretched out over the bone and are removed only after about 3 weeks. In more than 50 patients, a success rate of over 80% was achieved with this technique (see Zook's technique for treating pincer nail in Chapter 4B).

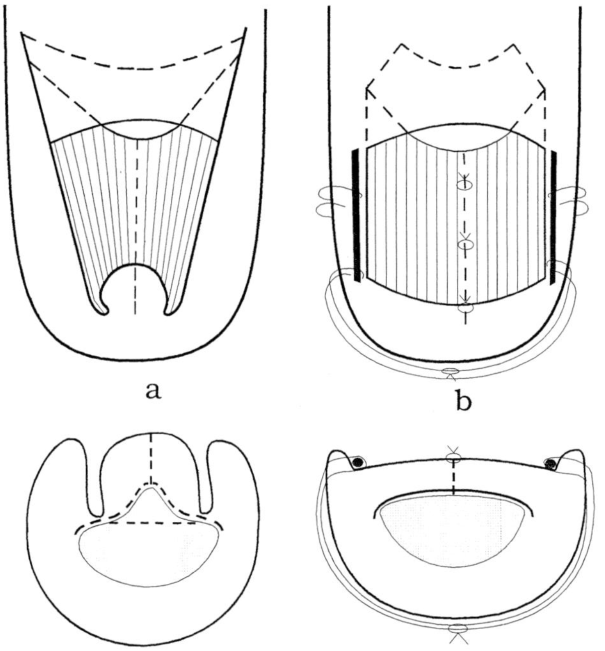

FIGURE 5D.2. Pincer nail treatment.

REFERENCES

1. Baran R, Broutard JC. Epidermoid cyst of the thumb presenting as pincer nail. *J Am Acad Dermatol* 1988;19:143–144.
2. Haneke E. Segmentale Matrixverschmälerung zur Behandlung des eingewachsenen Zehennagels. *Dtsch Med Wochenschr* 1984;109:1451–1453.
3. Haneke E. Etiopathogénie et traitement de l'hypercourbure transversale de l'ongle du gros orteil. *J Med Esth Chir Dermatol* 1992;19:123–127.
4. Chapman RS. Overcurvature of the nails—an inherited disorder. *Br J Dermatol* 1973;89:317.
5. Baran R. Pincer and trumpet nails. *Arch Dermatol* 1974;110:639–640
6. Steigleder GK, Stober-Münster J. Das Syndrom des eingewachsenen Nagels. *Z Hautkr* 1977;52: 1225.
7. Bureau H, Baran R, Haneke E. Nail surgery and traumatic abnormalities. In: Baran R, Dawber RPR, eds. *Diseases of the nails and their management.* Oxford: Blackwell Science, 1984:347–402.
8. El-Gammal S, Altmeyer P. Erfolgreiche konservative Therapie des Pincer-Nail-Syndroms. *Hautarzt* 1993;44:535–537.
9. Fraser AR. Orthonyx: theory and practice. *Br J Chirop* 1967;32:229–232.
10. Farnsworth FC. A treatment for convoluted nails. *J Am Podiatr Assoc* 1972;62:110.
11. Dagnall JC. The development of nail treatments. *Br J Chirop* 1976;41:165–168.
12. Effendy I, Ossowski B, Happle R. Zangennagel. Konservative Korrektur durch Aufkleben einer Kunststoffspange. *Hautarzt* 1993;44:800–802.
13. Zaias N. *The nail in health and disease.* Jamaica, NY: Spectrum Publications, 1980:171.
14. Haneke E, Baran R, Bureau H. Chirurgie der Nagelregion. *Z Hautkr* 1982;57:1107–1116.
15. Runne U. Operative Eingriffe am Nagelorgan. *Z Hautkr* 1983;58:324–332.
16. Fosnaugh RF. Surgery of the nail. In: Epstein E, Epstein E Jr., eds. *Skin surgery,* 5th ed. Springfield, IL: Charles C Thomas, 1977:981.
17. Douglass MC, Krull EE. Diseases of the nail. In: Rakel R, ed: *Conn's current therapy.* Philadelphia: WB Saunders, 1981:712–718.
18. Suzuki K, Yangi I, Kondo M. Surgical treatment of pincer nail syndrome. *Plast Reconstr Surg* 1979;63:570–573.
19. Baran R, Haneke E. Nail surgery. In: Epstein E, Epstein E Jr., eds. *Skin surgery,* 6th ed. Phildadelphia: WB Saunders, 1987:534–547.
20. Haneke E, Baran R. Nails: surgical aspects. In: Parish LC, Lask GP, eds. *Aesthetic dermatology.* New York: McGraw-Hill, 1991:236–247.

5E

CONSERVATIVE MANAGEMENT OF NAIL PROBLEMS

EDWARD A. KRULL

Circumstances may prevail that preclude nail surgery. Among the most commonly encountered problems are inadequate circulation and diabetes. Dermatologic surgeons must develop a discipline in examining the digits, especially the feet, for adequacy of circulation. Presence or absence of hair on the digits, quality of skin, adequacy of pulses, rapidity of refill of the pad of the digits after blanching pressure, pallor with elevation rubor with dependency, and a history of intermittent claudication are all part of that disciplined approach. If there is a question of the circulation being adequate for surgery, whether it is scalpel, electrosurgery, laser surgery, or chemodestruction, consultation with vascular surgeons or other appropriate physicians is indicated. Doppler assessment of peripheral circulation may be a necessity (1).

Sometimes the patient does not want to undergo surgery. Their age and immobility may influence the patient, caregivers, or relatives to forego surgery. Whatever the reasons, conservative management of nails may be the only option available to the physician (Figs. 5E.1 and 5E.2).

The key instruments in the conservative management of nail problems are a nail cutter, a foot or hand basin for soaking the digits to soften the nail plate, a nail file, a nail grinder, and packing material such as cotton or plastic tubing (2–9). The English nail splitter and the double-action nail clipper are especially useful nail cutters (Figs. 5E.3 and 5E.4) (2). A pear-shaped Carborundum burr that comes to a point facilitates the grinding down of the bulk of the nail plate and yet allows finely controlled nail grinding, especially along the edges of the nail (7,9). The burr can be driven by a number of power sources. Expensive equipment, such as the instruments used for dermabrasion, or inexpensive Dremel hand motors, can be used (Figs. 5E.5–5E.8). Masks and dust evacuators or extractors should be used to prevent inhalation of the powder from the grinding (9). The particle size of much of the nail dust allows inhalation into the alveoli. Diffuse infiltrative lung diseases and immunologic irregularities have been reported to affect significant numbers of podiatrists as a result of inhaling nail dust (10,11). Gloves may protect against skin infection from dermatophytes in the nail.

Nail plates usually are cut straight across and the distal free ends filed smooth so as not to catch on socks or stockings (9). Soaking the nails in water for 10 to 15 minutes facilitates clipping and filing. Many older patients cannot adequately see or take care of their nails. Patients with significant diabetes or circulatory disease are at risk from complications of improper nail care. Physicians, podiatrists, and other trained caregivers should be selected appropriately for the complexity of the patient's problems.

E. A. Krull: Department of Dermatology, Henry Ford Hospital, Detroit, Michigan 48202.

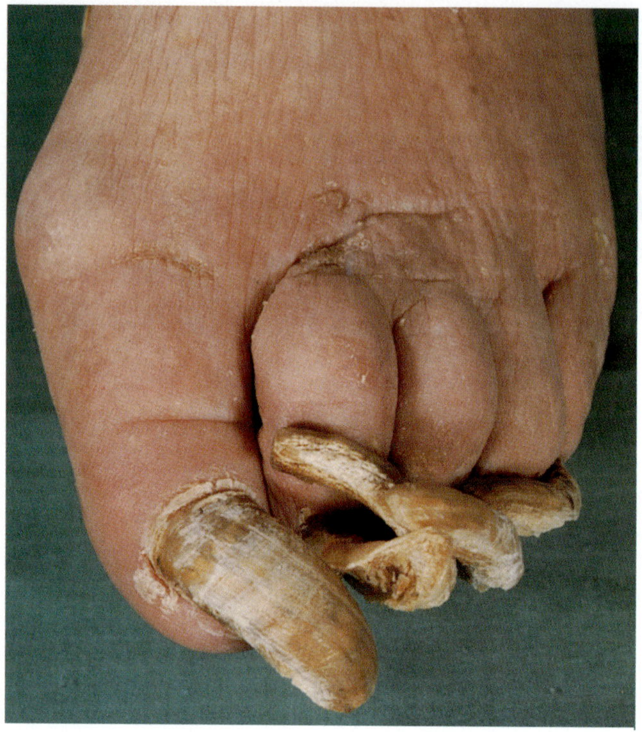

FIGURE 5E.1. Marked onychogryphosis. Conservative management with grinding and nail cutting was effective.

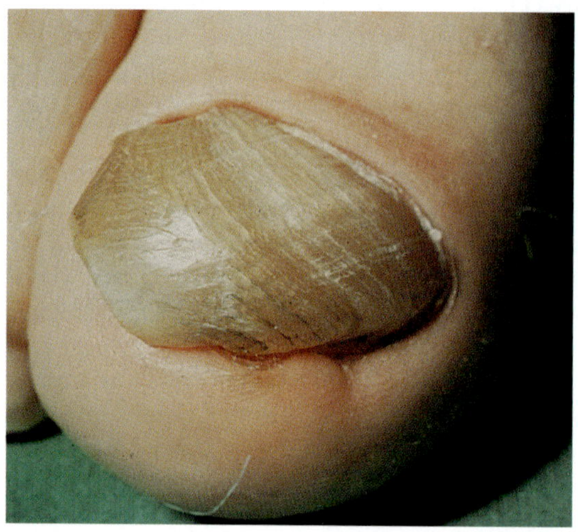

FIGURE 5E.2. Onychauxis causing pain when wearing shoes. Patient preferred grinding and nail cutting rather than surgery. Because of the thickness of the nail plate, the nail was slow growing in terms of length, but probably not in terms of volume (i.e., length × thickness).

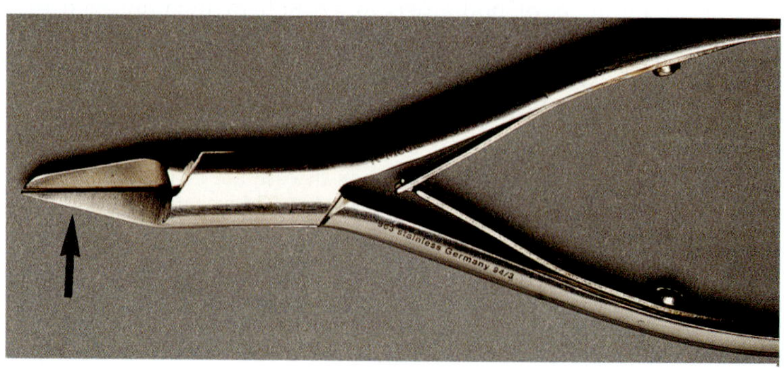

FIGURE 5E.3. English nail splitter. Anvil-like lower jaw *(arrow)* slides easily over nail bed and matrix and under nail plate, which facilitates nail cutting.

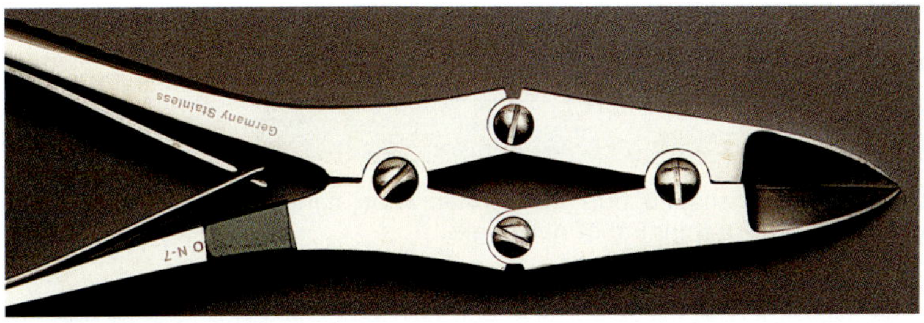

FIGURE 5E.4. Double-action nail clipper works well for onycholytic as well as thick nail trimming.

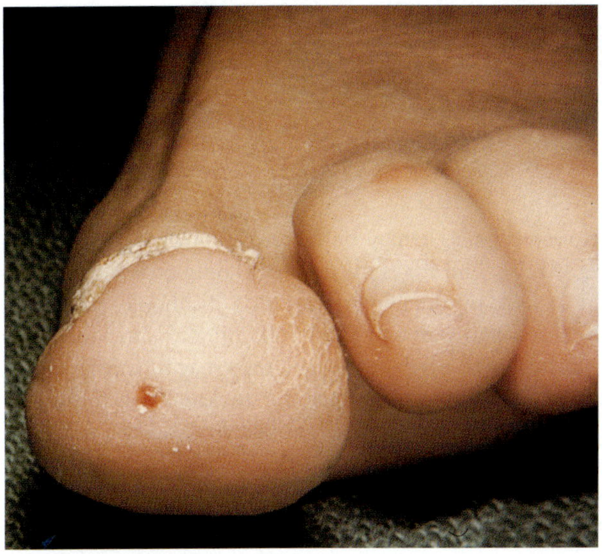

FIGURE 5E.5. Mildly uncomfortable thickened nail plate before grinding.

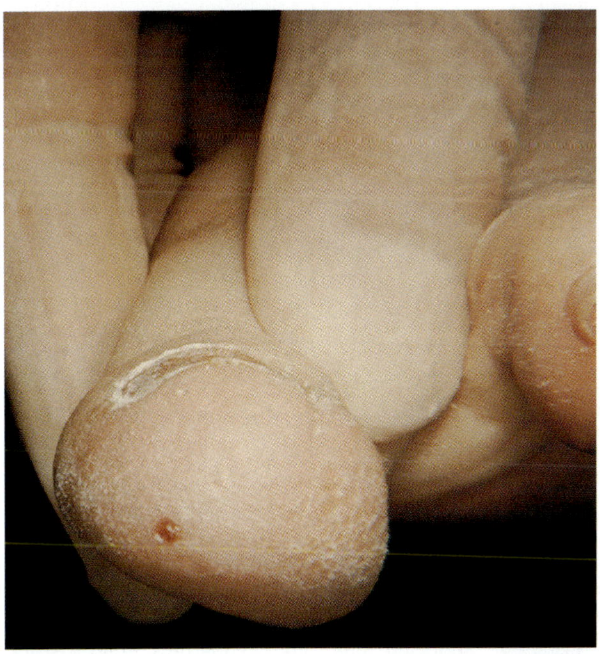

FIGURE 5E.6. Same nail as shown in Fig. 5E demonstrating marked thinning of the nail plate from grinding. Patient's symptoms disappeared.

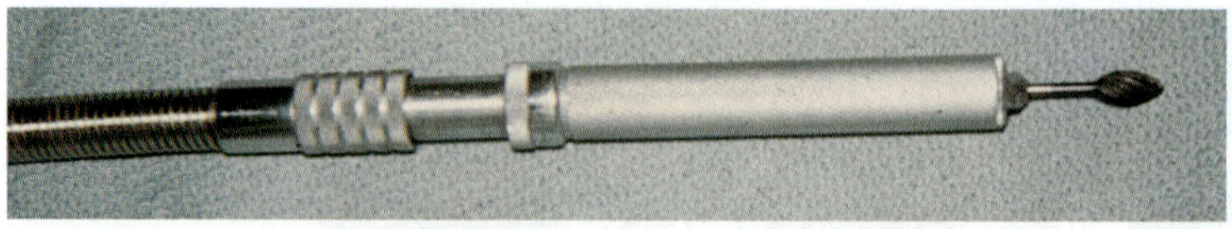

FIGURE 5E.7. Carborundum pear-shaped burr on dermabrasion handle. The shape of the burr allows careful thinning of nail edges.

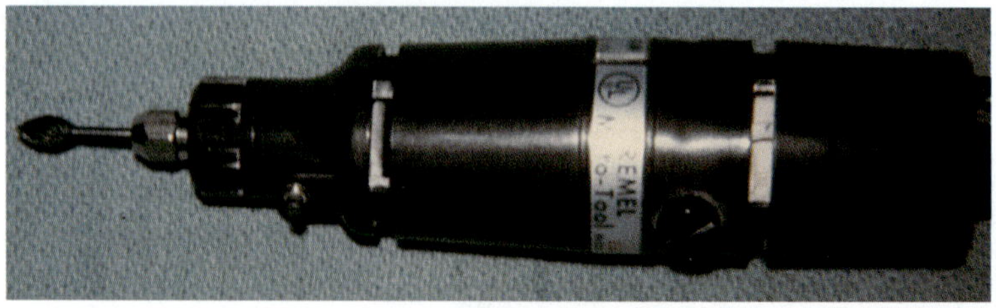

FIGURE 5E.8. Same burr used in inexpensive Dremel hand motor.

For pincer nails or other incurvated nails, it may be better to cut deeply along the edge instead of straight across (Fig. 5E.9). The edges must be cut completely through so as not to leave a pointed remnant of nail, producing the so-called "fishhook" nail. This splinter may become buried in the distal nail wall and cause inflammation and infection (Fig. 5E.10). To facilitate cutting of the lateral edges of the nail, they may be initially carefully ground down with the point of the Carborundum burr to a thickness that makes the nail flexible and easily cut (9). Sometimes an ointment such as salicylic acid 3% to 5% in petrolatum applied twice daily to the distal nail groove will help loosen the edge of the nail plate from the underlying tissues, making it easier to clip more deeply along the edges of the nail. Despite these measures, there may be more pain than the patient can tolerate. If the clinical situation permits, a very small amount of anesthesia of plain lidocaine, which is a vasodilator, is injected into the side. Usually 0.1 to 0.2 mL is an adequate amount to produce comfort for the trimming procedure.

Thickened nails are seen frequently with aging. Pincer nails may be thickened and markedly incurvated. Thinning the nail plate by grinding may make the patient more comfortable. With pincer nails, grinding may lessen the pressure from the shoe, decrease

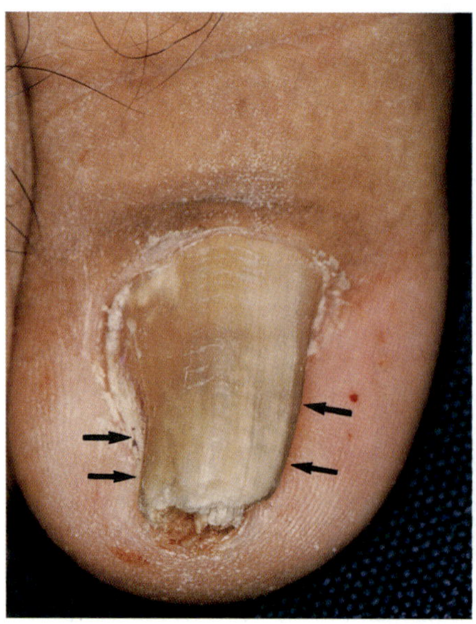

FIGURE 5E.9. Pincer nail. Cutting deeply down the sides of the nail may be more effective in reducing pain and tenderness than cutting straight across. *Arrows* point to edge of nail plate that was clipped.

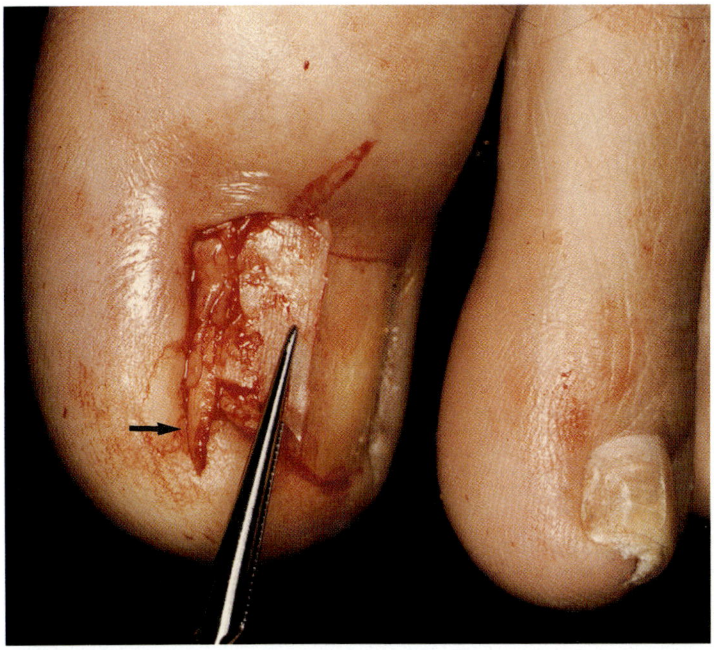

FIGURE 5E.10. Fishhook nail *(arrow)* resulting from failure to cut across the entire distal nail edge.

the tenderness originating from the incurvated edges, and reduce the likelihood of an ingrown toenail (7,9). Urea paste 40% applied under occlusion for several days allows only abnormal nail to be atraumatically removed from the surrounding nail bed, matrix, and nail folds. It is of limited value in normal nail plate integrity, even though these nails may be thick or incurvated (12,13).

Uncommonly the pressure of the shoe on a thickened nail may be enough to critically affect the precarious circulation to the nail tissue and initiate gangrene (Fig. 5E.11).

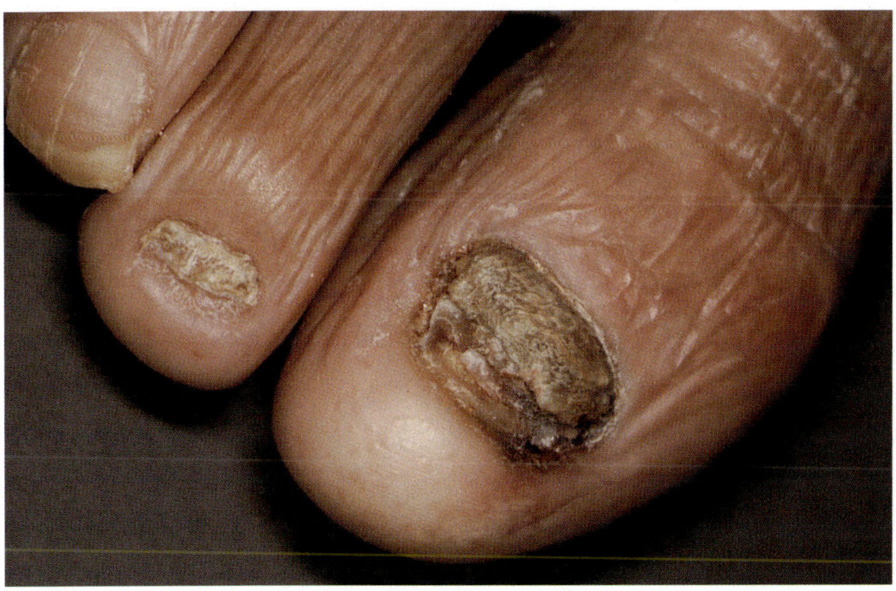

FIGURE 5E.11. Initial site of gangrene formation was beneath thickened nail in this patient with severe vascular insufficiency. Bacterial and fungal cultures of serous-colored oozing and residual nail repeatedly produced no growth. Note crinkly, smooth skin without hair reflecting arterial diseases. Necrosis occurred under the nail plate.

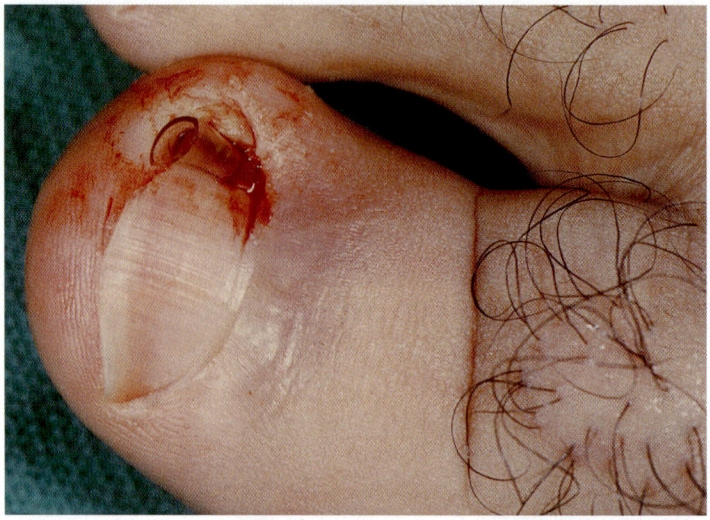

FIGURE 5E.12. Plastic intravenous tubing cut lengthwise to separate nail edge from nail groove in ingrown toenail.

Cotton, wool-cotton, and cotton coated with flexible collodion may be inserted between the distal nail edges and nail groove to decrease pain and injury of the nail tissue of lateral nail grooves from the edges of the nail (3–5). If the nail is ingrown, splinting the sides of the nail with plastic tubing that has been cut lengthwise may be less traumatic to the tissue than other surgery (Figs. 5E.12 and 5E.13); however, a small amount of local anesthesia is needed (6). Taping the lateral nail folds to retract from the edge of the nail plates also has been helpful (14).

Conservative management of nail care may be essential to patients suffering from vascular disease and diabetes. In general, careful handling of the nails and the surrounding tissue may prevent the initiation and development of serious problems that lead to chronic ulcerations, infection, and amputation. Sometimes foreseeing and predicting future problems may result in more aggressive curative procedures performed earlier in

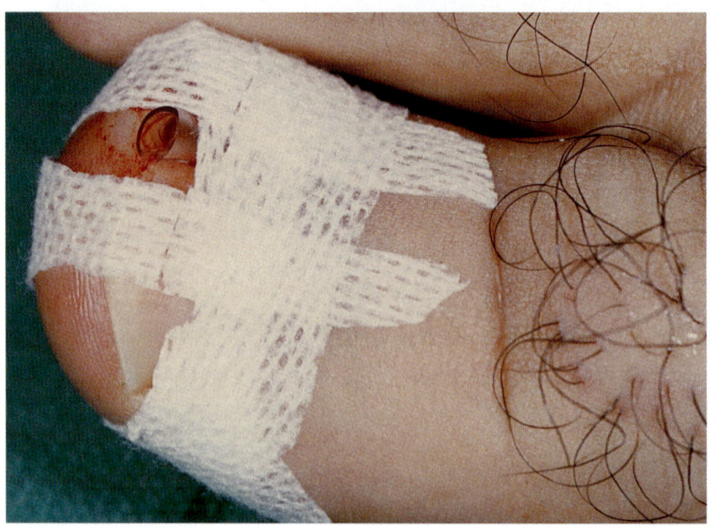

FIGURE 5E.13. Plastic tube taped in place. There was almost immediate relief of pain and tenderness. After about 4 weeks, the plastic tube splint may be permanently removed. Daily washing with povidone iodine or other sterilizing material is used.

life that would not be tolerated later when the circulation and other conditions become progressively worse.

REFERENCES

1. Ceilley RI, Collison DW. Matricectomy. *J Dermatol Surg Oncol* 1992;18:728–734.
2. Clark RE, Madani S, Bettencourt MS. Nail surgery. *Dermatol Clin* 1998;16:145–164.
3. Connolly B, Fitzgerald RJ. Pledget in ingrowing toenails. *Arch Dis Child* 1988;63:71–72.
4. Ileld FW. Ingrown toenail treated with cotton collodion insert. *Foot Ankle* 1991;11:312–313.
5. Murray WR. Onychocryptosis principles of non-operative and operative care. *Clin Orthop Rel Res* 1978;142:96–102.
6. Schulte KW, Neumann NJ, Ruzicka T. Surgical pearl: nail splinting by flexible tube—a new non-invasive treatment for ingrown toenail. *J Am Acad Dermatol* 1998;39:629–630.
7. Maeda N, Mizuno N, Ichikawa K. Nail abrasion: a new treatment for ingrown toenails. *J Dermatol* 1990;17:746–749.
8. Haneke E, Baran R. Nail surgery and traumatic abnormalities. In: Baran R, Dawber RPR, eds. *Disease of the nails and their management,* 2nd ed. London: Blackwell Scientific, 1994:345–364.
9. Master G, Kelly PF, Walinsky MD. Burring technique in modern podiatric care. *J Am Podiatr Med Assoc* 1990;80:649–652.
10. Abramson C, Wilton J. Inhalation of nail dust from onychomycotic toenails. Part I. Characterization of particles. *J Am Podiatr Med Assoc* 1985;75:563.
11. Abramson C, Wilton J. Inhalation of nail dust from onychomycotic toenails. Part II. Clinical and serological aspects. *J Am Podiatr Med Assoc* 1985;75:631.
12. Kechijian P. Nail biopsy vignettes. *Cutis* 1987;40:331–335.
13. South DA, Farber E. Urea ointment in nonsurgical avulsion of nail dystrophies: reappraisal. *Cutis* 1980;21:609–620.
14. Manca D. Taping toes—effective treatment for ingrown toenails. *Can Fam Physician* 1998;44:275.

CHAPTER

6

INFECTION OF THE NAIL APPARATUS

6A

FUNGAL NAIL INFECTION

ROBERT BARAN
ECKART HANEKE

Onychomycosis is a common condition, accounting for about 30% of all nail diseases. A fungus gains entry into a nail by one of four main routes, each resulting in a different clinical pattern of infection (1–4). The four entry routes are as follows.

1. *Via the lateral nail groove and distal area* (Fig. 6A.1), leading to distal and lateral subungual onychomycosis (DLSO). This is the most common form of fungal nail infection. It may be associated with three major clinical features whose contribution may vary with individual cases: subungual hyperkeratosis, onycholysis, and paronychia.

Causative organisms are

Dermatophytes: *Trichophyton* sp; *Epidermophyton floccosum*, rarely
Nondermatophytic molds: *Scopulariopsis brevicaulis*; *Scytalidium* sp producing onycholysis and paronychia
Yeasts: *Candida* sp.

2. *Via the dorsal surface of the nail plate* (Fig. 6A.2), producing superficial onychomycosis, usually white (SWO). These white patches (pseudoleukonychia) are easily removed by scraping. This type of infection is seen almost exclusively on toenails. In rare cases, superficial black onychomycosis (SBO) can be observed.

Causative organisms are

Dermatophytes: *Trichophyton mentagrophytes* var. *interdigitale* (in more than 90% of the cases) *Trichophyton rubrum* is found rarely.
SWO nondermatophytic molds: *Acremonium* sp, *Aspergillus, Fusarium*
SBO nondermatophytic molds: *Scytalidium dimidiatum, Trichophyton rubrum nigricans.*

3. *Via the undersurface of the proximal nail fold* (Fig. 6A.3A–B). In proximal white subungual onychomycosis (PWSO) due to *Trichophyton rubrum,* the proximal nail fold remains normal but becomes inflamed in chronic paronychia (*Candida* sp) (5) and with some nondermatophytes molds, such as *Fusarium, Scopulariopsis,* and *Aspergillus.* PWSO is most common after trauma or in individuals infected with the human immunodeficiency virus, in whom it may be considered a marker of this virus-induced condition. *Candida paronychia* usually is secondary to chronic paronychia from protein-contact dermatitis favored by "wet work" (6).
4. *Endonyx onychomycosis.* This form of infection involves invasion of the superficial surface as well as deeper penetration of the nail plate. It is caused notably by *Trichophyton. soudanense.*

R. Baran: Nail Disease Centre, 42, Rue des Serbes, 06400 Cannes, France.
E. Haneke: Klinikk Bunæs, Løkkeåsveien, 1337 Sandvika, Oslo, Norway.

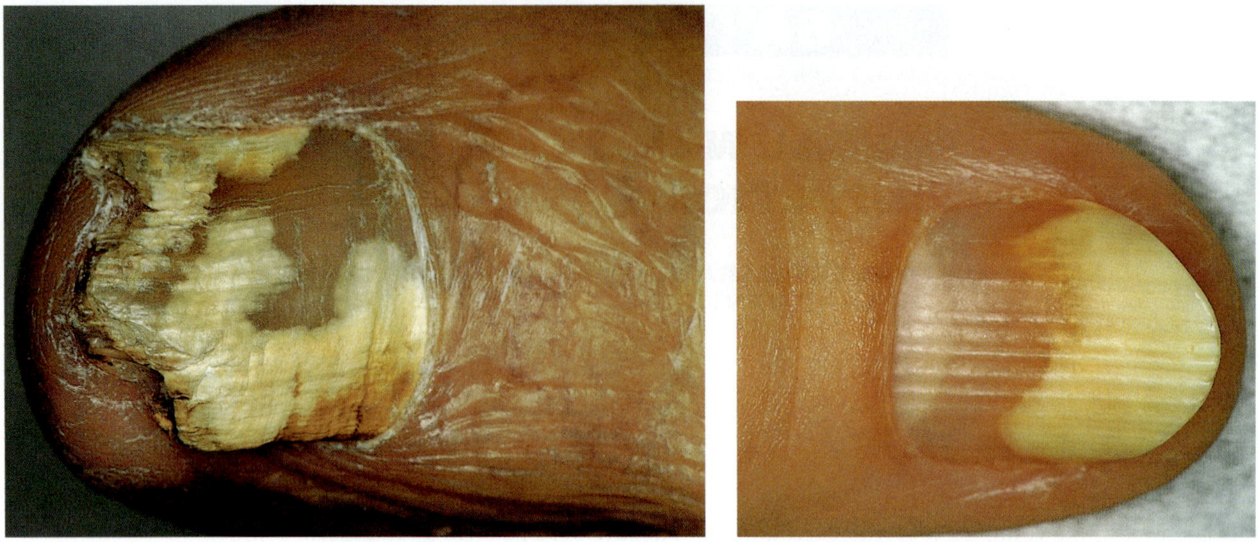

FIGURE 6A.1. A: Distal and lateral subungual onychomycosis (DLSO). **B:** Candida onycholysis (DLSO).

Total dystrophic onychomycosis may result from progression of any of these forms of infection or appear primarily in patients with chronic mucocutaneous candidiasis.

The clinical patterns of change in fungal nail disease give only a clue to the type of infection. Definitive diagnosis of fungal infection cannot be made on clinical grounds alone, because the clinical appearance of infection caused by one species of fungus usually is indistinguishable from that caused by another. However, there are some exceptions. For example, PWSO usually is due to *Trichophyton rubrum*; the same agent is responsible for the one-hand two-foot tinea syndrome in which 15 nails are normally infected.

If no local antifungal treatment has been applied recently, the diagnosis of onychomycosis must be confirmed by laboratory tests using the proper technique. In DLSO, samples should include clippings taken from the border of the affected nail segment and the superficial subungual tissue of the healthy nail portion. The ventral aspect of the nail should be vigorously scraped with a no. 15 scalpel blade for microscopic examination and culture.

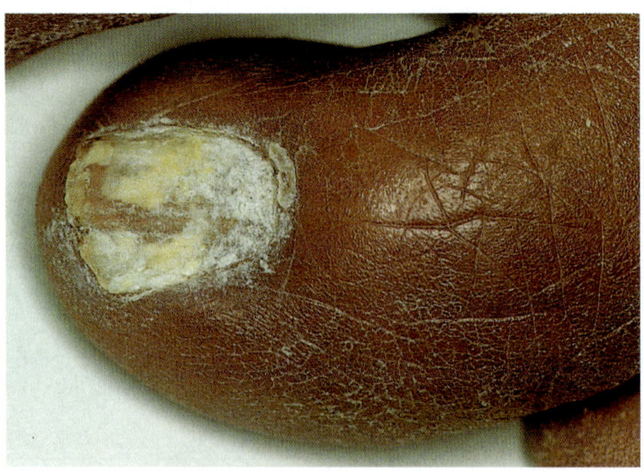

FIGURE 6A.2. Superficial white onychomycosis (SWO).

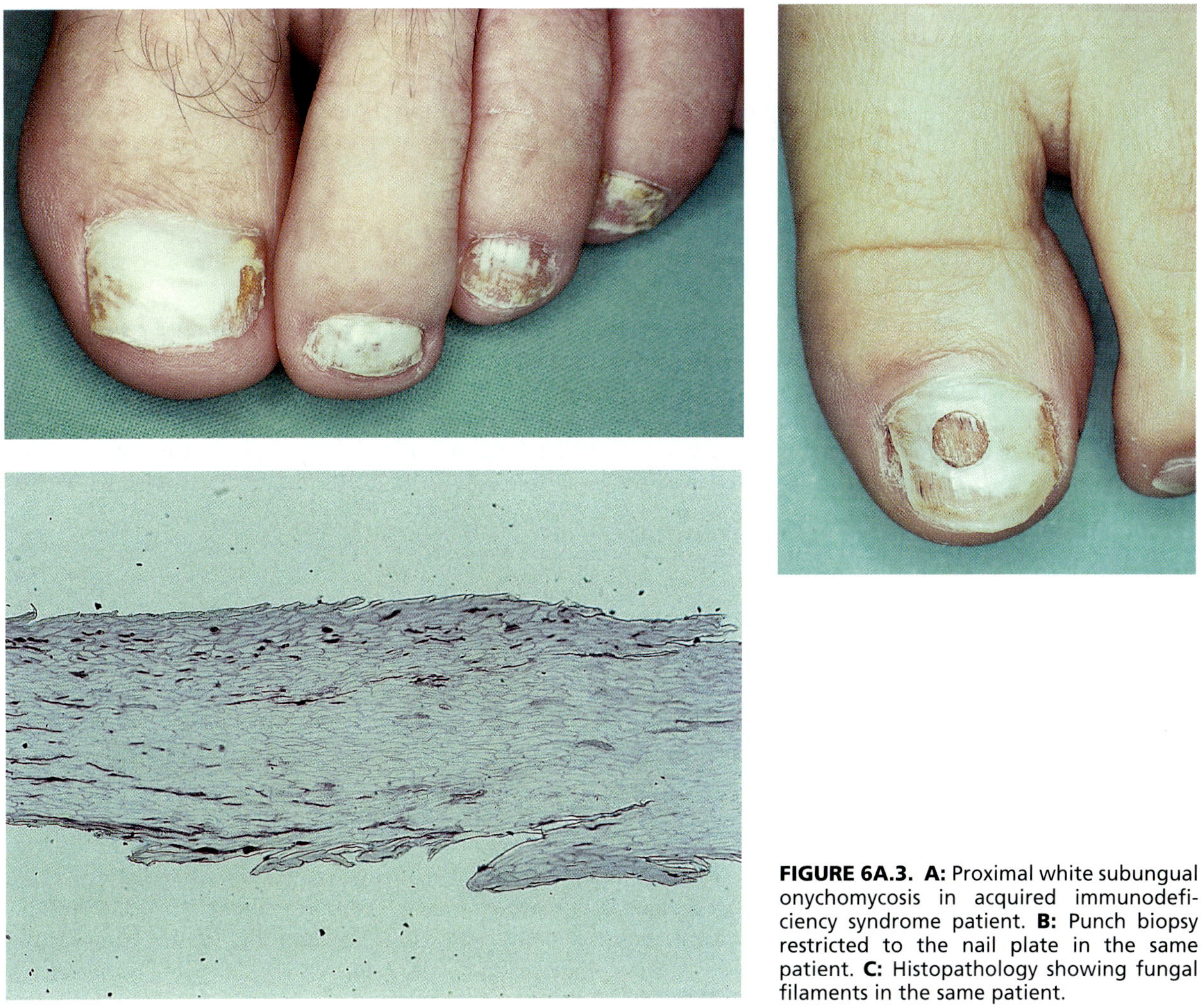

A

B

C

FIGURE 6A.3. A: Proximal white subungual onychomycosis in acquired immunodeficiency syndrome patient. **B:** Punch biopsy restricted to the nail plate in the same patient. **C:** Histopathology showing fungal filaments in the same patient.

In SWO, shallow shaving or scraping will permit microscopy and culture. In chronic paronychia, samples can be taken from beneath the proximal nail fold by passing a disposable loop underneath this affected area. Such samples may be used to prepare wet mounts or can be rubbed directly onto a culture plate.

MICROSCOPIC EXAMINATION

Direct microscopy can differentiate yeast cells from dermatophyte hyphae and other molds. Species identification usually cannot be made from wet mounts, except for *Scopulariopsis brevicaulis,* a common mold that may be suspected when thick-walled and lemon-shaped spores are seen, and *Scytalidium dimidiatum,* which can be identified by its dark coloration. Because different causative organisms may require different therapies, reliable identification of the causal agent is mandatory and can only be made by culture.

CULTURE

Cultures should be grown in a mycology laboratory and specimens sent in folded black paper. The report may take 2 to 5 days for yeasts, which will be subsequently identified by biochemical tests, and may take as long as 1 month for dermatophytes.

The drawbacks in culturing a sample suspected for onychomycosis are twofold:

1. Identification by culture of a particular fungus is not always proof of its etiologic role. It is only presumptive evidence of fungal infection. Some criteria are necessary to determine if a fungus is truly responsible for the dystrophic nail. These criteria should include

   - The absence of physical evidence or history of trauma; the absence of signs of psoriasis, lichen planus, or peripheral vascular disease;
   - KOH positivity;
   - If nondermatophyte molds were grown, the same fungus must repeatedly be isolated in the absence of dermatophytic growth when cultured on selective media.

2. It sometimes is difficult to isolate fungi in culture. From 30% to 50% of mycologic specimens can be negative, even from nails that are positive on direct microscopy. Nail biopsy is appropriate.

NAIL BIOPSY

Histologic observation of fungal invasion of keratin reverses false-negative results of culture in fungal infection where dermatophytes are not found (1,2,4,7).

Nail biopsy is indispensable for

1. A dystrophic nail, suggestive of onychomycosis, but with negative culture. In this case, two samples must be taken: one for a new culture and one for pathology. Periodic acid-Schiff staining will easily show the fungal elements. Histopathology reliably demonstrates whether a fungus is invasive or merely colonizes subungual debris. Because yeasts and nondermatophytic molds grow easily and quickly, the hyphae observed should belong to dermatophytes. This allows one to treat the patient systemically and/or with topical antifungal nail lacquers.
2. PSWO with nail changes, distant from the free edge. A 3-mm punch nail biopsy excluding the underlying tissue is essential for quick diagnosis (Fig. 6A.3C).
3. Psoriatic nails, resistant to local corticosteroid injections. Dual pathology should be sought and antifungal treatment given as described later.

TREATMENT

To date, therapy for onychomycosis has been unsatisfactory for several reasons.

Traditional Therapy

Topical therapy has been inefficient, despite anecdotal reports. The long duration of oral treatment (6 months for fingernails and 12 months or more for toenails) may have discouraged the patient, especially use of ketoconazole, which is active against dermatophytes and yeasts but needs continuous monitoring. Interference with some drugs (such as anticoagulants and griseofulvin, active only on dermatophytes) may be a problem. However, the high rate of relapse is probably the most common complication.

As a refinement to traditional therapy, to eradicate the pathogen the maximum possible amount of nail material containing fungi should be removed.

A. Mechanically by cutting, filing, or abrading. Nail abrasion is never sufficient; the remaining nail bed is still infected.

B. Chemically by keratinolysis, with 40% urea paste, used for nail avulsion. Only the clinically affected nail becomes detachable, but fungi may still be present in the margin of the clinically "normal" adherent portion of the nail keratin. It is difficult to ensure that infected material under the proximal nail fold is completely removed, even with a combination of 40% urea and 1% bifonazole.

C. Surgically by nail avulsion. In this case, counterpressure normally produced by the nail is lost, which allows progressive dorsal expansion of the distal soft tissue of toes. The free edge of the regrowing nail may embed itself in the unrestricted growth of the underlying distal nail bed. *Partial* surgical avulsion (8) of the nail is suitable for onychomycosis to a limited extent and results in reasonable remission rates while reducing the time needed for griseofulvin therapy. Partial nail avulsion by removing the *lateral* or *medial segment* of the nail plate permits, especially on the great toenail, enough normal nail to be left, which counteracts the upward forces exerted on the distal subungual tissue when walking. When the affected portion of the nail plate has been separated from the nail bed and the proximal nail fold using a Freer septum elevator, it is incised from distal to proximal with the nail splitter and the freed nail segment is avulsed.

In *distal subungual onychomycosis* with hyperkeratosis, elevation of the nail plate from the nail bed permits easy removal of the affected portion of the nail with a double-action bone rongeur. In such cases, the distal edge of the remaining normal portion of the nail plate has only a short distance to grow to reach the distal nail bed, which prevents the slow development of a possible distal nail wall.

In *Candida onycholysis,* thorough clipping away of as much of the detached nail as possible facilitates the daily application of antifungal drugs until normal nail growth is achieved.

In *proximal subungual onychomycosis,* removal of the loose base of the nail plate is ideal. The lunula region is cut transversely with a nail splitter by inserting the instrument beneath the lateral edge of the nail plate. Leaving the distal adherent portion of the nail in place decreases discomfort.

In chronic paronychia, protection of the hands for any "wet work" is an indispensable part of management. Topical antifungal agents active against *Candida* must be applied twice daily to the groove between the nail plate and the proximal nail wall when this yeast has been cultured. Intralesional steroids given once or twice often solve the problem. In recalcitrant cases due to a foreign body for example (9), the thickened nail fold may be excised *en bloc* as a crescent-shaped full-thickness skin 4 mm at its greatest width, which extends from one lateral nail fold to the other (Fig. 6A.4) (10). A crescentic superficial excision of the proximal nail fold is an alternative (11). In patients who experience repeated acute painful flares associated with chronic paronychia, removal of the base of the nail plate is useful (12).

New Antifungal Drugs

The fungicidal effect of some drugs (terbinafine) and the mode of penetration of these new drugs (itraconazole, fluconazole, and terbinafine) that reach the nail not only through the matrix, but also through the nail bed, explain their efficacy. In addition, retention for a long time of the drug in the nail keratin explains the ability to prescribe drugs for short periods (2 weeks to 3 months, or intermittent therapy).

Efficient new antifungal transungual delivery systems (5% amorolfine and 8% ciclopirox) may not only bring new life to old systemic drugs such as griseofulvin, but they can be associated with the new systemic therapy. If they are used periodically after cure of onychomycosis, they may reduce the risk of relapse.

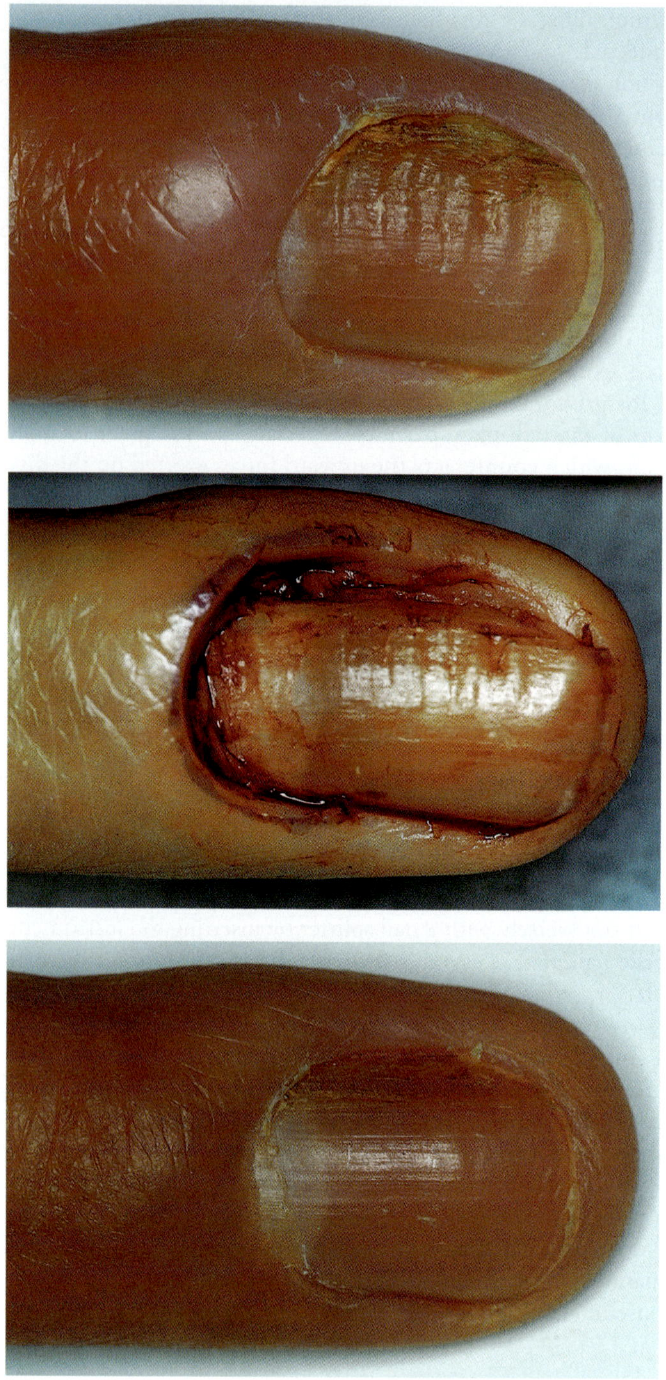

A

B

C

FIGURE 6A.4. A: Recalcitrant chronic paronychia. **B:** Crescentic excision of the proximal nail wall. **C:** Same patient shown postoperatively.

REFERENCES

1. Achten G, Wanet-Rouard J. *Onychomycosis (mycology no. 5).* Brussels: Cilag Ltd., 1981.
2. Haneke E. Fungal infections of the nail. *Semin Dermatol* 1991;10:41–51.
3. Zaias N. Clinical manifestations of onychomycosis. *Clin Exp Dermatol* 1992;17[Suppl 1]:6–7.
4. Hay RJ, Baran R, Haneke E. Fungal onychomycosis and other infections involving the nail apparatus. In: Baran R, Dawber RPR, eds. *Diseases of the nails and their management,* 2nd ed. London: Blackwell Science, 1994:97.

5. Hay RJ, Baran R, Moore MK, et al. Candida onychomycosis—an evaluation of the role of Candida in nail disease. *Br J Dermatol* 1988;118:47–58.
6. Tosti A, Fanti PA, Guerra L, et al. Role of foods in the pathogenesis of chronic paronychia. *J Am Acad Dermatol* 1992;27:706–710.
7. Suarez SM, Silvers DN, Scher RK, et al. Histologic evaluation of nail clippings for diagnosing onychomycosis. *Arch Dermatol* 1991;127:1517–1519.
8. Baran R, Hay R. Partial surgical avulsion of the nail in onychomycosis. *Clin Exp Dermatol* 1985;10: 413–418.
9. Stone OJ. Chronic paronychia in which hair was a foreign body. *Int J Dermatol* 1975;14:661–663.
10. Baran R, Bureau H. Surgical treatment of recalcitrant chronic paronychia of the fingers. *J Dermatol Surg Oncol* 1981;7:106–107.
11. Keyser JJ, Eaton RG. Surgical cure of chronic paronychia by eponychial marsupialization. *Plast Reconstr Surg* 1976;58:66.
12. Baran R. Nail growth direction revisited (why do nails grow out instead of up?). *J Am Acad Dermatol* 1981;4:78–82.

6B

CHRONIC BACTERIAL INFECTION

ECKART HANEKE
ROBERT BARAN

TUBERCULOSIS CUTIS VERRUCOSA

Butcher's nodule or prosector's wart is a secondary infection due to (auto) inoculation of *Mycobacterium hominis* or, less frequently, *bovis* (Fig. 6B.1). When an occupational infection is suspected, a culture is recommended for type determination. Body parts that tend to be mechanically injured are affected more often; a periungual location is rare (1). Verrucous tuberculosis commonly presents as a round-to-oval, bluish-red nodule with a rough hyperkeratotic surface. Common periungual warts as well as verrucous Bowen's disease must be differentiated. Treatment is conservative, although a small circumscribed lesion may be excised and sutured.

SWIMMING POOL GRANULOMA

Atypical mycobacteriosis is mainly seen in fish mongers and hobby pisciculturists, but several epidemics were observed in people attending public communal swimming pools (Fig. 6B.2). The infection remains restricted to the site of inoculation. Dissemination is rare and may follow a sporotrichoid pattern along the lymphatic vessels. Swimming pool infections were mainly seen at the site of superficial lacerations from the concrete wall of the pool. These lesions usually resemble tuberculosis cutis verrucosa, whereas "sporadic" infections look like lupus vulgaris. Further differential diagnoses are leishmaniasis, blastomycosis, sporotrichosis, and tularemia.

Mycobacterium marinum infections tend to heal spontaneously within 1 to 2 years. The treatment of choice is surgical removal of solitary primary lesions. Antibiotic treatment may be necessary for multiple lesions but requires previous cultural identification and sensitivity studies. Minocycline, doxycycline, rifampicin, cotrimoxazole, and ciprofloxacin usually are effective (2,3).

LEPROSY

Leprosy causes a wide spectrum of nail changes that are mainly due to peripheral neuropathy and consequent chronic repeated injury to the finger tips as well as unnoticed

E. Haneke: Klinikk Bunæs, Løkkeåsveien, 1337 Sandvika, Oslo, Norway.
R. Baran: Nail Disease Centre, 42, Rue des Serbes, 06400 Cannes, France.

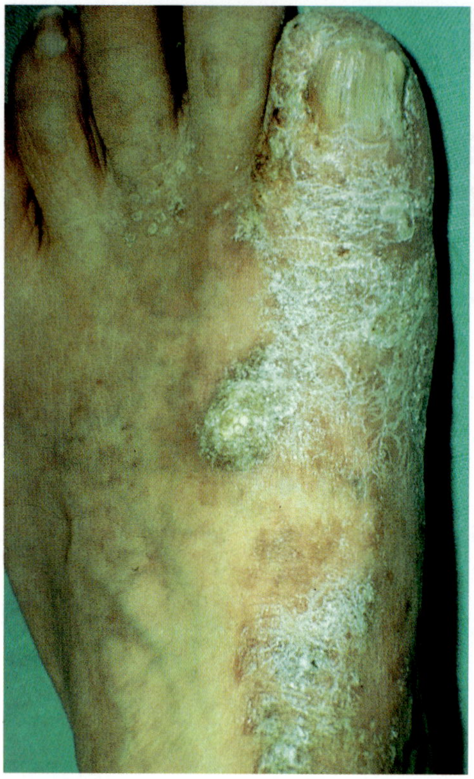

FIGURE 6B.1. Tuberculosis cutis verrucosa.

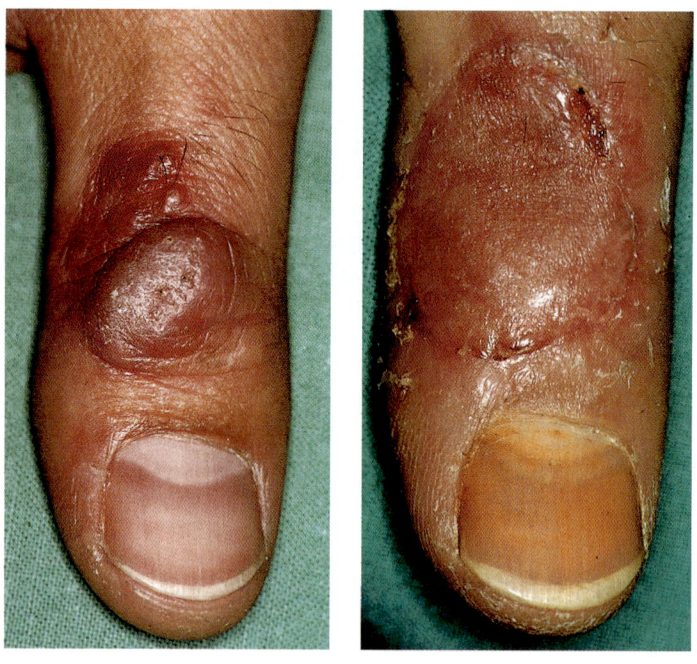

A B

FIGURE 6B.2. A: Swimming pool granuloma. **B:** Graft following surgical treatment in same patient.

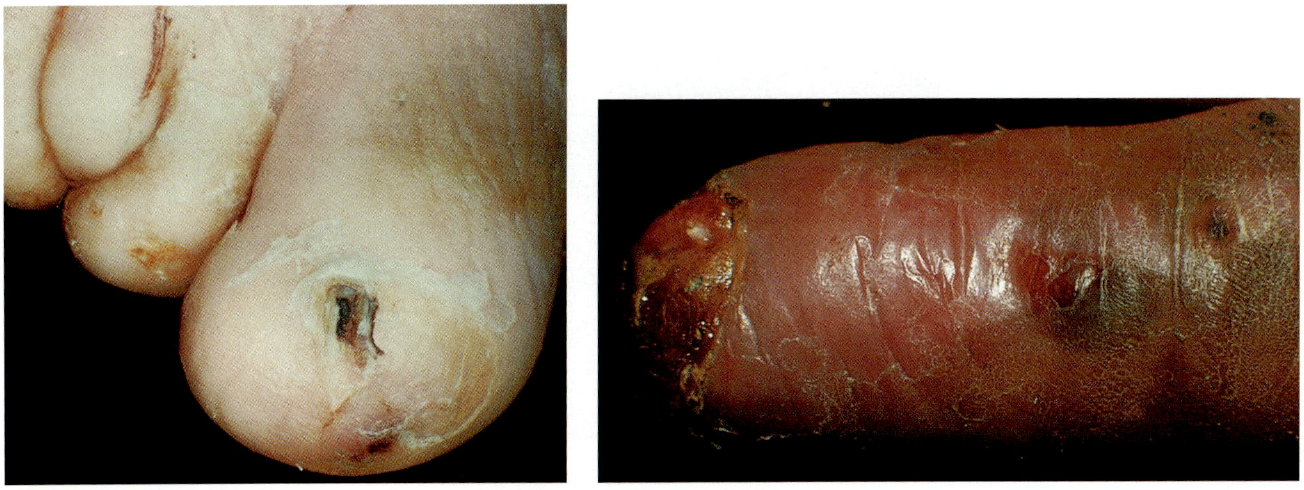

A B

FIGURE 6B.3. A: Dystrophic nail with acroosteolysis in leprosy (T). **B:** Morvan's syndrome.

infections (Fig. 6B.3 and Table 6B.1). Nail dystrophy, discoloration, secondary micronychia, claw nails, acroosteolysis, and loss of digits are common and usually not amenable to treatment (4). The differential diagnosis includes Morvan's syndrome, in which analgesic whitlow with dermal changes may be associated wtih leprosy, syringomyelia, and hereditary sensory neuropathy.

TABLE 6B.1. NAIL CHANGES IN LEPROSY

Neuropathy and unnoticed trauma
Subungual hematoma
Onycholysis
Onychauxis
Onychogryphosis
Claw nails
Pterygium unguis
Nail spicules
Ectopic nail
Anonychia
Acquired racket nail from distal acroosteolysis
Trauma and infection
Pyogenic infection
Acute paronychia
Onychomycosis
Impaired peripheral blood supply
Nail plate thinning
Ridging and splits
Atrophy
Slow growth
Onychauxis
Pterygium unguis

Data from Patki A, Baran R. Significance of nail changes in leprosy: a clinical review of 357 cases. *Semin Dermatol* 1991;24:158.

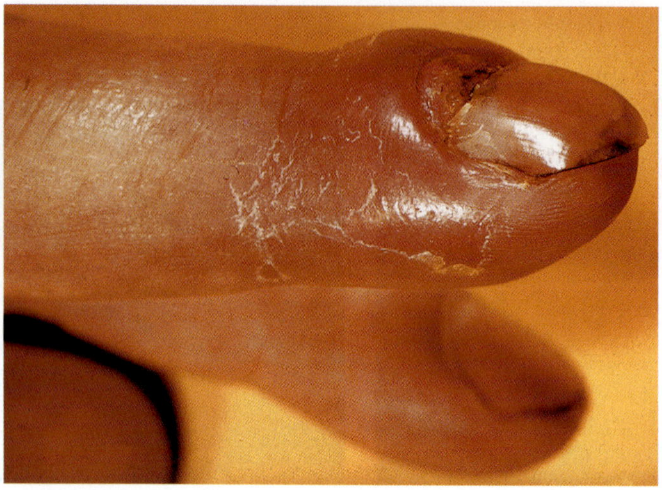

FIGURE 6B.4. Primary syphilis (Hutchinson's syphilitic whitlow). Courtesy of J. Garcia-Silva, et al., Spain.

SYPHILIS

Syphilitic chancres of the fingers due to occupational infection or sexual contact may present as periungual erosion or ulceration that may involve the nail bed, as in the syphilitic whitlow of Hutchinson or paronychia, or it may resemble a pyogenic granuloma (Fig. 6B.4). A crusty ulceration covering the free edge of the nail in a half-moon shape or developing in one of the lateral nail folds has been reported. Chancres in this location usually are painful and have a more chronic clinical course than chancres located elsewhere. Regional lymphadenopathy accompanies the primary lesion. Chancres of fingers are associated with painless unilateral, epitrochlear, and/or axillary nodes. The affected lymph nodes are discretely enlarged, hard, and nonsuppurating.

In Starzycki's clinic, the incidence of primary syphilis of the fingers was 14% of extragenital chancres (5). Of his 6 cases, two had chancres on both fingers and genitals resulting from sexual foreplay.

REFERENCES

1. Heller J. Die Krankheiten der Nägel. In: Jadassohn HJ, ed. *Handbuch der Haut-und Geschlechtskrankheiten. Bd IX/2.* Spezielle Dermatologie, Springer, Berlin 1927:S.150–172.
2. Loria PR. Minocycline hydrochloride treatment for atypical acid-fast infection. *Arch Dermatol* 1976; 112:517–519.
3. Brown JW, Sanders CV. Mycobacterium infections. *Arch Intern Med* 1987;147:817–818.
4. Patki A, Baran R. Significance of nail changes in leprosy: a clinical review of 357 cases. *Semin Dermatol* 1991;24:158.
5. Starzycki Z. Primary syphilis of the fingers. *Br J Vener Dis* 1983;59:169–171.

6C

PARONYCHIA

ELVIN G. ZOOK

A paronychia is a staphylococcus infection of the paronychium (1). The dermal and epidermal layers of the paronychium are arranged in an overlapping fashion much like shingles on a roof. A paronychia is usually the result of an infection in the area of a "hangnail." When one of these layers is pulled up, a "hangnail" is created and bacteria starts to grow.

Confusion with infections caused by the herpes simplex virus (Fig. 6C.1) in which surgical drainage is contraindicated must be avoided. If herpes blisters beneath the nail cause significant pain, nail removal may provide relief.

Localized bacterial infections superficial to the nail can be treated by pushing back the paronychium to drain the pus (Fig. 6C.2), followed by soaking the finger to maintain adequate drainage and administering antibiotics when there is cellulitis. If the purulence extends beneath the nail (Fig. 6C.3A), it is essential to remove 3 to 4 mm of the side of the nail to adequately drain the abscess. This usually can be done with a finger block, but occasionally a general anesthetic is required. A Kutz elevator or fine scissors are (Fig. 6C.3B) placed between the nail and nail bed at the hyponychium and the nail freed along the involved paronychium. Dissection between the eponychium and the nail plate to free the entire lateral surface of the nail also is necessary. Using a scissors, the nail is split longitudinally and the lateral fragment removed (Fig. 6C.3C–D). A small piece of Adaptic or other water-soluble gauze is placed under the nail fold to maintain drainage (Fig. 6C.3E). The finger is soaked in warm soapy water several times a day to facilitate drainage. In a neglected paronychia, purulence may undermine the nail around the nail fold and under the opposite paronychium (Fig. 6C.4). This is called a "runaround" abscess. Both sides of the nail and the nail in the nail fold are removed to drain the abscess. Sometimes the entire nail must be removed. It is not uncommon for pressure to cause ischemia of the germinal matrix so that the nail does not regrow. An incision in the eponychium for drainage is unnecessary (Fig. 6C.5) and only leads to notching of the eponychium.

Chronic paronychia are typified by intermittent inflammation and tenderness proximal to the eponychium and deformity of the nail surface (Fig. 6C.6A) (2–4). It usually responds to antibiotics and soaks, but recurs with cessation of treatment. Chronic infection usually is caused by gram-negative organisms or fungus that invades the nail wall. Treatment with a 4 to 5 mm wide, "quartermoon"-shaped excision of the nail wall proximal to the eponychium has been shown to adequately treat chronic paronychia (Fig. 6C.6B). The skin and underlying soft tissue are excised full thickness down to the nail and allowed to heal, with frequent soapy water soaks. The eponychium contracts proximally to some degree lengthening the visible lunula and the nail (Fig. 6C.6C).

E. G. Zook: Division of Plastic Surgery, Southern Illinois University, Springfield, Illinois 62708.

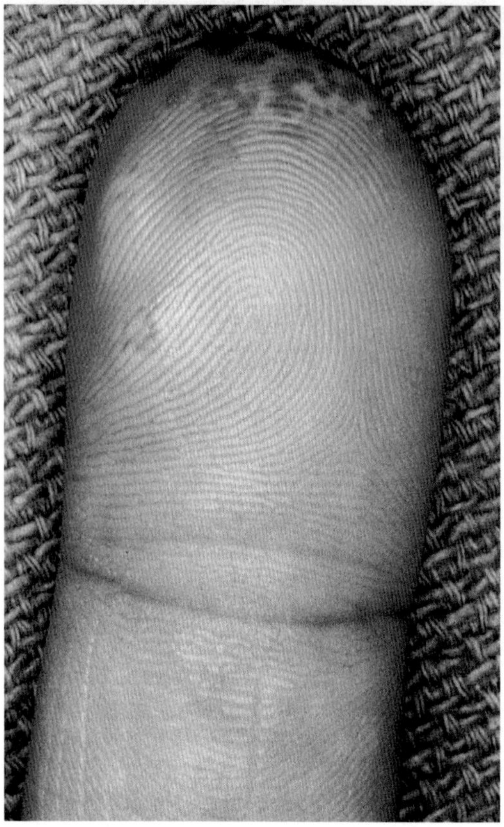

FIGURE 6C.1. Typical blisters of a herpes simplex skin infection of finger.

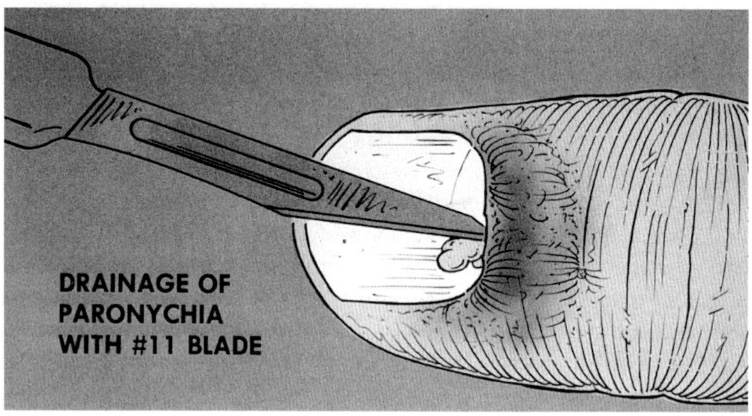

FIGURE 6C.2. The soft tissue superficial to the nail may be elevated with an instrument to drain the abscess.

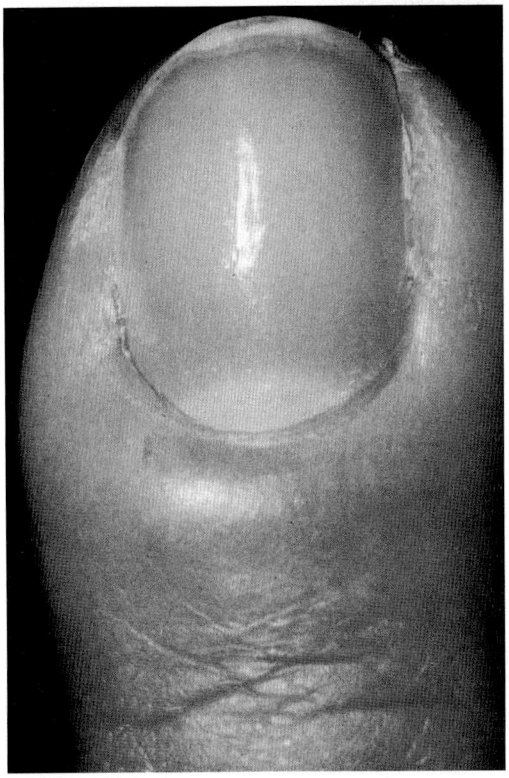

A

FIGURE 6C.3. A: Red, tender, painful, swollen paronychium indicating a paronychia. **B:** Scissors used to free soft tissue from the nail. **C:** The lateral nail is cut off and removed to allow drainage.

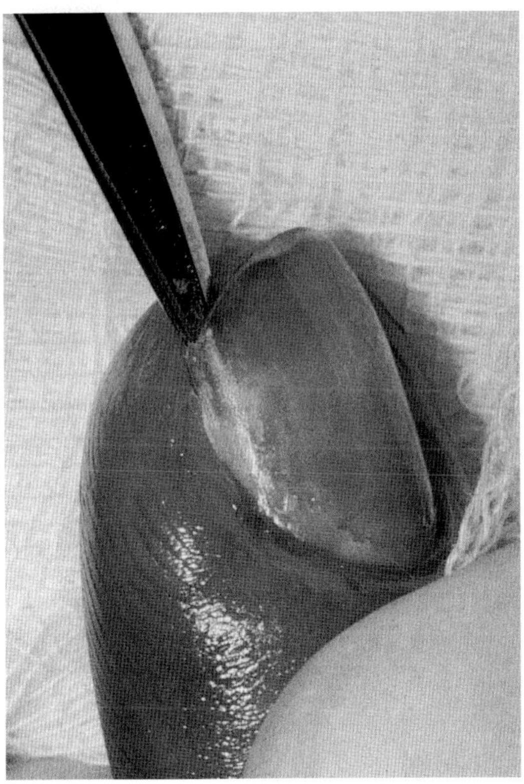

B

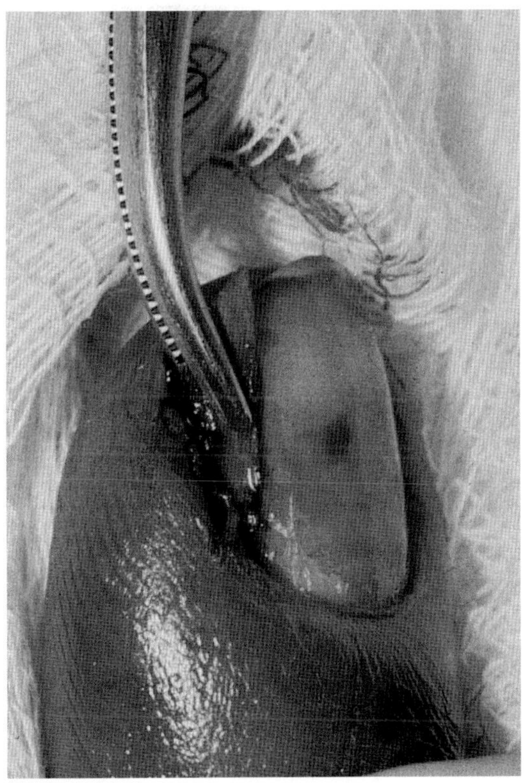

C

(continued)

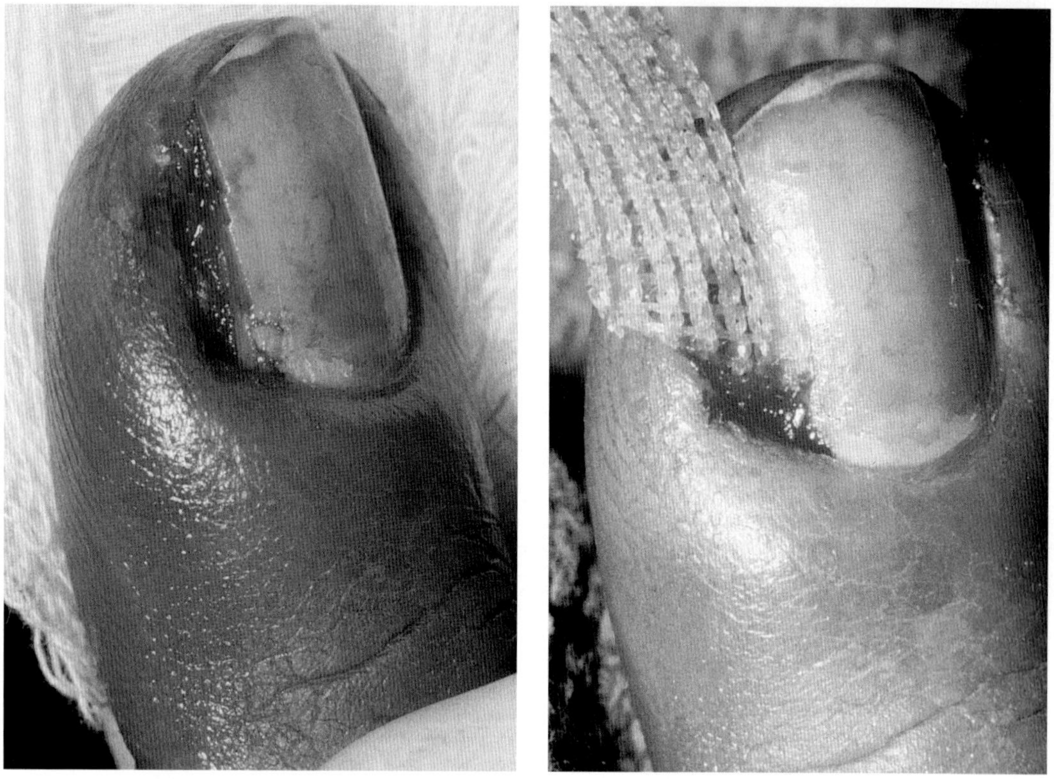

D E

FIGURE 6C.3 ***(continued).*** **D:** Paronychium adequately drained. **E:** Water-soluble gauze drain.

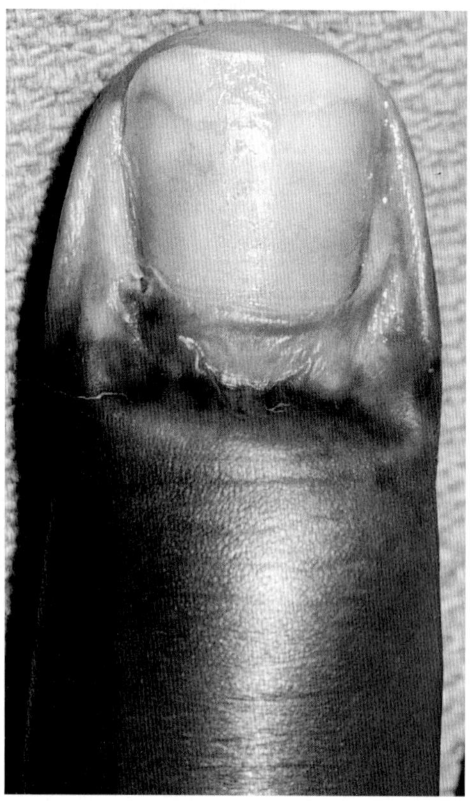

FIGURE 6C.4. Abscess of both paronychium and eponychium.

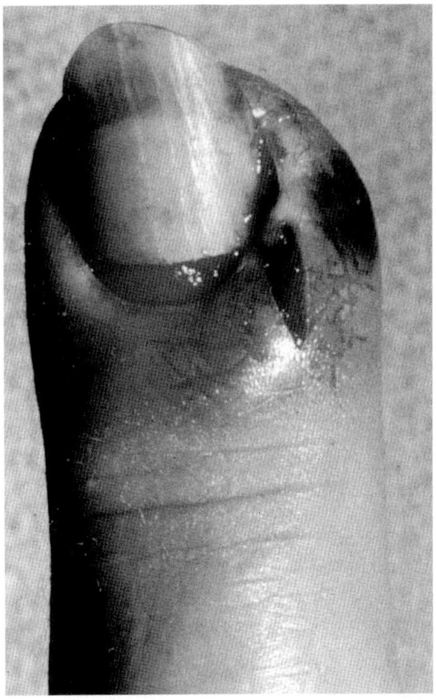

FIGURE 6C.5. Paronychia can be drained adequately without eponychial incisions.

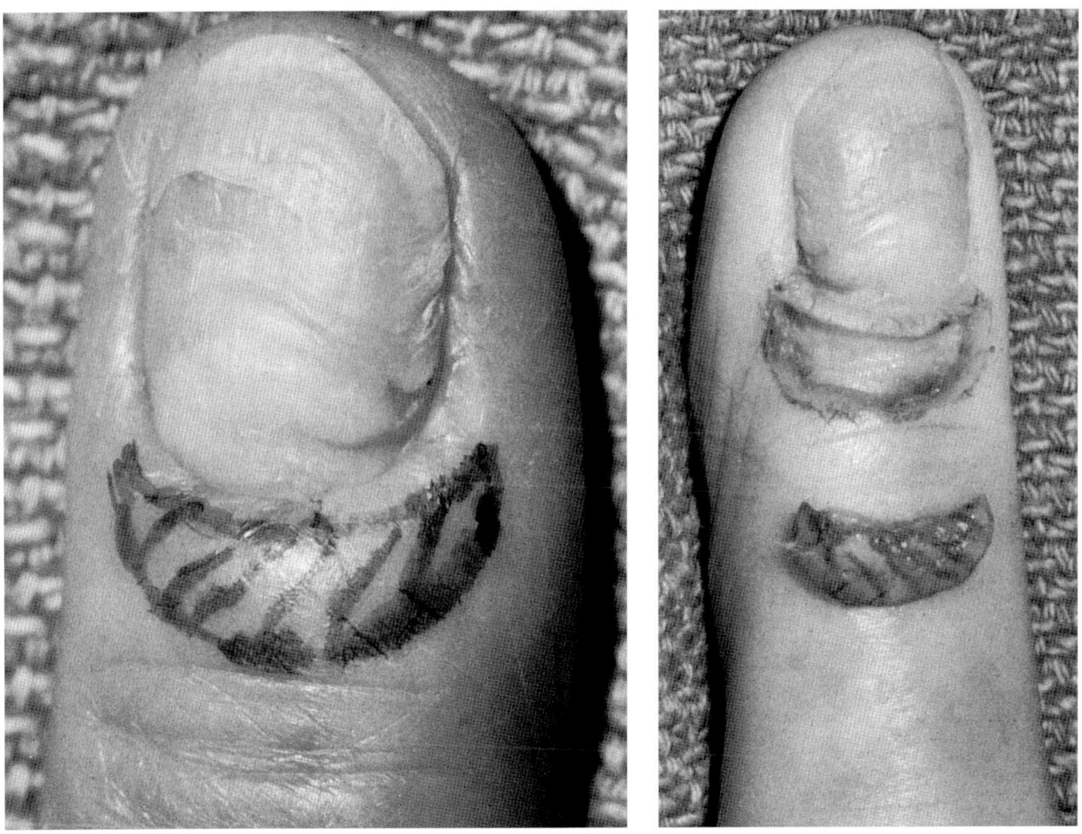

A B

FIGURE 6C.6. A: Chronic paronychia with swollen and tender nail wall and nail deformity. **B:** An arc-shaped portion of the nail wall is excised to the nail.

(continued)

C

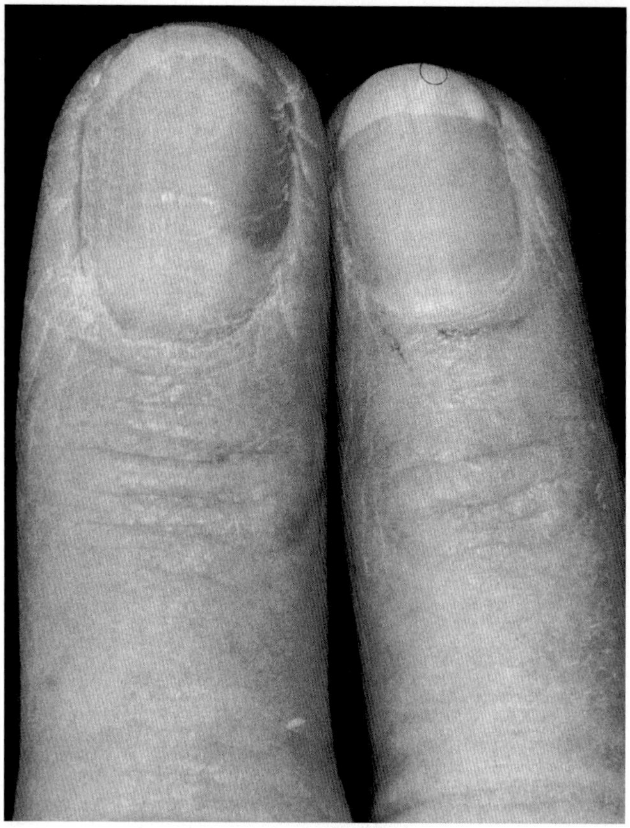

FIGURE 6C.6 *(continued)*. C: At 1 year after surgery, there is no further problem. Note lengthened lunula.

REFERENCES

1. McGinley KJ, Larson EL, Leyden JJ. Composition and density of microflora in the subungual space of the hand. *J Clin Microbiol* 1988;26:950–953.
2. Bednar MS, Lane LB. Eponychial marsupialization and nail removal for surgical treatment of chronic paronychia. *J Hand Surg* 1991;16A:314–317.
3. Keyser JJ, Eaton RG. Surgical care of chronic paronychia by eponychial marsupialization. *Plast Reconstr Surg* 1976;58:66–70.
4. Keyser JJ, Littler JW, Eaton RG. Surgical treatment of infections and lesions of the perionychium. *Hand Clin* 1990;6:137–157.

6D

VIRAL INFECTION

ROBERT BARAN
ECKART HANEKE

HERPES SIMPLEX

Herpetic whitlow is less rare than often estimated (Fig. 6D.1). Herpes simplex virus (HSV) types 1 and 2 occur at almost equal frequency. It is seen in primary herpetic gingivostomatitis when the child sucks the thumb or finger and thus can inoculate HSV into periungual skin (1). In adults, medical and dental personnel are prone to become infected from direct contact with saliva and oral mucosa of patients shedding active HSV. Examination gloves should be worn to avoid HSV infection of the fingers. HSV type 2 often is acquired from genital herpes simplex of the partner.

Usually the index finger or thumb is affected (2–4), but any finger, and rarely a toe or several fingers, may be involved. After an incubation period of 3 to 8 days, erythema and swelling develop and finally give rise to confluent small clear blisters. They rapidly become putrid and cloudy. Because the periungual epidermis has a resistant horny layer, the blister roof usually remains intact, although the small vesicles may merge to form a large bulla. Small new vesicles may appear for 8 to 12 days. Lymphangitis virtually always is present and, in recurrent herpetic whitlow, often precedes vesicle eruption. Tender regional lymph nodes are found in about two-thirds of patients with periungual herpes simplex. Pain that is disproportionally severe is another consistent symptom. Numbness and hypesthesia are rare sequelae (3).

Diagnosis of herpetic whitlow depends on considering this viral disease. Grouped vesicles with clear contents, eruption of new vesicles, lack of response to systemic antibiotics, pain, and lymphangitis in the early course all help to make the clinical diagnosis. A Tzanck test of fresh vesicles will reveal multinucleated keratinocytes with steel-blue homogeneous karyoplasm. Virus culture and immunofluorescence using type-specific monoclonal antiherpes simplex antibodies allow differentiation of HSV types (4).

Treatment of herpetic whitlow is essentially conservative. Because pain may significantly interfere with professional and sports activities and virus shedding may be a risk for spreading the infection from medical and dental personnel to patients, systemic acyclovir may be indicated at a dose of five times 200 mg daily for about 5 days. Unfortunately, many cases are misdiagnosed as a paronychial felon and are treated with incisions, systemic antibiotics, and splinting, which delay (spontaneous) healing.

ORF AND MILKER'S NODULE

Orf (ecthyma contagiosum) is due to a pox virus causing infection of the mouth and lips in sheep, goat, and reindeer (Fig. 6D.2). Lesions in humans are commonly seen on the

R. Baran: Nail Disease Centre, 42, Rue des Serbes 06400 Cannes, France.
E. Haneke: Klinikk Bunæs, Løkkeåsveien, 1337 Sandvika, Oslo, Norway.

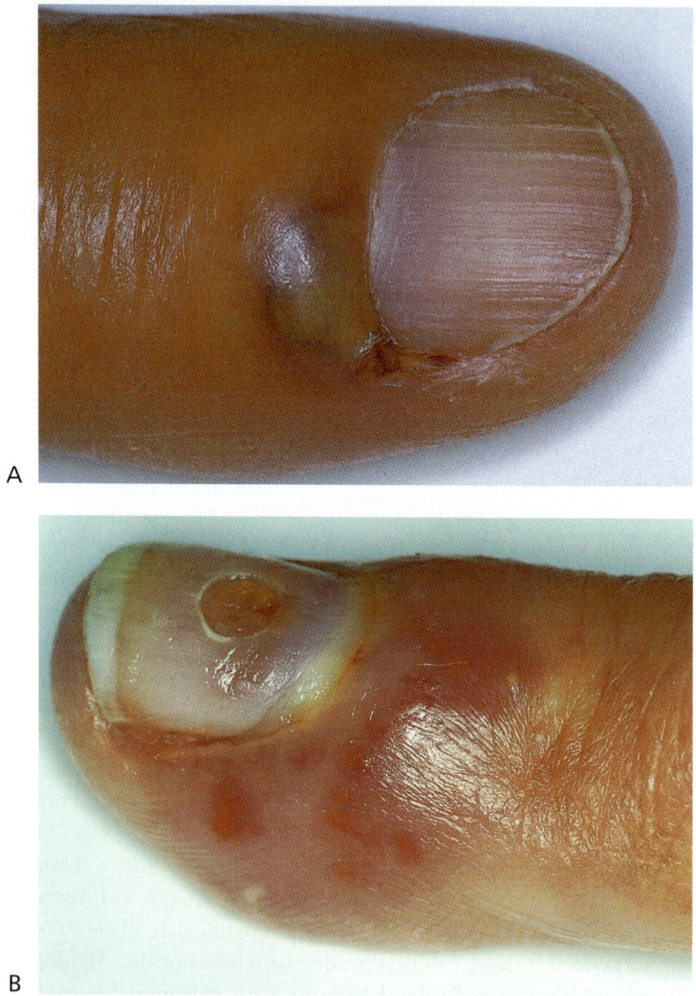

FIGURE 6D.1. A: Herpes of the proximal nail fold. **B:** Periungual herpetic manifestation after punch biopsy of the nail bed.

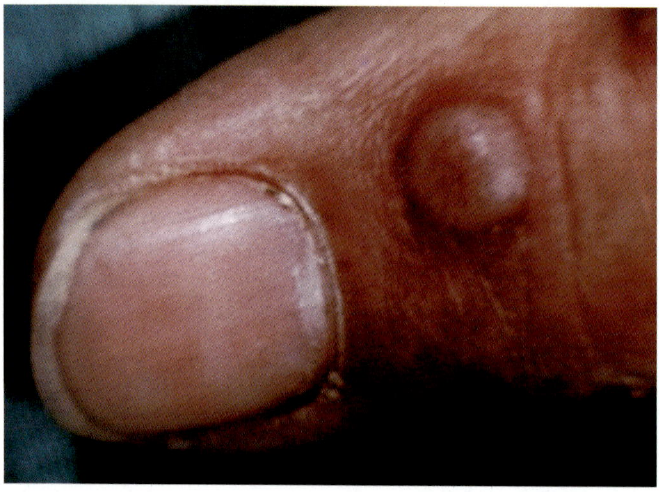

FIGURE 6D.2. Orf affecting the proximal nail fold.

middle phalanx of the third finger of the dominant hand. If the diagnosis is correct, a shave excision supplemented by antimicrobial topical treatment may accelerate healing.

Milker's nodule is another pox virus infection transmitted from cattle. It takes 1 to 3 months to heal, but healing is faster after removal by epidermal subsections (5).

WARTS

Periungual verrucae vulgares are due to human papillomaviruses (HPVs) of different strains, mainly HPV-2 and HPV-4, less frequently HPV-1, and, in butchers HPV-7 (Fig. 6D.3). Warts are benign fibroepithelial tumors that are weakly contagious. Autoinoculation in nail biters and pickers appears to be a frequent cause for this particular localization and may cause spread to the lips and oral mucosa.

Periungual warts are skin-colored to grayish nodules with a rough hyperkeratotic surface. Usually they are asymptomatic, but fissuring may cause pain and secondary inflammation. They may appear as single or multiple lesions and merge to form a large verrucous mass surrounding the whole nail. Involvement of the matrix causes nail plate deformation. Subungual warts sometimes are quite painful, mimicking a subungual glomus tumor. Bone erosion has been observed in some cases (6–8), although it is not absolutely clear that subungual keratoacanthoma had been excluded in all of these cases.

The differential diagnosis of periungual and subungual warts is Bowen's disease, squamous cell carcinoma, keratoacanthoma, mucinous syringometaplasia (9), onychophosis, amyloidosis, subungual corn, subungual filamentous tumor and fibrokeratoma, verrucous epidermal nevus, inflammatory linear verrucous epidermal nevus, verrucous stage of incontinentia pigmenti in infants and subungual papillomatosis of incontinentia pigmenti in pregnant women, multicentric histiocytosis, and subungual exostosis.

Treatment of periungual and subungual warts remains a challenge. Although a single wart usually has a natural lifespan no longer than 2 to 5 years, the warts may spread and resist treatment for years.

Treatment should be graduated, beginning with the least harmful conservative treatment, such as a wart paint containing about 20% salicyclic acid. In many countries, there is a wealth of different wart remedies providing evidence that none is universally active. If this therapy fails, cantharidine 0.007% (Cantharone) is applied to the warts, allowed to dry, and covered with an occlusive tape for 24 hours. The resultant blister with its subepidermal split should contain the wart in its roof; however, retreatment may be necessary after 2 or more weeks for up to four times if the wart recurs. Tkach (10) suggested that

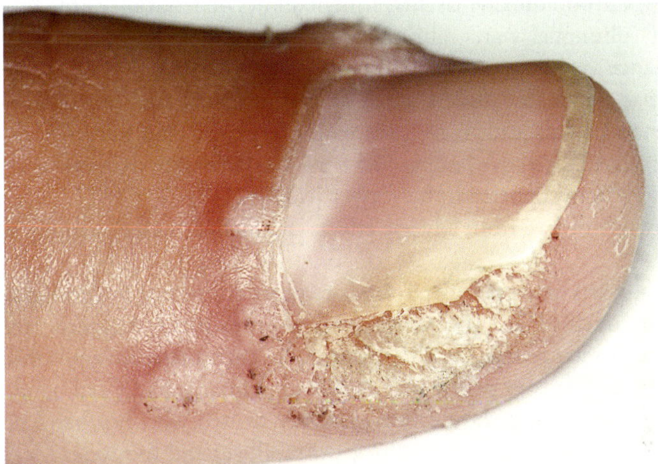

FIGURE 6D.3. Subungual and periungual warts.

blister formation might spread the warts and recommends avoiding blister formation by wiping off the cantharidine after 2 hours.

Bleomycin has been recommended for resistant warts (11), but both transitory and permanent nail dystrophy were observed after injection of bleomycin into periungual warts (12,13). Using a multiple puncture technique with 1 U/mL bleomycin sulfate with a bifurcated vaccination needle, a 92% healing rate was achieved in a series of 258 periungual warts after a single treatment (14). This may be the best treatment, unless the promising results with cimetidine 25 to 40 mg per kg per day for 2 months are confirmed (15).

Topical immunotherapy using diphencyprone or squaric acid dibutylester may be alternatives for adult patients (16,17).

The success rate of most surgical approaches is not better than that of conservative treatment. Because, in genital warts, HPV DNA was detected by hybridization as far as 15 mm around the lesion, it must be assumed that DNA of common warts also may be found in healthy adjacent skin. Excision with a safety margin of 15 mm or more for a totally benign, self-limited lesion is absolutely inadequate.

Cryosurgery as a simple procedure often is performed (18). A single freeze of 20 seconds using the dipstick technique is sufficient to raise a blister, the roof of which contains the epidermal component (with HPV DNA) of the wart. Some authors prefer to only slightly freeze the warts to a subblister level and repeat this treatment in biweekly intervals until the warts are gone. The latter causes localized hyperemia of the inflamed treatment area, which may be the mechanism of response. Side effects from cryosurgery are postcryosurgical edema with intense throbbing pains, later leukonychia, Beau's lines, onychomadesis, and nail loss. Permanent nail dystrophy may develop (19). To avoid untoward sequelae, freezing times must be carefully controlled, particularly when treating the proximal nail fold. Prophylactic antiinflammatory and analgesic treatment with aspirin, indomethacin, or another nonsteroidal antiinflammatory agent helps to prevent excessive postoperative pain and swelling.

Curettage, both alone and followed by electrodesiccation, rarely is successful and may cause extensive scarring. Infrared light coagulation commonly used for hemorrhoid therapy has been attempted, but it is difficult to target exactly the infrared beam to periungual warts.

Carbon dioxide laser treatment has been recommended as the treatment of choice (20). Two treatment modalities can be used: (i) defocused mode to vaporize the epithelial portion of the wart, including its capillary vessels reaching high up in the papillae; and (2) focused mode to carbonize the warts. Both modalities often are used consecutively. The wart itself is carbonized and the surrounding epidermis is treated with the defocused laser beam to necrose it. The wart then is wiped off from the underlying dermis, which does not contain viral DNA. Carbon dioxide laser treatment allows penetration of the nail plate and carbonization of subungual warts, although we prefer to cut the plate away before laser therapy to enhance visibility and accuracy.

Particular problems arise when the undersurface of the proximal nail fold also is affected. Treatment of both sides will lead to necrosis of the proximal nail fold. A two-stage operation then is necessary. Pulsed dye laser at 585 nm, 450 μs, 5-mm diameter spot size, and fluences of 6.25–7.5 J/cm^2 appears to be very efficient (21).

Laser treatment of warts is not free from adverse effects; permanent nail dystrophy is possible (22).

Treatment of periungual warts often is frustrating. It requires consistent medical care and enormous compliance to eliminate the warts, and posttreatment measures are extremely important to prevent recurrences. Although common warts often are amenable to suggestion therapy, this usually fails in periungual as well as plantar warts. Cool hands and feet are virtually always present and should be vigorously treated with hot hand or foot baths, abstention from nicotine, and wearing thick warm gloves and footwear.

The incubation period of HPV is up to 6 months or more; therefore, new warts may appear even after successful treatment. Regular follow-up visits are recommended.

REFERENCES

1. Muller SA, Hermann EC. Association of stomatitis and paronychias due to herpes simplex. *Arch Dermatol* 1970;101:394.
2. LaRossa D, Hamilton R. Herpes simplex infections of the digits. *Arch Surg* 1971;102:600.
3. Chang T, Gorbach SL. Primary and recurrent whitlow. *Int J Dermatol* 1977;16:752.
4. Haneke E. Klinik der Herpes simplex-Virusinfektionen. In: Hornstein OP, ed. *Virale and bakterielle Infektionskrankheiten der Haut.* Nürnberg: Perimed-Spitta, 1993:35–43.
5. Shelley WB, Shelley ED. Surgical treatment of farmyard pox. *Cutis* 1983;31:191–192.
6. Shapiro L, Flushing LY, Stoller NM. Erosion of phalanges by subungual warts. *JAMA* 1961;176: 379.
7. Plewig G, Christophers E, Braun-Falco O. Mutilierende subunguale Warzen: Abheilung durch Methotrexat. *Hautarzt* 1973;24:338–341.
8. Kumar B, Shazema SC, Kaur S. Phalangeal erosions with subungual warts. *Indian J Dermatol Vener* 1980;46:166–168.
9. Scully K, Assaad D. Mucinous syringometaplasia. *J Am Acad Dermatol* 1984;11:503–508.
10. Tkach JR. Finding and inventing alternative therapies. How I do it. *Dermatol Clin* 1989;7:1–18.
11. Shumer SM, O'Keefe EJ. Bleomycin in the treatment of recalcitrant warts. *J Am Acad Dermatol* 1983;9:91–96.
12. Baran R. Onychodystrophie induite par injection intralésionnelle de bléomycine pour verrue périunguéale. *Ann Dermatol Venereol* 1985;112:463–464.
13. Urbina-Gonzales F, Cristobal-Gil M, Aguilar Martinez A. Cutaneous toxicity of intralesional bleomycin administration in the treatment of periungual warts. *Arch Dermatol* 1986;122:974–975.
14. Shelley WB, Shelley ED. Intralesional bleomycin sulfate therapy for warts: a novel bifurcated needle puncture technique. *Arch Dermatol* 1991;127:234–236.
15. Orlow ST, Paller A. Cimetidine therapy for multiple viral warts in children. *J Am Acad Dermatol* 1993;28:794–796.
16. Orrecchia G, Douville H, Santagostino L, et al. Treatment of multiple relapsing warts with diphenciprone. *Dermatologica* 1988;177:225–231.
17. Iijima S, Otsuka F. Contact immunotherapy with squaric acid dibutylester for warts. *Dermatology* 1993;187:115–118.
18. Dawber RPR, Colver GB, Jackson A. *Cutaneous cryosurgery.* London: Martin Dunitz, 1992:30–38.
19. Baran R. Brachytéléphalangie révélée à l'occasion de dystrophies unguéales induites par cryothérapie. *Ann Dermatol Venereol* 1985;112:365–367.
20. Street MI, Roenigk RK. Recalcitrant periungual verrucae: the role of carbon dioxide laser vaporisation. *J Am Acad Dermatol* 1990;23:115–120.
21. Tan OT, Hurwitz RM, Stafford TJ. Pulsed dye laser treatment of recalcitrant verrucae: a preliminary report. *Lasers Surg Med* 1993;13:127–137.
22. Olbright SM, Stern RS, Tang SV, et al. Complications of laser surgery: a survey. *Arch Dermatol* 1987;123:345–349.

TUMORS AND CYSTS

7A

GANGLIONS OF THE DISTAL INTERPHALANGEAL JOINT

ELVIN G. ZOOK

Distal interphalangeal (DIP) ganglions (mucous cysts, clear cysts, and many other names have been given to this entity) have been an enigma to physicians for years (1,2). Myriad methods of treatment, including injections, radiation, aspiration, and excision, have all shown a high recurrence rate.

Kleinert et al. (3) showed with careful microdissection that subcutaneous cysts over the dorsum of the DIP joint and the eponychium always communicated with the joint. Newmeyer et al. (4) and Eaton et al. (5) showed communication between the cyst and the joint and between the joint and the cyst with injections of dye into the joint.

The pressure of the ganglion may press volarly on the forming nail and germinal matrix, creating a groove or disruption of the nail (Fig. 7A.1A], or it may dissect along the periosteum to elevate the germinal matrix and disrupt the nail (Fig. 7A.1B].

In 1991, Brown et al. (6) reported on the treatment of ganglions via a "T" or "H" incision over the DIP joint, resecting the cyst and debriding the joint of osteoarthritic spurs. Ganglions originate from the area of the arthritic spurs (Fig. 7A.2). All patients had deformities of the nail bed secondary to pressure of the ganglion on the germinal matrix area. They showed a 60% return to normal, with good improvement of the nail deformity in the other 40%. In 1995, the same group of authors reported on the treatment of 20 patients with ganglion-caused nail bed deformities by debriding the DIP joint osteophytes and only decompressing the ganglion through the incision (7). Ninety percent of the nails returned to normal, with only slight deformity in 10% even when the deformity had been long-standing. Of the total of 45 nail deformities, there was no recurrence of ganglions. Almost all of these patients had evidence of DIP joint osteophytes noted on x-ray film, and all had osteophytes found at exploration. The joint osteophytes are removed with a fine rongeur. The extensor tendon is gently raised and the osteophytes removed from beneath it. Care must be taken to not detach the tendon from the distal phalanx. It is not uncommon for the tendon to be stretched by the osteophyte proliferation. When removed, the tendon remains long and may give a few degrees of lag at the DIP joint. Postoperative splinting in extension for several weeks may be needed in the injured or greatly stretched extensor tendon.

There is no need to excise the thinned skin over the cyst, because it will collapse and thicken with decompression. There is no reason to remove the cyst wall, as it is a compressive-type lining and obliterates when decompressed.

E. G. Zook: Division of Plastic Surgery, Southern Illinois University, Springfield, Illinois 62708.

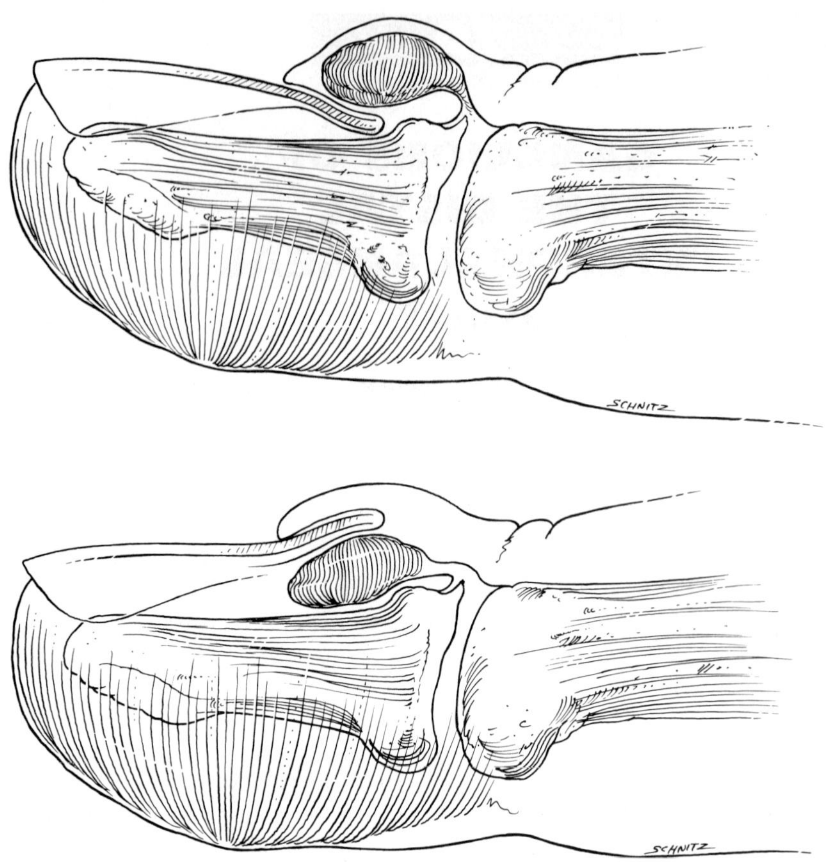

A

B

FIGURE 7A.1. A: Ganglion dissecting from the joint and putting downward pressure on the nail and germinal matrix. **B:** The ganglion may dissect along the periosteum, elevating the germinal matrix and nail and causing disruption of the nail.

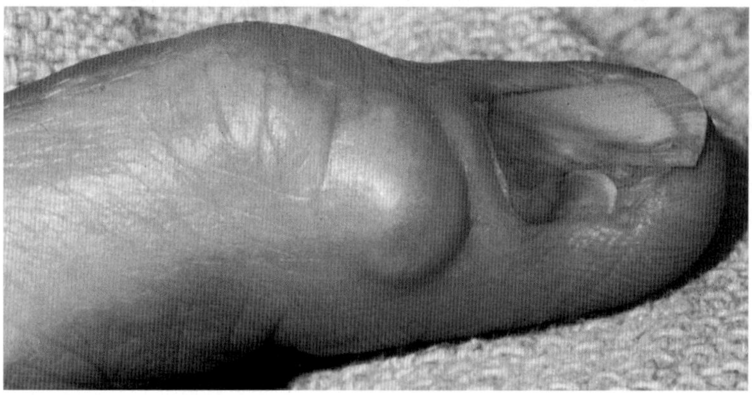

A

FIGURE 7A.2. A: Ganglion is seen causing disruption of the nail.

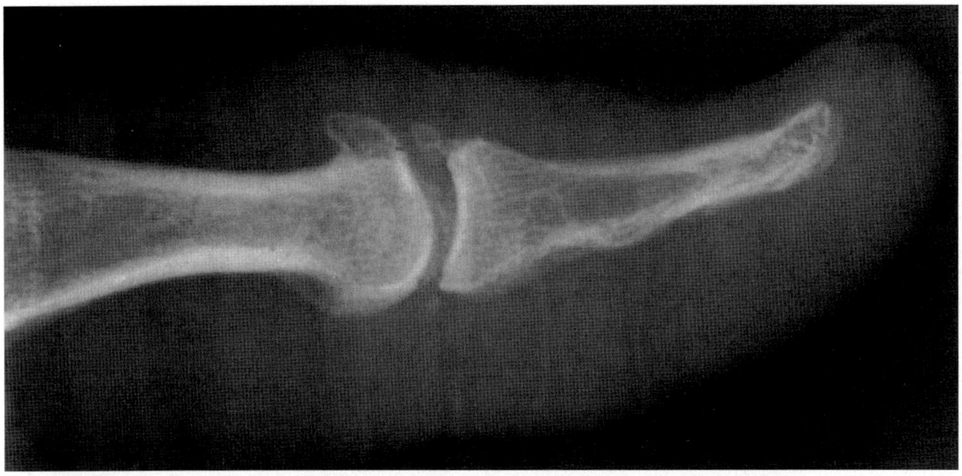

B

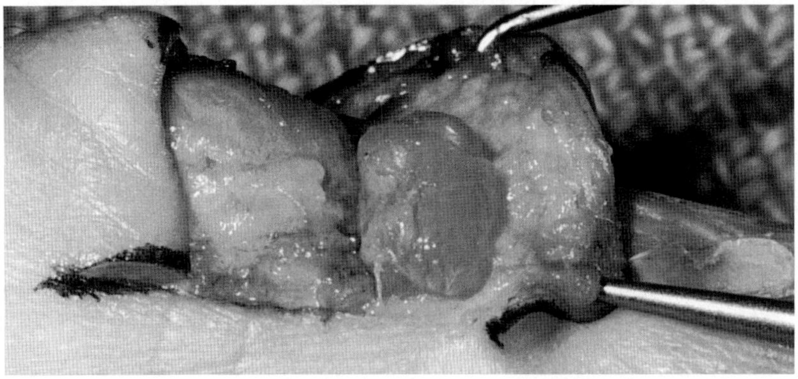

C

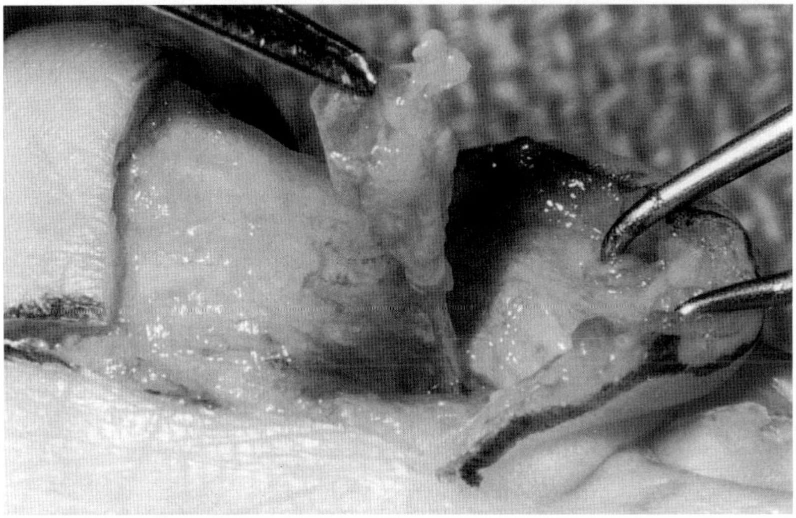

D

FIGURE 7A.2 *(continued).* **B:** Osteophytes on the distal interphalangeal joint shown on x-ray film. **C:** With the flaps raised, the ganglion is identified. **D:** Ganglion was resected. Identification and dissection of the ganglion is no longer recommended.

(continued)

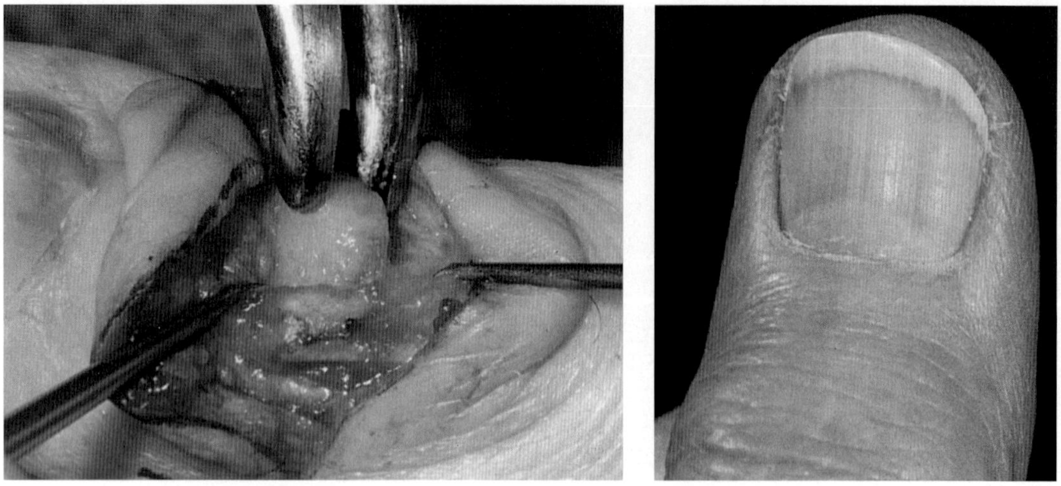

E F

FIGURE 7A.2 *(continued).* E: Osteophytes on both the distal and middle phalanges are removed with a rongeur. **F:** Essentially normal nail is seen 1 year later.

REFERENCES

1. Newmeyer WL, Kilgore ES. Common injuries of the fingernail and nail bed. *Am Fam Physician* 1977; 16:93–95.
2. Zaias N. *The nail in health and disease.* Jamaica, NY: Spectrum Publications, 1980.
3. Kleinert HE, Kutz JE, Fishman JH, et al. Etiology and treatment of the so-called mucous cyst of the finger. *J Bone Joint Surg* 1972;54A:1455–1458.
4. Newmeyer WL, Kilgore ES Jr., Graham WP III. Mucous cysts: the dorsal distal interphalangeal joint ganglion. *Plast Reconstr Surg* 1974;53:313.
5. Eaton RG, Dobranski AI, Littler JW. Marginal osteophyte excision in treatment of mucous cysts. *J Bone Joint Surg* 1973;55A:570.
6. Brown RE, Zook EG, Russell RC, et al. Fingernail deformities secondary to ganglions of the DIP joint (mucous cyst). *Plast Reconstr Surg* 1991;87:14:718–725.
7. Gingrass MK, Brown RE, Zook EG. Treatment of fingernail deformities secondary to ganglions of the DIP joint. *J Hand Surg* 1995;20A:502–505.

7B

FIBROUS AND FIBROEPITHELIAL TUMORS

ROBERT BARAN
ECKART HANEKE

A large variety of clinical presentations of fibrous tumors may develop in the subungual and periungual areas. These tumors may represent true entities in themselves or merely be variants of the same process, ranging from fibrous dermatofibroma to digital fibrokeratoma. This contrasts markedly with the uniformity of the histology of some fibrous tumors and is an argument for a single and identical pathologic process including Koenen's tumor, acquired fibrokeratoma, and dermatofibroma of the nail apparatus.

FIBROUS DERMATOFIBROMAS OR "TRUE" FIBROMAS

Fibromas usually develop as painless slow-growing nodular tumors. They may be spherical or oval in shape, and either firm or elastic in consistency. They can develop in any subepidermal structure of the nail apparatus and may move freely or remain in place (1).

The tumor is smooth on the dorsal aspect of the proximal fold, or on the nail bed, and more often spherical, resembling a small pea, than ovoid. The fibroma may be spherical on the lateral nail fold, but without the collar of slightly elevated skin seen in acquired periungual fibrokeratoma. Fibromas may become cherry shaped or polypoid and lift the nail in the distal subungual area (Fig. 7B.1).

Fibroma of the matrix results in nail dystrophy. Clinical features are variable according to the anatomic site, ranging from simple thinning of the nail plate to a longitudinal canal (Fig. 7B.2), which sometimes is partially covered by nail keratin to form a tunnel-like structure. This grooved nail results from the pressure of the tumor on the matrix. Beyond just a deformity of the nail, displacement of the finger pulp and erosion of the distal phalanx may have led to unnecessary amputation, because of misdiagnosis as a malignant tumor.

Two types of dermatofibroma are described histologically (2): fibroma (or fibrous dermatofibroma) and histiocytoma. In the latter, which is uncommon in the nail apparatus, "histiocytic" proliferation sometimes is associated with an angiomatous component, most often referred to as sclerosing hemangioma.

Three main histologic clues distinguish isolated "true" fibroma from acquired fibrokeratoma and other tumors of the nail apparatus (Table 7B.1) (3). (i) The lesions are composed of areas of very thick, hypocellular, hyalinized collagen bundles, in haphazard array. (ii) There is an ill-defined nodule situated principally in the reticular dermis. (iii) The elastic fibers usually are absent or scarce.

R. Baran: Nail Disease Centre, 42, Rue des Serbes, 06400 Cannes, France.
E. Haneke: Klinikk Bunæs, Løkkeåsveien, 1337 Sandvika, Oslo, Norway.

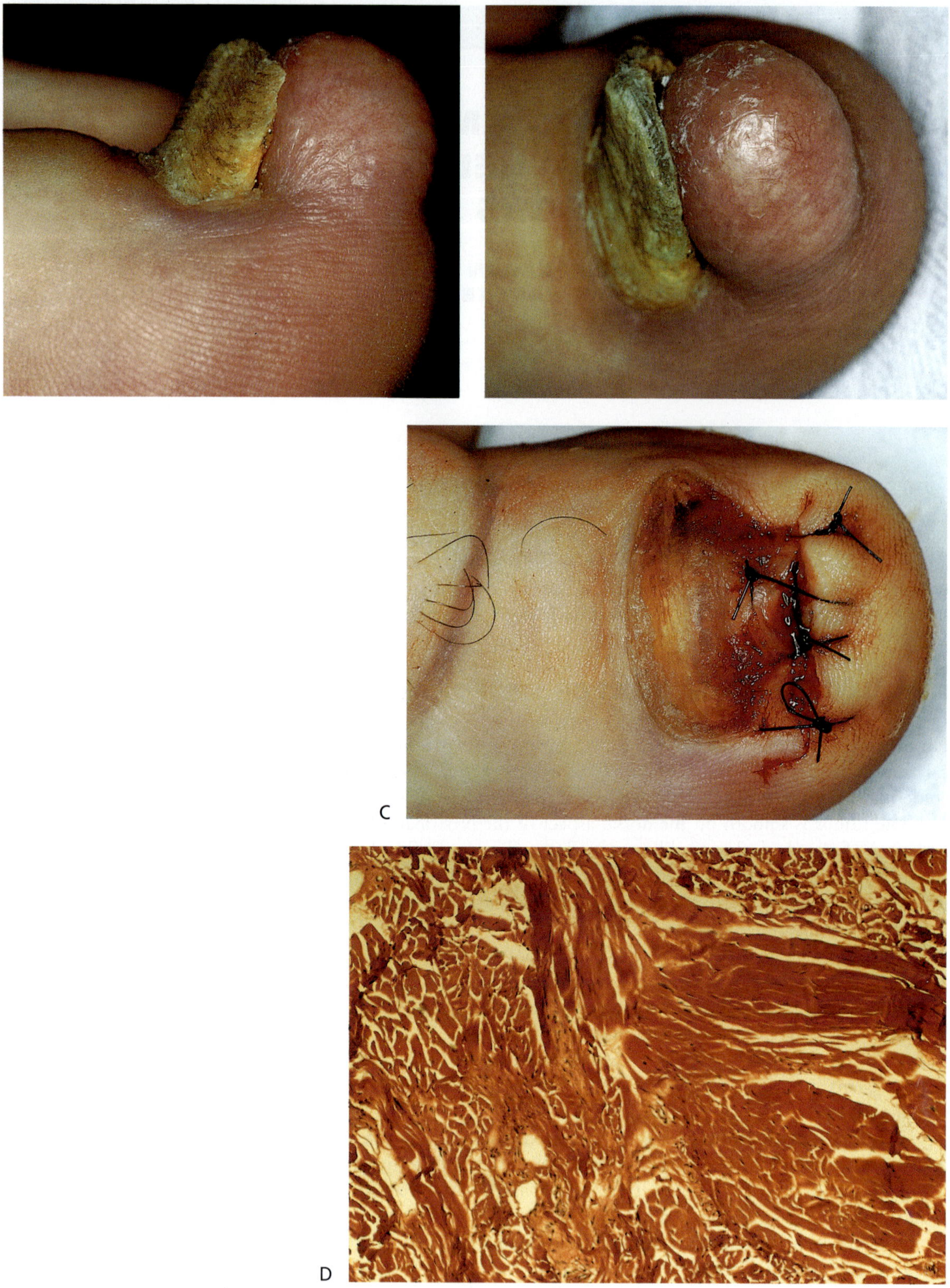

FIGURE 7B.1. A: Fibroma lifting the nail in the distal subungual area. **B:** Front view. **C:** After partial nail avulsion and removal of the tumor. **D:** Histology of the fibroma.

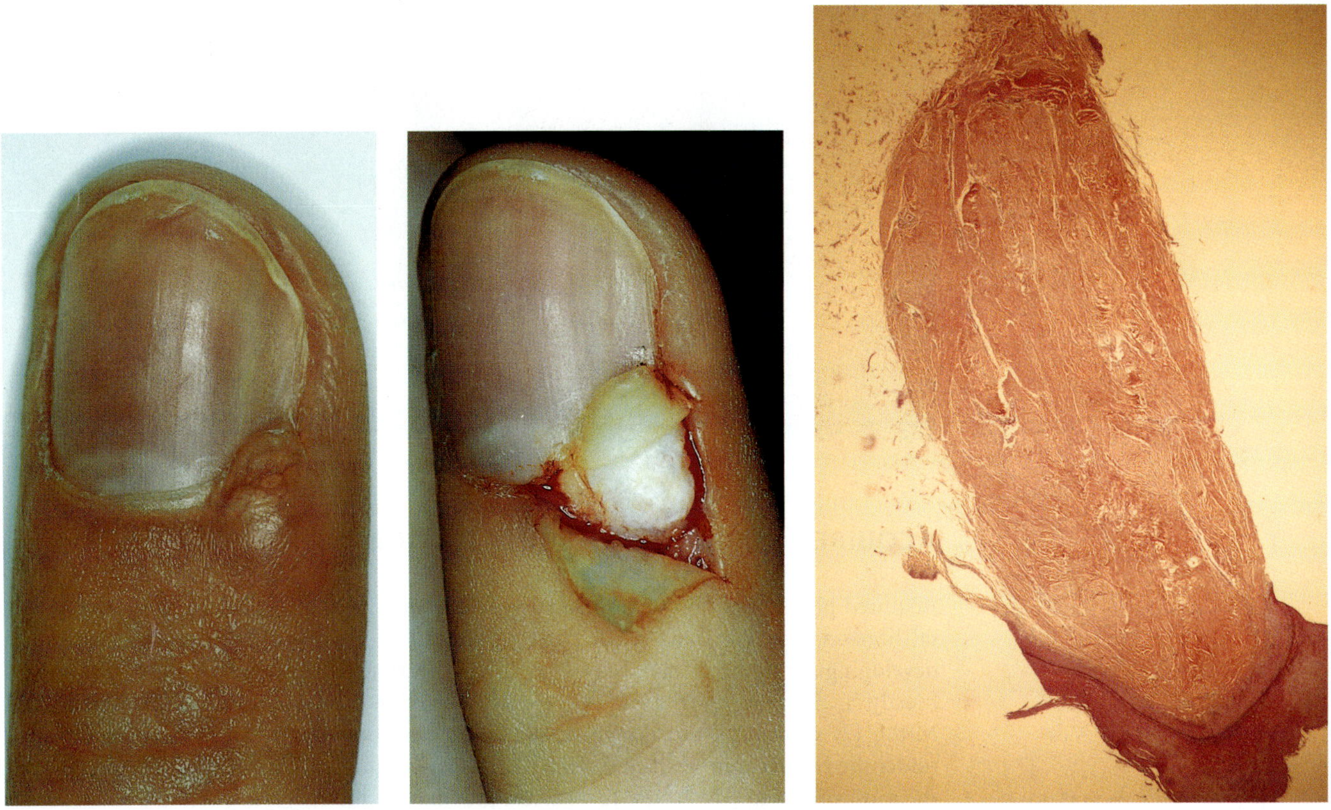

A B,C

FIGURE 7B.2. A: Fibroma grooving longitudinally along the nail plate. **B:** Appearance of the tumor after reflecting of the proximal nail fold. **C:** Histology of the tumor.

TABLE 7B.1. DIFFERENTIAL DIAGNOSIS OF TRUE FIBROMA FROM OTHER FIBROUS TUMORS OF THE NAIL APPARATUS

- Sclerotic fibroma (28)
  - Well circumscribed
  - Overlying epidermis thin
  - Collagen bundle in a "whorllike" pattern
- Fibroma of tendon sheath
  - Well circumscribed
  - Attachment to tendon or tendon sheath
  - Gradual transition between cellular and more hyalinized areas
- Pleomorphic fibroma
  - Multinucleated cells with large hyperchromatic nuclei
- Keloid (Fig. 7B.3)
  - Well circumscribed
  - Hyalinized collagen bundles
  - Papillary dermis normal
  - Hypocellular areas mixed with more cellular areas
- Acquired digital fibrokeratoma
  - External fingerlike projection
  - Hyperkeratosis at the distal portion (9)
- Leiomyoma (29,30)
  - Muscle stains red with Masson trichrome (in contrast, fibroma stains blue)
  - Leiomyoma is labeled by smooth muscle actin and desmin (31)
- Recurrent infantile digital fibromas
  - First year of life
  - Characteristic inclusion bodies visualized with phosphotungstic acid hematoxylin stain (30)
- Rudimentary supernumerary digits
  - Present at birth
  - Involvement of the fifth digit
  - Multiple nerve bundles within the dermal core
- Hereditary digital fibrokeratoma (Koenen's tumor)
  - Similar or less keratotic, dome shaped, clinically identical to true fibroma

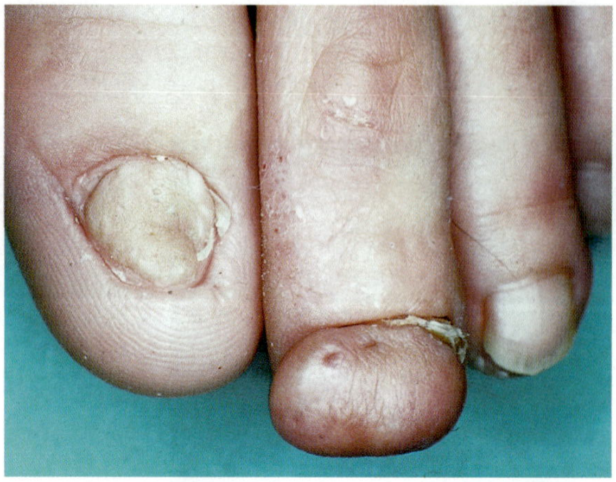

FIGURE 7B.3. Keloid involving the nail bed.

ACQUIRED PERIUNGUAL FIBROKERATOMAS

Acquired periungual fibrokeratomas (Figs. 7B.4 and 7B.5) are identical to acquired digital fibrokeratomas (4) and garlic clove fibromas (5,6). They are benign, spontaneously developing asymptomatic nodules with a hyperkeratotic tip and a narrow base that occur mostly in the periungual area. There may be double or even triple lesions. Takino and Mitoh (7) reported a case in which the lesion was located beneath the nail and visible under the free margin of the great toe nail. Most periungual fibrokeratomas emerge from the most proximal part of the nail sulcus, growing on the nail and causing a sharp longitudinal depression (8). Trauma is thought to be a major factor initiating acquired periungual fibrokeratoma.

Biopsy is mandatory for diagnosing nail tumors, especially as Bowen's disease may simulate acquired ungual fibrokeratoma.

Histologic examination of 50 cases of acquired digital fibrokeratoma disclosed three variants of these lesions (9): (i) a tumor composed of thick, dense, and closely packed collagen bundles; (ii) a variant with an increased number of fibroblasts in the cutis; and (iii) a type demonstrating an edematous and poorly cellular structure.

A 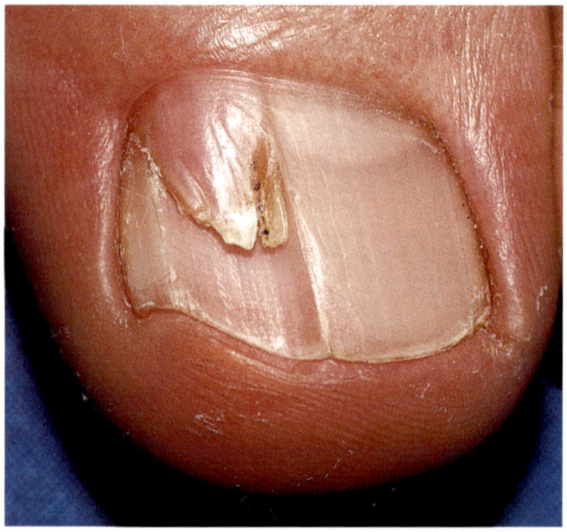 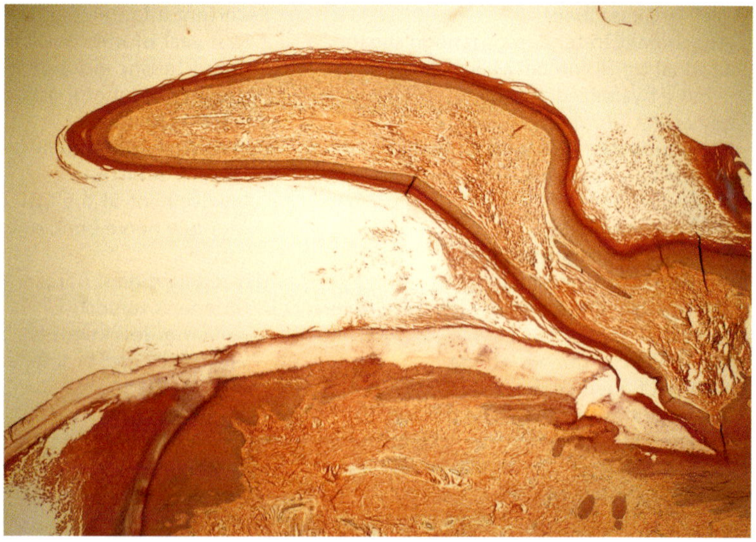 B

FIGURE 7B.4. A: Acquired fibrokeratoma causing a large longitudinal depression of the nail plate. (Courtesy of S. Goettmann, Paris, France.) **B:** Histology of fibrokeratoma.

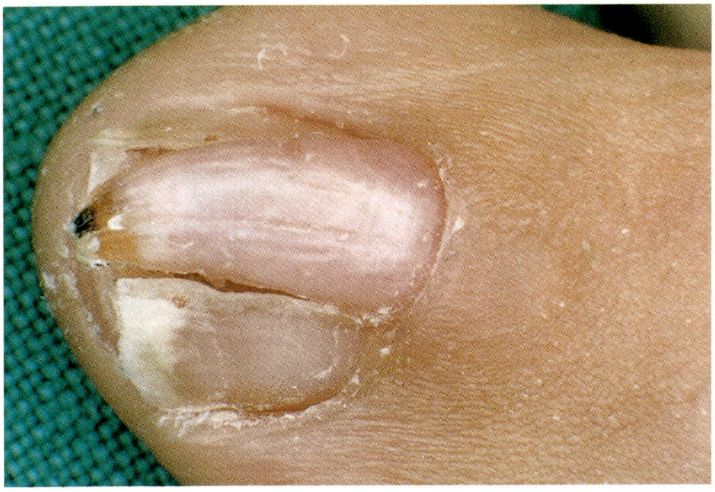

FIGURE 7B.5. Garlic clove fibroma.

Surgical treatment is the same as for Koenen's tumors and depends on the size and location of the tumor.

The differential diagnosis of acquired periungual fibroma requires exclusion of fibroma, keloid, Koenen's tumor, recurring digital fibrous tumors of childhood, dermatofibrosarcoma, fibrosarcoma, cutaneous horn, Bowen's disease (10,11), eccrine poroma, pyogenic granuloma, verruca vulgaris, and exostosis (12).

SUBUNGUAL FILAMENTOUS TUMORS

Subungual filamentous tumors are threadlike horny subungual lesions growing with the nail plate and emerging from under its free edge. They may cause a longitudinal rim. This entity is probably a narrow, extremely hyperkeratotic fibrokeratoma; it pares down painlessly when the nail is cut. Radical treatment requires nail bed exposure and excision of the tether of the lesion.

Histology reveals a subungual rim of keratinous substance in an irregular whorled arrangement. The nail bed may show a single slightly papillomatous projection with marked hypergranulosis.

KOENEN'S TUMORS

Koenen's periungual fibromas develop in 50% of tuberous sclerosis cases (epiloia or Bourneville-Pringle disease) and, thus, occur as frequently as renal hamartomas (Figs. 7B.6 and 7B.7). They usually appear in patients between 12 and 14 years of age and increase progressively in size and number with age. Individual tumors are small, round, flesh colored, and asymptomatic, with a smooth surface, but the tip of the tumor may be slightly hyperkeratotic, resembling fibrokeratoma. These hereditary periungual fibrokeratomas grow out of the nail fold, eventually overgrowing the nail bed and destroying the nail plate. Depending on their location, they may thin the nail or cause a longitudinal depression in the nail plate. Even tiny hyperkeratotic lesions in the cuticle area may produce similar longitudinal nail grooves and have the same significance as Koenen's tumors (13). Very large tumors often are painful.

In the Koenen's tumors we examined (14), two compartments could be distinguished: a small distal segment with loose collagen and many blood vessels, and a larger

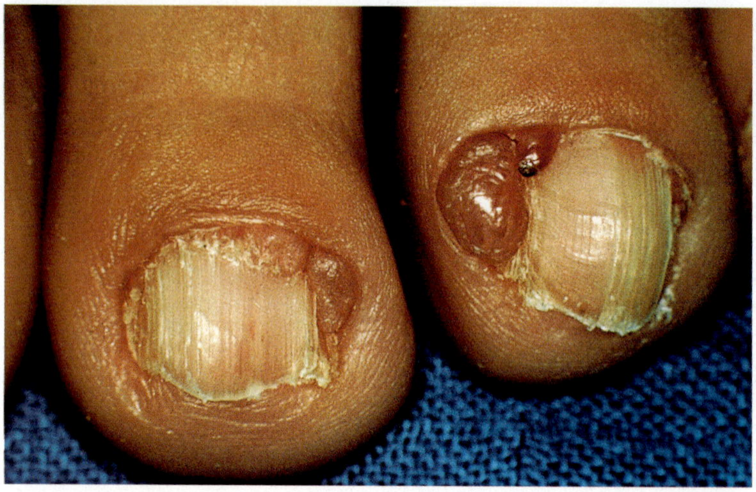

A

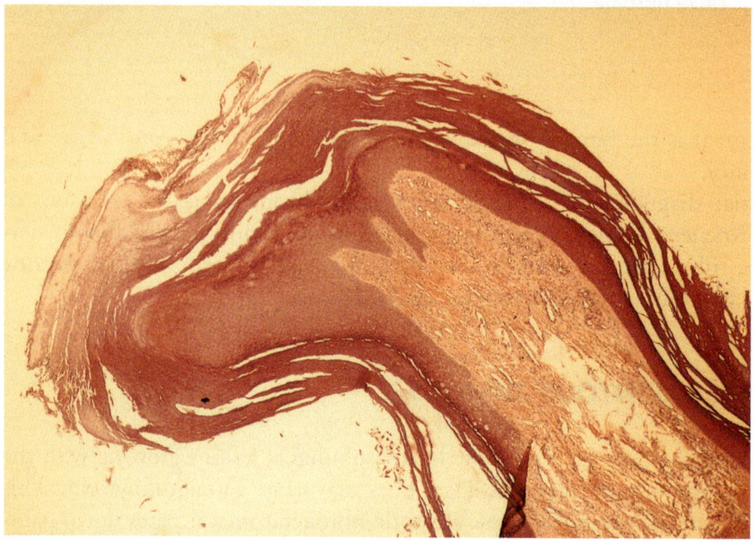

B

FIGURE 7B.6. A: Koenen's tumors. **B:** Histology of Koenen's tumor.

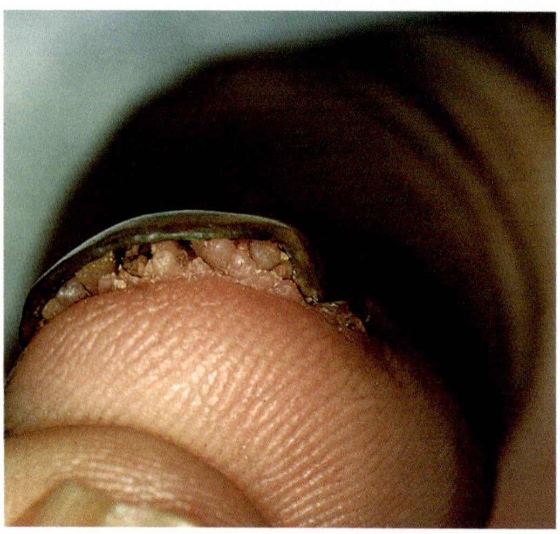

A

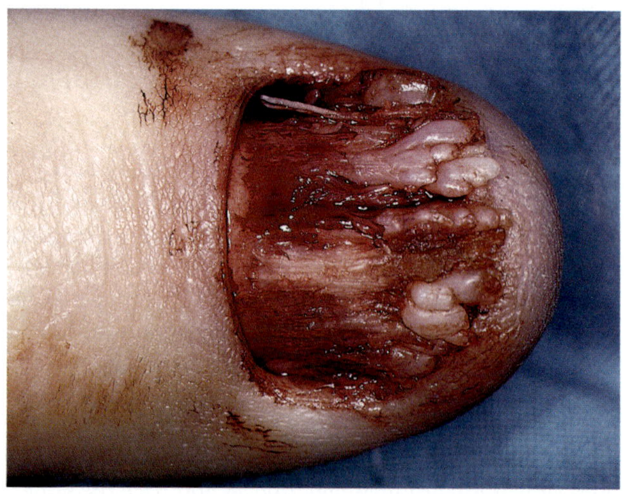

B

FIGURE 7B.7. A: Subungual Koenen's tumor. **B:** Appearance of Koenen's tumor after nail avulsion. (Courtesy of B. Richert, Liege, Belgium.)

proximal part with dense collagen bundles but fewer capillaries. Neither neural or glial appearance (15) nor arteriovenous anastomoses (16) could be found.

Koenen's tumors are treated by simple excision at their base. Usually, no suture is necessary. Tumors growing from under the proximal nail fold are removed after bending back the proximal nail fold by making a lateral incision perpendicular to the lateral curve of the proximal nail fold on one or both sides. Subungual tumors sometimes can be removed after avulsion of the corresponding portion of the nail plate.

LOCALIZED DISTAL SUBUNGUAL KERATOSIS

Distal subungual keratosis is a small horny lesion resembling a forme fruste of the subungual filamentous tumor originating in the hyponychium region (Fig. 7B.8) (17). It may be followed clinically and histologically as far as the lunula and is well demonstrated by magnetic resonance imaging (MRI). Multinucleate cells may be found without any sign of Darier's disease.

INFANTILE DIGITAL FIBROMATOSIS (RECURRING DIGITAL FIBROUS TUMORS OF CHILDHOOD, BENIGN JUVENILE DIGITAL FIBROMATOSIS)

Recurring digital fibrous tumors (Fig. 7B.9) (18) are round, smooth, dome-shaped, shiny (19), firm, tense dermal nodules, with reddish or livid-red color that appear on the dorsal and axial surfaces of the fingers and toes, although they characteristically spare the thumbs and great toes. They may be present at birth or develop during infancy. A single case has been described in an adult (20). Fingers are affected more often than toes. The tumors often are multicentric, occurring on several digits. On reaching the nail unit, they may elevate the nail plate, leading to dystrophy but not destruction, but they may cause considerable distortion of the digits.

Multiple lesions occur in 50% of patients, more often on the fingers, and especially on adjacent fingers. Bilateral lesions may be observed.

Because there is recurrence in 60% of cases after excision, surgery should be attempted only if there is functional impairment. The surgery necessitates going to the

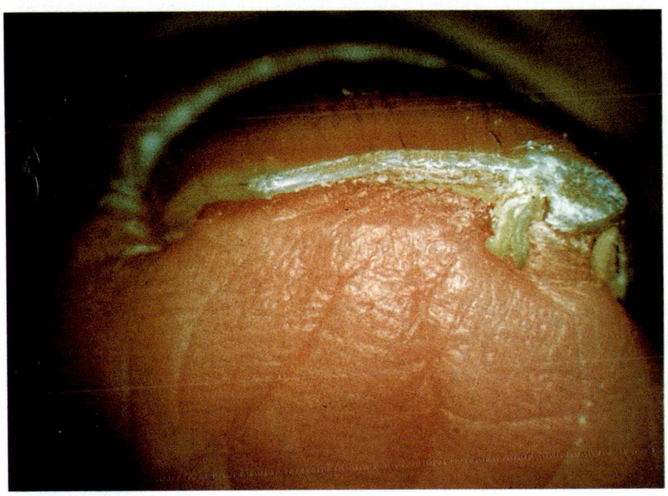

FIGURE 7B.8. Localized distal subungual keratosis.

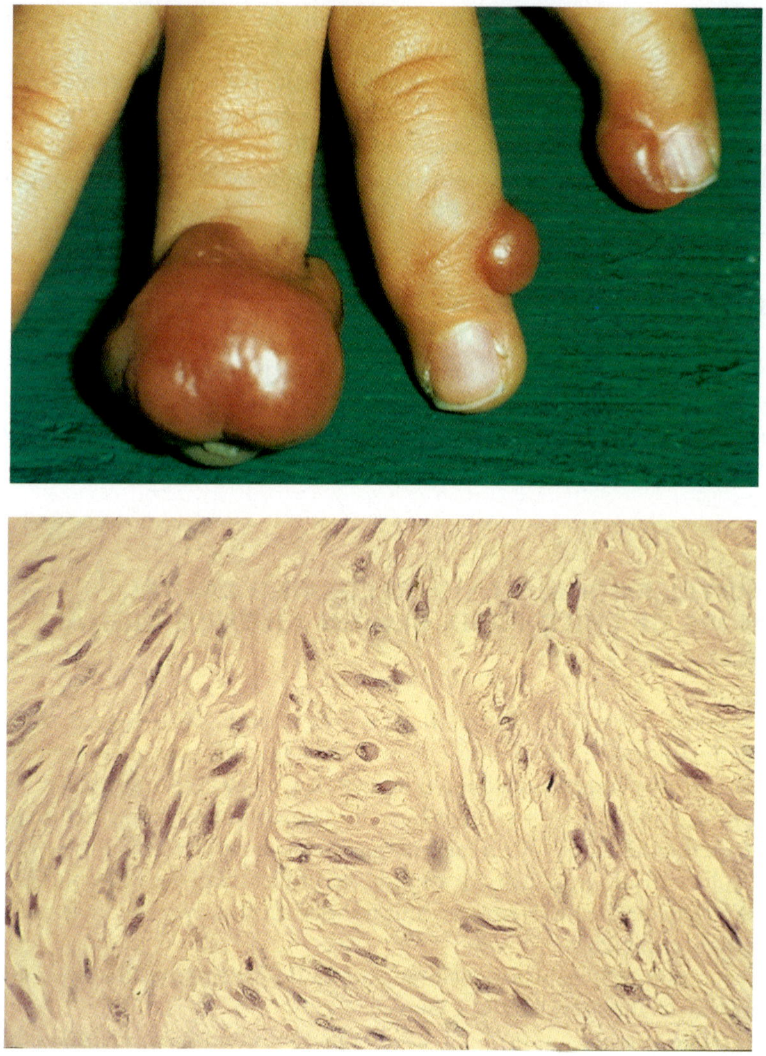

A

B

FIGURE 7B.9. A: Infantile digital fibromatosis. **B:** Histology of the tumor. (Courtesy of Pr. Menni, Milan, Italy.)

fascial and tendinous areas to avoid recurrences and can be mutilating. As metastasis has never been recorded, amputations (38 of 115 cases) can no longer be justified (21).

Conservative treatment is best because the lesions follow a natural course. A tumor stage followed by spontaneous resolution explains why the lesion is not found in adults (22). Cryosurgery may accelerate this natural course.

Paranuclear inclusion bodies, 3 to 10 μm in diameter, can be seen histologically in properly fixed specimens using stains such as iron hematoxylin, methyl green–pyronin, and phosphotungstic acid-hematoxylin, in about 2% of the fibroblasts (23). Under electron microscopy, tumor cells appear as typical myofibroblasts, containing inclusion bodies and bundles of microfilaments. Immunohistochemistry demonstrates that the paranuclear inclusions consist of actin fibers.

Differential diagnosis requires exclusion of pseudoinfantile digital fibromatosis with hypertrophic lateral lips of the great toe in early infancy, fibrosarcoma, and neurofibrosarcoma. Cerebriform connective tissue nevus (24), and juvenile hyaline fibromatosis

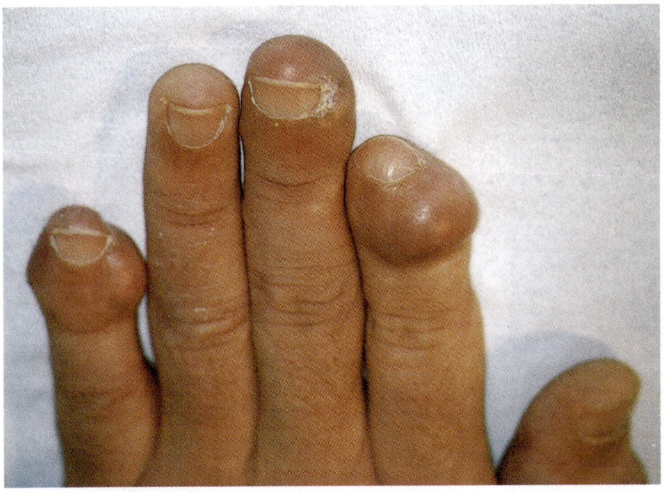

FIGURE 7B.10. Juvenile hyaline fibromatosis involving the dorsal aspect of the distal phalanges.

(Fig. 7B.10) should also be ruled out. Histopathologically, Reye's fibrous tumor may be confused with dermatofibroma, fibroma, and scar tissue.

ONYCHOMATRICOMA (FILAMENTOUS TUFTED TUMOR IN MATRIX OF FUNNEL-SHAPED NAIL)

Clinically identical lesions of the nails of one finger in four patients and on three fingers in another subject were disclosed as a new entity (Fig. 7B.11) (25–27). Four main clinical signs are striking enough to make the diagnosis or to arouse suspicion of the condition.

1. Yellow coloration of variable width along the entire length of the nail plate, leaving a single or double portion of normal pink nail, with splinter hemorrhages in the yellow area involving the proximal nail region.
2. More prominent ridging of the affected nail than of a normal nail from the same patient.
3. Tendency toward transverse overcurvature of the affected nails that becomes increasingly pronounced when the yellow color covers a more extensive surface.
4. Nail avulsion exposes a tumor emerging from the matrix region while the nail appears as a flat funnel, storing filamentous digitations in holes in its proximal section. Retraction of the proximal nail fold when the tumor is small allows its superficial removal from the matrix without suturing. If the root of the tumor is not clearly exposed, the proximal nail fold should be pulled back.

We observed a recurrence in two digits. This may be explained by the MRI features of the lesion that thickens the matrix from its most proximal portion in the ungual cul-de-sac (27).

The tumors all have the same histologic features and consist of epithelial proliferations related to the nail matrix or to the surrounding epidermis. The lobules are delineated by normal basal cells and keratinocytes. A narrow column of dense elongated cells with dark nuclei that are similar in structure to keratinizing matrix cells can be seen centrally.

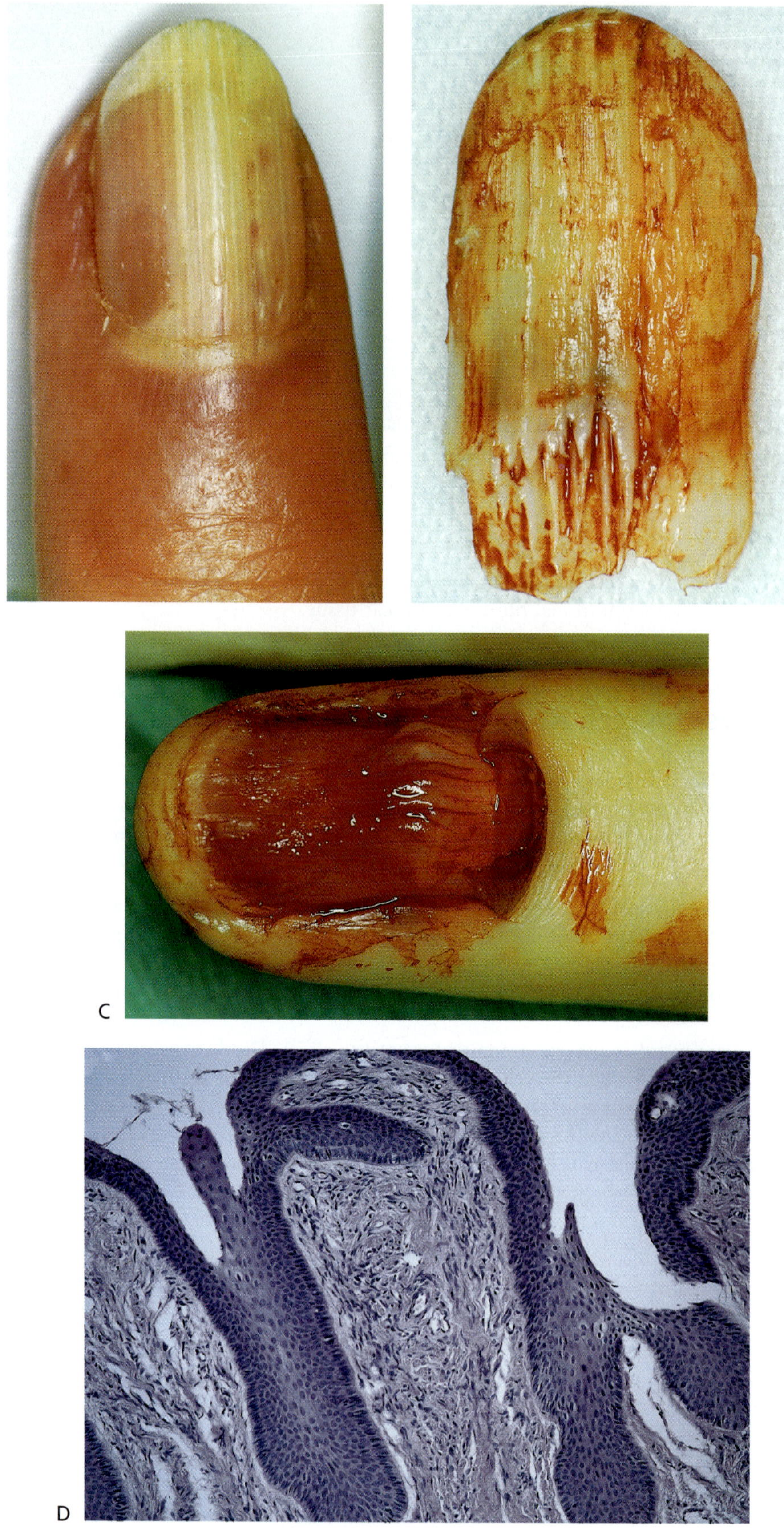

FIGURE 7B.11. A: Onychomatricoma (longitudinal xanthonychia). **B:** Ventral aspect of the nail plate. **C:** Filamentous tumor after nail avulsion. **D:** Histology of the same tumor.

REFERENCES

1. Butler ED, Hamill JP, Seipel RS, et al. Tumors of the hand. *Am J Surg* 1960;100:293–302.
2. Lever WF, Schaumburg-Lever G. *Histopathology of the skin,* 7th ed. Philadelphia: JB Lippincott Co., 1983.
3. Baran R, Perrin C, Baudet J, et al. Clinical and histological patterns of dermatofibromas (true fibromas) of the nail apparatus. *Clin Exp Dermatol* 1994;19:31–35.
4. Bart RS, Andrade R, Kopf AW, et al. Acquired digital fibrokeratomas. *Arch Dermatol* 1968;97: 120–129.
5. Steel HH. Garlic-clove fibroma. *JAMA* 1965;191:1082–1083.
6. LoBuono P, Jothikumar T, Kornblee L. Acquired digital fibrokeratoma. *Cutis* 1979;24:50–51.
7. Takino C, Mitoh Y. A case of acquired ungual fibroma located beneath the nail. *Jpn J Clin Dermatol* 1983;37:57–62.
8. Kikuchi I, Ishii Y, Inoue S. Acquired periungual fibroma. *J Dermatol* 1978;5:235–237.
9. Kint A, Baran R, De Keyser H. Acquired (digital) fibrokeratoma. *J Am Acad Dermatol* 1985;12: 816–821.
10. Haneke E. Epidermoid carcinoma (Bowen's disease) of the nail simulating acquired ungual fibrokeratoma. *Skin Cancer* 1991;6:217–221.
11. Baran R, Perrin C. Pseudo-fibrokeratoma of the nail apparatus with melanocytic pigmentation. A clue for diagnosing Bowen's disease. *Acta Dermatol Venereol* 1994; 74:449–450.
12. Cahn RL. Acquired periungual fibrokeratoma. *Arch Dermatol* 1977;113:1564–1568.
13. Colomb D, Racouchot J, Jeune R. Les lésions des ongles dans la sclérose tubéreuse de Bourneville isolées ou associées aux tumeurs de Koenen. *Ann Dermatol Syphil* 1976;103:431–437.
14. Kint A, Baran R. Histopathologic study of Koenen tumors. *J Am Acad Dermatol* 1988;18:369–372.
15. Nickel WR, Reed JR. Tuberous sclerosis. Special reference to the microscopic alterations in the cutaneous hamartomas. *Arch Dermatol* 1962;85:209–224.
16. Knoth W, Meyhöfer W. Zur Nosologie des Adenoma sebaceum Typ Balzer, der Koenenschen Tumoren und des Morbus Bourneville-Pringle. *Hautarzt* 1957;8:359–366.
17. Baran R, Perrin C. Localized multinucleate distal subungual keratosis. *Br J Dermatol* 1995;133: 77–82.
18. Reye RDK. Recurring digital fibrous tumors of childhood. *Arch Pathol* 1965;80:228–231.
19. Oñate-Cuchet MJ, Vargas Castrillon J, Sanchez Gomez-Coronada P, et al. Fibromatosis digital recurrente infantil. *Acta Dermo-Sif* 1989;80:15–18.
20. Sarma DP, Hoffman ED. Infantile digital fibroma-like tumor in an adult. *Arch Dermatol* 1980;116: 578.
21. Ryman W, Bale P. Recurring digital fibromas of infancy. *Aust J Dermatol* 1985;26:113–117.
22. Sangueza P, Jove N. Fibromatosis digital infantil recidivante. *Med Cut ILA* 1983;11:307–310.
23. Mehregan AM. *Pinkus' guide to dermatohistopathology,* 4th ed. Norwalk, CT: Appleton-Century-Crofts, 1986.
24. Bauer EA, Eisen AZ. Biochemical changes in certain genodermatoses. *Clin Dermatol* 1985;3: 135–142.
25. Baran R, Kint A. Surgical treatment of a filamentous tufted tumor in the matrix of a funnel-shaped nail—a new entity. *Zbl Haut Geschl Kr* 1991;159:337.
26. Baran R, Kint A. Onychomatrixoma. Filamentous tufted tumor in the matrix of a funnel-shaped nail: a new entity. *Br J Dermatol* 1992;126:510–515.
27. Goettmann S, Drapé JL, Baran R, et al. Onychomatricome: deux nouveaux cas-Intérêt de la résonance magnétique nucléaire. *Ann Dermatol Venereol* 1994;121[Suppl 1]:S145.
28. Rapini RP, Golitz LE. Sclerotic fibromas of the skin. *J Am Acad Dermatol* 1989;20:266–271.
29. Requena L, Baran R. Digital angioleiomyoma: an uncommon neoplasm. *J Am Acad Dermatol* 1993;29:1043–1044.
30. Fitzpatrick JE, Mellette JR, Hwang RJ, et al. Cutaneous angiolipoleiomyoma. *J Am Acad Dermatol* 1990;23:1093–1098.
31. Lever WF, Schaumburg-Lever G. *Histopathology of the skin,* 7th ed. Philadelphia: JB Lippincott Co., 1990.

7C

KERATOACANTHOMA

ROBERT BARAN

Characteristically keratoacanthoma, which has a predilection for the three first fingers, especially the thumb, is painful and occurs in the distal nail bed, where its growth separates the nail plate from the nail bed. Partial onycholysis precedes the appearance of a keratotic crusted nodule at the distal nail bed (Fig. 7C.1). The distal digit sometimes is red and swollen. The tumor itself is drastically altered clinically by the confining effect of the overlying nail plate (1). The pain initially is intermittent but increases in intensity and duration (1); however, some lesions cause only slight pain.

In the proximal subungual tissue, the tumor may produce a painful, periungual, paronychialike lesion. When present, drainage from the nail bed varies from firm keratinaceous plugs (2) to cheesy or rarely purulent exudate secondary to bacterial infection (1).

The hallmark of distal digital keratoacanthoma (DKA) is its rapid growth, which is within a period of weeks. This serves to differentiate it from squamous cell carcinoma. There may be a preceding history of trauma.

Whether trauma plays a role in DKA has not been shown conclusively. Spontaneous resolution has been noted in only three patients without incontinentia pigmenti (IP). In subungual tumors incontinentia pigmenti (STIP), it was observed in nine cases that we collected, but regression may occur only during pregnancy.

The radiographic appearance of subungual keratoacanthoma consists of a punched-out area of bone destruction in the terminal phalanx and well-circumscribed margins of the radiologic defect (Fig. 7C.2). This crescent-shaped lytic defect without accompanying sclerosis or periosteal reaction (3) is limited to the bone underlying the tumor and is the result of pressure erosion.

The histologic changes of DKAs have the same general configuration as keratoacanthomas located elsewhere, but they tend to be more vertically oriented. They have more dyskeratotic cells, fewer neutrophils and eosinophils in their epithelium, and little or no fibrosis at their bases (4).

Keratoacanthoma should be considered in the differential diagnosis of painful lesions of the distal phalanx. Clinically, there are three main diagnoses of practical significance, epidermoid implantation cyst, verruca vulgaris, and squamous cell carcinoma (SCC); and histologically, SCC.

Local excision and curettage are sufficient and are the initial treatment of choice (5). Mohs' micrographic surgery performed at the time of curettage to determine adequacy of surgical margins may reduce the potential risk of persistent or recurrent tumor and allow preservation of the involved fingertip.

Amputation should be reserved for those cases of multiple recurrence that do not respond to etretinate or acitretin, cases of extensive bony destruction, or cases in which a clear differentiation from SCC cannot be made (6,7).

R. Baran: Nail Disease Centre, 42, Rue des Serbes, 06400 Cannes, France.

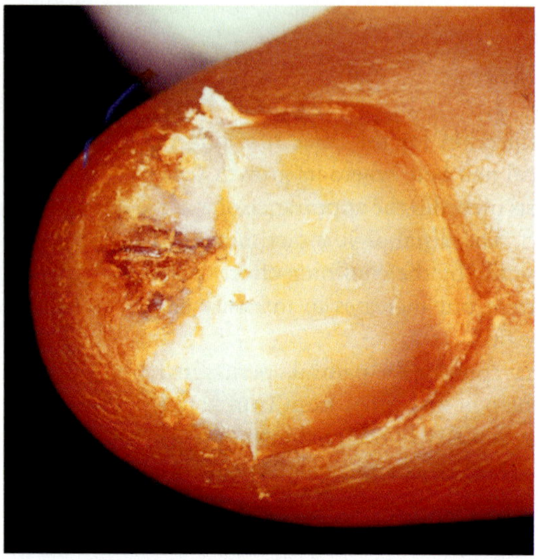

FIGURE 7C.1. Keratoacanthoma of the distal nail bed.

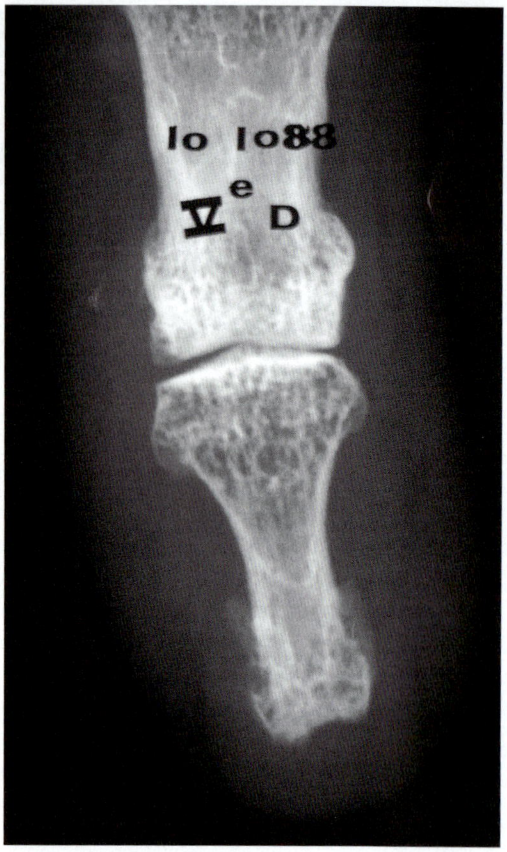

FIGURE 7C.2. Well-circumscribed defect due to keratoacanthoma.

REFERENCES

1. Lamp JC, Graham JH, Urbach F, et al. Keratoacanthoma of the subungual region. *J Bone Joint Surg* 1964;46A:1721–1731.
2. Fisher AA. A distinctive, destructive, digital disease: follow-up. *Arch Dermatol* 1968;98:98.
3. Levy DW, Bonakdarpour A, Putong PB, et al. Subungual keratoacanthoma. *Skeletal Radiol* 1985;13: 287–290.
4. Gimenez Moreno JC, Lerma Puerta E, Sanchez Conejo-Mir J, et al. Queratoacantoma subungueal. *Acta Dermo-Sif* 1987;78:561–564.
5. Patel MR, Desai SS. Subungual keratoacanthoma in the hand. *J Hand Surg* 1989;14A:139–142.
6. Lovett JA, Haines TA, Bentz ML. Subungual keratoacanthoma masquerading as a chronic paronychia. *Plast Surg* 1995;38:84–87.
7. Baran R, Goettmann S. Distal digital keratoacanthoma: a report of 12 cases. *Br J Dermatol* 1998;139: 512–515.

7D

GLOMUS TUMORS

EDWARD A. KRULL

Glomus tumors, first well defined by Barré and Masson (1), most commonly occur in the subungual region of fingers. Females are affected more often than males. In my experience, most of the subungual glomus tumors have been in females.

The most characteristic history is an acutely tender point in the nail tissue in which pain is triggered by pressure, trauma, and cold. The pain is intense, lancinating, and pulsating, and it frequently radiates up the arm, sometimes up to the shoulder (2–5). Because of minor trauma initiating the symptomatology, the patient may take a guarded, protective attitude with the involved fingernail. Because of these symptoms and the fact that the tumor may be hidden under the proximal nail fold, causalgia and neuroma may be diagnosed incorrectly and may result in unfortunate and unfavorable attempts at therapy, such as posterior rhizotomy and amputation (2).

The history alone should alert the physician to the possibility of a glomus tumor. A visible bluish-red tumor of the nail tissue, usually of the matrix, is further evidence (Figs. 7D.1 and 7D.2). However, the tumor may be hidden by the proximal nail fold (Figs. 7D.3 and 7D.4). The matrix may be distorted by the tumor within it (Fig. 7D.3). The nail plate may show changes of ridging and distal splitting (Figs. 7D.4–7D.7) (3). These changes result from the tumor compressing and replacing the nail matrix, which subsequently produces a thin nail plate. As the nail plate grows out, normal wear and tear causes the nail plate to split. An important sign is reproduction of the pain with pressure. Applying gentle pressure with the wooden end of an applicator stick will initiate severe pain (Fig. 7D.8). The same pressure on surrounding tissue will not trigger pain. This method is also an excellent way to localize the tumor if it is hidden by the nail fold.

Various studies, such as ultrasound, magnetic resonance imaging, and arteriography, have been suggested for diagnosis and location of the lesion (4,6). Ultrasound of a glomus tumor is shown (Fig. 7D.9), but these diagnostic tests are expensive and I believe they are not usually necessary. Pressure test, history, and nail plate changes, even if the bluish tumor is hidden by the nail fold, should prompt the surgeon to perform confirmatory and definitive exploration and surgery of the nail tissue.

Histopathology of glomus tumor reveals an encapsulated circumscribed lesion. Large epithelioid tumor cells surround vascular spaces that are lined with cuboidal endothelial cells. The stroma is edematous and contains mucin. Nerve fibers may be seen and may be related to the pain. The immunochemistry of glomus cells is positive for vimentin, a 42-kDa muscle actin, and a smooth muscle actin, but negative for desmin and factor VIII-related antigen, whereas endothelial cells stain with factor VIII-related antigen (3,7). Nerve fibers contain S-100 protein, Leu 7 antigen, nonspecific enalase, and neurofilaments (3,7).

E. A. Krull: Department of Dermatology, Henry Ford Hospital, Detroit, Michigan 48202.

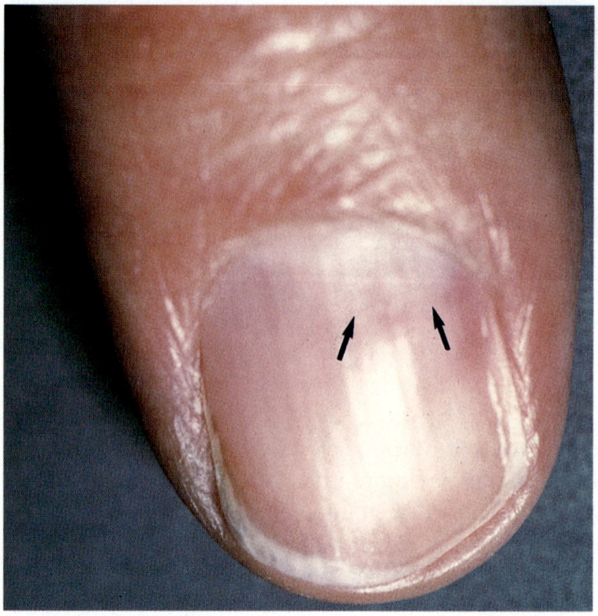

FIGURE 7D.1. Bluish glomus tumor of matrix that seems to extend into the nail bed *(arrows)*. There is history of pain and area being tender to touch.

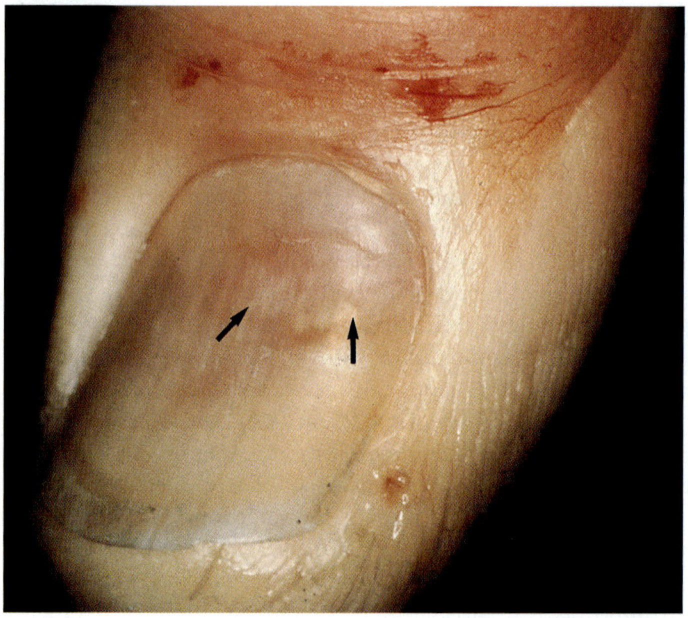

FIGURE 7D.2. Bluish glomus tumor of matrix that was painful and tender.

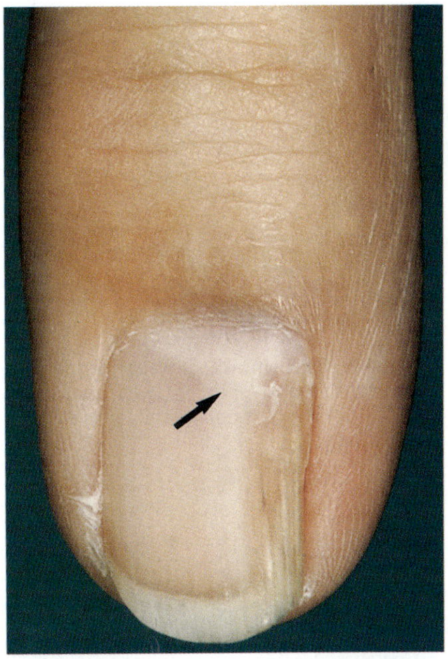

FIGURE 7D.3. Nail dystrophy with painful and tender glomus tumor. Distal matrix edge is fuzzy and distorted *(arrow)*.

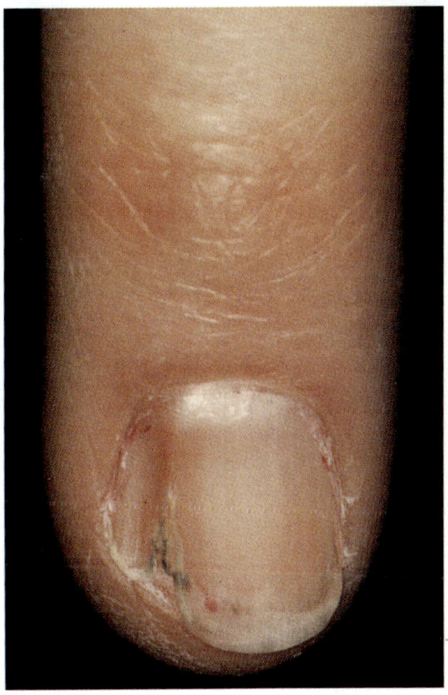

FIGURE 7D.4. Distally split nail with proximal ridge. Painful and tender site beneath proximal nail fold without visible bluish glomus tumor.

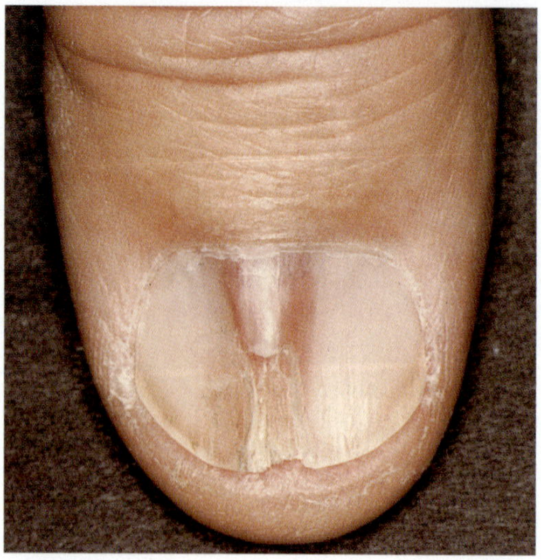

FIGURE 7D.5. Bluish glomus tumor that was tender and painful, with associated partial splitting of nail distally and ridging proximally.

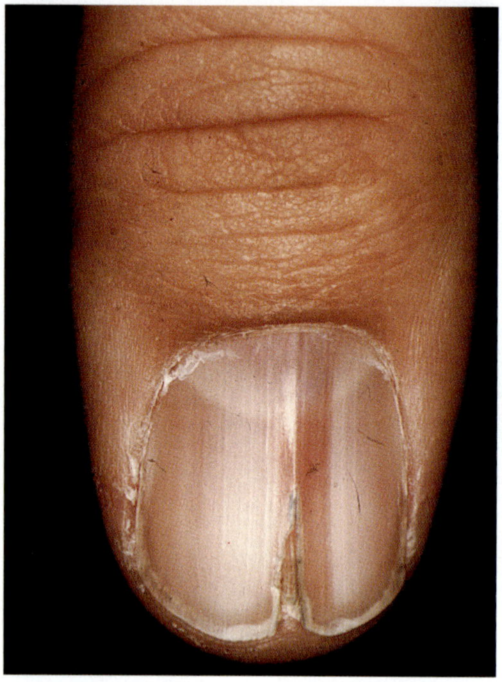

FIGURE 7D.6. Bluish glomus tumor of matrix that was painful and tender, with proximal nail ridging and distal nail splitting.

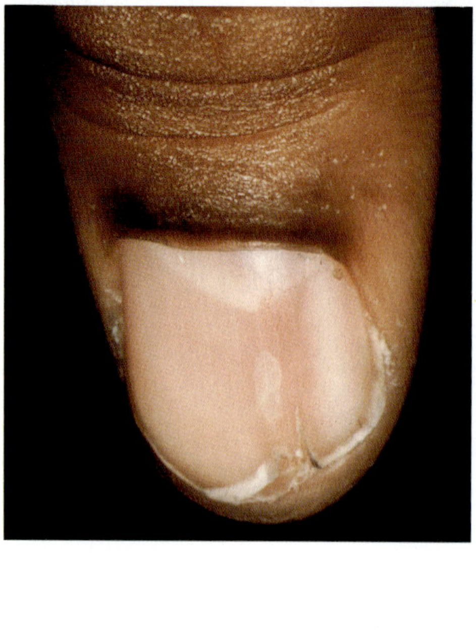

FIGURE 7D.7. Faint bluish glomus tumor of proximal lunula that was painful and tender. There was some pink ridging of proximal nail plate and small split of nail distally.

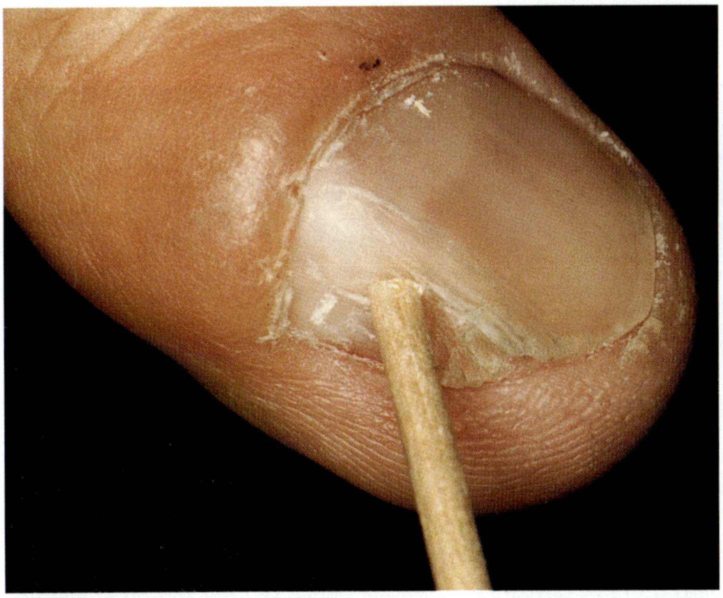

FIGURE 7D.8. Glomus tumor that was severely tender when pressure was applied with the wooden end of a cotton-tipped swab.

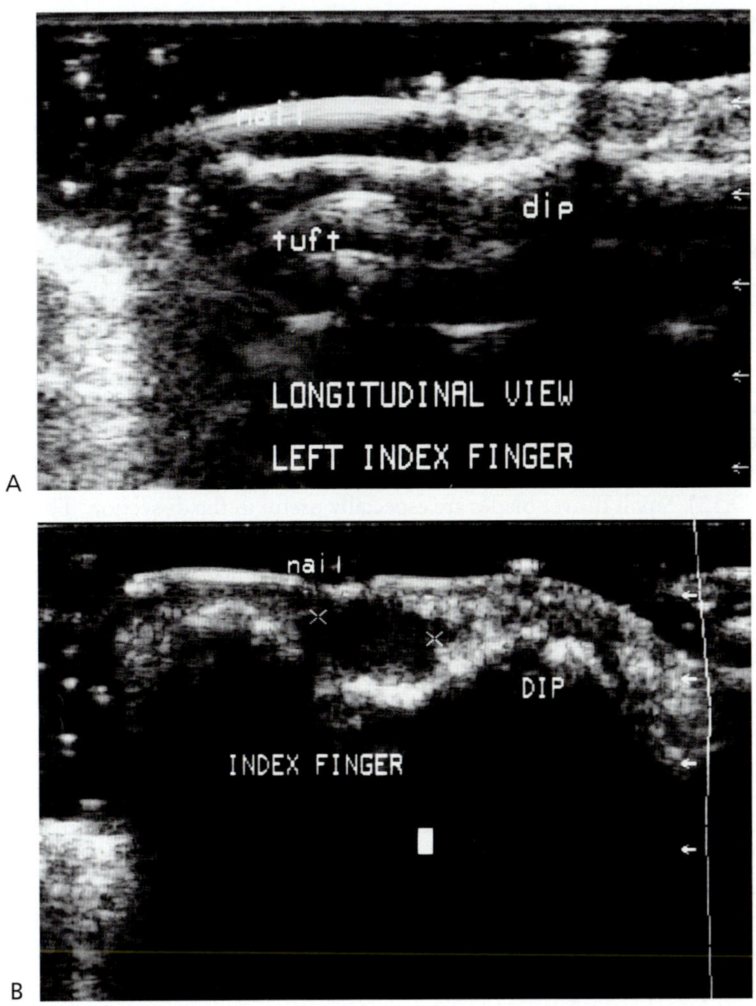

FIGURE 7D.9. Ultrasound of glomus tumor. **A:** Longitudinal section of uninvolved finger. **B:** Longitudinal view of involved finger. Tumor marked by *X*. *(continued)*

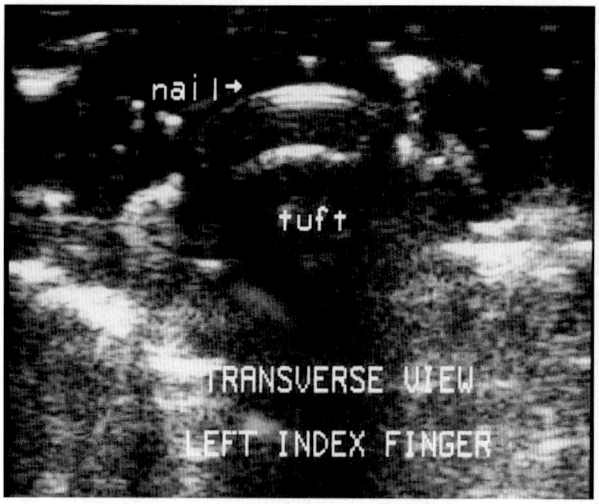

C

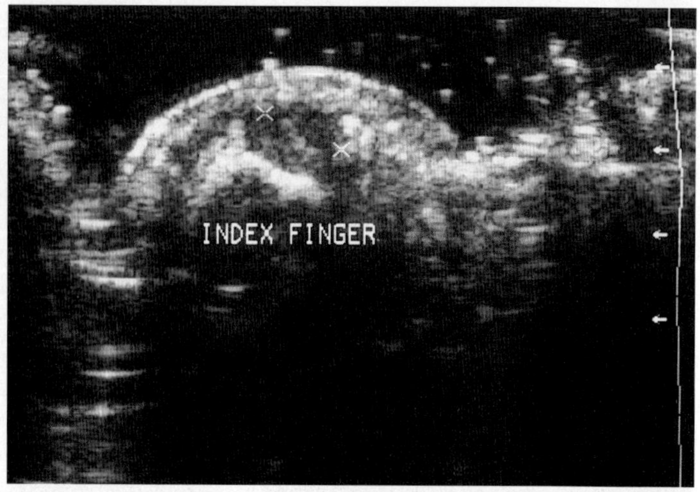

D

FIGURE 7D.9 *(continued)*. **C:** Cross section of uninvolved finger. **D:** Cross section of involved finger. Tumor marked by *X*.

SURGERY FOR REMOVAL OF GLOMUS TUMOR

After localization of the tumor, the finger is anesthetized in the usual manner and a tourniquet applied. The nail plate is removed in the manner described in the section on avulsion of the nail. If the lesion has been identified as being in the matrix, incisions are made in the proximal nail folds detailed in biopsy and exploration technique (Fig. 7D.10). This allows the proximal nail fold to be reflected to visualize the entire matrix. If the tumor lies totally beneath the proximal nail fold, the glomus tumor cannot be visualized until this fold is reflected (Fig. 7D.10C). The tumor is identified by its reddish-blue color or bulge in the matrix. In principle, incisions over the tumor preferably are made in a transverse direction to avoid injury to the proximal matrix and possible resultant nail deformity (Fig. 7D.11). However, in reality, an incision in the proximal matrix without removal of matrix tissue probably will have no effect on the regrowing nail plate. Incisions in the distal matrix do not have to be closed but are preferentially closed by the author.

The tumor in the tissue, which may be reddish-blue or gray, is identified and excised (Figs. 7D.11C and 7D.12). Small Beaver blades are especially useful in the dissection. The defect in the matrix is closed with 5-0 absorbable sutures, such as polyglactin (Fig. 7D.13). The incisions in the skin of the proximal nail fold are closed with 5-0 nylon sutures, beginning at the cuticular border.

On occasion, the tumor may be located in the nail bed (Fig. 7D.14). A longitudinal incision, parallel to the rete ridges, is made through the nail bed. The tumor is excised. The author prefers to close the wound with 5-0 absorbable sutures, although little, if any, adverse effect to the adherence of the nail plate would result from not closing the incision of the nail bed. The nail bed tears more easily than the matrix and must be handled more gently. The wound is dressed with antibiotic ointment, petrolatum gauze, fine-mesh gauze cut to size, and a self-adhering dressing such as Kling, and completed with tubular elastic net.

Other techniques can be used to remove the tumor. If the tumor is small, a 6-mm punch through the nail plate alone will expose the tissue with the tumor, which then can be excised through the opening of the nail. The circle of nail plate can be used as a dressing of the surgical wound (3). However, the tumors may be larger than anticipated, requiring removal of the entire or proximal part of the nail plate to obtain enough exposure for resection of the tumor. If the glomus tumor is located in the lateral part of the matrix, it may be removed through an incision in the lateral nail wall (3). However, the size of the tumor may make this approach more difficult.

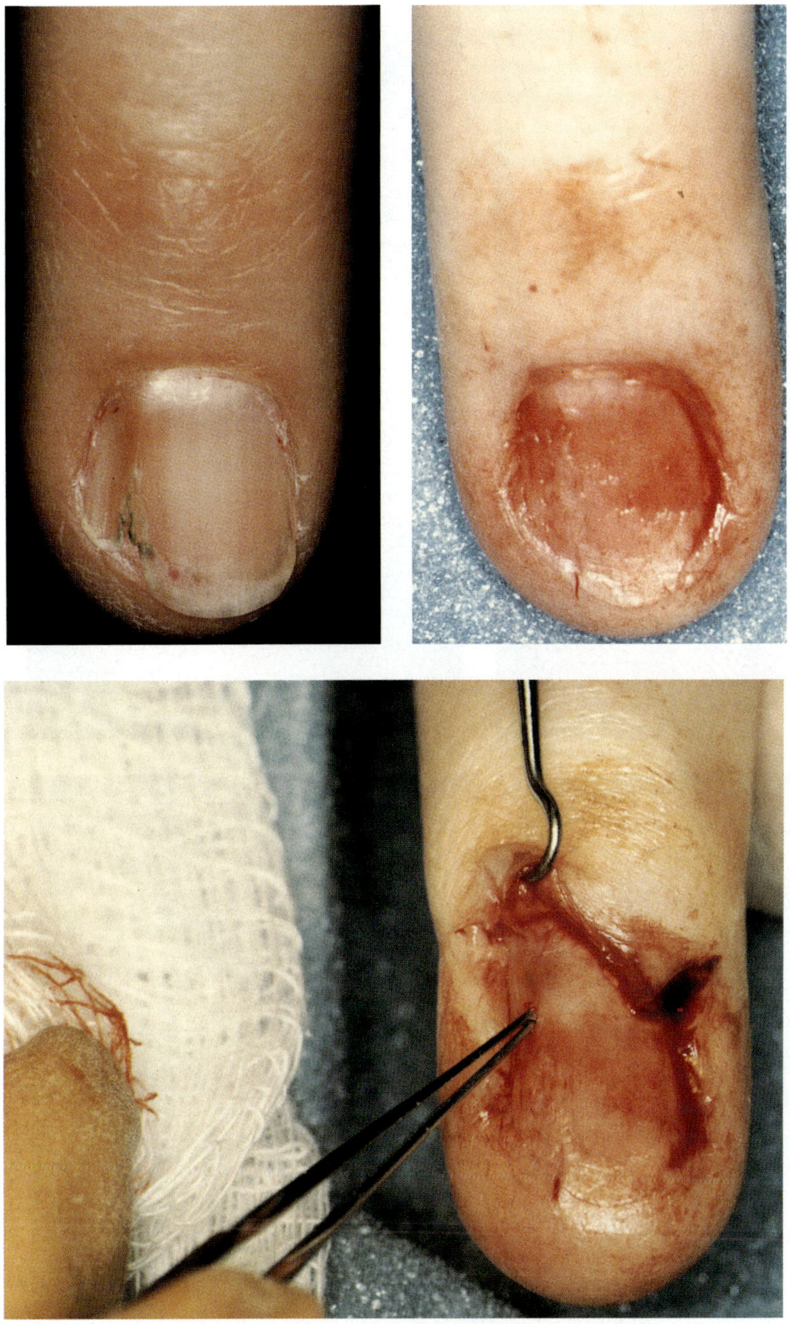

FIGURE 7D.10. A: Glomus tumor identified by history, point tenderness, and nail changes. Proximal nail fold obscures tumor. **B:** Nail plate removed. Glomus tumor is not visible. **C:** Incision through proximal nail folds reveals blue glomus tumor of the midmatrix.

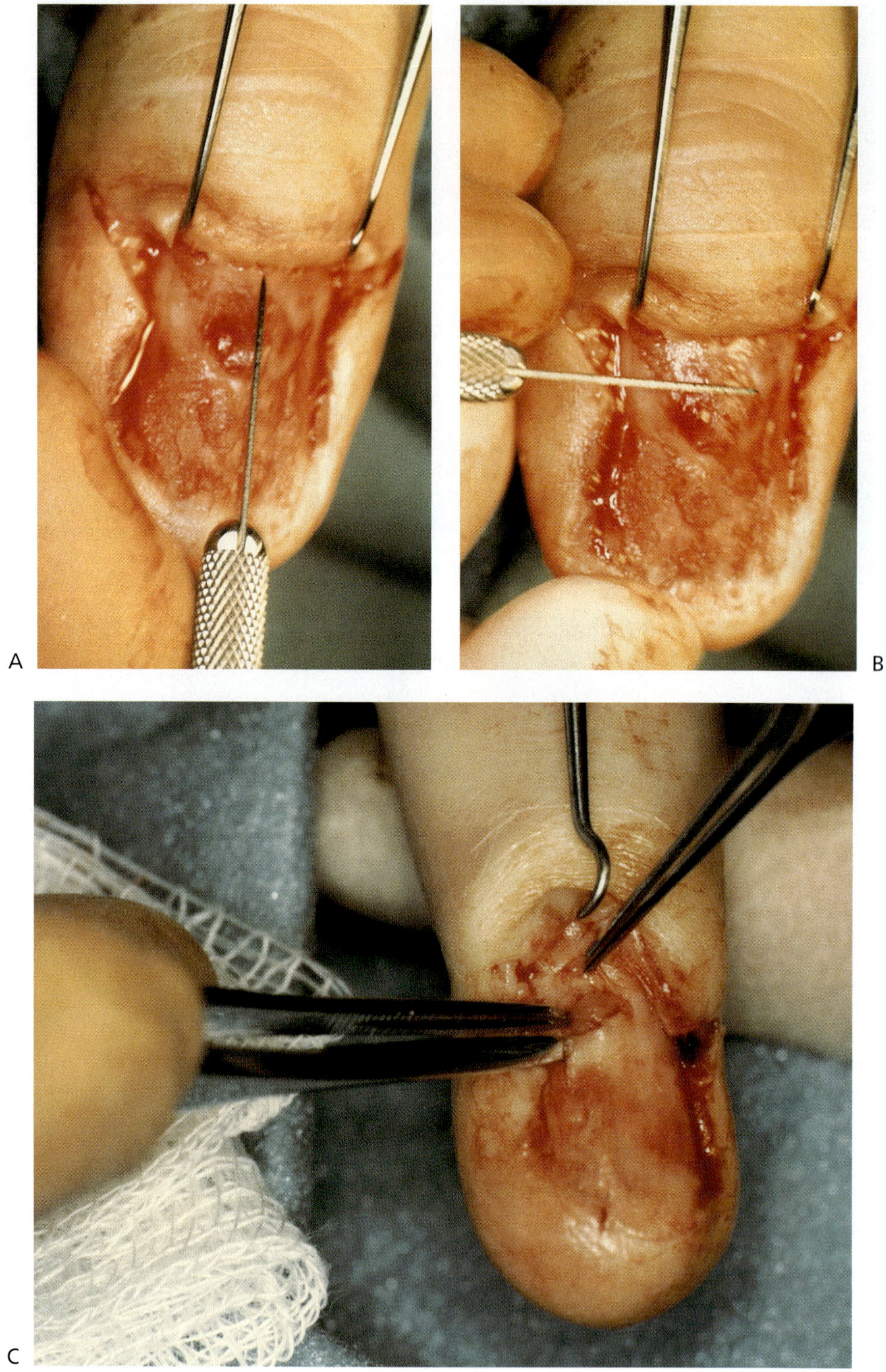

FIGURE 7D.11. A: Incisions for removal of glomus tumor. Glomus tumor identified. Longitudinal incision for excision considered. **B:** Transverse incision preferred to minimize possibility of proximal matrix injury and resultant nail dystrophy. **C:** Transverse incision revealing gray glomus tumor.

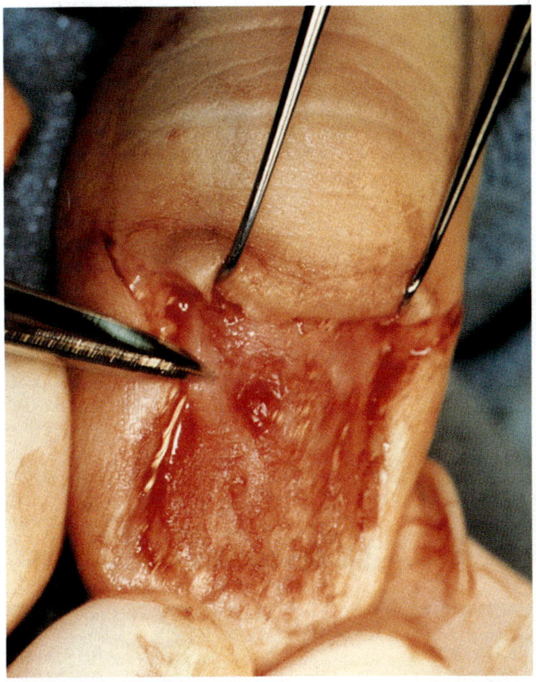

A

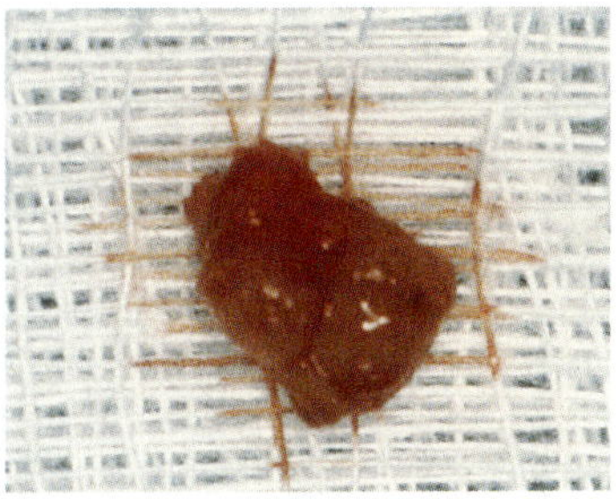

B

FIGURE 7D.12. A: Reddish-blue color of glomus tumor. Some tumors are blue and some are gray. **B:** Gray color of excised glomus tumor.

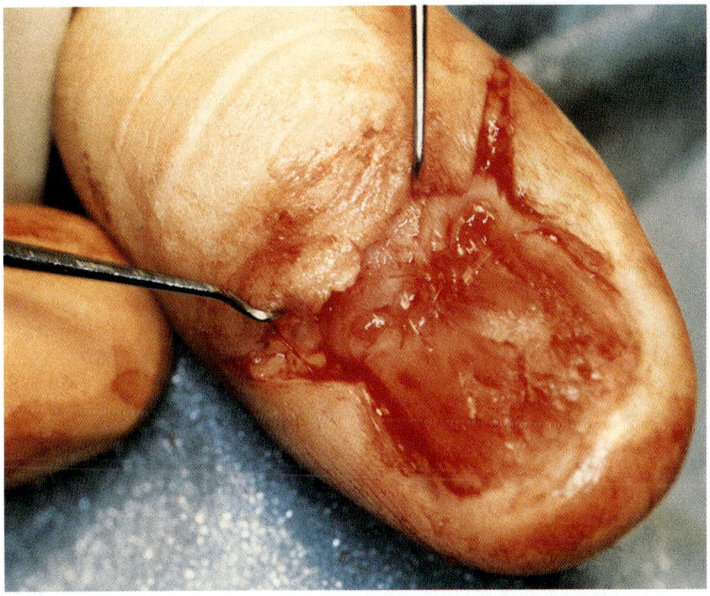

FIGURE 7D.13. Transverse incision for removal of a glomus tumor, closed with 5-0 polyglactin suture. The author prefers to close incisions in the distal matrix, although it is generally not necessary.

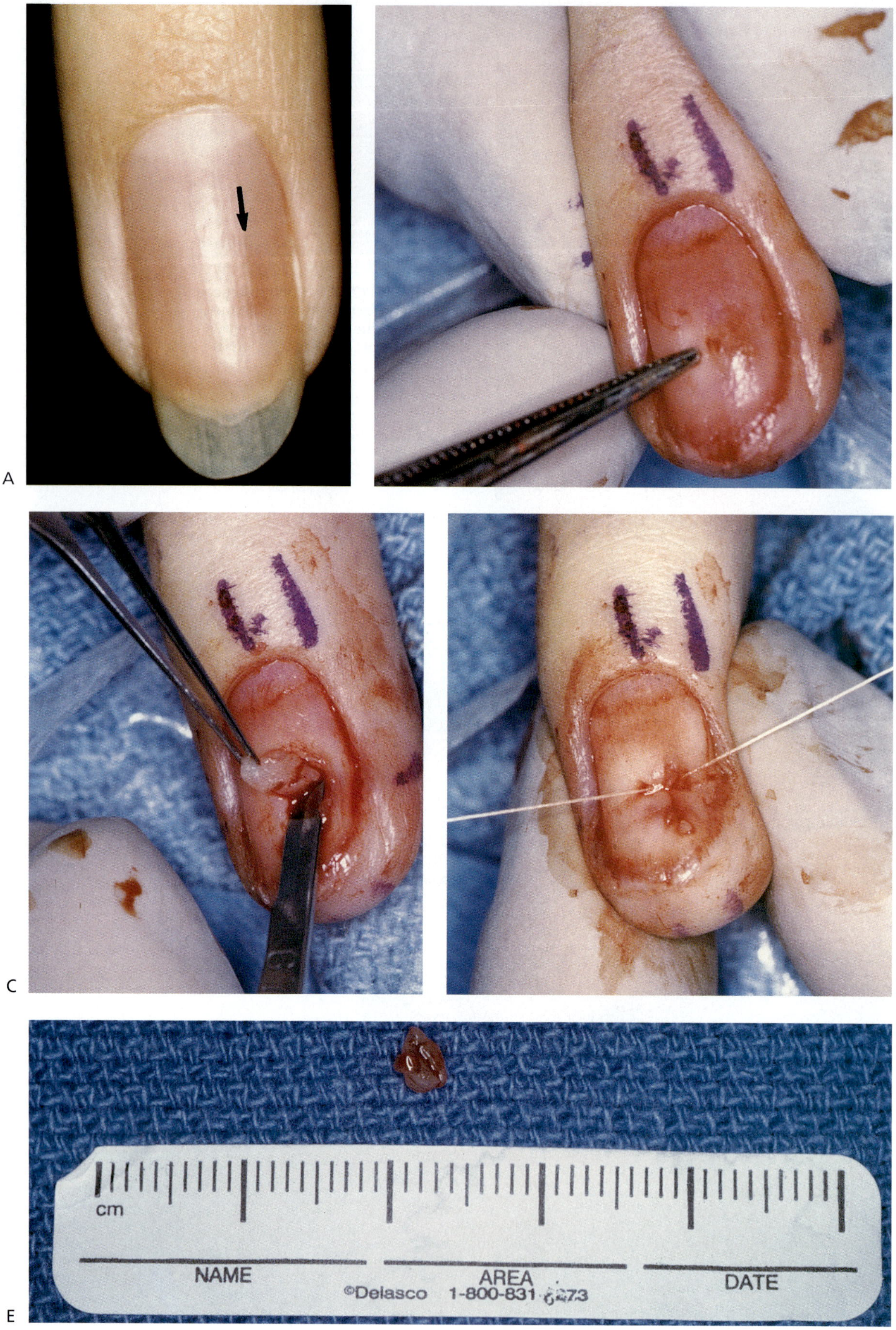

FIGURE 7D.14. Glomus tumor of the nail bed. **A:** Reddish-blue glomus tumor of nail *(arrow)*. Tumor was painful and tender. **B:** Nail plate removal reveals reddish-blue tumor. Purple lines were used to demarcate tumor before the nail plate was removed in case the tumor was difficult to locate. **C:** Tumor is partially excised. Longitudinal incision parallels rete ridges of the nail bed. **D:** Wound closed with polyglactin. **E:** Gray tumor specimen.

REFERENCES

1. Barré JA, Masson P. Etude anatomo-clinique de certaines tumeurs sous-unquéales douloureuses (tumeurs du glomus neuro-myoarterial des extémités). *Bull Soc Fr Dermatol Syphil* 1924;31:149–159.
2. Rettig AC, Strickland JW. Glomus tumor of the digits. *J Hand Surg* 1977;2:261–265.
3. Baran R, Dawber RPR, eds. *Diseases of the nails and their management,* 2nd ed. London: Blackwell Science, 1994:455–457.
4. Jablon RP, Howowith A, Bernstein DA. Magnetic resonance imaging of a glomus tumour of the fingertip. *J Hand Surg* 1990;15A:507–509.
5. Heys SD, Brittenden J, Atkinson P, et al. Glomus tumour: an analysis of 43 patients and review of the literature. *Br J Surg* 1992;79:345–346.
6. Holzberg M. Glomus tumor of the nail. A "red herring" clarified by magnetic resonance imaging. *Arch Dermatol* 1992;128:160–162.
7. Herbst WM, Nakoyama K, Hornstein OP. Glomus tumors of the skin: an immunohistochemical investigation of the expression of marker proteins. *Br J Dermatol* 1991;124:172–176.

7E

LONGITUDINAL MELANONYCHIA

EDWARD A. KRULL

Longitudinal melanonychia (LM) is caused by melanin pigment in the nail plate. It is observed most commonly in African-Americans and other dark-skinned races, reflecting the close relationship of nail melanogenesis and skin color (1). A number of medical conditions and drugs may produce LM (Table 7E.1 and Fig. 7E.1). Subungual hematomas, especially of the nonmigratory variant, foreign bodies, and microbial infections that produce pigmentation may simulate LM (Figs. 7E.2–7E.5). Of greatest importance is to exclude atypical melanocytic hyperplasia and subungual melanoma as the cause of LM.

The management of LM is subject to much controversy (Table 7E.2) This controversy is derived from a number of factors, which include the high prevalence of LM, the unresolved criteria and clinical characteristics of LM in terms of melanoma development, the undetermined likelihood of the evolution of LM into melanoma, and the specific site on the pigmented streak that could become malignant. These issues raise the question of when and how to biopsy LM.

The incidence of LM is significantly increased in dark-skinned people as compared to whites. Nearly 100% of African-Americans have LM by the age of 50 years (1–4). Longitudinal melanonychia occurs in 20% of Japanese individuals and frequently is observed in Hispanics (1). In whites the incidence of LM is 1%, although a recent study suggests the incidence is higher (Table 7E.3) (4). LM increases with age and probably is associated with trauma. The thumb and index finger as well as the first toe are the most frequent sites of LM, possibly reflecting trauma to these digits (5).

There is evidence suggesting that LM is a precursor lesion of actual melanoma. Some studies indicate that 30% or more of subungual melanomas begin with LM (6,7). Other reports support the observation that subungual melanomas may start as LM many years before the appearance of melanomas (9–13). Secondary melanomas may present as pigmented streaks (8).

In addition, features of LM and subungual melanoma are intriguingly similar, suggesting the relationship of LM and melanoma (Table 7E.4) Both predominantly occur on the thumb, index finger, and first toe. Longitudinal melanonychia and subungual melanoma have a disproportionately greater incidence in dark-skinned people. Both seem to have a relationship to trauma.

In the face of the information associating LM and subungual melanoma, what should be done? Should all LM be biopsied or excised? Based on the data of the incidence of LM in the population of the United States, there are probably millions of people, possibly more than 60 million, with LM, whereas subungual melanoma is not common (Tables 7E.3, 7E.5, and 7E.6) (14). It would seem impractical and unnecessary to biopsy or excise all LMs any more than one should excise all nevi of the skin to reduce the number of melanomas. What is needed are reliable and clinically useful criteria and characteristics of LM that are more indicative of early subungual melanoma or more likely to

E. A. Krull: Department of Dermatology, Henry Ford Hospital, Detroit, Michigan 48202.

TABLE 7E.1. CAUSES AND SIMULATION OF LONGITUDINAL MELANONYCHIA

Trauma
Foreign body
Subungual non-migratory hematoma
Acquired and congenital melanocytic nevi
Proliferation of melanocytes, normal and atypical
Melanoma *in situ,* invasive, and metastatic melanoma
Bowen's disease
Laugier-Hunziker syndrome
Lichen planus
Drugs, such as adrenocorticotropic hormone, ketoconazole, psoralens, minocycline, zidovudine, cancer chemotherapeutic agents
Pregnancy
Microbiologic agents: chromogenic bacteria coexisting with dermatophytes and *Candida,* acquired immunodeficiency syndrome
Addison's disease
Malnutrition
Radiotherapy of digits
Breast carcinoma without therapy

Modified from Baran R, Kechijian P. Longitudinal melanonychia (melanonychia striata): diagnosis and management. *J Am Acad Dermatol* 1989;21:1165–1175.

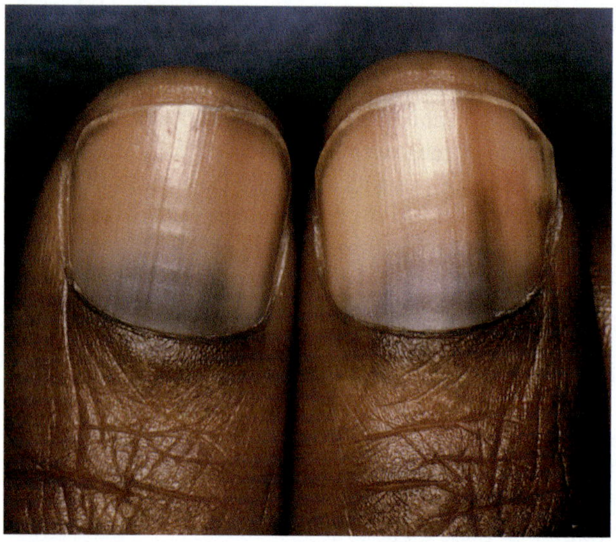

FIGURE 7E.1. Bluish-gray pigmentation of lunula and associated longitudinal melanonychia (LM) from zidovudine (AZT).

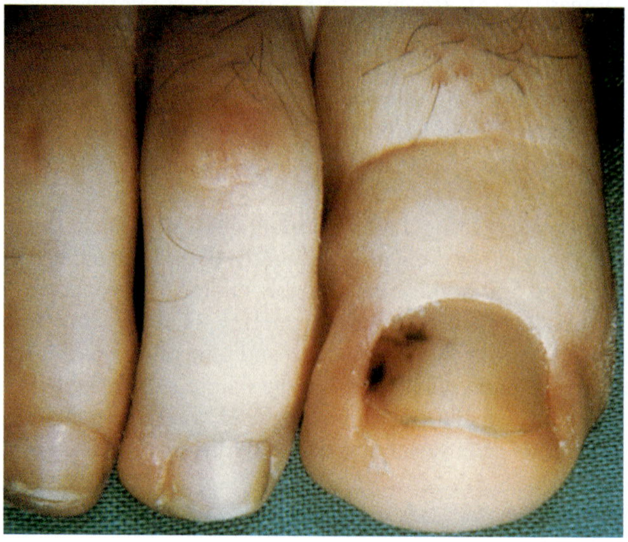

FIGURE 7E.2. Nonmigratory subungual hematoma.

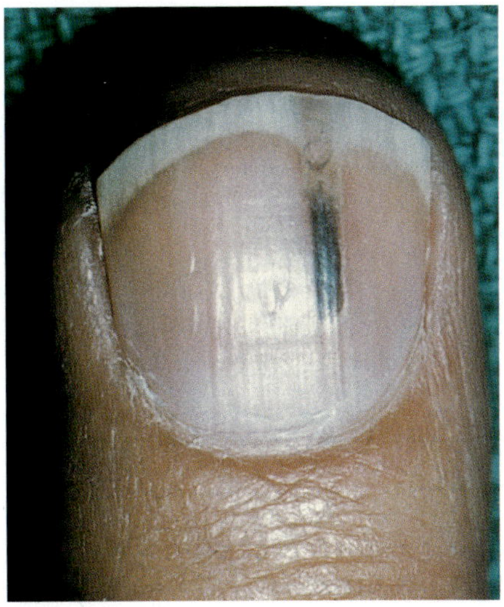

FIGURE 7E.3. Subungual lesion from foreign body that might raise concern for longitudinal melanonychia or melanoma.

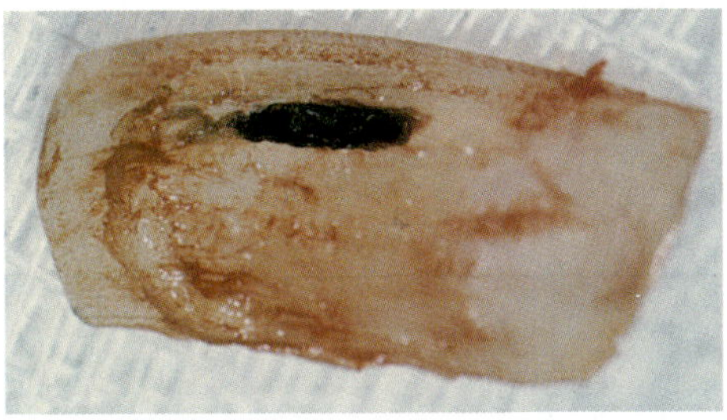

FIGURE 7E.4. Nail plate removed, revealing foreign body attached to undersurface of the nail.

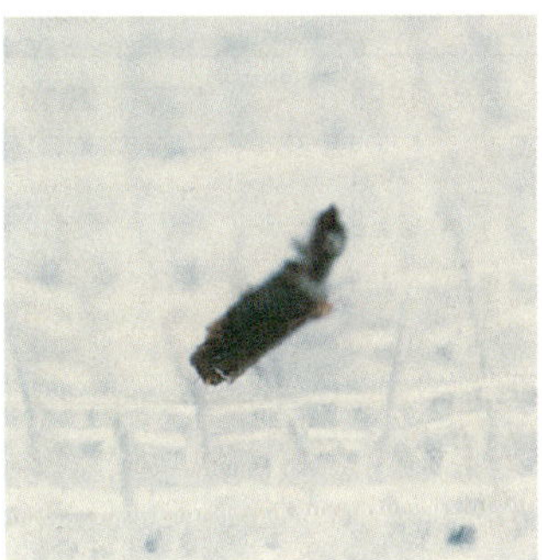

FIGURE 7E.5. Foreign body suggestive of piece of wood.

TABLE 7E.2. ISSUES OF LONGITUDINAL MELANONYCHIA

1. Incidence
2. Diagnosis
3. Relationship to subungual melanoma
4. Biopsy
   A. When
   B. How

TABLE 7E.3. INCIDENCE OF LONGITUDINAL MELANONYCHIA

1. African-American
   A. 77% over age 20 years
   B. 100% over age 50 years
2. Japanese 20%
   A. Chinese about the same
3. White 1%
4. Hispanic origin 10% or more

TABLE 7E.4. COMPARISON OF SUBUNGUAL MELANOMA AND LONGITUDINAL MELANONYCHIA

	Subungual melanoma	Longitudinal melanonychia
Toe, thumb, index finger	+	+
Increased incidence in African-Americans	+	+
Black or darkly pigmented streak	+	+ especially in African-Americans
Childhood	Rare	±
Trauma	+	+
Nail plate destruction	+	–

TABLE 7E.5. BUREAU OF THE CENSUS PROJECTED UNITED STATES POPULATION (IN MILLIONS) FOR 2010 BASED ON CURRENT TRENDS

	Population	
Census	1995	2010
African-American	33	40
White	218	240
Asian and Pacific Islander	9.7	17
Hispanic origin	26.7	40.5

TABLE 7E.6. POTENTIAL UNITED STATES POPULATION WITH LONGITUDINAL MELANONYCHIA (IN MILLIONS) BASED ON INCIDENCE AND CENSUS PROJECTIONS

Year	1995	2010
African-American	29.7	36.2
White	21.8	24
Asian and Pacific Islander	1.9	3.4
Hispanic origin	2.6	4.0
Potential total no. of people with longitudinal melanonychia	56	67.6

progress to subungual melanoma. Although these criteria are fairly well-established for nevi for the rest of the skin, they are not well-defined or well-evaluated for LM.

At one time, this author felt that only very dark or black nail streaks, especially those undergoing change, needed to be biopsied or excised (Fig. 7E.6). Although this is essentially correct, i.e., that black or dark expanding LMs are of more concern, some lighter-brown LMs have been reported to be melanoma (15–17). There is additional difficulty and confusion because 20–25% of the subungual melanomas are amelanotic (11,18) and even higher percentages have been reported in some series (19).

The criteria to determine which LMs are more likely to be melanoma or precursors of melanoma are not precise (6). At present, even with the recommended criteria, the decision as to the significance of the LM often becomes a question of judgment, intuition, and experience. In the future, the criteria and characteristics of LM that are precursors to melanoma need to be better established than is currently the case (6).

Some criteria that may reflect the significant LM as precursors to melanoma are as follows (Table 7E.7):

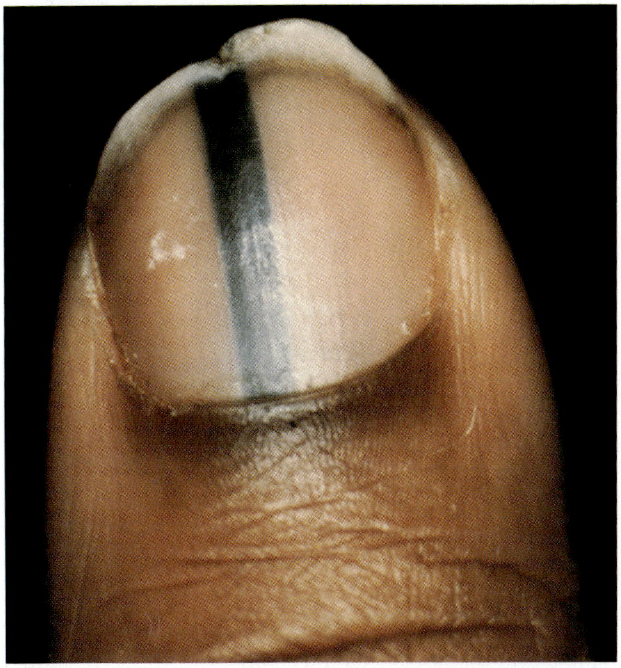

FIGURE 7E.6. Blue-black longitudinal melanonychia. Biopsy of the lesion revealed melanocytic hyperplasia. No recurrence was seen 5 years after excision.

TABLE 7E.7. POSSIBLE INDICATIONS FOR BIOPSY OF LONGITUDINAL MELANONYCHIA

1. Becomes suddenly wider and/or darker
2. Border is blurred rather than sharply well-demarcated
3. Irregular color, usually in bands
4. Black bands, especially in white patients, and solitary
5. Nail plate dystrophy/destruction
6. Wide band

1. Black or very dark brown pigmented streaks (Figs. 7E.6–7E.8). However, heavily pigmented streaks in dark-skinned people are not uncommon (Figs. 7E.6, 7E.9, and 7E.10). Despite this observation, it has been the standard of my practice to explore the intensely black pigmented streaks in African-Americans and other dark-skinned patients.
2. Pigmented nail streaks that are expanding. This is especially important if the pigment is dark brown or black and is rapidly enlarging (Fig. 7E.11).
3. Fuzzy or indistinct edges of the pigmented nail streak (Fig. 7E.7).
4. The width of the LM has been considered an indication for concern (Figs. 7E.11–7E.14). There is no agreement on the exact width of the pigmented streak that has importance to melanoma development, but 6 mm has been proposed (5). Whether wider bands that are common in African-Americans have significance if not associated with darker pigment or fuzzy borders is not known at this time.
5. Variegated shades of pigment may be important in determining which LMs are suspicious of being related to melanoma (Figs. 7E.11–7E.16). Yet, in African-Americans variegations of color of the pigmented nail streaks are common (Figs. 7E.11–7E.13) (5). What may be more significant is when the variegation has black and darker colors.

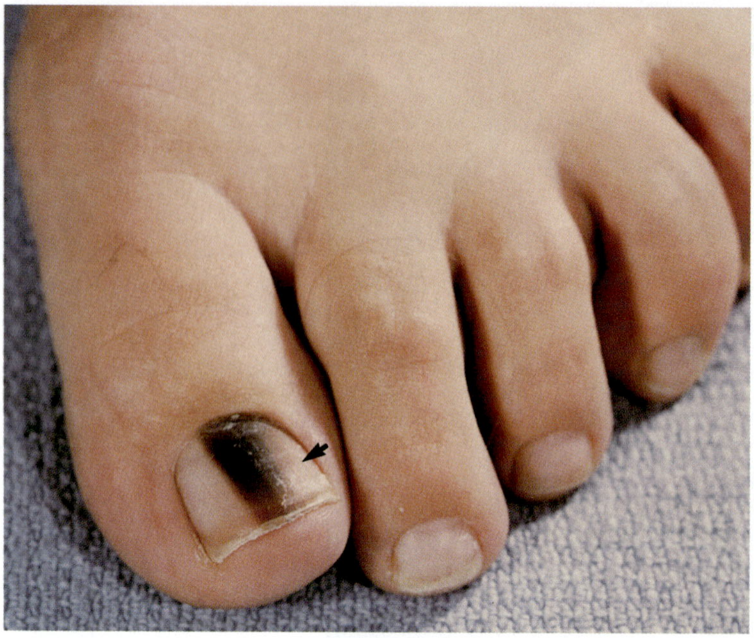

FIGURE 7E.7. Sudden onset of longitudinal melanonychia during pregnancy. Lateral edge is fuzzy *(arrow)*. Biopsy excision revealed atypical melanocytic hyperplasia. No recurrence was seen 7 years after biopsy excision.

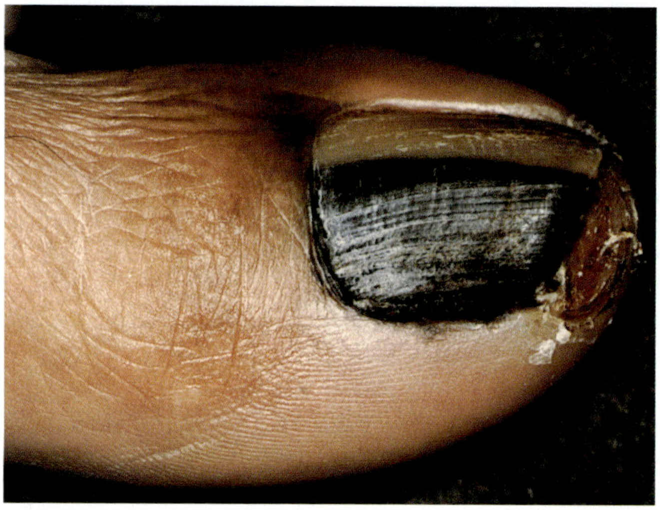

FIGURE 7E.8. Deep blue-black, wide longitudinal melanonychia and ulceration of the tip of the toe. There is early pigment spread to the lateral nail fold and probably cuticular region. Skin around ulcer also showed pigment spread consistent with Hutchinson's sign. Biopsy excision revealed thick melanoma.

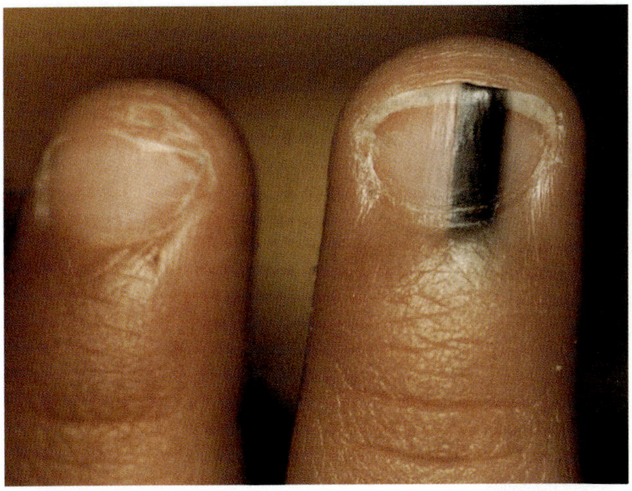

FIGURE 7E.9. Blue-black longitudinal melanonychia in tip of nail from an African-American patient reveals pigment through entire thickness of the nail plate, suggesting its origin from proximal, middle, and distal matrix. Pigment is seen through cuticle and distal proximal nail fold representing pseudo-Hutchinson's sign. Histopathologic interpretation was melanocytic hyperplasia. No recurrence was seen after biopsy excision.

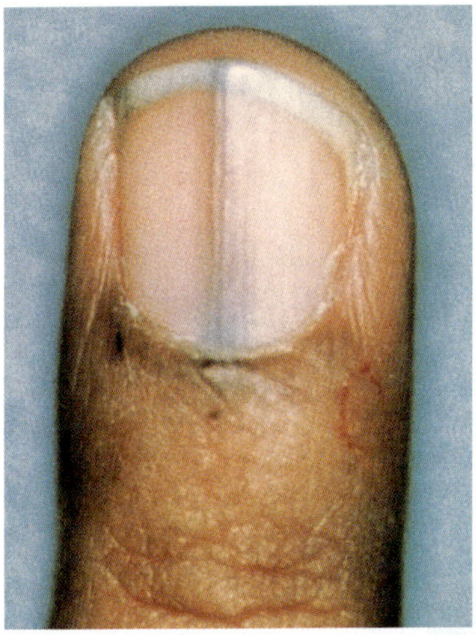

FIGURE 7E.10. Deep-blue longitudinal melanonychia in an African-American patient. Biopsy was interpreted as melanocytic hyperplasia.

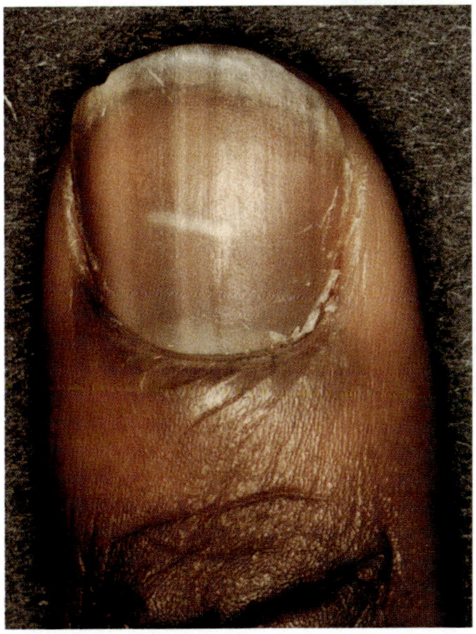

FIGURE 7E.11. Wide longitudinal melanonychia involving most of the nail and of variegated colors. Histopathology was interpreted as melanocytic hyperplasia.

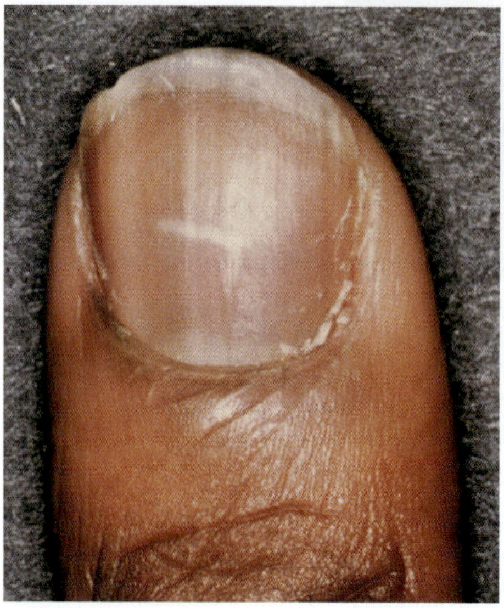

FIGURE 7E.12. Wide longitudinal melanonychia with variegated shades of dark brown. Histopathology was interpreted as melanocytic hyperplasia.

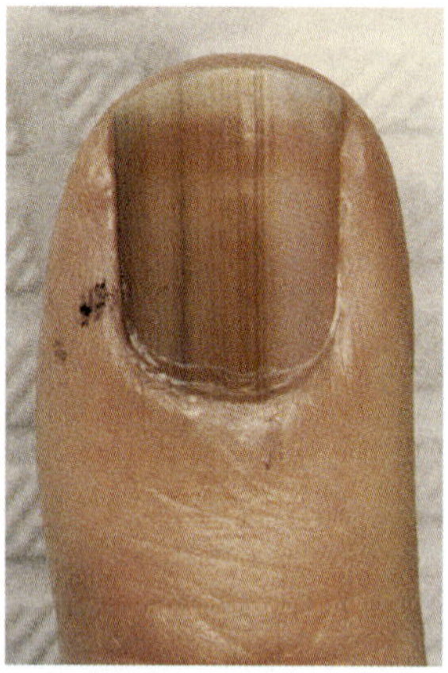

FIGURE 7E.13. Wide longitudinal melanonychia with variegated shades of brown. Histopathology was interpreted as melanocytic hyperplasia.

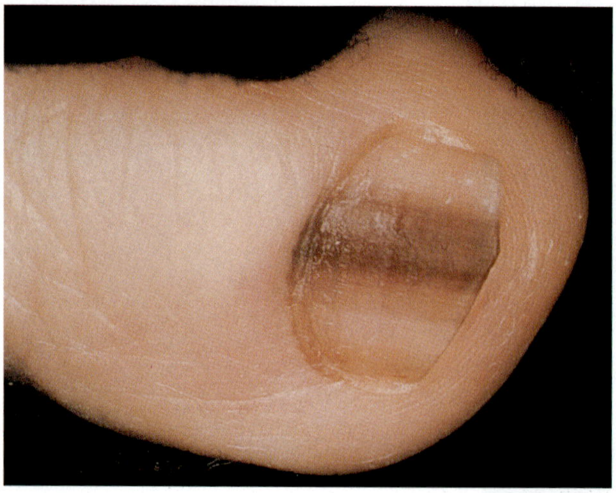

FIGURE 7E.14. Wide longitudinal melanonychia with variegated colors. Histopathology was interpreted as melanocytic hyperplasia.

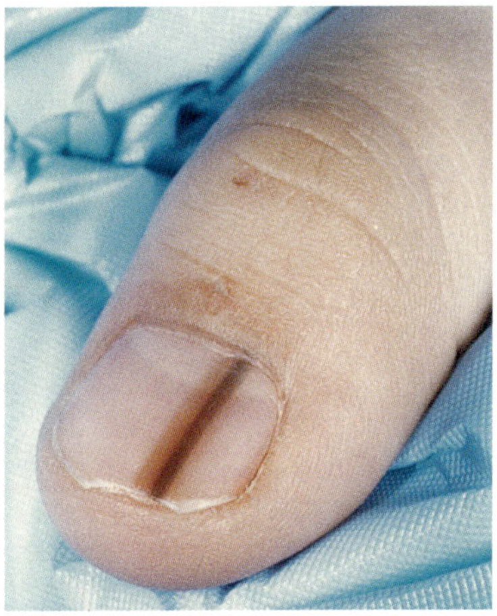

FIGURE 7E.15. Irregularly pigmented streak in an 11-year-old Hispanic boy, appearing 1 year previously. No other pigmented nail streaks were present in the other nails. There was no history of specific trauma.

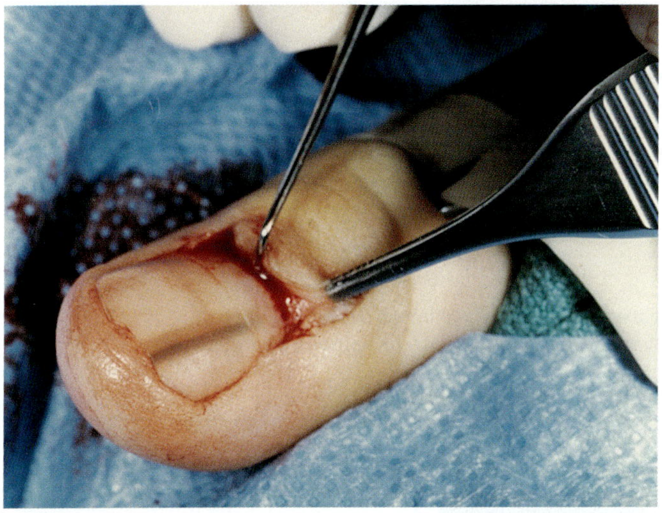

FIGURE 7E.16. Origin of pigment of the patient's nail shown in Fig. 7E.15 is distal matrix. Histopathology, reviewed by a number of dermatopathologists, was melanoma *in situ.* This is especially unusual for his age. Local excision with 5mm margins resulted in no recurrence in 1 year.

6. Destruction or deterioration of a nail plate. This suggests a tumor that is attacking or affecting the nail plate itself. However, this is a late sign.
7. Patients with higher risks of developing melanoma, such as those with dysplastic nevus syndrome (20).
8. The presence of Hutchinson's sign, which is the extension of pigment into the soft tissues surrounding the nail (Figs. 7E.8, 7E.17). It is almost pathognomonic of subungual melanoma (21), but there are other infrequent causes of periungual pigmentation (Table 7E.8) (22). It usually is not observed in pigmented streaks with early melanoma, but usually in more advanced cases of melanoma (Fig. 7E.8). Regardless of the circumstances under which it is seen, the presence of Hutchinson's sign is presumptive of melanoma and should be investigated and managed as such. Occasionally with a pigmented nail streak there is a suggestion of a pigment spread to the cuticle and distal part of the proximal nail fold (Figs. 7E.9 and 7E.18). Careful examination, frequently requiring a magnifying lens, should differentiate pigment in the soft tissue of the proximal nail fold from the transmission of the underlying pigmented nail steak through the thin cuticle and distal part of the proximal nail fold. This latter finding is

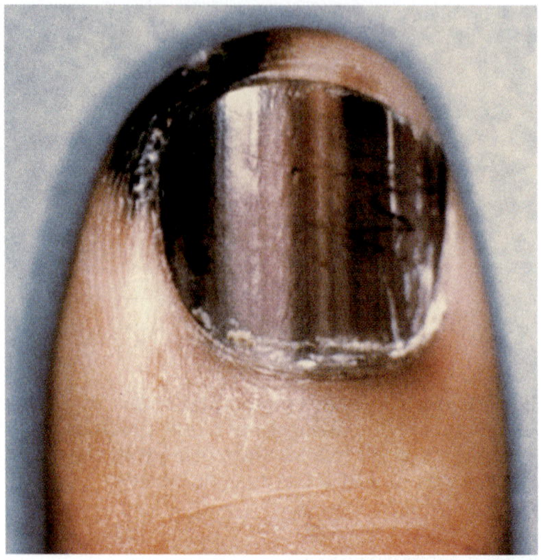

FIGURE 7E.17. Deep black-blue tumor involving entire nail plate and extending onto the surrounding skin-Hutchinson's sign. Biopsy confirmed thick melanoma.

TABLE 7E.8. OTHER MIMICKERS OF HUTCHINSON'S SIGN

1. Laugier-Hunziker syndrome
   A. Longitudinal melanonychia
   B. Sometimes pigment spread to surrounding nail tissue
   C. Lip pigmentation in Laugier-Hunziker syndrome
2. Congenital nevi of nail unit
3. Radiation therapy
4. Malnutrition
5. Ethnic nail pigment changes, especially in dark-skinned people
6. Acquired immunodeficiency syndrome patients with or without treatment
7. Drugs such as minocycline
8. Regressive nevoid nail melanosis in Japanese children

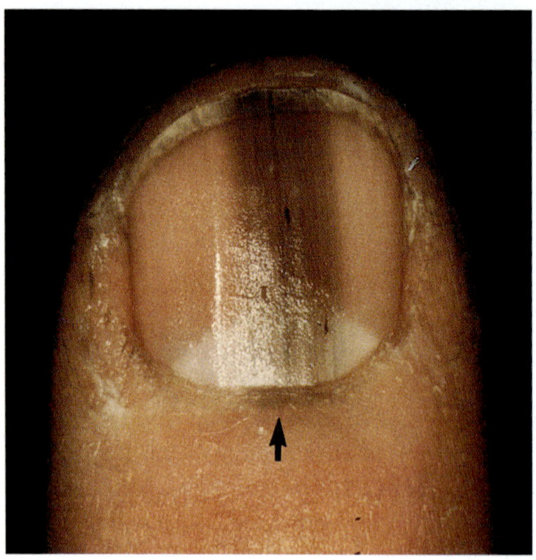

FIGURE 7E.18. Pseudo-Hutchinson's sign with pigmentation of the nail becoming visible because of the thin, transparent proximal nail fold and cuticle *(arrow).*

referred to as pseudo-Hutchinson's sign and has no significance to subungual melanoma (5).

Less helpful criteria are as follows (Table 7E.9):

1. Longitudinal melanonychia of the thumb, index finger, and first toe. These are common sites for all LM as well as subungual melanoma.

TABLE 7E.9. LESS VALUABLE INDICATION FOR BIOPSY OF LONGITUDINAL MELANONYCHIA

1. Abrupt onset
2. Single digit in adult
3. Thumb, index fingers, or great toe involvement
4. History of trauma

2. Single pigmented nail streaks in dark-skinned people, especially those who are younger. This may simply be the early onset of LM that is common in some races.
3. Trauma has been implicated in benign LM and subungual melanoma. A distinctly specific trauma associated with the other findings in LM may be more significant (5,23).
4. Usually multiple LMs, especially if all are similarly colored, suggest a benign lesion. However, the multiplicity of nail involvement does not completely exclude one or rarely multiple streaks developing a malignant course (6). Darkening of the color, changes in the borders of the pigment, or significant pigment spread of one or more of the pigmented nail streaks should raise the suspicion of melanoma even in patients with multiple LM.

Patients with LM should be directed to return to the physician if the pigmented nail streak changes. Polaroid prints or other photographs are useful in monitoring changes. Photographs of the LM given to the patient may be helpful in the patient's own early recognition of nail changes. We give a copy of these photographs to the patient as well as including them in the chart.

The site of pigment formation frequently can be determined by careful examination of the distal nail plate edge, especially if it is freshly cut. The proximal matrix gives rise to the dorsal nail plate and the distal matrix to the ventral nail plate (Fig. 7E.19). Thus, if the ventral part of the nail plate alone is involved, then the origin of the pigmentation should be the distal matrix; if only the dorsal nail plate demonstrates pigment, then the source is the proximal matrix. If the entire thickness of the nail plate reveals pigmentation, then the LM involves the entire longitudinal length of the matrix (Fig. 7E.9).

If the nail bed alone is involved, the nail plate will not develop a pigmentary streak because the nail bed essentially does not contribute to the nail plate. However, a melanoma arising in or metastatic to the nail bed will result in a pigmented lesion that may cause destruction of the nail plate by the tumor. It will not involve the layers of the nail plate as occurs with matrix lesions and will not extend to the end of the nail plate unless it does so by local tumor destruction and extension (24).

In my initial studies of LM, I removed the nail plate to identify the point of origin of the pigmented streak (Figs. 7E.20–7E.22). Most of the pigmented streaks originated in the

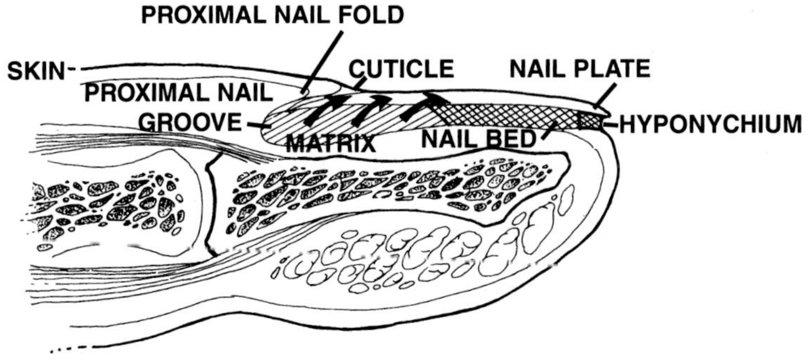

FIGURE 7E.19. Longitudinal section of nail unit illustrating that proximal matrix gives rise to dorsal nail plate, mid matrix to intermediate nail plate, and distal matrix to the ventral nail plate.

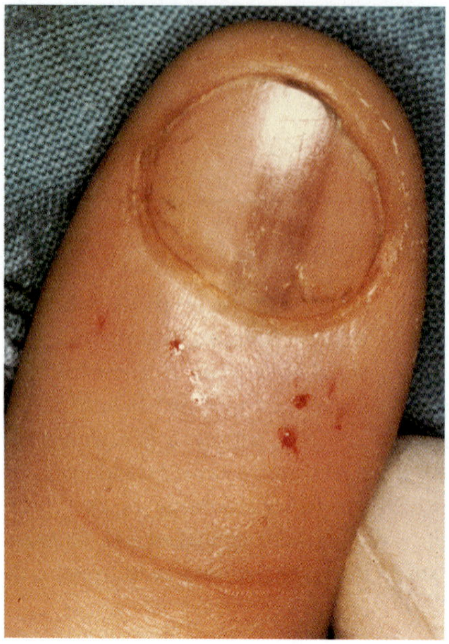

FIGURE 7E.20. Longitudinal melanonychia of more than 1 year's duration.

distal half of the matrix (Fig. 7E.23). This may reflect the a higher concentration of melanocytes in the distal matrix as compared to the proximal matrix. However, rather than the increased melanocyte density, LM may occur more frequently beginning in the distal matrix because only the distal matrix contains an active melanin synthesis compartment; in the proximal matrix, the dormant melanocyte compartment predominates (25). However, despite this condition, there were significant numbers of pigmented streaks that began in the most proximal part of the matrix in my patients (Figs. 7E.20–7E.22, 7E.24, 7E.25). It was interesting that the pigmentation in the matrix was often lighter in color than that

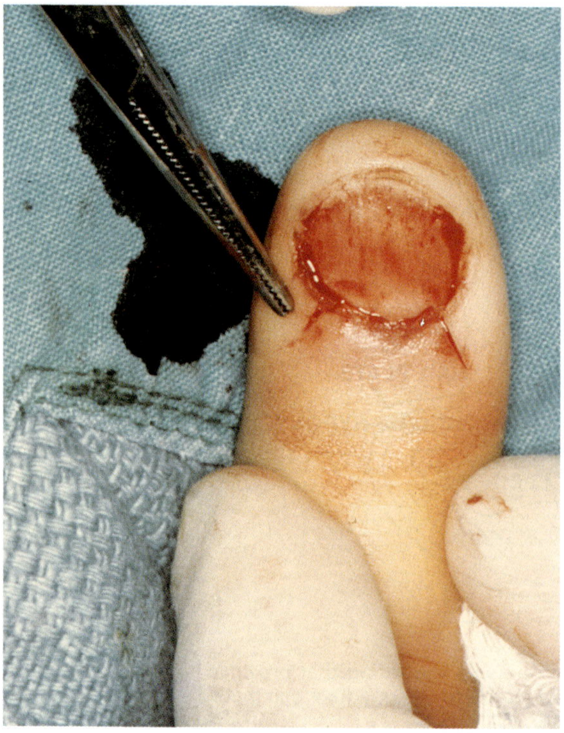

FIGURE 7E.21. Nail plate avulsed. Incision was made in the proximal nail fold to visualize the origin of longitudinal melanonychia.

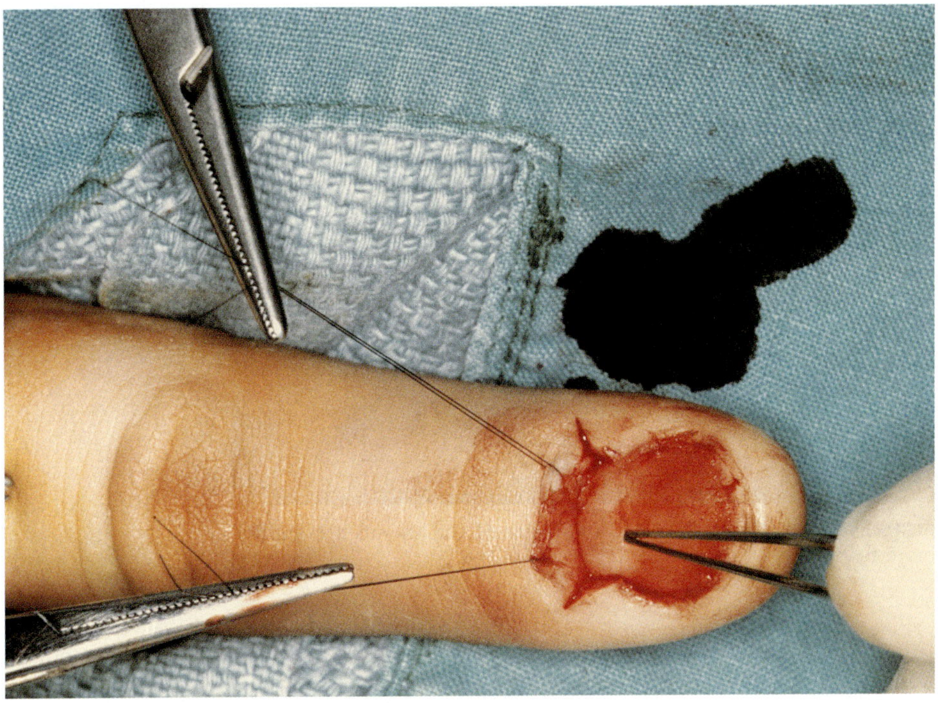

FIGURE 7E.22. Longitudinal melanonychia involves the proximal as well as the distal matrix.

of the nail plate (Figs. 7E.20–7E.22, 7E.24, 7E.25). The color of the pigmentation of the matrix frequently resembles the color of coffee-and-cream or light brown. In fact, the pigmented matrix streak often was difficult to identify. Therefore, I began marking the soft tissues at the distal end of the LM before removing the nail plate (Fig. 7E.25). I also left the plate on the surgical tray in case I wished to temporarily replace it if I had great difficulty determining the borders of the pigmentation in the matrix. Although nevi are seen in the matrix (26), it was interesting to note how few of the lesions resembled a nevus.

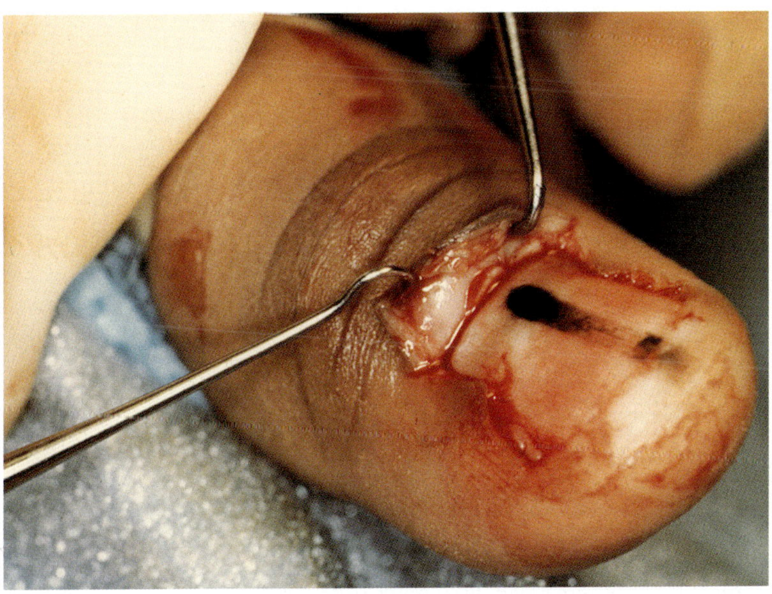

FIGURE 7E.23. Longitudinal melanonychia begins in the distal half of the matrix.

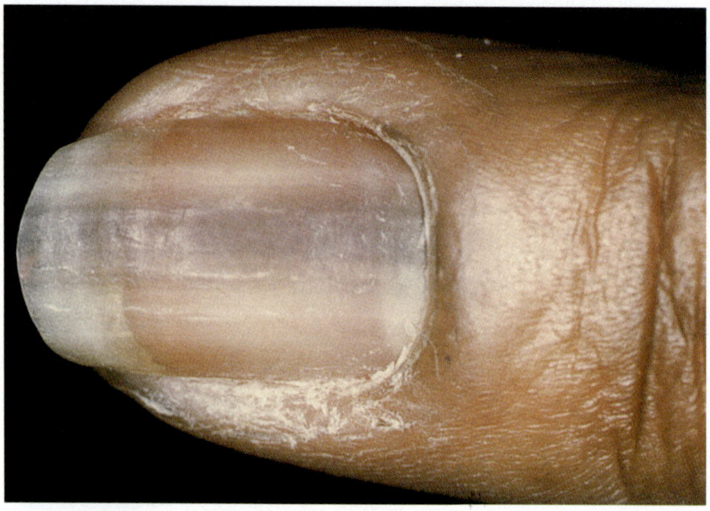

FIGURE 7E.24. Longitudinal melanonychia of the thumb present for more than 2 years.

Instead, there were varying shades of brown or black pigmentation. The histopathology was usually melanocytic hyperplasia. Although removal of the nail plate helped visualize the point of origin of the pigmentation and facilitated the biopsy or excision of the LM, the process of nail removal produced some destruction of the epithelial tissue and melanocytes. This made the histopathologic interpretation of the tissue more difficult and less reliable. As a result, I changed my procedures so that the involved nail plate of LM is left attached to its underlying matrix and nail bed, even for punch biopsies. For those who wish to perform punch or ellipse biopsies for LM with partial or total avulsion of the nail plate, this is described in Chapter 3c in the section on nail unit biopsies.

A number of methods have been recommended to biopsy LM (5,27–30). To perform a punch biopsy, the proximal nail wall is reflected from the underlying nail plate. The entire proximal nail wall or just enough wall to visualize the entire extent of the streak and perform the planned procedure is reflected (Figs. 7E.26 and 7E.27). Then a 2- to 3-mm

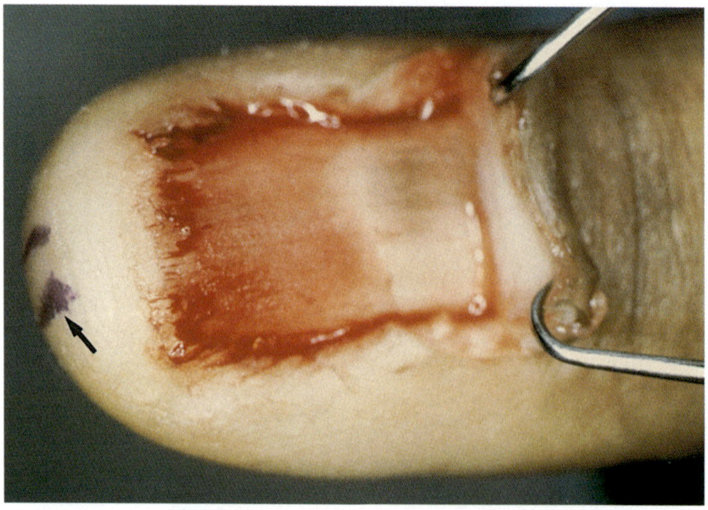

FIGURE 7E.25. Most intensely pigmented part of the matrix giving rise to the longitudinal melanonychia shown in Fig. 7E.24 is in the middle third, but pigment also is seen in the proximal and distal thirds of the matrix. *Arrow* points to marking of edges of pigmented streak because pigmented steak of matrix, in contrast to that of nail plate, may be very faint.

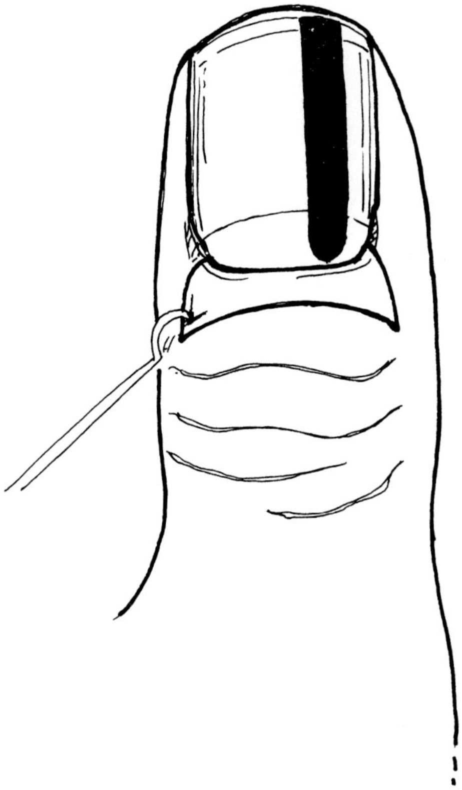

FIGURE 7E.26. Incision through entire width of proximal fold.

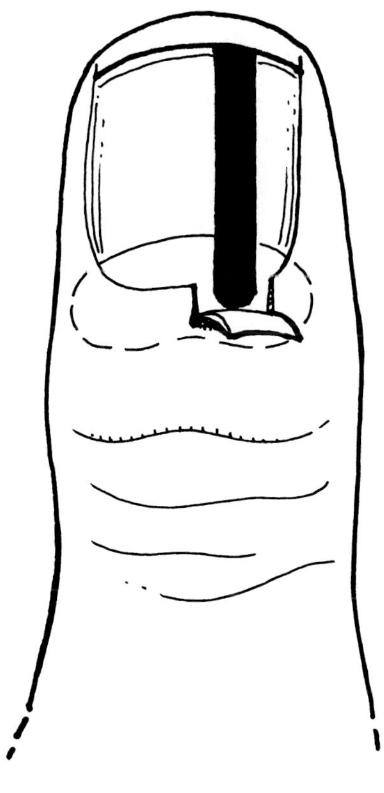

FIGURE 7E.27. Incisions in proximal nail fold wide enough to explore source of lesion and perform planned surgical procedure.

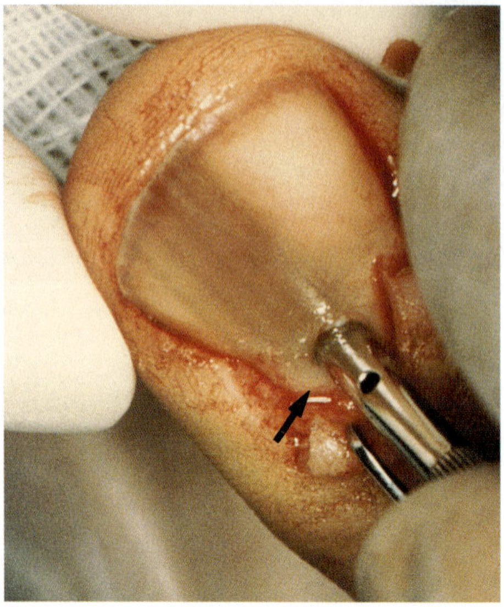

FIGURE 7E.28. Punch biopsy through the nail plate at the proximal point of the longitudinal melanonychia. The origin of the longitudinal melanonychia is distal matrix. Proximal matrix appears uninvolved *(arrow)*.

punch biopsy is carried out at the proximal part of the pigmented nail streak through the nail plate and matrix to the underlying bone (Figs. 7E.28 and 7E.29). It is not easy to remove the specimen with the nail plate attached to the matrix using this biopsy method because of the small opening and the surrounding nail plate (Fig. 7E.30). Bleeding can be controlled by inserting a piece of Gelfoam into the biopsy site, which also acts as a tamponade. This method does not permit determination of whether further melanocytic activity and pigmentation remains, unless more nail plate is removed after completion of the biopsy to explore the rest of the matrix. Because of the difficulty in removing the tissue and the inability to determine if there is residual pigmentation, Baran and Kechijian (5) recommend a different approach (Fig. 7E.31). They used a punch biopsy incision of the proximal part of the pigmented streak, leaving the circle of nail plate attached to the underlying matrix (Fig. 7E.32). The nail plate was transversely incised with a scalpel or nail clipper at

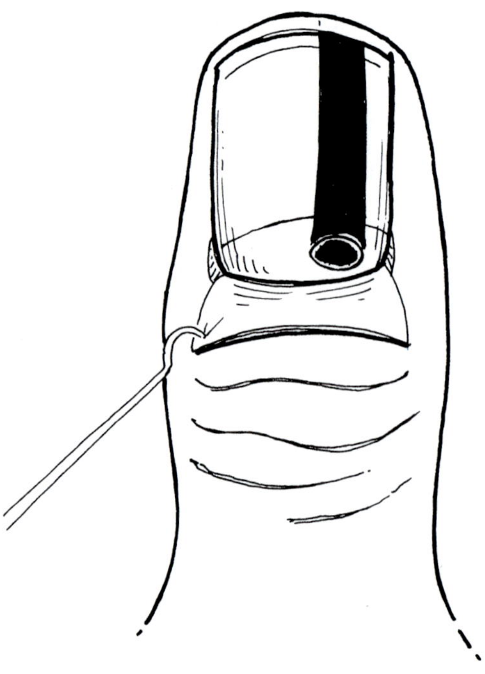

FIGURE 7E.29. Punch biopsy through nail plate to underlying bone.

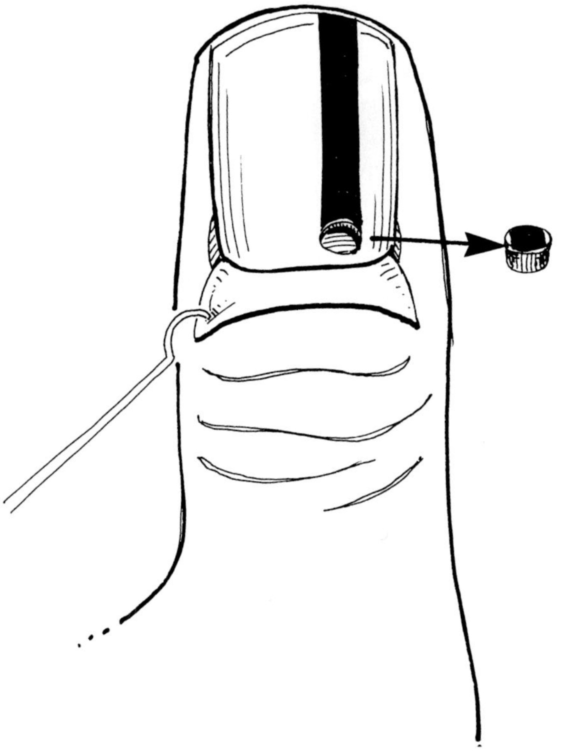

FIGURE 7E.30. Removal of punch biopsy.

about the level of the junction of the proximal and middle third of the nail (Fig. 7E.32). The nail plate proximal to the transverse incision was avulsed, leaving only a circle of nail plate attached to the cylinder of underlying pigmented matrix (Fig. 7E.33). The punch biopsy specimen was then easily excised from the underlying bone (Figs. 7E.34–50).

If the punch biopsy does not involve the most proximal 10% of the matrix or the punch diameter is less than 3 mm, either no closure or partial closure is used. If the biopsy

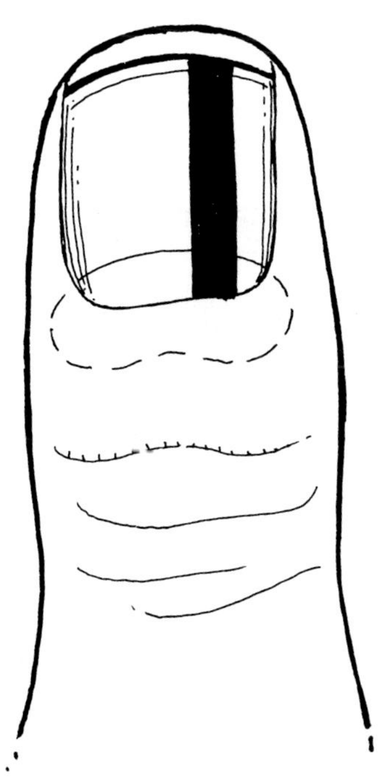

FIGURE 7E.31. Pigmented streak.

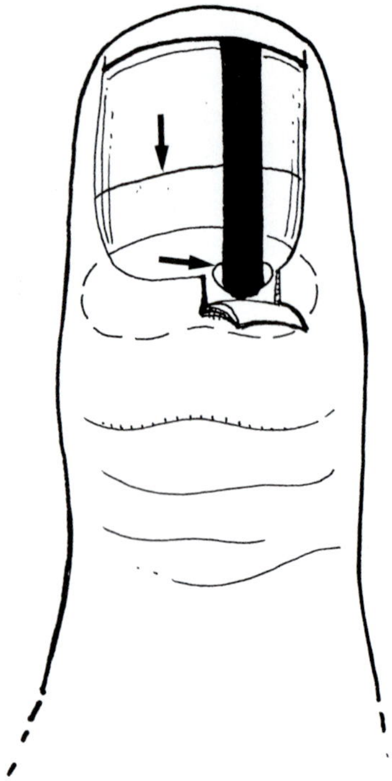

FIGURE 7E.32. Punch biopsy through the nail plate to the underlying bone at the proximal point of the longitudinal melanonychia *(arrow)*. Transverse semicircular incision is made through only the nail plate at about the junction of the proximal and middle thirds of the nail *(arrow)*, distal to the lunula. (Adapted from Baran R, Kechijian P. Longitudinal melanonychia (melanonychia striata): diagnosis and management. *J Am Acad Dermatol* 1989;21:1165–1175.)

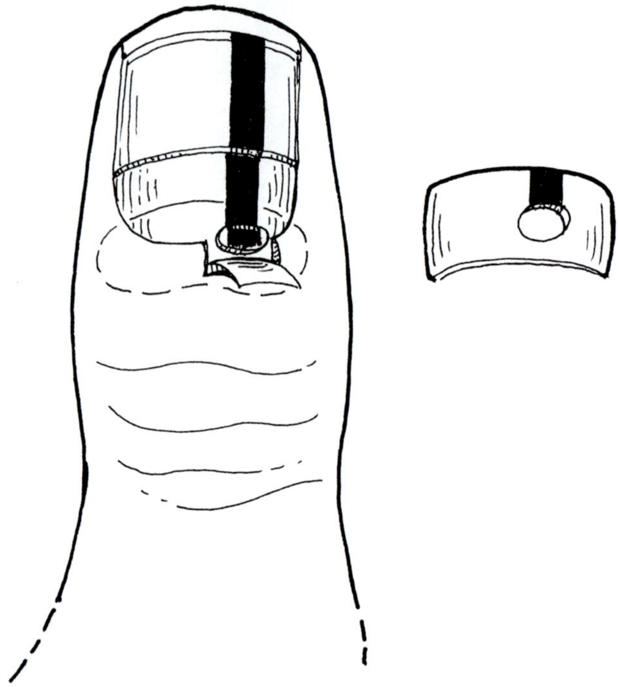

FIGURE 7E.33. Proximal part of nail plate is avulsed, leaving only the punch biopsy with its corresponding adherent nail. (Adapted from Baran R, Kechijian P. Longitudinal melanonychia (melanonychia striata): diagnosis and management. *J Am Acad Dermatol* 1989;21:1165–1175.)

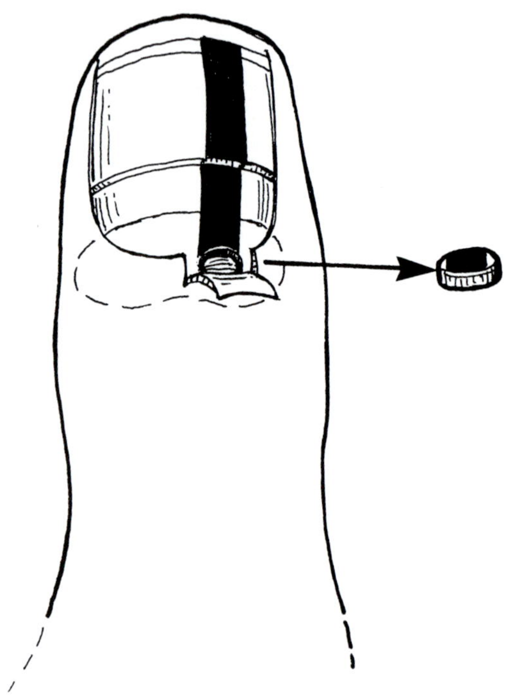

FIGURE 7E.34. With surrounding nail plate removed, the punch biopsy with the attached nail is now easily removed. (Adapted from Baran R, Kechijian P. Longitudinal melanonychia (melanonychia striata): diagnosis and management. *J Am Acad Dermatol* 1989;21:1165–1175.)

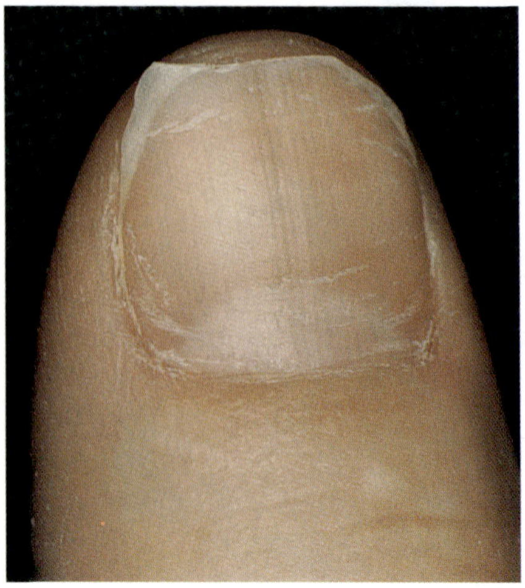

FIGURE 7E.35. Pigmented nail streak in mother of boy with melanoma *in situ* (patient shown in Figs. 7E.15 and 7E.16).

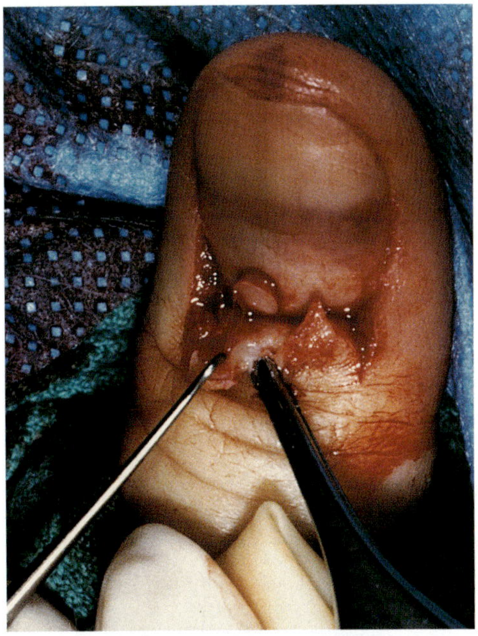

FIGURE 7E.36. Proximal nail wall reflected. Punch biopsy of proximal part of pigmented streak.

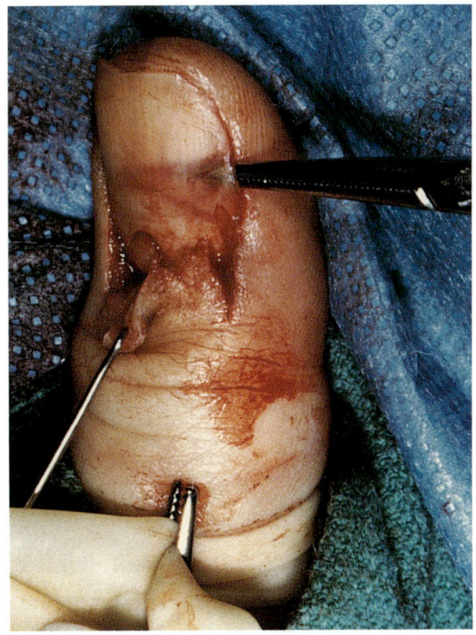

FIGURE 7E.37. Mosquito hemostat frees nail distal to biopsy site to allow cutting of nail transversely. This separation of the nail plate should avoid the punch biopsy site.

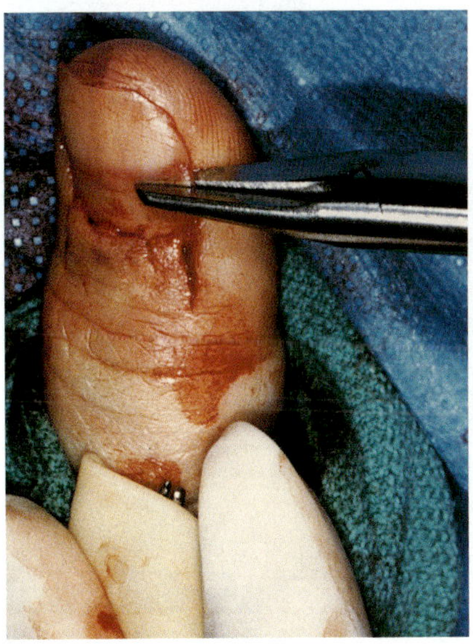

FIGURE 7E.38. Transection of nail plate with an English nail splitter.

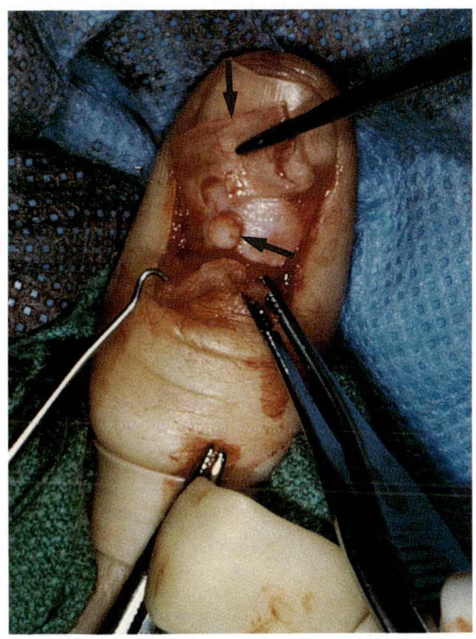

FIGURE 7E.39. Proximal nail plate removed *(arrow)*, leaving punch biopsy with attached nail in place *(arrow)*. The distal retained nail plate acts as a long-term dressing for the nail bed.

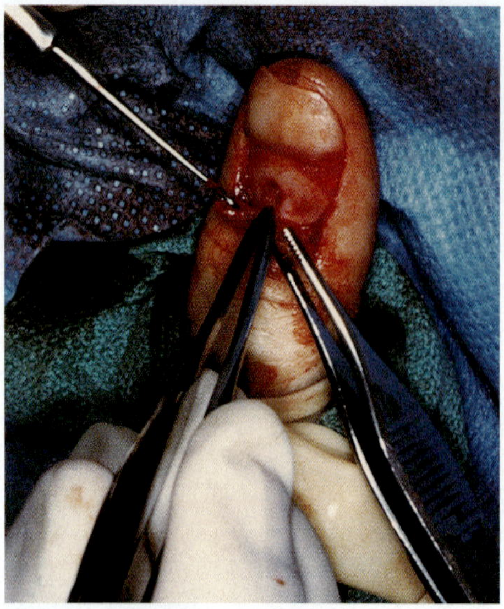

FIGURE 7E.40. Removal of biopsy with attached nail plate. Care must be used to avoid grasping the biopsy specimen by the attached nail plate alone. It will separate from the underlying tissue and possibly damage the epithelial cells and melanocytes, which in turn may make the histopathology more difficult and less accurate.

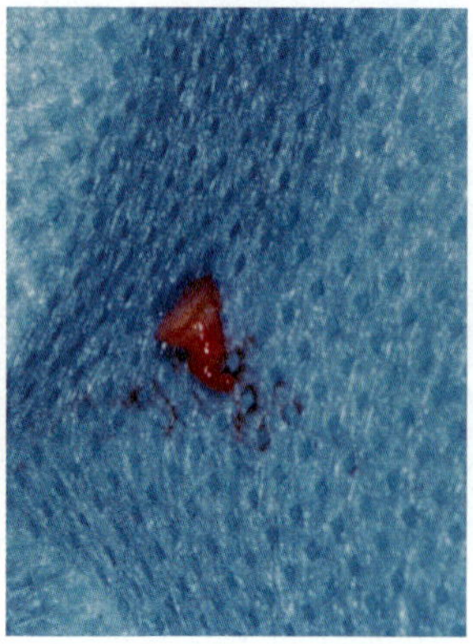

FIGURE 7E.41. Specimens with nail attached.

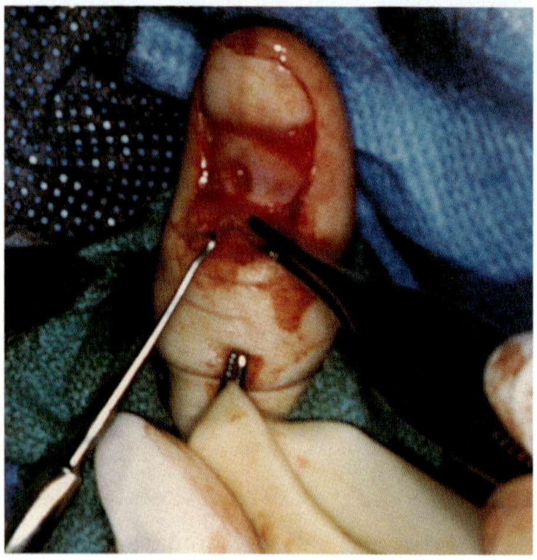

FIGURE 7E.42. Residual defect.

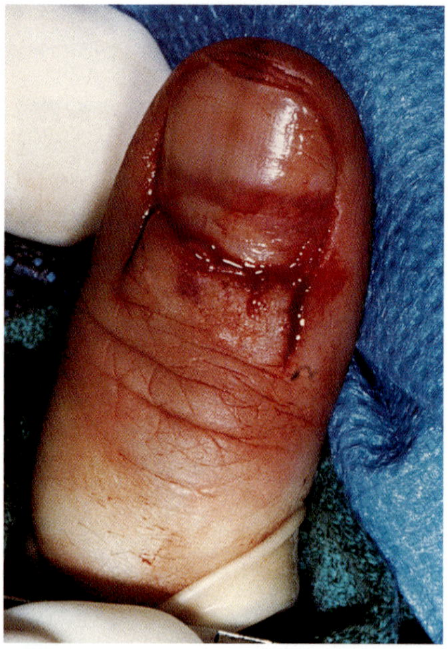

FIGURE 7E.43. Although not necessary in the distal matrix, this defect was closed with 5-0 polyglactin.

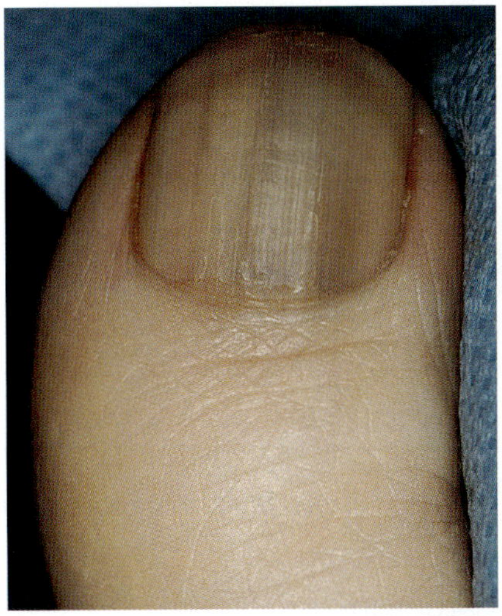

FIGURE 7E.44. The Baran-Kechijian procedure may be applied to broad bands of pigment as well as wide irregular pigmented streaks.

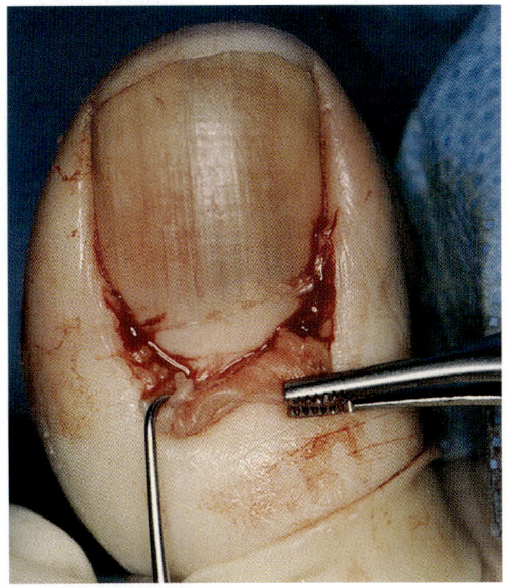

FIGURE 7E.45. Pigmentation originates from distal matrix.

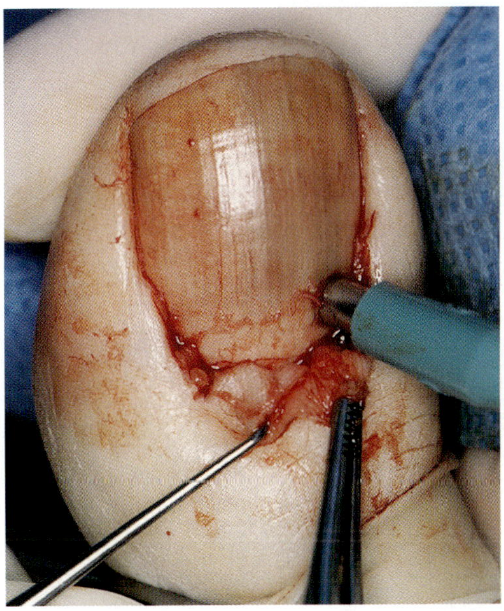

FIGURE 7E.46. First of multiple biopsies.

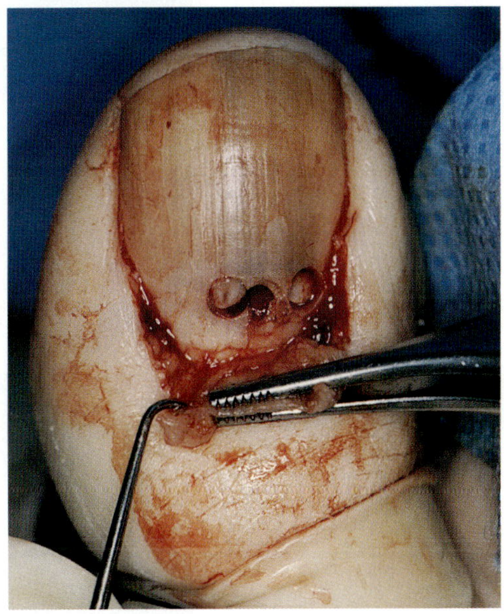

FIGURE 7E.47. Three punch biopsies performed.

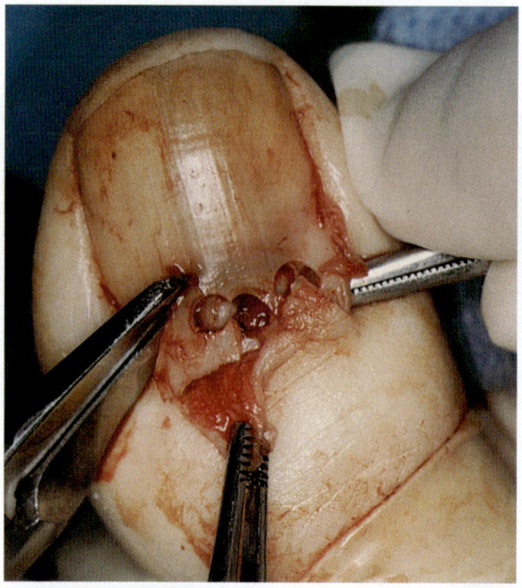

FIGURE 7E.48. Transverse splitting of the nail.

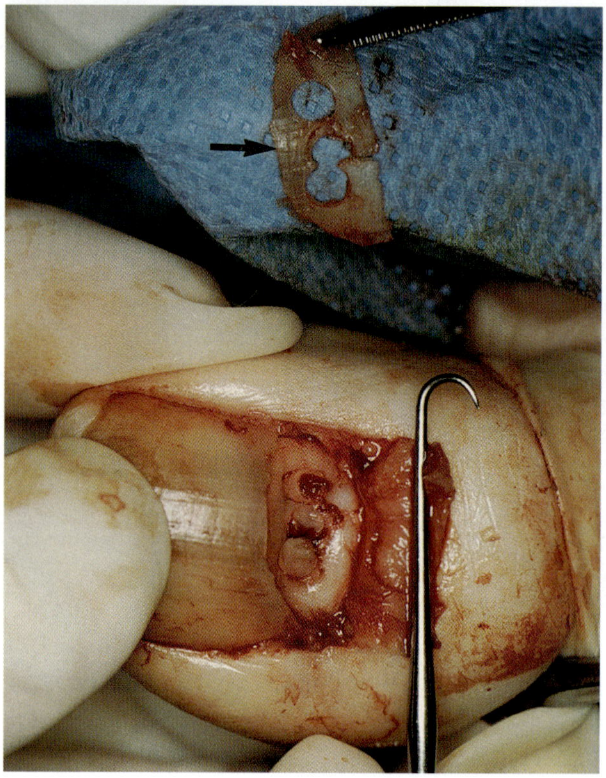

FIGURE 7E.49. Proximal nail plate removed *(arrow)*, leaving punch biopsies with nail plate attached. Residual nail plate over nail bed acts as a long-term dressing.

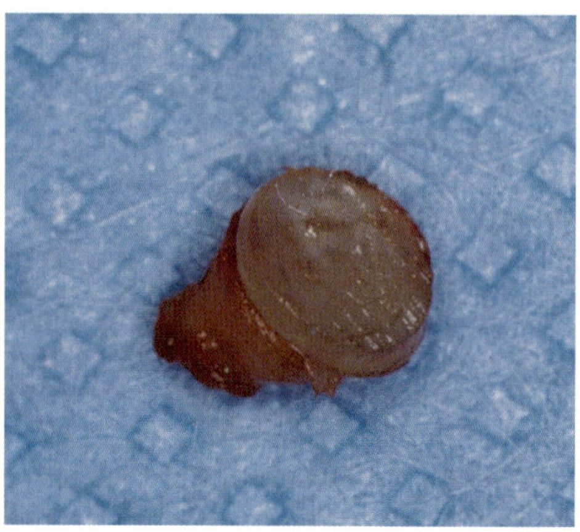

FIGURE 7E.50. One of the specimens with the nail plate attached.

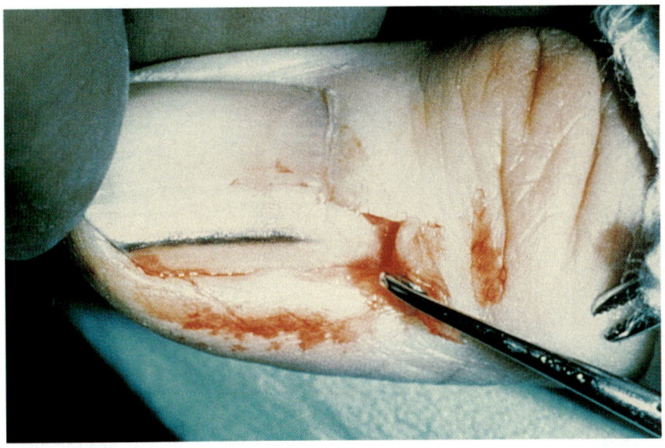

FIGURE 7E.51. Thin black longitudinal melanonychia showing histopathologic changes of melanoma. (Courtesy of Paul Kechijian.)

is larger than 3 mm in width and involves the proximal matrix, an attempt to repair the matrix should be made. Because the dorsal nail plate is made by the proximal matrix, injury or surgery to proximal 10% of the matrix may result in a longitudinal nail dystrophy or split nail (Fig. 7E.19). Zaias (31) found that biopsies smaller than 3 mm in width through the proximal part of matrix might result in nail dystrophy, and uncommonly results in a split nail. However, this author tends to close biopsies involving the proximal matrix to reduce the possibility of development of split or dystrophic nails.

Underlying the recommendation to biopsy the proximal part of the pigmented streak in the matrix is an assumption that the melanoma will develop only in the proximal part of the pigmented streaks. Most of the pigmented streaks that begin in the proximal matrix extend through most of the length of the matrix (Fig. 7E.22). There is no reason why the malignant change could not begin anywhere along the pigmented streak of the matrix. In fact, I am aware of three established cases in which the melanoma did not originate in the proximal part of the pigmented streak, but more distally so the melanomas would have been missed by a 2- to 3-mm punch biopsy of the proximal part of the pigmented streak (Figs. 7E.51–7E.53). How many other such cases exist is unknown. But this author feels that patients in whom punch biopsies are performed on the proximal part of LM leaving residual pigment in the matrix should be monitored carefully to determine whether the

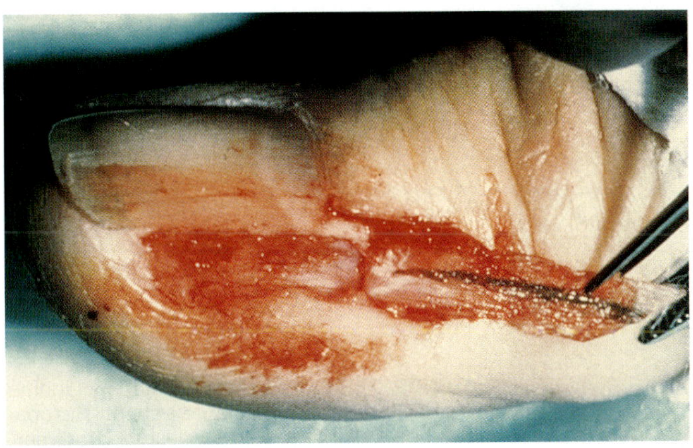

FIGURE 7E.52. The site of the melanoma was 4 to 5 mm distal to the proximal point of the longitudinal melanonychia. (Courtesy of Paul Kechijian.)

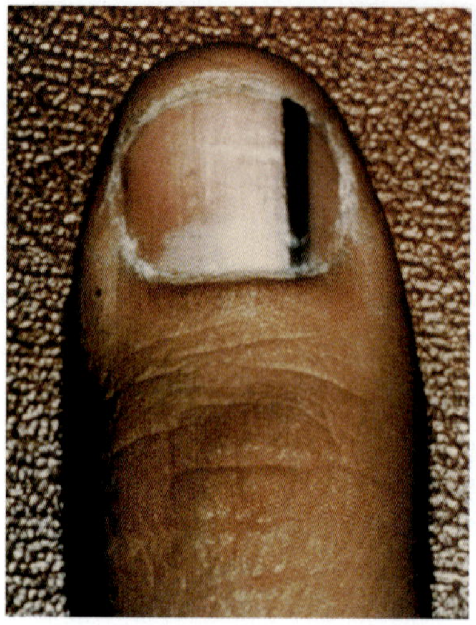

FIGURE 7E.53. This longitudinal melanonychia revealed histopathologic evidence of melanoma about 4 mm distal to the most proximal point of the pigmentation. (Courtesy of Alejandro Morales.)

pigmented streak disappears in time after the biopsy. If it does not, then removal of the remaining part of the pigmented streak should be strongly considered.

Because of the uncertainty about where the melanoma may begin in LM and the desire to obtain complete histopathologic examination of all the pigmented tissue, the author prefers to excise the entire suspicious streak. As stated earlier, the author used to remove the entire nail plate before excising the pigmented streak. Because of epithelial tissue destruction with nail plate removal, biopsy excisions of LM now are performed to include the nail plate attached to underlying tissues. The steps in the procedure are as follows.

A. Longitudinal melanonychia of the central nail plate with involvement of the proximal matrix or extending through the longitudinal length of matrix (Fig. 7E.54).

1. Incisions are made in the proximal nail fold. Retraction of the fold allows visualization of the extent of the pigmented streak (Fig. 7E.55).
2. An incision is made through the nail plate, matrix, and nail bed to underlying bone, with a 1- to 2-mm margin on the sides of the pigmented streak (Figs. 7E.55 and

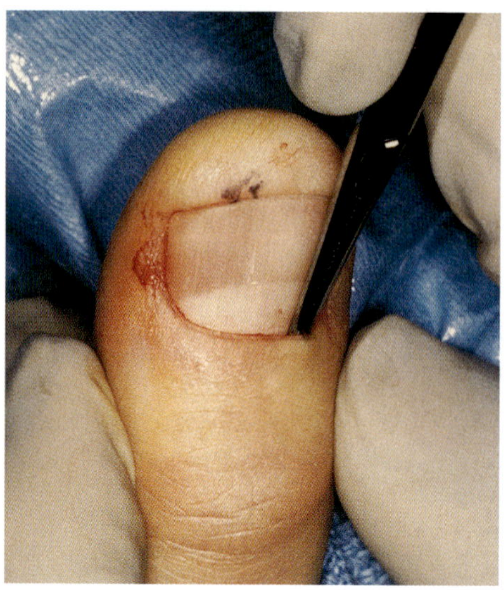

FIGURE 7E.54. New longitudinal melanonychia of the thumb in a white patient. Histopathology was interpreted as melanocytic hyperplasia. One year later, the patient developed a primary melanoma of the thigh. The thumb lesion has not returned in 5 years.

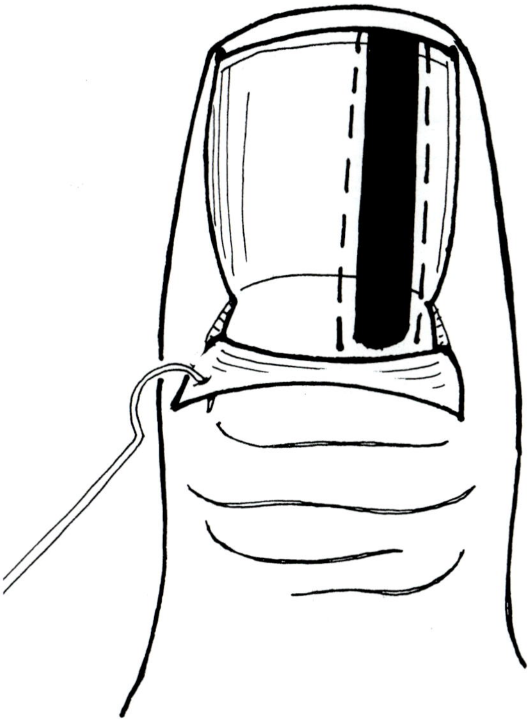

FIGURE 7E.55. Proximal nail fold retracted to reveal source of pigmentation of longitudinal melanonychia. Incisions through nail plate with about 2-mm margins for the longitudinal melanonychia.

7E.56). The nail can be softened by soaking the digit in water for 10 to 15 minutes before surgery. Because of the hardness of the nail plate, a rocking motion of the scalpel, back and forth in the direction of incision, may be necessary to get through the nail plate. Care should be taken to prevent the scalpel from slipping off the hard nail and cutting the surgeon especially at the distal end of the nail plate. The no. 15 blade cuts through the nail better than the small Beaver blades. The incision is extended to the proximal part of the pigmented streaks.

When the streak involves the proximal matrix, the end of the incision is somewhat rectangular or blunt (Figs. 7E.55 and 7E.57).

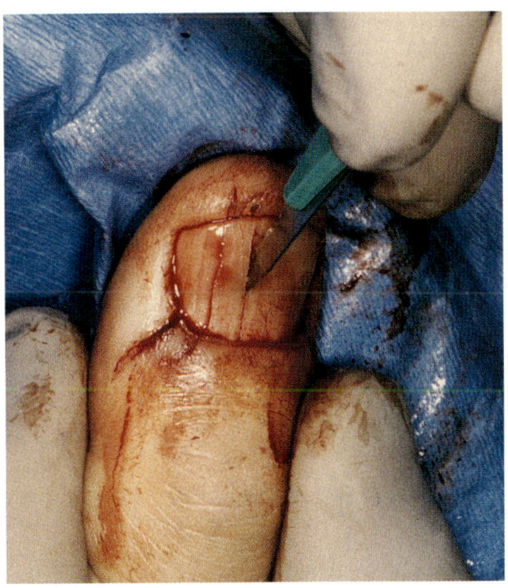

FIGURE 7E.56. Incisions in the proximal nail folds for visualization of the matrix. Incision through the nail plate made with no. 15 blade and disposal scalpel.

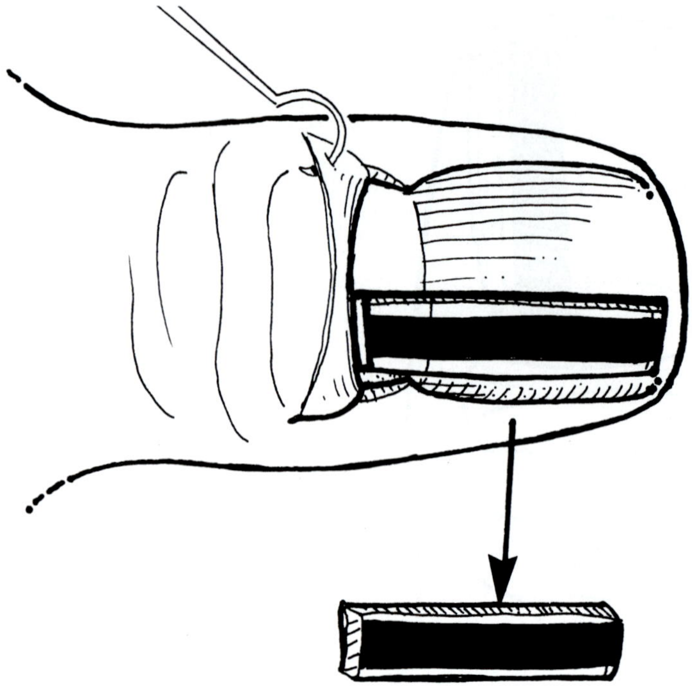

FIGURE 7E.57. Nail plate avulsed on both sides of biopsy specimen, which has nail plate still attached. This facilitates removal of the specimen and closure.

3. A small-nosed mosquito hemostat or similar instrument is inserted under the uninvolved nail plate on either side of the pigmented streak. This frees the uninvolved nail plate from its underlying matrix and nail bed. The uninvolved nail plate on both sides of the pigmented streak is avulsed, leaving only the nail plate with the pigmented streak attached to the underlying tissues (Figs. 7E.57 and 7E.58).
4. Using a Beaver blade and starting distally, first the nail bed and then the matrix are excised from the underlying bone to the most proximal point of the pigmented streak (Fig. 7E.59). The nail bed is included in the excision because, in doing so, it facilitates

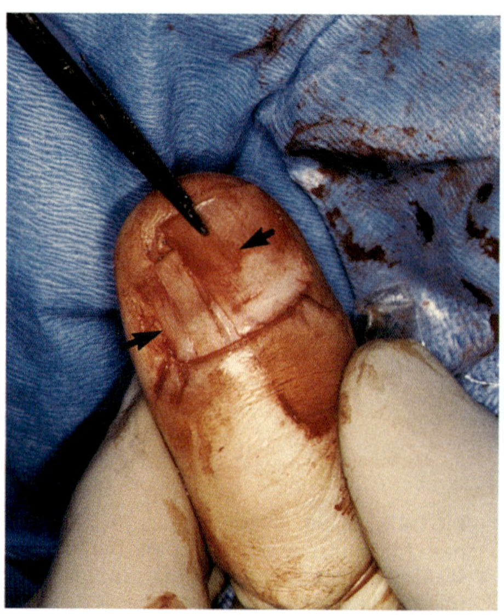

FIGURE 7E.58. Avulsion of nail plate on both sides of the biopsy (*arrows* point to planned avulsed nail) leaving only biopsy specimen with nail attached.

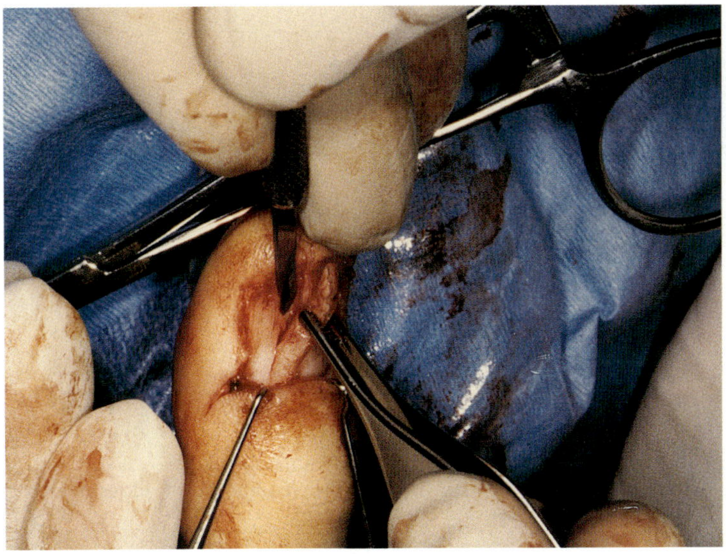

FIGURE 7E.59. Biopsy specimen with nail plate attached is being excised *en bloc.*

the procedure and it is not always easy to determine the precise demarcation of matrix and nail bed.

5. The excision site is closed. The sides of the excision must be mobilized to close the defect. The matrix and the nail bed are firmly bound to the underlying bone. If the width of the excision is narrow, then a few millimeters of undermining may be all that is necessary. More undermining is indicated with increasing width of the defect. With greater widths, the undermining of the matrix and nail bed from the underlying bone usually has to extend to the fibrofatty tissue of the lateral nail folds (Fig. 7E.60). This is performed initially on the narrower side of the excision. If the matrix and nail bed cannot be closed by this one-sided mobilization, then the other side of the excision also is undermined to the fibrofatty tissue of the lateral nail wall.

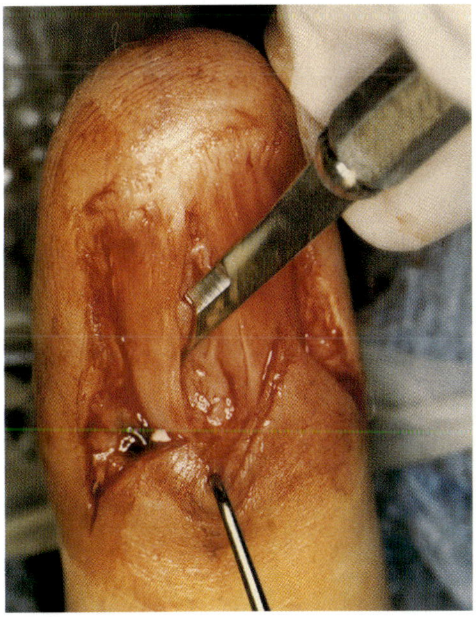

FIGURE 7E.60. Undermining edges of the excision to the fibrofatty tissue of the lateral nail fold to close the defect.

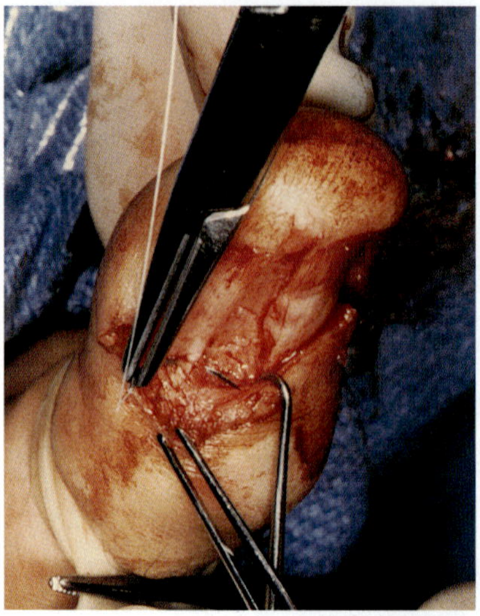

FIGURE 7E.61. Approximation of edges of defect using 5-0 polyglactin.

6. The closure is performed with 5-0 absorbable sutures such as polyglactin using a fine needle such as a P-3 (Figs. 7E.61 and 7E.62). The most critical suture placement is in the most proximal part of the matrix; this is also usually the first suture placed (Fig. 7E.62). The matrix is fairly tough tissue and will hold sutures under significant tension, unlike the nail bed, which tends to tear more easily. Careful closure of the matrix, especially the proximal part, affords the best opportunity to avoid a split nail or significant nail dystrophy. The rest of the matrix and nail bed are closed. The incisions in the proximal nail fold are closed with 5-0 or 6-0 nylon, with the first suture placed at the distal end of the incision at the cuticle to reform the cuticular margin for the best cosmetic results. The wound is dressed with an antibiotic ointment, petrolatum gauze, a fine-mesh gauze (Sof-Wick), a self-adhering dressing such as Kling, and finished with tubular elastic net (Se Pro Net). The dressing should be applied gently but firmly; it should not impair circulation. The dressing is changed in 48 hours and as often as necessary subsequently. The nylon sutures in the proximal nail fold are removed in 5 to 7 days, and Steri-Strips or similar material may be applied for another 2 to 3 days.

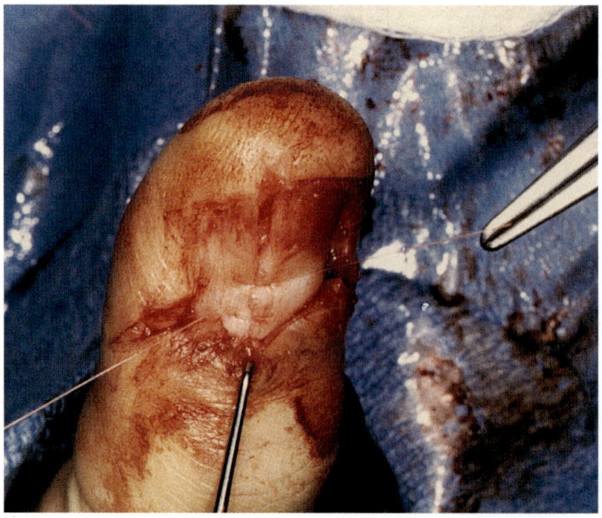

FIGURE 7E.62. First suture approximating the proximal matrix is key suture. Materials used are 5-0 polyglactin on a P-3 needle.

B. The biopsy of LM involving only the distal matrix differs in only a few ways from that involving the proximal matrix. The incision ends in a wide angle taper at the most proximal part of the pigmented streak, thus avoiding injury to the proximal matrix and reducing the possibility of a split or dystrophic nail plate (Figs. 7E.63–74). In addition, the first suture in the matrix is easier to place and is not as critical to the final appearance of the nail as it is when the pigmented streak involves the proximal matrix. In fact, the matrix may not have to be closed under these circumstances.

C. If the LM involves the lateral edge of the nail plate, then the procedure is essentially the same as for a lateral longitudinal biopsy performed in five steps. The procedure is illustrated in Figs. 7E.75–88.

We prefer that tissue be submitted for serial sectioning along the long axis of the tissue block. This permits better examination of the changes in the pigmented streak and localization of atypical cellular or tumor changes. The tissue may be marked with dyes for better tissue orientation. The dyes are those used in Mohs surgery. The dyes also can be used to differentiate the matrix from nail bed to assist the less experienced pathologist in examining nail tissue.

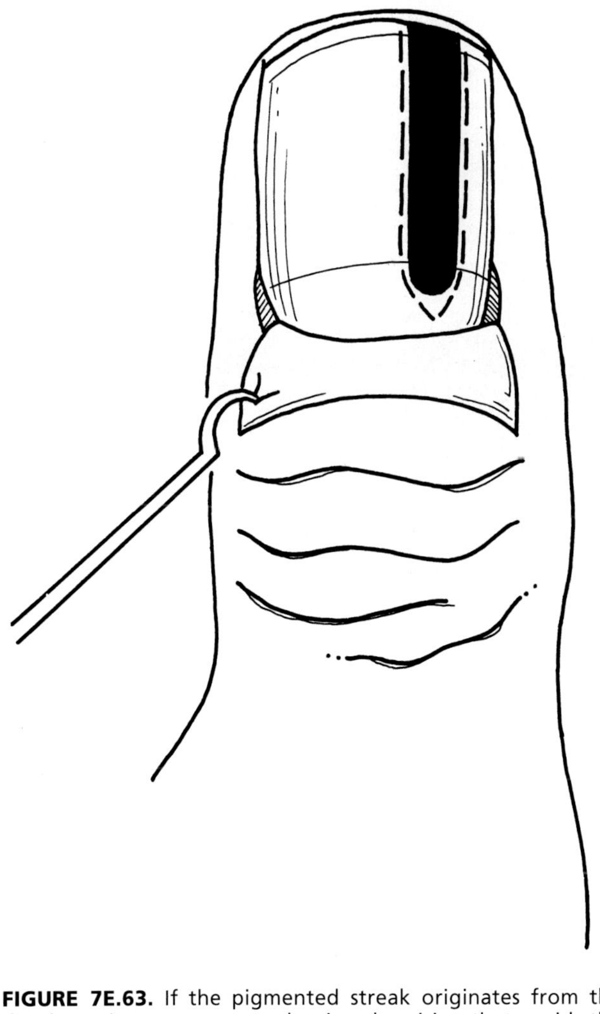

FIGURE 7E.63. If the pigmented streak originates from the distal matrix, a more tapered pointed excision that avoids the proximal matrix can be made.

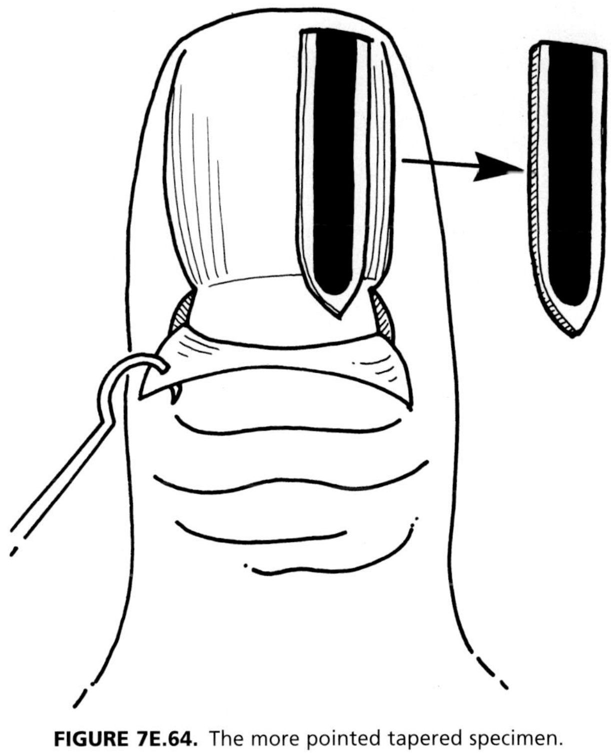

FIGURE 7E.64. The more pointed tapered specimen.

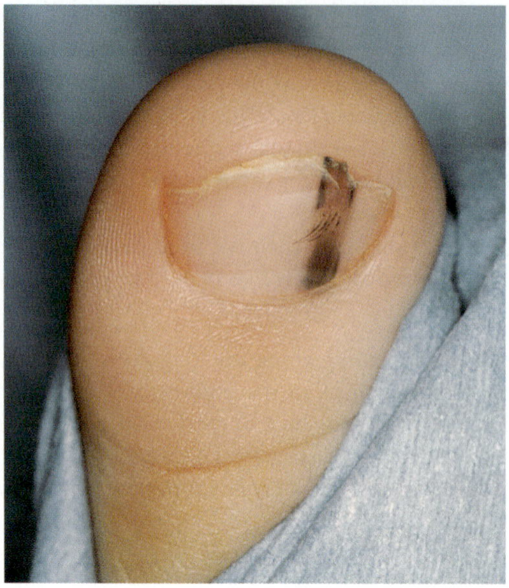

FIGURE 7E.65. Heavily pigmented nail streak.

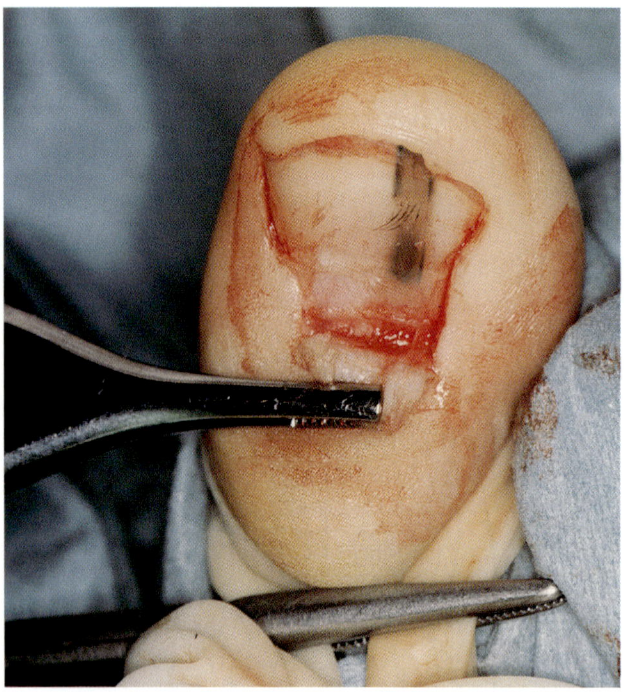

FIGURE 7E.66. Origin of pigment is the distal matrix.

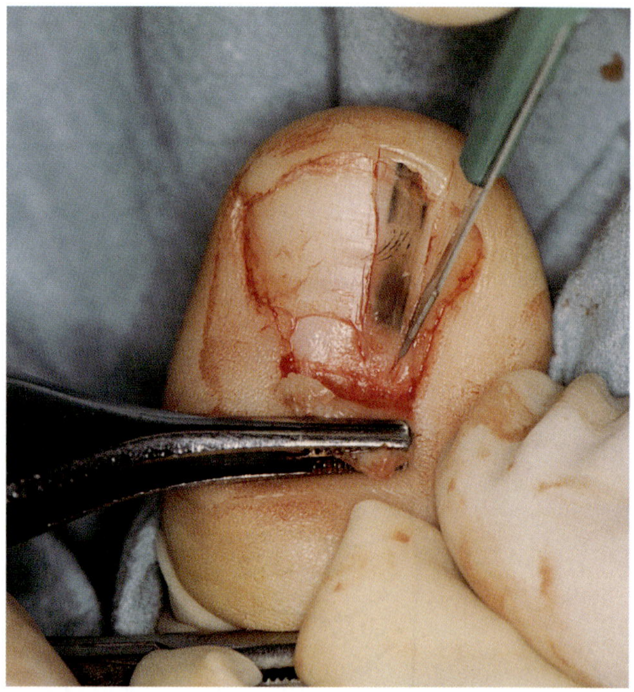

FIGURE 7E.67. A pointed tapered excision avoiding the proximal matrix.

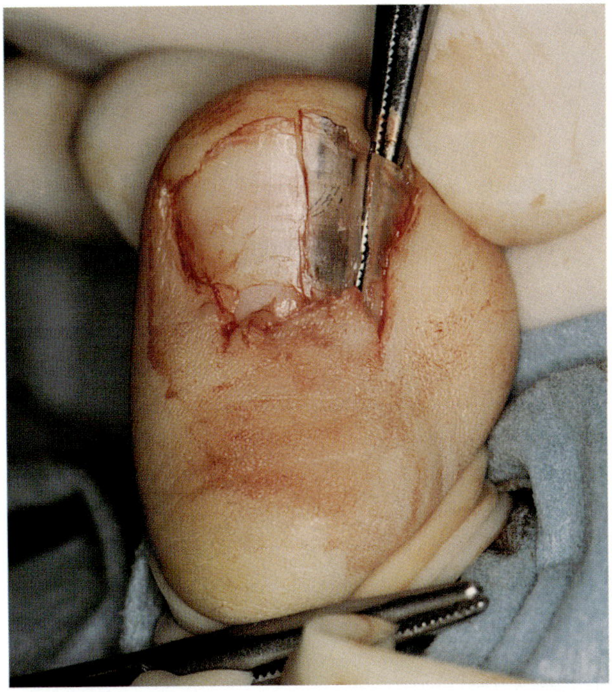

FIGURE 7E.68. The narrower rim of nail plate lateral to the pigmented streak is separated from the underlying nail tissues.

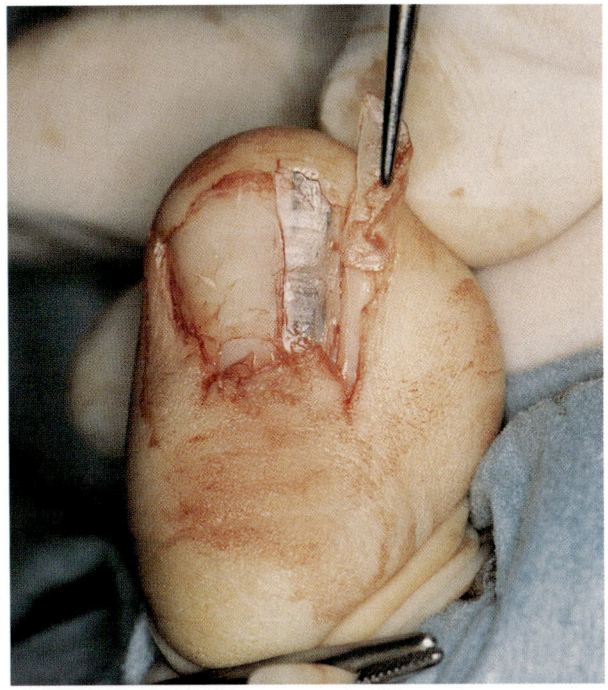

FIGURE 7E.69. The narrower rim of nail is removed to facilitate the excision of the pigmented streak.

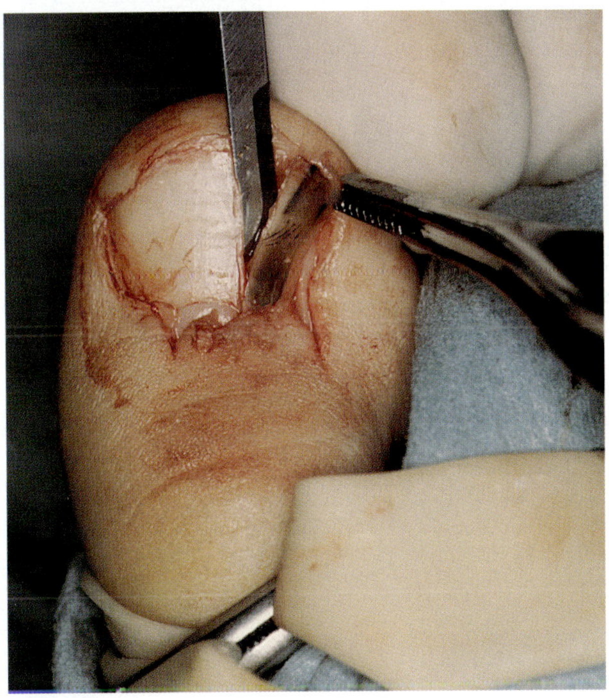

FIGURE 7E.70. Excision of the pigmented streak with the nail plate attached.

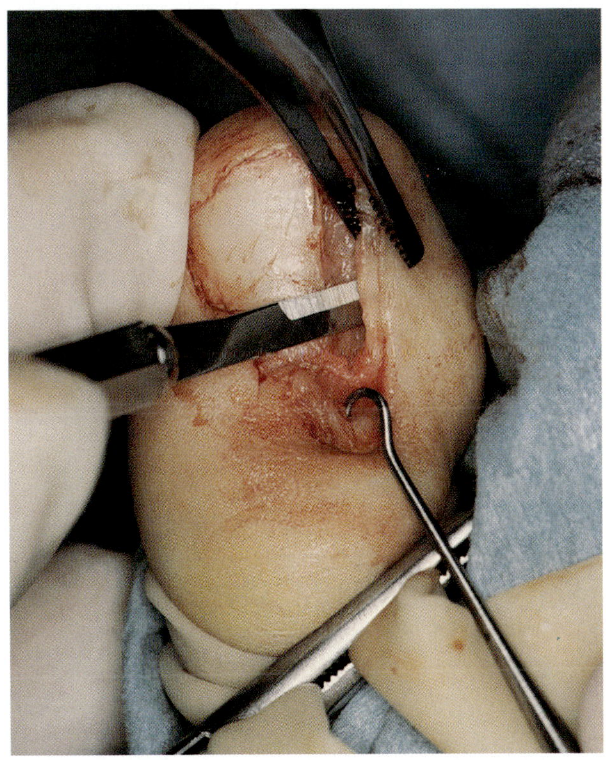

FIGURE 7E.71. The matrix and nail bed on the narrower side is undermined into the fibrofatty tissue of the lateral nail wall to mobilize the tissue and close the defect.

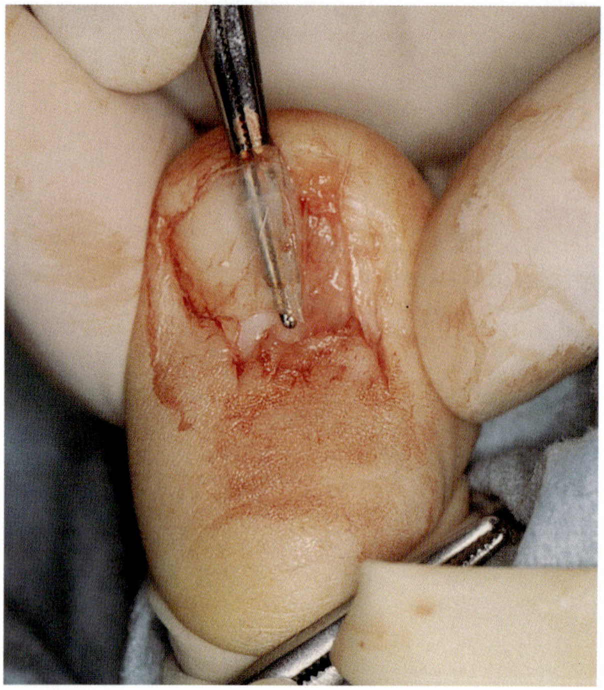

FIGURE 7E.72. An edge of nail plate is removed from the larger piece of nail plate.

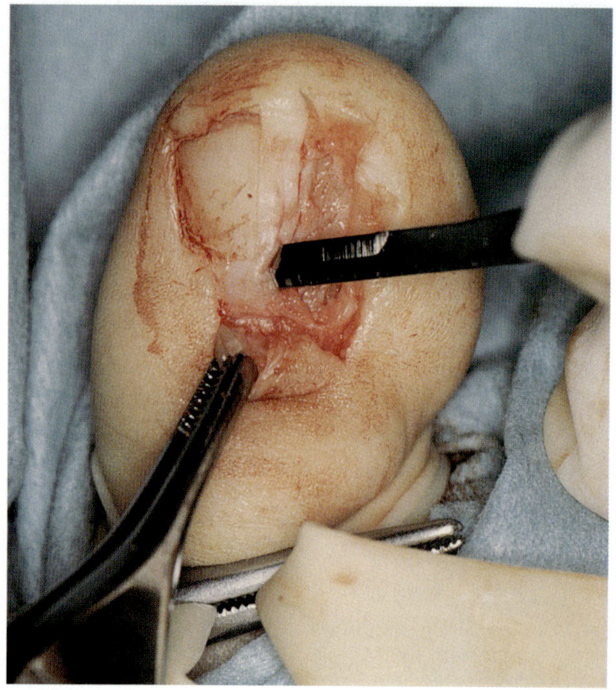

FIGURE 7E.73. A modest amount of undermining on the wider sides allows easier closure.

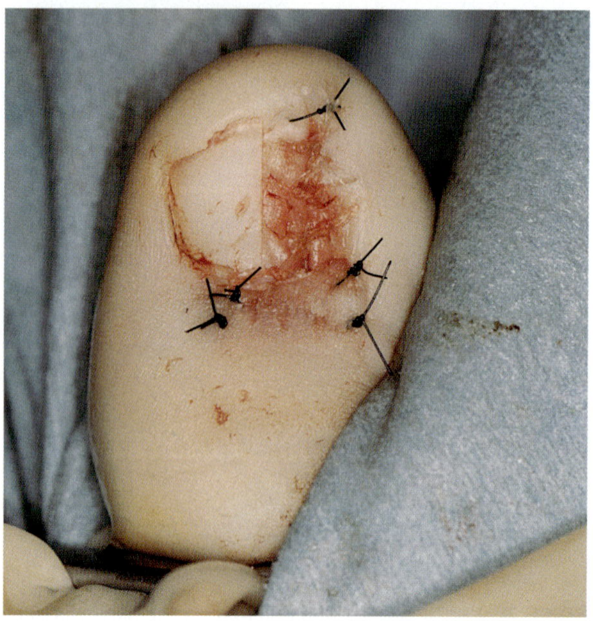

FIGURE 7E.74. Final closure with the defect sutured with 5-0 polyglactin. The proximal matrix suture placement is less critical because the proximal matrix has been avoided; the origin of the pigmented streak was in the distal matrix. Another approach would have been to remove the pigmented streak and the lateral nail and corresponding matrix, which would have left a narrower nail plate.

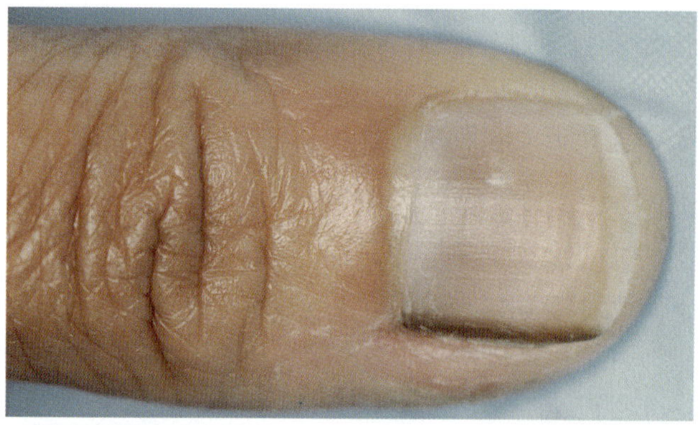

FIGURE 7E.75. Single longitudinal melanonychia of recent onset. Location is the lateral nail plate.

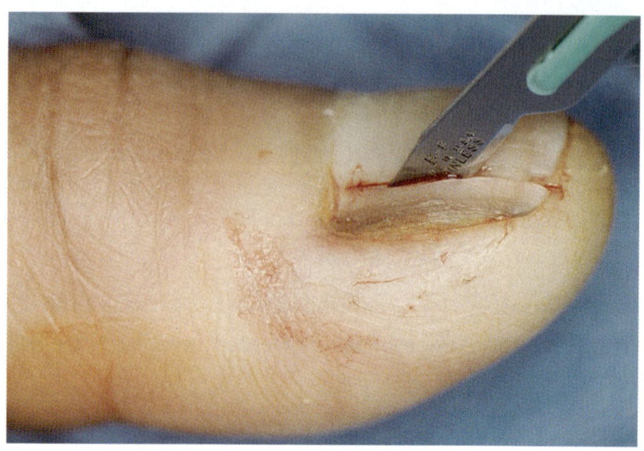

FIGURE 7E.76. Step 1: Incision in progress through nail plate of nail bed and visible matrix (lunula) stopping at the cuticle.

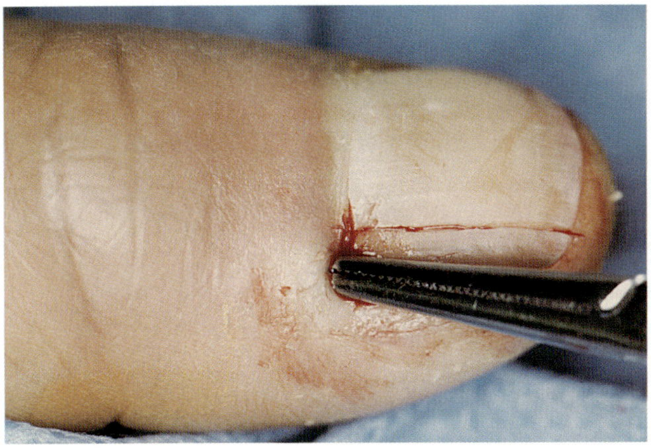

FIGURE 7E.77. Step 2: Straight part of incision of proximal nail fold is determined by depth of insertion of fine-nosed mosquito hemostat. The incision of the proximal nail fold extends only to the depth of the nail groove.

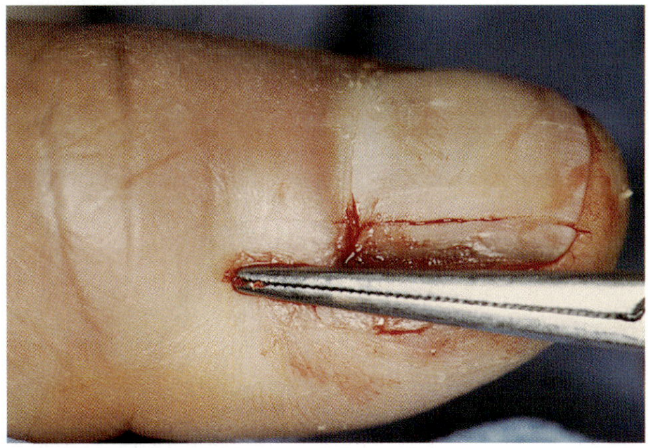

FIGURE 7E.78. The length of the straight part of incision of the proximal nail fold is being measured by previously inserted hemostat.

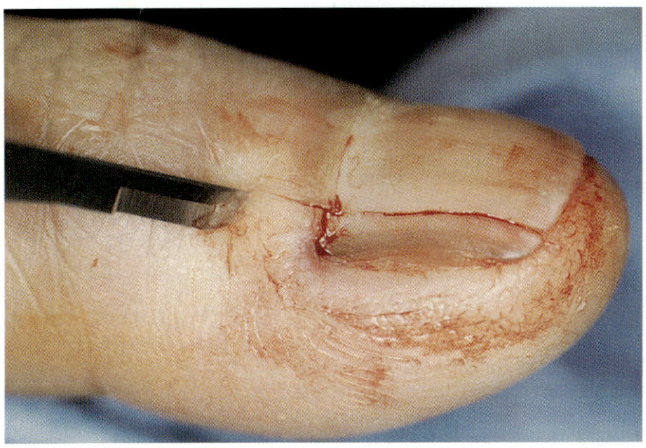

FIGURE 7E.79. Laterally curved excision permits visualization of lateral horn of the nail and matrix.

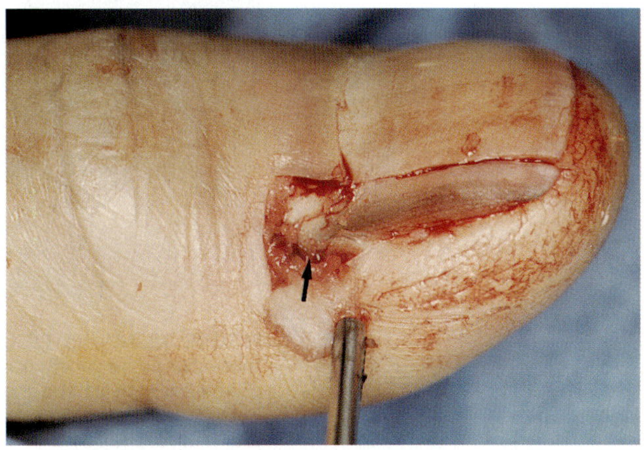

FIGURE 7E.80. Lateral horn of nail identified *(arrow)*.

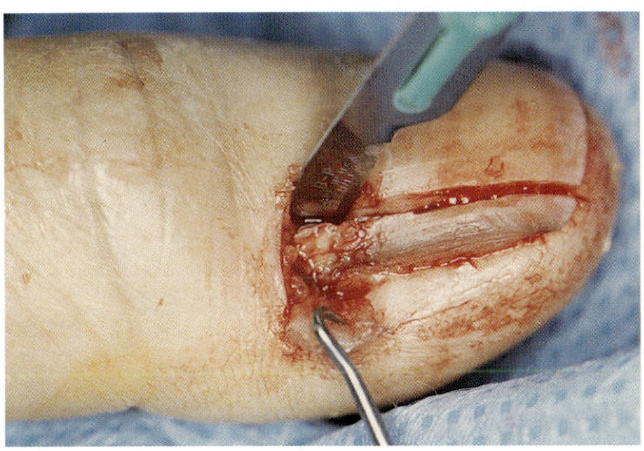

FIGURE 7E.81. Step 3: Direct continuation of incision in step 1. This completes the incision through the entire matrix.

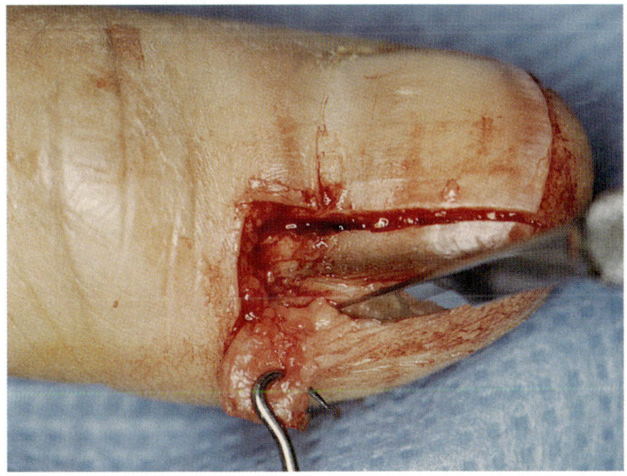

FIGURE 7E.82. Step 4: Incision in lateral nail groove dorsal to the nail plate.

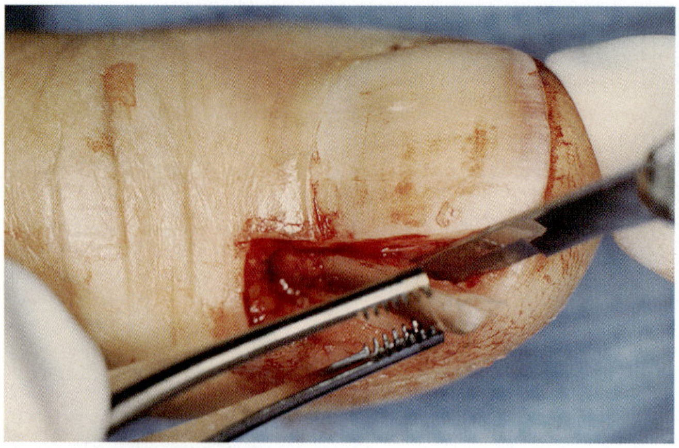

FIGURE 7E.83. Step 5: Excision of tissue begins distally and medially. The underlying bone defines the depth of the excision.

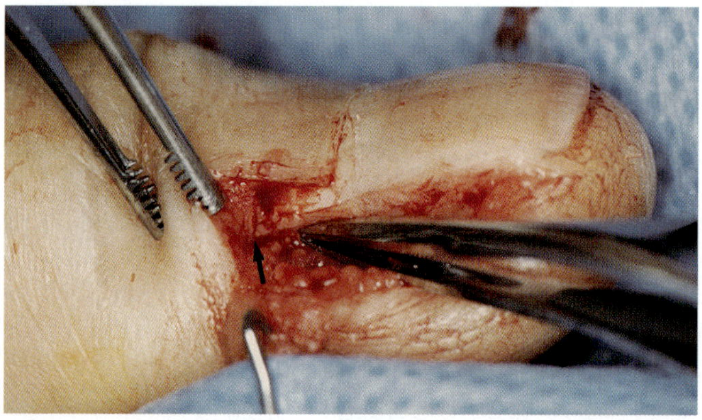

FIGURE 7E.84 and **FIGURE 7E.85.** Critical landmarks for which there should be careful examination to determine the completeness of the excision of the matrix. Figure 7E.84**,** most proximal part of the nail plate and matrix; and Figure 7E.85, fibrofatty tissue of lateral nail fold *(arrows)*.

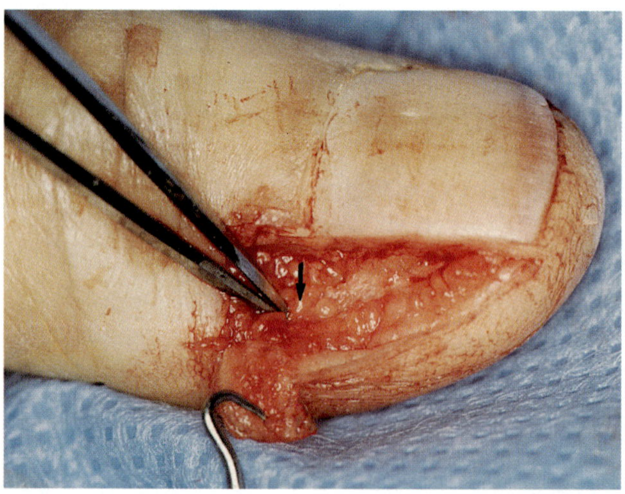

FIGURE 7E.85.

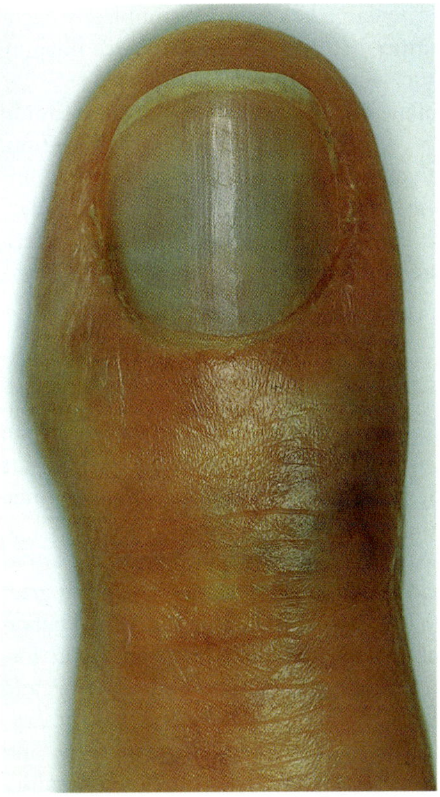

FIGURE 7F.3. Venous malformation showing swelling of the soft tissue of the distal phalanx and a blue nail bed.

ARTERIOVENOUS FISTULA (ANEURYSMAL BONE CYST)

Arteriovenous fistulas may occur in the distal phalanx of young persons. They are rapidly growing, painful lesions with marked bulbous enlargement of the tip of the digit. Aneurysmal bone cysts are a clinical entity distinct from hemangiomas of bone and from other tumors in which giant cells are a prominent feature (Fig. 7F.4). It was named aneurysmal bone cyst because the contour of the affected bone suggests a blowout type of distention resembling the saccular protrusion of the walls of an aneurysm and because cystic blood-filled spaces are encountered at surgery.

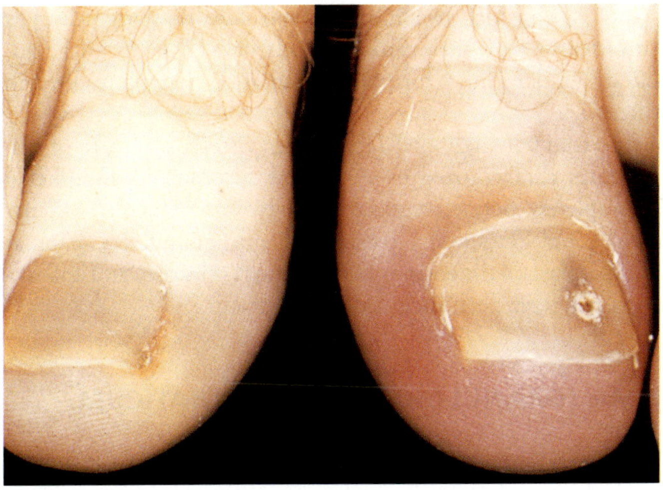

FIGURE 7F.4. Aneurysmal bone cyst. (Courtesy of J.L. Schmutz, Nancy, France).

On x-ray film, the phalanx may be excessively enlarged and almost completely occupied by osteolytic tissue simulating a malignant tumor (5). Magnetic resonance imaging permits specific diagnosis of aneurysmal bone cysts (6).

"If radiographic imaging is indicated in cutaneous vascular anomalies, one should first consider ultrasonography with Doppler flow. It is the least expensive examination, although highly operator dependent. MRI with gadolinium is now the standard. Computed tomography with contrast may be useful if there is a question of osseous involvement by vascular malformation. Angiography is used only to document or treat arteriovenous malformations and arteriovenous fistulae" (7).

EPITHELIOID HEMANGIOMA

Weiss and Enzinger (8) proposed the term "epithelioid hemangioendothelioma" to describe an unusual tumor of soft tissue having an epithelioid appearance. These tumors pursue a clinical course between that of a hemangioma and a conventional angiosarcoma. Similar neoplasms occur in others sites such as the lung, liver, and bone.

Epithelioid hemangioma masquerading as a paronychia was described in a 42-year-old woman with a 6-month history of progressive swelling and some tenderness of the left great toe. The toe was diffusely swollen, the pulp was bluish-red, and there was increased curvature of the nail. Large lytic lesion of the distal phalanx without any reactive new bone formation was observed on x-ray film, as was expansion of the proximal end of the phalanx and an associated large soft tissue mass. The tumor was multicentric, as is almost always the case.

Histopathology of curetted tissue disclosed numerous vessels of varying size and development, many with large epithelioid endothelial cells. An associated inflammatory infiltrate contained foci of eosinophils (9).

Because of its rare occurrence, epithelioid hemangioma may be misdiagnosed as a metastatic carcinoma or other neoplasm (10).

PSEUDOPYOGENIC GRANULOMA (HISTIOCYTOID HEMANGIOMA)

Avenel et al. (11) described a 40-year-old woman with angiomatous nodules affecting the fingertip, lateral nail folds, and nail bed. Histologic and ultrastructural changes of their patient were consistent with a diagnosis of pseudopyogenic granuloma (12–15). Baran reported a case of histiocytoid hemangiomas with simultaneous cutaneous and bone involvement. The patient, a 31-year-old Mexican-American man, presented with nail changes including onycholysis of the distal area, longitudinal splitting, subungual and periungual erythema, and paronychial swelling with purulent discharge. Biopsy specimens demonstrated a proliferation of histiocytoid endothelial cells with intracytoplasmic vacuoles and associated vascular lumen formation. Radiation therapy resulted in significant clinical improvement.

The collective term "histiocytoid hemangioma" (13) encompasses a spectrum of diseases that share histologic features characterized by distinctive histiocytoid endothelial cells (Fig. 7F.5). Several incompletely defined cutaneous and extracutaneous vascular tumors, including atypical pyogenic granuloma, pseudopyogenic granuloma, papular angioplasia, angiolymphoid hyperplasia with eosinophilia, Kimura's disease, and inflammatory arteriovenous hemangioma, have been included in this group.

PYOGENIC GRANULOMA (GRANULOMA TELANGIECTATICUM, BOTRYOMYCOMA)

Pyogenic granuloma is a benign eruptive hemangioma that typically occurs after a minor penetrating skin injury (Figs. 7F.6–7F.9). It first appears around the nail as a minute red

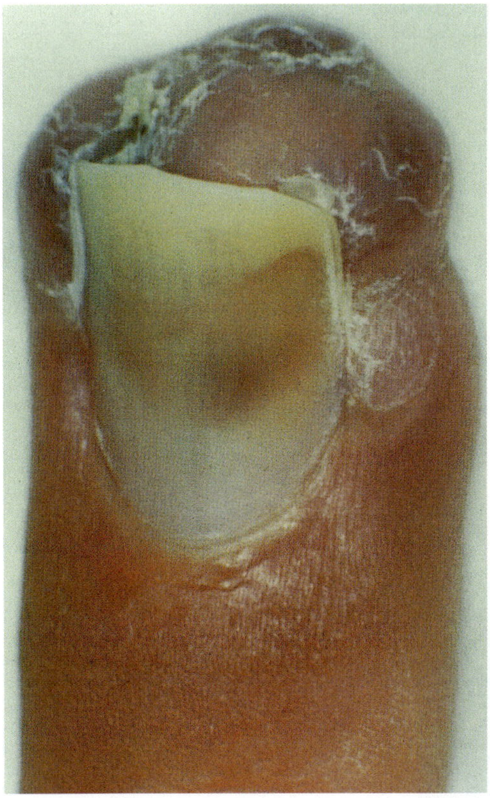

FIGURE 7F.5. Histiocytoid hemangioma. (Courtesy of J.L. Verret, Angers, France.)

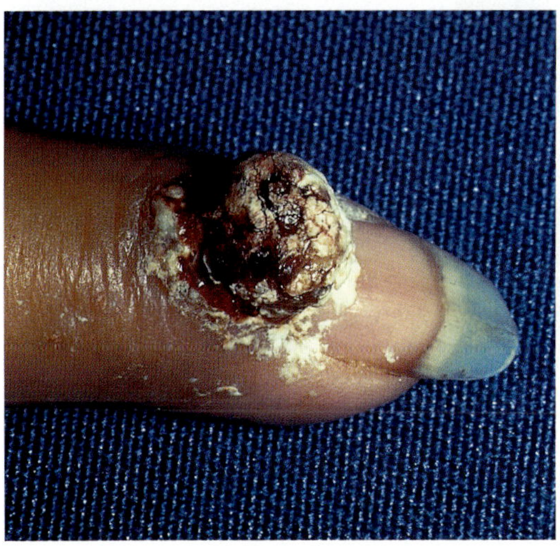

FIGURE 7F.6. Pyogenic granuloma on the proximal nail fold.

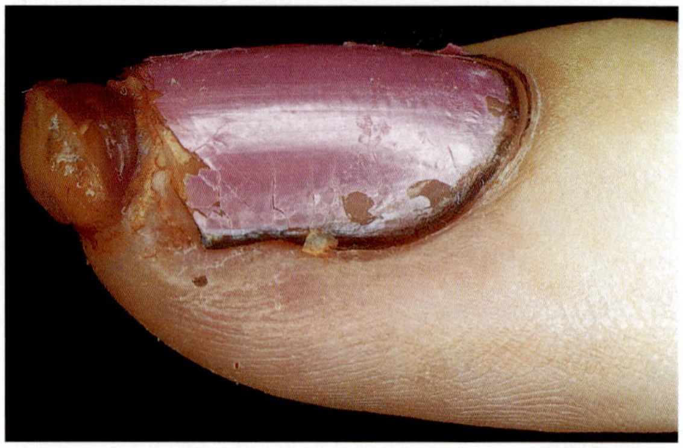

FIGURE 7F.7. Pyogenic granuloma involving the tip of the distal phalanx.

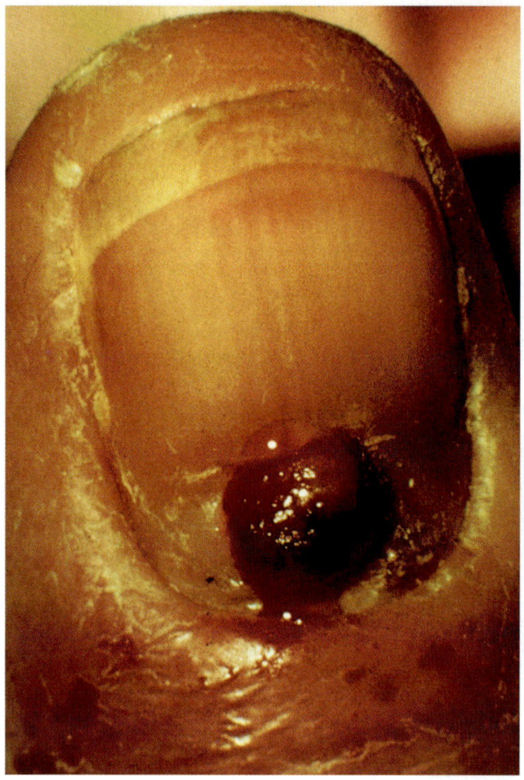

FIGURE 7F.8. Pyogenic granuloma involving the nail bed after a penetrating wound of the nail plate.

papule that rapidly grows to the size of a pea or even a cherry. Its surface may become eroded from necrosis of the overlying epidermis. Crusting may mimic a malignant melanoma, although a typical collarette usually can be seen. Pyogenic granuloma is commonly located at the proximal nail fold, but it may develop distally in the hyponychium region with onycholysis or in the nail bed after a penetrating wound of the nail plate. Tenderness and a ready tendency to bleed are characteristic features. Extensive granulation tissue due to an ingrowing toenail may mimic a periungual pyogenic granuloma, and it has been observed in patients treated with aromatic retinoids (16). Differential diagnosis must rule out cavernous angioma, pseudopyogenic granuloma, hemangiosarcoma, and especially amelanotic melanoma.

Histologic investigation of the specimen is essential. Therapy should be as simple as possible to avoid disfiguring scars or nail deformity, The easiest treatment method is to paint the tumor with a silver nitrate stick. Pyogenic granuloma can be removed at its base by excision followed by electrodesiccation. Application of Monsel's solution usually is not

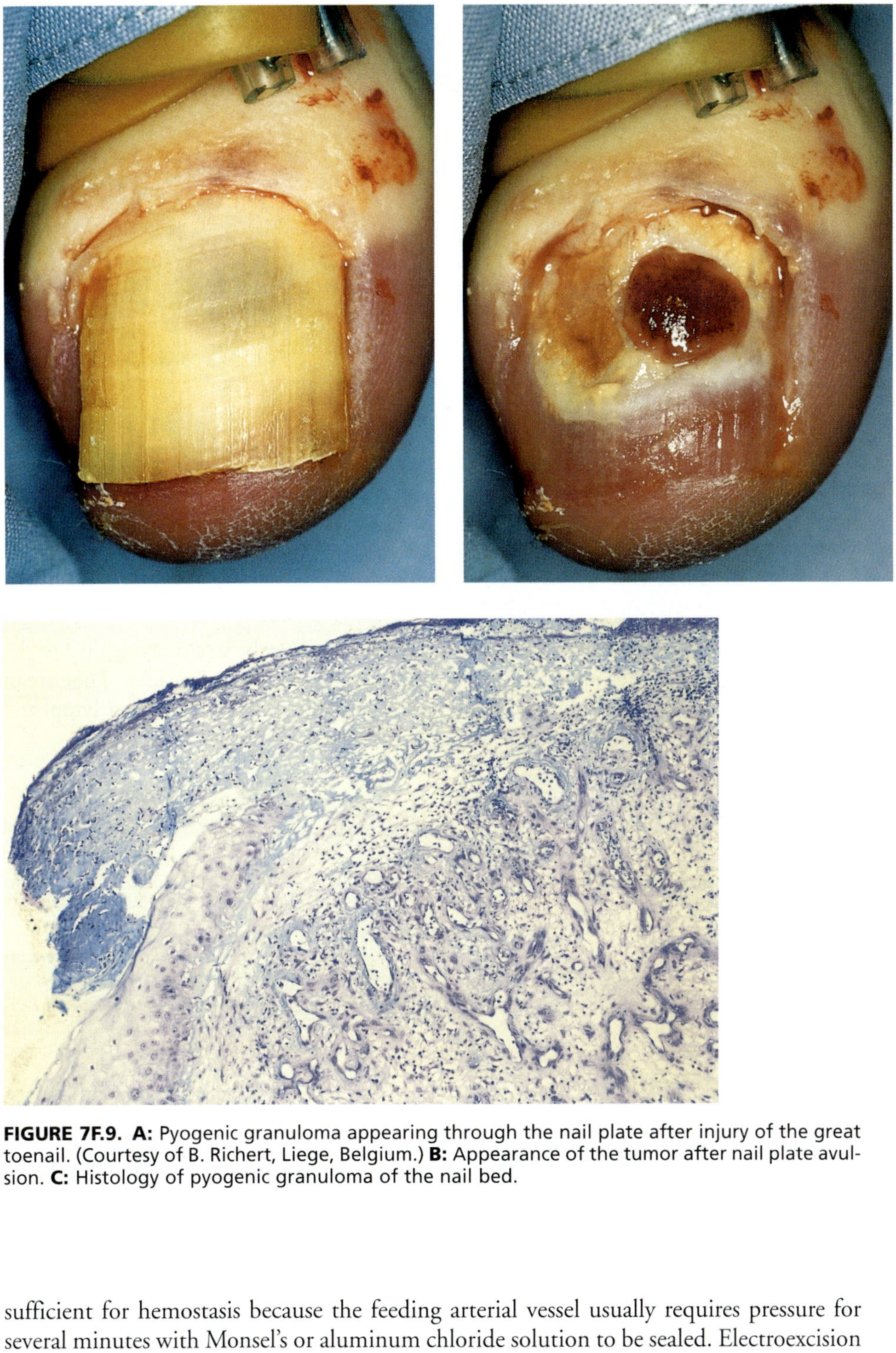

FIGURE 7F.9. A: Pyogenic granuloma appearing through the nail plate after injury of the great toenail. (Courtesy of B. Richert, Liege, Belgium.) **B:** Appearance of the tumor after nail plate avulsion. **C:** Histology of pyogenic granuloma of the nail bed.

sufficient for hemostasis because the feeding arterial vessel usually requires pressure for several minutes with Monsel's or aluminum chloride solution to be sealed. Electroexcision (Chapter 5c), carbon dioxide laser excision, and cryotherapy are curative; however, a tissue specimen for histopathologic examination is mandatory.

LYMPHANGIOMA CIRCUMSCRIPTUM

Appearing as a cluster of vesicles resembling frog spawn (some of the vesicles can be filled with fresh or altered blood), lymphangioma circumscriptum on the distal digit is rare (Fig. 7F.10).

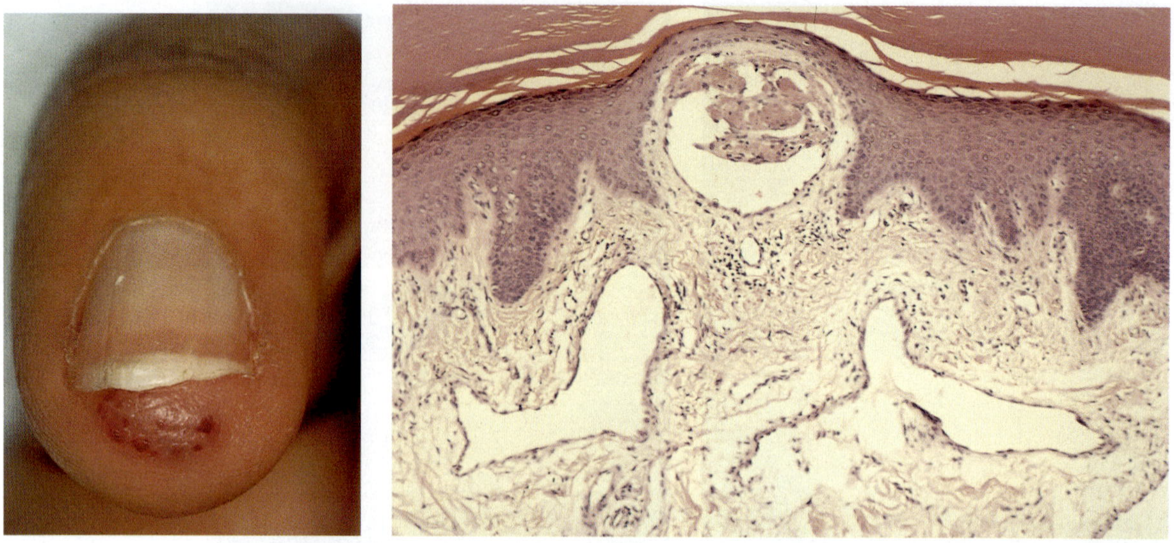

FIGURE 7F.10. A: Lymphangioma circumscriptum involving the tip of the digit. (Courtesy of B. Richert, Liege, Belgium.) **B:** Histology of the same lesion. (Courtesy of G. Pierard, Liege, Belgium.)

Histopathology shows endothelium-lined spaces that may be so closely applied to the lower surface of the epidermis that they may simulate intraepidermal vesicles. They are in communication with large subcutaneous cisterns consisting of muscle-coated lymphatic vessels that must be excised to effect cure (17). Excision has to be wide and deep.

REFERENCES

1. Enjolras O, Riché MC. *Hémangiomes et malformations vasculaires superficielles.* Paris: Medsi/McGraw-Hill, 1990.
2. Monses B, Murphey WA. Distal phalangeal erosive lesions. *Arthritis Rheum* 1984;27,449–455.
3. Heller J. Die Krankheiten der Nägel. In: Jadassohn J, ed. *Handbuch der Haut-und Geschlechtskrankheiten. Bd IX/2.* Spezielle Dermatologie. Berlin: Springer 1927:150–172.
4. Burge SM, Baran R, Dawber RPR, et al. Periungual and subungual arteriovenous tumors. *Br J Dermatol* 1986;115:361–366.
5. Schajowicz F, Aiello C, Slullitel I. Cystic and pseudocystic lesions of the terminal phalanx with special reference to epidermoid cysts. *Clin Orthop Rel Res* 1970;68:84–92.
6. Schmutz JL, Cuny JF, Duprez A, et al. Kyste osseux anévrismal d'un orteil. *Rech Dermatol* 1988;1:679–681.
7. Mulliken JB. A biologic approach to cutaneous vascular anomalies. *Pediatr Dermatol* 1992;9:356–357.
8. Weiss SW, Enzinger FM. Epithelioid haemangioendothelioma. A vascular tumor often mistaken for a carcinoma. *Cancer* 1982;50:970–981.
9. Kennedy CTC, Burton PA, Cook P. Swollen toe due to epithelioid haemangioma of bone. *Br J Dermatol* 1990;123[Suppl 37]:85–89.
10. Tsuneyoshi M, Dorfman HD, Bauer TW. Epithelioid hemangioendothelioma of bone. A clinicopathologic, ultrastructural, and immunohistochemical study. *Am J Surg Pathol* 1986;10:754–756.
11. Avenel M, Verret JL, Fortier P. Finger localisation of Wilson-Jones pseudo-pyogenic granuloma. In: *Case Presentations in XVI Congressus Int Derm Tokyo.* Tokyo: University of Tokyo Press, 1982:38.
12. Wilson-Jones E, Bleehen SS. Pseudo-pyogenic granuloma. *Br J Dermatol* 1969;81:804–816.
13. Rosai J, Gold J, Landy R. The histiocytoid hemangiomas. A unifying concept embracing several previously described entities of skin, soft tissue, large vessels, bone, and heart. *Hum Pathol* 1979;10:707–730.
14. Verret JL, Avenel M, François H, et al. Hémangiomes histiocytoides des pulpes digitales. *Ann Dermatol Venereol* 1983;110:251–257.
15. Dannaker C, Piacquadio D, Willoughby CB, et al. Histiocytoid hemangioma: a disease spectrum. *J Am Acad Dermatol* 1989;21:404–409.
16. Baran R. Retinoids and the nails. *J Dermatol Treat* 1990;1:151–154.
17. Mehregan A. *Pinkus' guide to dermatohistopathology.* Norwalk, CT: Appleton & Lange, 1986:558.

7G

KERATIN CYSTS

ROBERT BARAN
ECKART HANEKE

IMPLANTATION EPIDERMOID CYSTS

Epidermoid cysts (keratin, squamous epithelial, or traumatic cysts) in the terminal phalanx of the digits usually are secondary to heavy trauma, with implantation of epidermis into subcutaneous tissue or even into the bone. The trauma may have occurred decades ago and may not always be remembered. Postoperative epidermoid cysts may occur in the proximity of scars (Fig. 7G.1) (1).

The distal phalanx gradually enlarges and clubbing becomes evident. Pain is of late onset and is caused by compression of the bone. Eventually a fracture may occur. Shooting pain without bone involvement was described in one case (2). Acquired pincer nail is an unusual presentation (3).

Intraosseous epidermoid cysts are solitary lesions seen twice as frequently in men as in women. Pain and swelling of the terminal phalanx are the most frequent clinical signs. We observed a case of painful discharge from an open cyst through both great toenails in a patient treated 15 years previously for ingrown nails (Fig. 7G.2) (4). The lesion appears as a round, osteolytic zone embedded in the bone, without trabecules and sclerosis (cf. enchondroma), or as a marginal defect of cortical substance of bone (5). The clinical and radiologic differential diagnosis may be very difficult (6). Histopathology shows a simple epidermoid cyst filled with orthokeratin and lined with a thin epidermis (Fig. 7G.3). However, if remnants of matrix epithelium are displaced into the subcutaneous tissue, the cyst may contain areas resembling trichilemmal cyst epithelium (7). Treatment is by enucleation of the lesion including its entire membrane. This usually can be achieved with a lateral L-shaped incision. Bone transplants are not necessary.

SUBUNGUAL EPIDERMOID INCLUSIONS

Subungual epidermoid inclusions (epidermal buds) develop from the ridges of the nail bed epithelium (8). Although histologically indistinguishable from subungual epidermoid cysts, they usually remain microscopic. In rare cases they become large enough to produce symptoms, such as swelling of the nail bed. Trauma is a possible cause (9). They occur especially with finger clubbing but without associated dystrophy (10).

In addition to the two main varieties, a new form of subungual epidermoid inclusions was reported in eight cases (11). The most striking clinical features of this form is subungual hyperkeratosis associated with shortened and dystrophic nail plate. Onycholysis was observed in one case. A history of trauma is frequent, but the condition is symp-

R. Baran: Nail Disease Centre, 42, Rue des Serbes, 06400 Cannes, France.
E. Haneke: Klinikk Bunæs, Løkkeåsveien, 1337 Sandvika, Oslo, Norway.

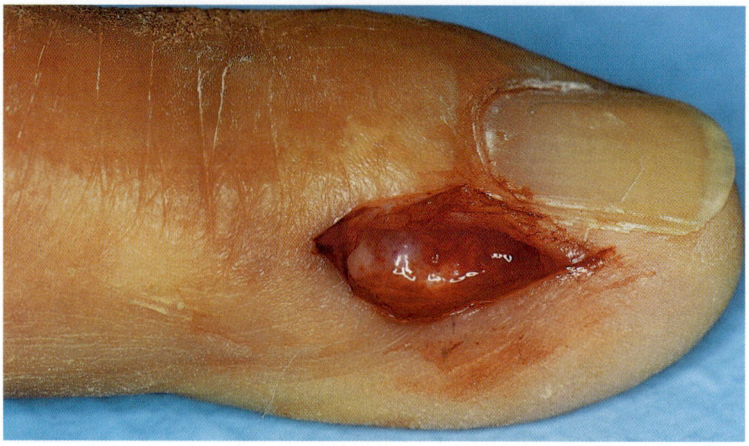

FIGURE 7G.1. Implantation epidermoid cyst in the proximity of surgical scar.

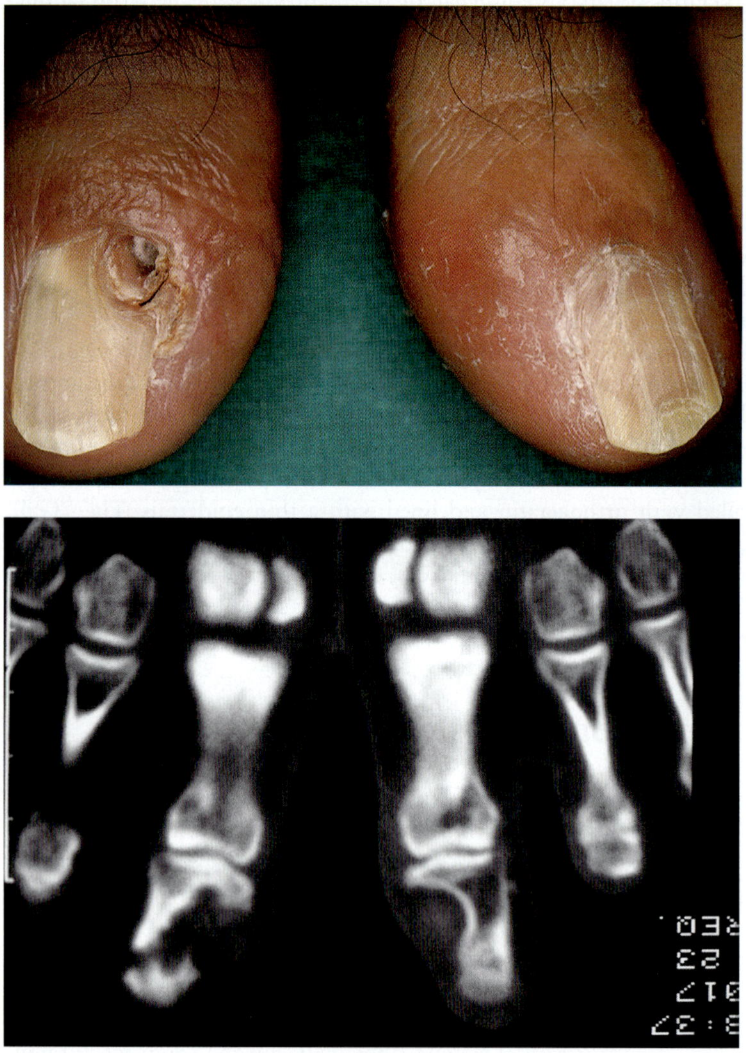

A

B

FIGURE 7G.2. A: Epidermoid implantation cysts involving both great toenails after several operations for ingrown toenails. (Courtesy of O. Binet, Paris, France.) **B:** X-ray film of the great toenails in the same patient.

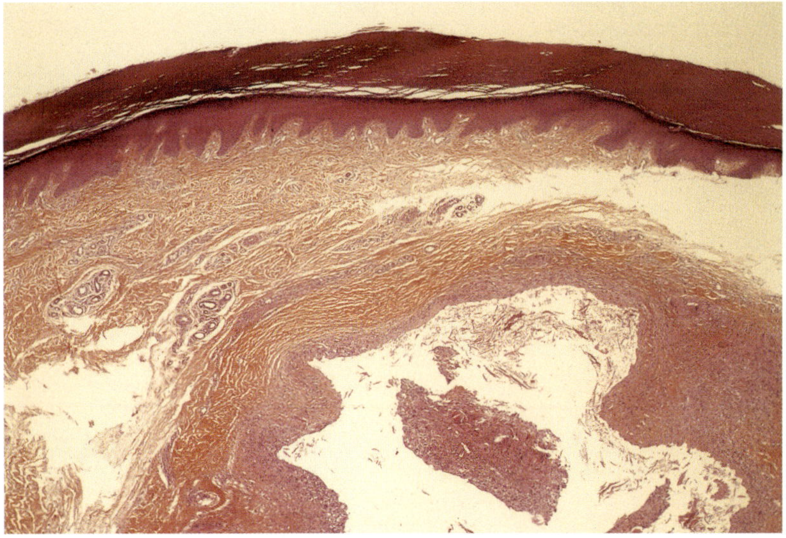

FIGURE 7G.3. Histology of epidermoid implantation cyst.

tomless. In all cases, the biopsy reveals marked hyperplasia of the nail bed and epidermoid cysts in the dermis. Onychomycosis and psoriasis are the main differential diagnosis.

Subungual epidermoid inclusions usually remain asymptomatic and therefore do not require treatment.

REFERENCES

1. Baran R, Bureau H. Two post-operative epidermoid cysts following realignment of the hallux nail. *Br J Dermatol* 1989;119:245–47.
2. Yung W, Estes SA. Subungual epidermal cyst. *J Am Acad Dermatol* 1980;3:599–601.
3. Baran R, Broutard JC. Epidermoïds cyst of the thumb presenting as a pincer nail. *J Am Acad Dermatol* 1989;19:143–144.
4. Challier L, Binet O, Baran R, et al. Kystes épidermoids d'implantation intra-osseuse de la phalange distale des deux gros orteils. *Ann Dermatol Venereol* 1994;121[Suppl 1]:S89–S90.
5. Drewes J, Günther D, Nolden HH. Intraossäre Epidermiszystem der Finger und Zehen. *Akt Chir* 1985;20:171–177.
6. Schajowicz F, Aiello Ca, Slullitel I. Cystic and pseudo cystic lesions of the terminal phalanx with special reference to epidermoid cyst. *Clin Orthop Rel Res* 1970;68:84–92.
7. Haneke E. Onycholemmal horn. *Dermatologica* 1983;167:155–158.
8. Zaias N. *The nail in health and disease,* 2nd ed. Norwalk, CT: Appleton & Lange, 1990:218.
9. Samman PD. The human toenail, its genesis and blood supply. *Br J Dermatol* 1959;71:296–302.
10. Lewin K. Subungual epidermoid inclusion. *Br J Dermatol* 1969;81:671–675.
11. Fanti PA, Tosti A. Subungual epidermoid inclusions: report of 8 cases. *Dermatologica* 1989;178: 209–212.

7H

BONE AND CARTILAGE TUMORS

ECKART HANEKE

SURGICAL TREATMENT OF SUBUNGUAL EXOSTOSIS

Dupuytren's subungual exostosis is a relatively common lesion of the nail region. It is seen mainly in youngsters, adolescents, and young adults. It occurs more often in females. Almost half of these reactive lesions involve the dorsomedial aspect of the terminal phalanx of the hallux; only 10% to 15% occur in fingers (1–8). Trauma, often repeated minor injuries, appears to be the main cause (9), although recent cytogenetic investigations in subungual exostosis revealed clonal abnormalities, suggesting that at least some of these heterotopic ossifications could be neoplastic in nature (10).

Clinically, most subungual exostoses arise from the laterodorsal aspect of the tip of the distal phalanx. They grow insidiously to a rock-hard, often tender mass that elevates the free margin of the nail plate. The overlying skin usually is overstretched, hyperkeratotic, and shiny, making the distal groove disappear. Occasionally, the surface is ulcerated and infected. The initial symptom often is pain, sometimes paronychia.

Clinical diagnosis is easy when one remembers to palpate this hard tissue mass, and the diagnosis is confirmed by x-ray film. This confirmation allows differentiation of subungual exostosis from bone spikes due to crush injury or turret exostosis due to a penetrating injury. Nevertheless, the diagnosis often is missed. Differential diagnosis includes ingrown toenail, viral wart, pyogenic granuloma, glomus tumor, epidermal cyst, enchondroma, granulomatous inflammations, fibroma, keratoacanthoma, ungual melanoma, squamous cell carcinoma, and many other rare nail tumors and conditions. Subungual location in multiple hereditary exostoses is very rare. It is not clear if subungual osteochondroma is really a different entity (11,12). Malignant degeneration does not appear to occur (3,13).

Histopathologic examination shows an exostosis capped by cartilage merging with the overlying fibrous tissue. The epidermis usually is hyperkeratotic (14).

Treatment is aimed at radical extirpation to avoid recurrence and saving as much tissue as possible to combine good function with cosmesis. Because most exostoses are located under the free edge of the nail rather than in the more proximal nail bed or even matrix region, the direct approach is recommended (15).

A ring block is performed at the base of the digit and a tourniquet can be applied. If the exostosis is visible as a tumor mass under the free margin of the nail plate, a fusiform incision with the scalpel beveled according to the growth of the lesion is performed. The soft tissue is dissected from the exostosis down to its base, which is clipped off with a sturdy nail clipper or bone rongeur. Slightly more of the bone is taken to safely avoid a recurrence. The wound is closed with simple 3-0 or 4-0 stitches that can be removed after 10 to 14 days (Figs. 7H.1 and 7H.2). If the lesion extends under the nail plate, part of it is cut away to allow access to the base of the exostosis. If the tumor is located under the

E. Haneke: Klinikk Bunæs, Løkkeåsveien, 1337 Sandvika, Oslo, Norway.

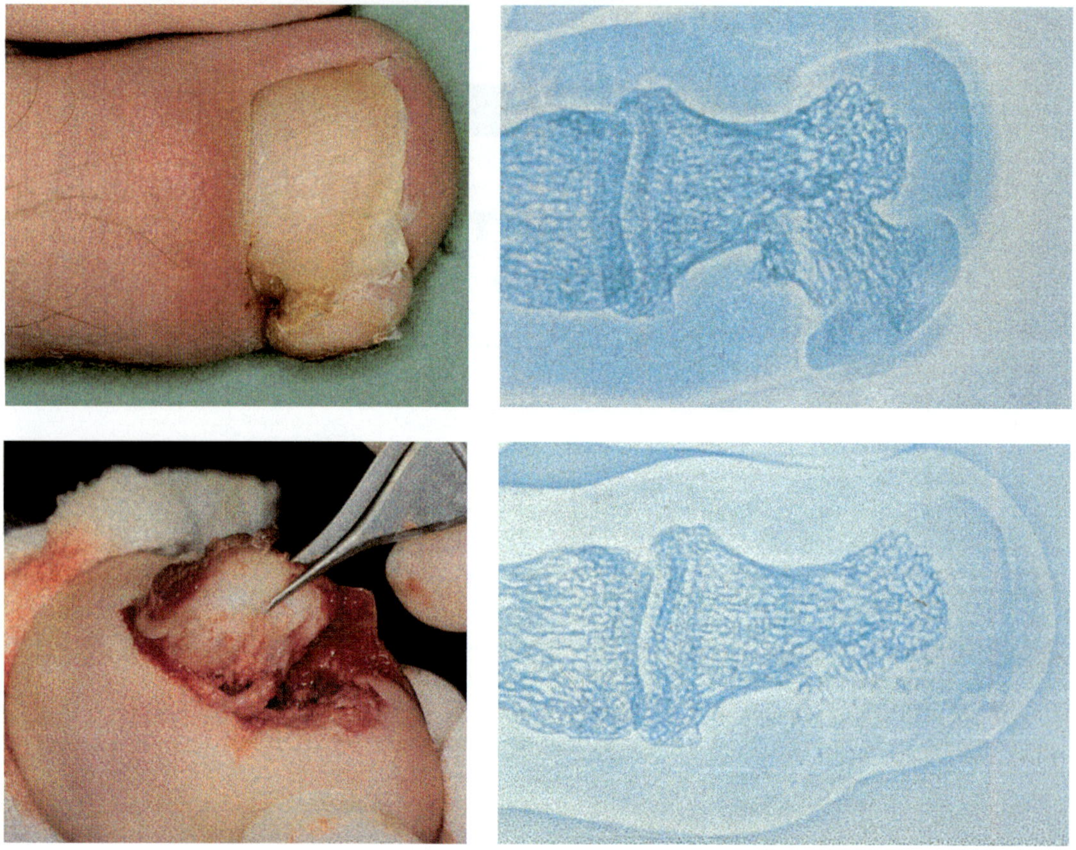

FIGURE 7H.1. Large subungual exostosis. **A:** Before operation. **B:** Radiograph before operation. **C:** Exostosis clipped off at its base and slightly lifted to show its connection with the terminal phalanx and its cancellous bone core. **D:** Radiograph after surgery.

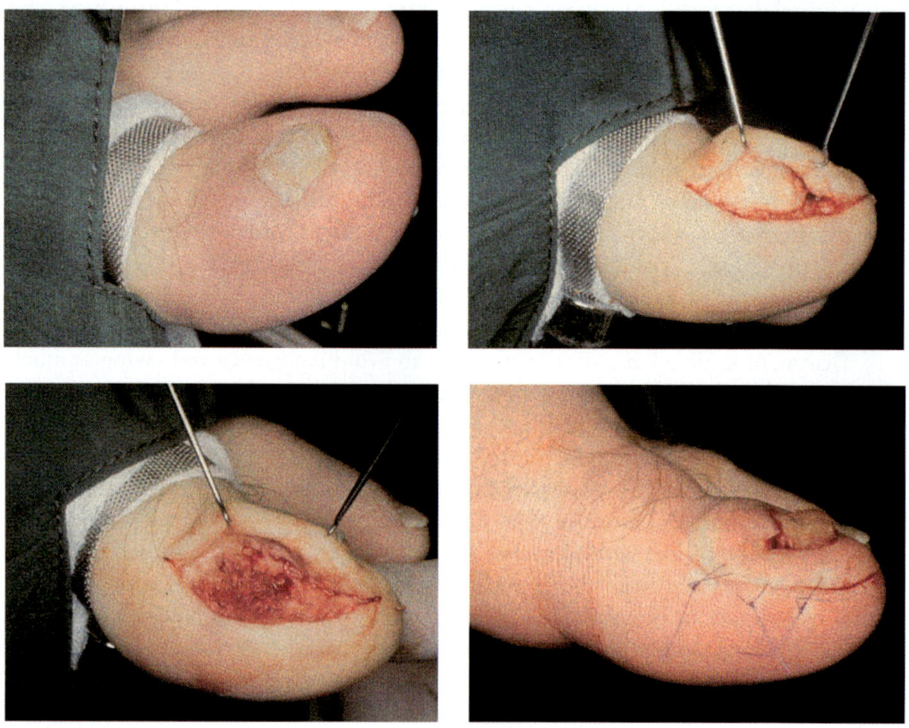

FIGURE 7H.2. Exostosis under the proximal medial aspect of the great toe nail. **A:** Before surgery. **B:** Lateral incision allows visualization of the exostosis after dissection. **C:** Exostosis is removed, showing the cancellous bone at its base. **D:** End of operation.

nail plate in a more central position, the nail plate is freed from the nail bed, incised transversely proximal from the lesion, and elevated to the other side but left in contact with the rest of the nail bed and plate. A longitudinal incision is made in the nail bed, and the tumor dissected and resected. The nail bed is sutured with fine resorbable material and the nail plate reflected again to give an optimal physiologic dressing (Fig. 7H.3). It rarely is necessary to excise part of the nail bed, although even then healing usually gives a good functional and esthetic result (16).

Recurrences are not seen provided the lesion was completely removed, particularly if no cartilage was left. A radiograph may prove the total excision.

SURGICAL TREATMENT OF SUBUNGUAL ENCHONDROMA

Enchondromas are cartilaginous tumors arising at a site where normally there is no cartilage, whereas (ec)chondromas are hyperplastic growths originating from normal cartilage; however, the terminology often is not strictly used. Enchondromas may occur as solitary or multiple tumors frequently arising from acral sites (17,18). Trauma is thought to be the cause in some cases. Solitary enchondroma has a low risk of malignant degeneration, which is common in enchondromatosis (19–21).

Subungual enchondroma is very rare. It usually presents as an insidious enlargement of the distal phalanx of the digit often mimicking chronic paronychia. The nail may be deformed and enlarged. Pain and tenderness are characteristic symptoms. Pathologic frac-

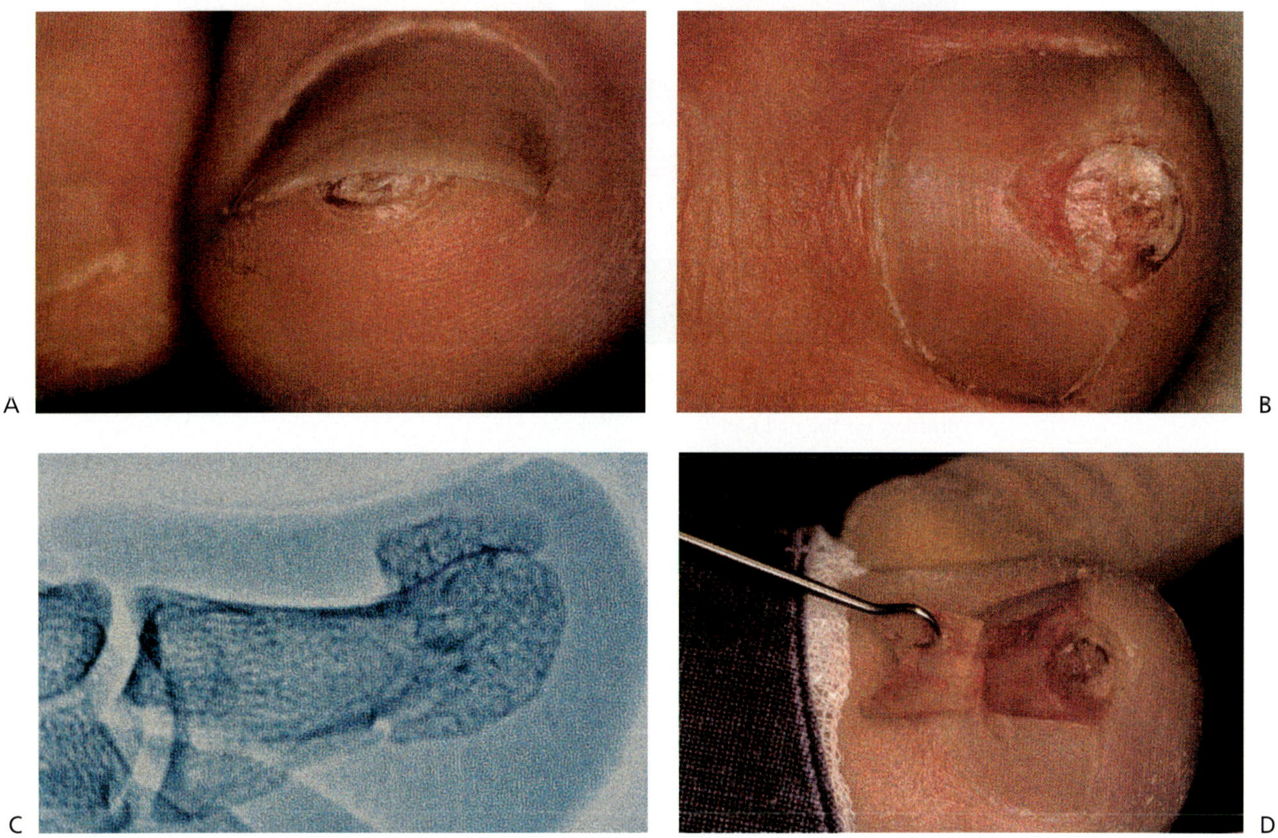

FIGURE 7H.3. Subungual exostosis in the median part of the distal nail bed. **A:** Seemingly keratotic tumor seen under the nail plate margin. **B:** Removal of the overlying nail plate renders the lesion visible. **C:** Preoperative xeroradiography shows a dorsal exostosis. **D:** Distal median portion of the nail is incised and reflected.

(continued)

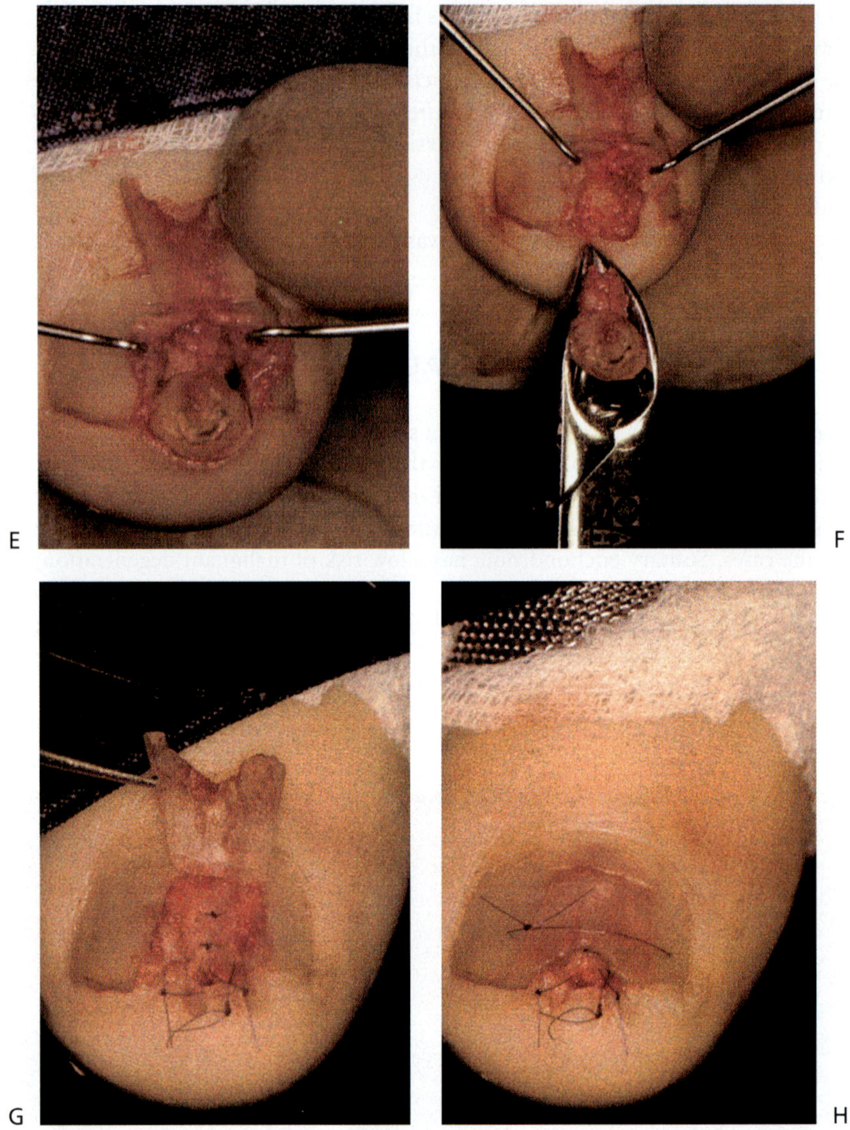

FIGURE 7H.3 *(continued)*. E: Nail bed is incised around the exostosis. **F:** Lesion is clipped off. **G:** Suture of the nail bed. **H:** Nail plate is laid back as a physiologic dressing for the nail bed.

tures and rupture of the tendon insertions may occur from extreme thinning of the cortical bone (22–26). Radiology shows an enlarged distal phalanx with expanded, paper-thin cortical bone often with a weblike appearance. Fine calcium deposits and bone destruction are characteristic for chondrosarcoma (21). Subungual chondroma clinically resembles subungual exostosis, but this lesion is radiolucent (18). Histology shows mature cartilage without cellular atypy. Differential diagnosis includes chronic paronychia, ingrown nail, acroulcerative neuropathy with bone enlargement and osteolysis, aneurysmal bone cyst, intraosseous epidermal cyst, and chondromyxoid fibroma (27), as well as any other subungual tumor mass.

Treatment of subungual enchondroma is surgical. The best approach is a longitudinal palmar incision that permits dissection of the pulp tissue and access to the bone. The latter is incised and curetted. The wound cavity is filled with cancellous bone or bone-morphogenetic protein (28,29). Tendon avulsion due to enchondroma may require

71

OSTEOID OSTEOMAS

ELVIN G. ZOOK

Osteoid osteoma is an uncommon lesion in the distal phalanx and frequently is difficult to diagnose. The most common finding in 92% of patients is pain. Swelling and clubbing occur in 75% of patients. Aspirin has been successful in relieving pain in 42% of individuals.

Radiographs frequently are nonspecific and, as a rule, lag significantly behind the symptoms. The successful use of computed tomography and bone scans for diagnosis has been reported.

Over time, the typical clubbing of the nail and enlargement of the tip occur (Figs. 71.1–71.5).

Treatment is surgical removal, which involves accessing the distal phalanx in the area of the osteoid osteoma and curettage of the tumor. The nail usually returns to a normal shape over time.

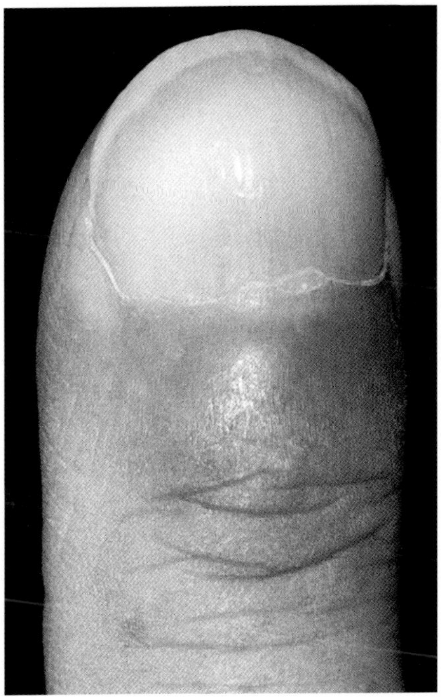

FIGURE 71.1. A 28-year-old woman with a painful tender fingertip that enlarged over the course of 1 year.

E. G. Zook: Division of Plastic Surgery, Southern Illinois University, Springfield, Illinois 62708.

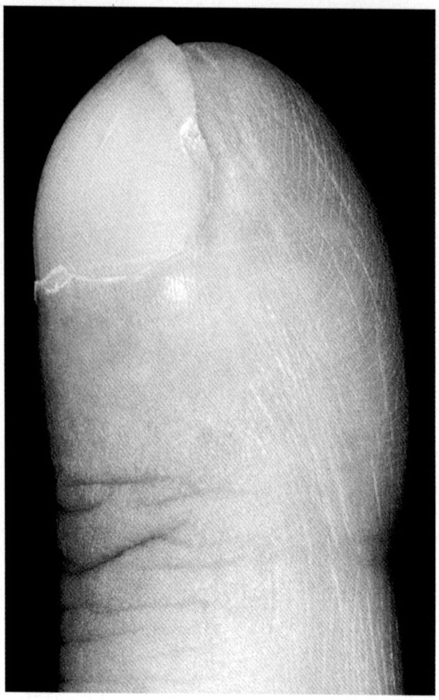

FIGURE 71.2. Lateral view of patient shown in Fig. 71.1.

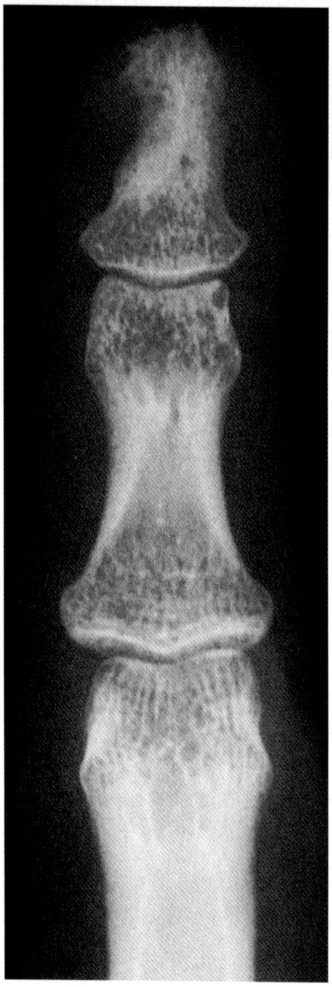

FIGURE 71.3. Anteroposterior x-ray film of distal phalanx. Note bony change on *left*.

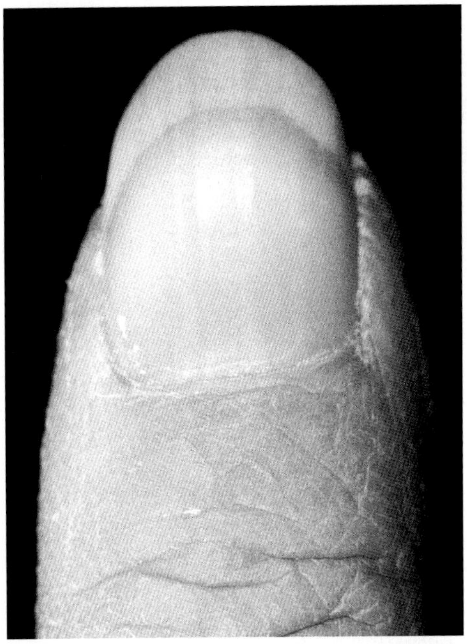

FIGURE 71.4. One year after operation. Patient is asymptomatic.

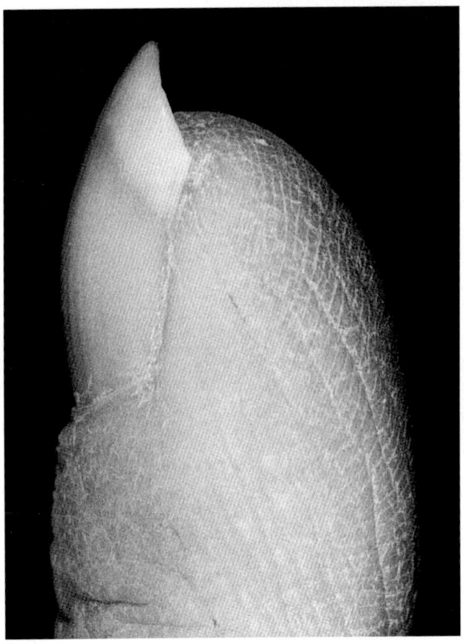

FIGURE 71.5. Lateral view taken at same time as Fig. 71.4.

8A

EPIDERMOID CARCINOMA OF THE SUBUNGUAL AND PERIUNGUAL TISSUES

EDWARD A. KRULL

Bowen's disease (squamous cell carcinoma *in situ*) and invasive squamous cell carcinoma of the nail tissue frequently are combined into the term "epidermoid carcinoma" for a number of reasons (1–3):

1. Difficulty in clinical differentiation of Bowen's disease from invasive squamous cell carcinoma of the nail tissues
2. More aggressive behavior of Bowen's disease in the nail tissues (2,3)
3. Histologic evidence of invasive squamous cell carcinoma in some parts of the lesions of Bowen's disease (1).

Epidermoid carcinoma of the subungual and periungual tissues is uncommon but not rare (1,4). Since the first description of squamous cell carcinoma of the nail unit in 1850 (5), more than 200 cases have been reported in the literature (1,4,6–8). There are probably many more cases that were treated in physicians' offices and clinics and remain unreported.

Epidermoid carcinoma may involve the periungual and subungual tissues. Most commonly it originates in the nail folds and groove extending under the nail (1,2). Clinically, epidermoid carcinoma presents as verrucous, hyperkeratotic, papillomatous lesions (Figs. 8A.1 and 8A.2) (1–4,9–13). Erosions, crusting, erythema, fissuring, scaling, granulation tissue, and ulcerations are common associations. Periungual swelling and erythema may be seen with deeper involvement in the nail folds. A "whitish band" appearance of the proximal nailfold is thought to be characteristic of invasion of this tissue (14). Onycholysis reflects nail bed involvement (2). The nail bed may show erythema, granulation tissue, and ulceration. Nail dystrophy and destruction are evidence of matrix involvement (1,3). Longitudinal melanonychia has been observed with subungual epidermoid carcinoma (15–17). Nodularity, ulceration, and bleeding suggest tumor invasion (Fig. 8A.3) (1).

The tumor frequently is misdiagnosed as verruca vulgaris (1,6,9). However, pyogenic granuloma, nail dystrophies, chronic paronychia, chronic inflammation, ingrown toenail, subungual exostosis, subungual melanoma, glomus tumor, and acquired fibrokeratoma have been diagnostic considerations (9–12). It is imperative to biopsy chronic lesions of the periungual and subungual tissues, especially those suggestive of verruca vulgaris (Fig. 8A.4). Partial or total nail avulsion may be necessary to obtain an adequate specimen and for treatment (Fig. 8A.5).

The greatest reported incidence of epidermoid carcinoma of the nail tissue is in the 50- to 69-year age group (1), but some reports indicate that as many as one-half of the cases have been in patients under the age of 50 years (6,18). The thumb is the most com-

E. A. Krull: Department of Dermatology, Henry Ford Hospital, Detroit, Michigan.

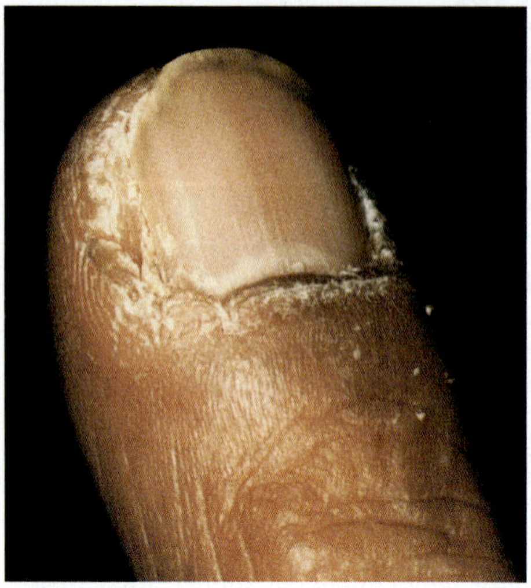

FIGURE 8A.1. Bowen's disease involving the periungual tissue. Mild changes of lateral edges.

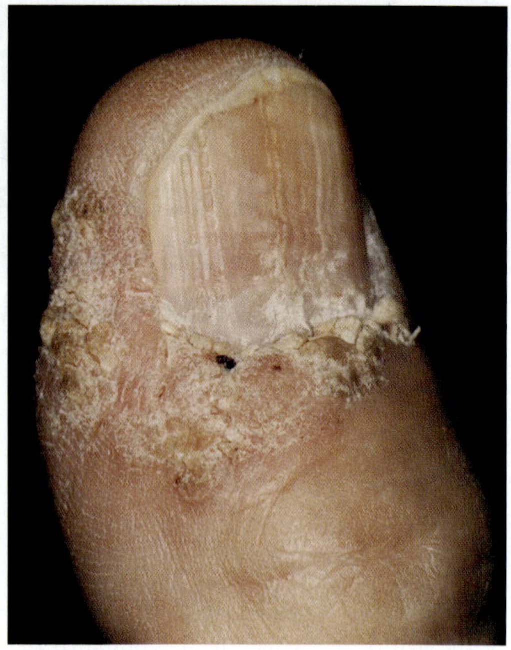

FIGURE 8A.2. Bowen's disease of thumb. This had been treated as warts for more than 2 years.

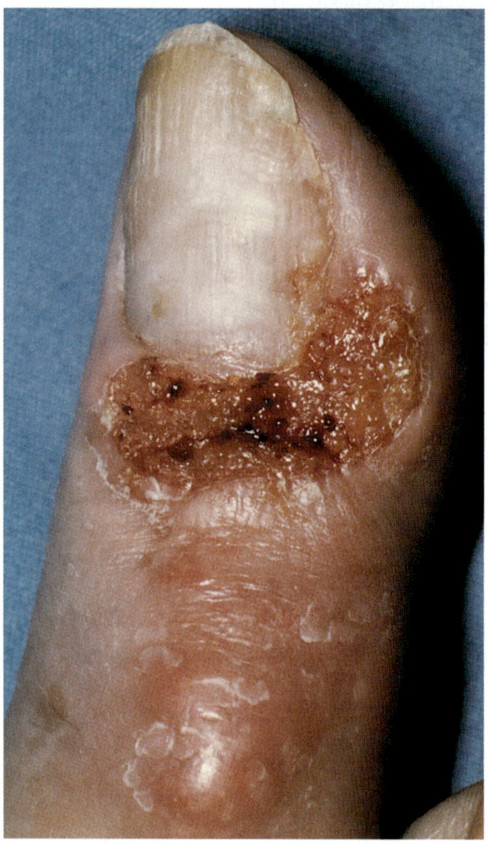

FIGURE 8A.3. Invasive squamous cell carcinoma with deep ulcerations. Tumor nodule is seen more proximally. Nodules of tumor in sporotrichoid fashion extended the entire length and base of the thumb.

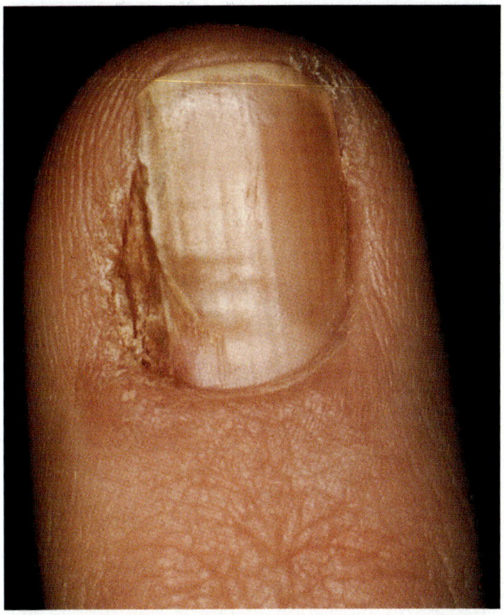

A

FIGURE 8A.4. A: Bowen's disease with hyperkeratotic sliver of tissue along edge of nail. Edge of lateral nail fold shows hyperkeratosis. Lateral edge of nail is missing, narrowing nail plate. **B:** Biopsy of involved tissue was histopathologically Bowen's disease. **C:** Closure of biopsy sites. Lesion was treated with Mohs' micrographic surgery.

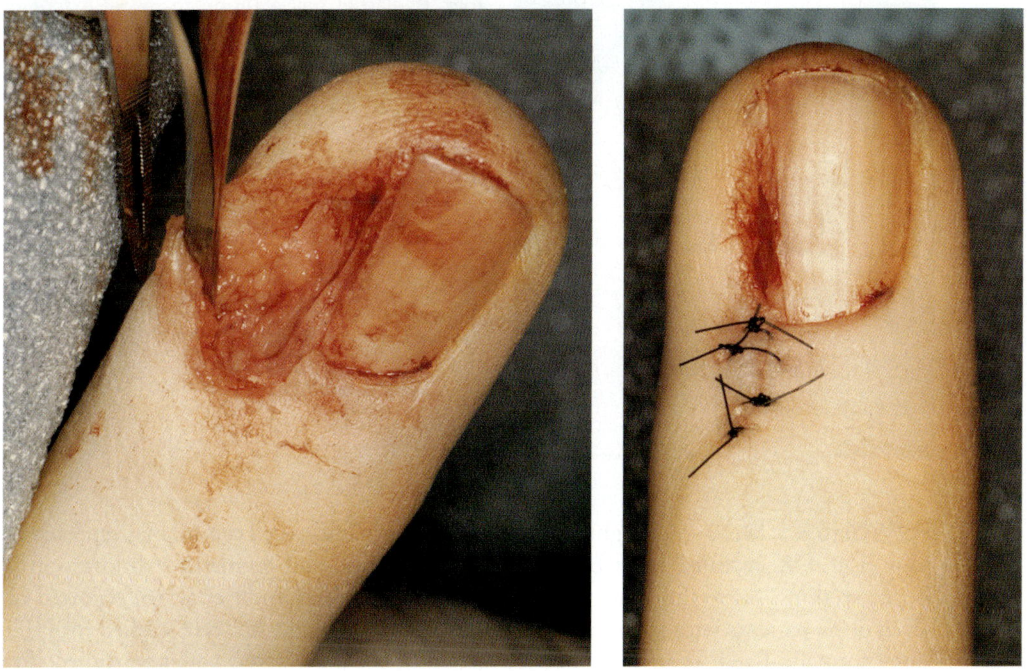

B C

monly involved finger (1,4), but some recent reports have noted more frequent involvement in other digits (6,7). Partly because epidermoid carcinoma of the nail tissues frequently is misdiagnosed as verruca vulgaris and partly related to the infrequency of biopsies, some tumors are not diagnosed for a prolonged period of time, extending to 15 to 30 years (4,18) or more. Even with bone involvement metastasis is uncommon (19–21). It is interesting that squamous cell carcinoma *in situ* of the nail tissues (4,8,19,20) has greater likelihood to become invasive squamous cell carcinoma than at other skin sites (1); however, invasive squamous cell carcinoma, even with bone involvement, is less likely to metastasize than a corresponding squamous cell carcinoma of other parts of the skin (1).

Risk factors attributed to the development of epidermoid carcinoma of the nail tissue include radiation treatment, arsenic ingestion, trauma, and chronic paronychia (2,3). However, more recent reports have implicated human papillomavirus (HPV), especially

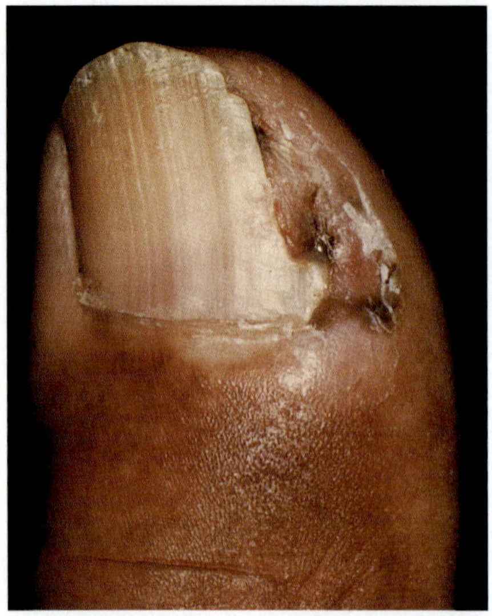
A

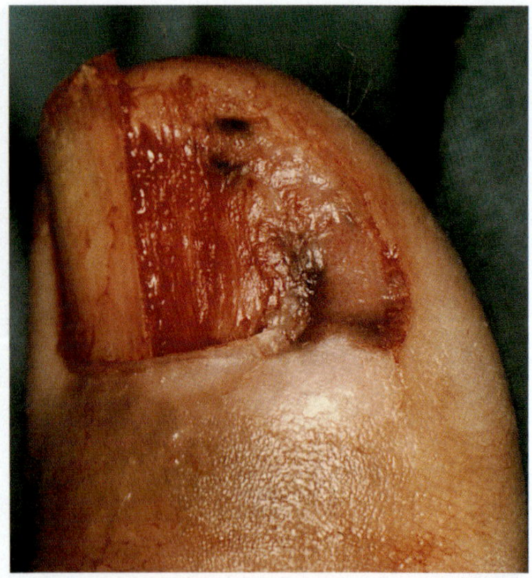
B

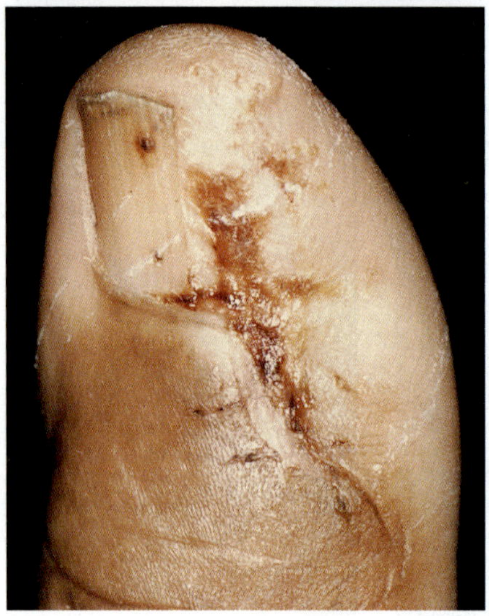
C

FIGURE 8A.5. A: Squamous cell carcinoma of the thumb with loss and onycholysis of nail edge. Suggestion of ulceration in lateral nail groove and fold. **B:** Partial nail avulsion to reveal extent of tumor. **C:** Ten days after repair of defect produced by Mohs' micrographic surgery for resection of tumor. Note how much thumb tissue was spared.

HPV-16 (6,8,9,18,22–24). Eight of ten periungual squamous cell carcinomas reported by Moy et al. (18) revealed HPV by dot blot hybridization and Southern hybridization in frozen sections; six of the eight were related to HPV-16. Ashinoff et al. (6) reported on seven of 12 patients with squamous cell carcinoma of the nail who were studied for HPV using formalin-fixed, paraffin-embedded specimens. Of the seven patients studied, five were found to contain HPV-16 with the polymerase chain reaction technique; *in situ* hybridization failed to identify HPV in any of the tumors (6). Rüdlinger et al. (24) reported on a 42-year-old woman with HPV-35 in a bowenoid papulosis and long-standing pruritus of the anogenital area who developed bowenoid dysplasia in a HPV-35–positive verruca of the periungual area 11 years later. The bowenoid papulosis of the anogenital region was never cured and may have been the source of antoinoculation. Eliezri et al. (25) found that 60% of squamous cell carcinomas of the genital areas and approximately the same percentage of squamous cell carcinoma of the fingers contained HPV-16. Kettler et al. (26) found HPV-16 in six cases of nongenital Bowen's disease; five of the sites were hands and feet but not periungual. All these studies suggest the possible autoinoculation of the finger from the genital area or the converse autoinoculation of the genital area from the fingers. McHugh et al. (8)detected HPV-35 RNA in both a periungual tumor and axillary lymph node metastasis. Metastases to lymph nodes from an HPV-associated digital squamous cell carcinoma is very uncommon (8). Therefore, there seems to be an etiologic role of HPV, predominantly HPV-16 but less commonly HPV-34 (23) and HPV-35

(8,24), in the development of epidermoid carcinoma, with a peculiar topographic localization of HPV-16 on the genital and periungual tissues and fingers.

The preferred treatment of epidermoid carcinoma of the nail tissues is surgical removal. This is especially true if risk factors of more aggressive tumors exist, such as etiologic relationship to radiation or congenital epidermal dysplasia. However, a number of other treatments methods have been attempted in selected cases, with variable results. The use of 5-fluorouracil 5% with a keratolytic agent has been successful in very few cases. Most reports indicated there is failure to cure the lesion and that the deeper parts of the tumor may be buried under the healing of the surface (7,28,29). Curettage and electrosurgery have been successful in a few patients (1,30). The nail plate should be removed, partially or totally, to ascertain the degree of spread of the tumor in the nail bed and matrix. Curettage should be aggressive and may extend to the bone. The cure rate with this modality of treatment has not been established because there have been too few cases. However, because many of the patients frequently were treated with curettage and electrosurgery because of misdiagnosis of wart, there is a presumption that the recurrence rate is high for this method. Cryosurgery performed by a physician familiar with its use for skin malignancies may be successful in selected cases (31).

Surgical removal of periungual and subungual epidermoid carcinoma is the treatment of choice (Fig. 8A.6). Local excision may be adequate to manage smaller lesions, yet clin-

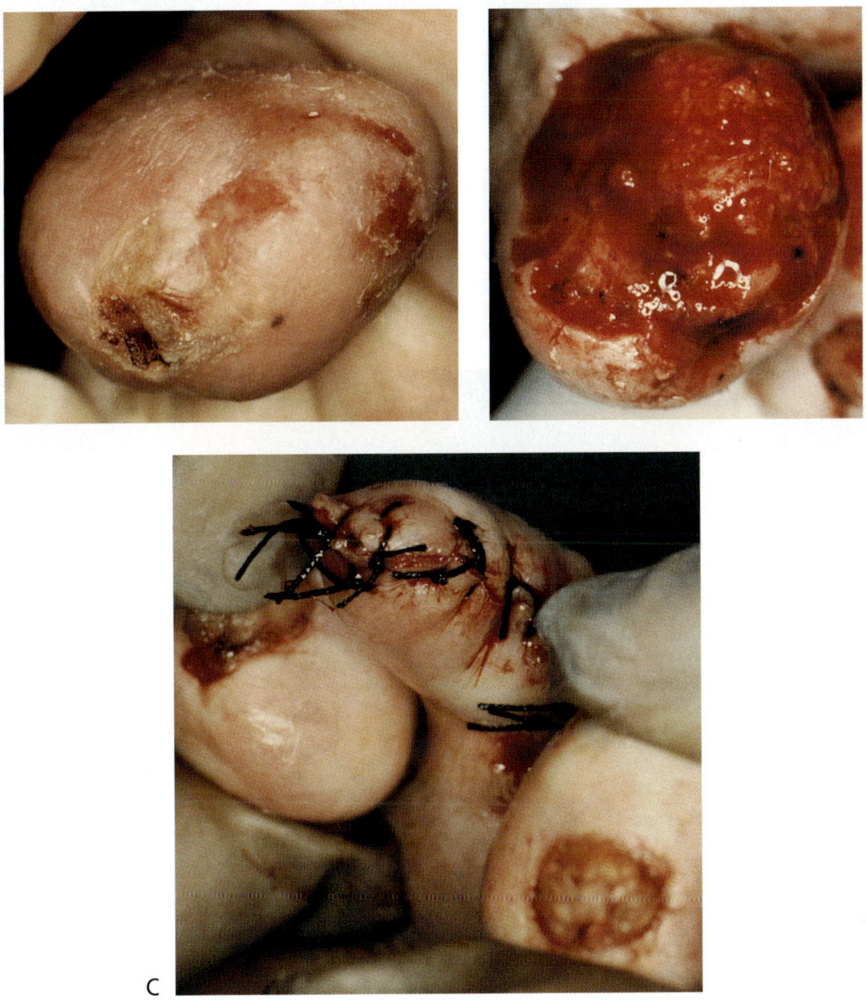

A B C

FIGURE 8A.6. A: Squamous cell carcinoma involving entire end of toe in elderly man who had received high doses of radiation for dyshidrosis of the feet before 1940. He already had multiple squamous cell carcinomas of the soles; some were simply excised, whereas others were treated with Mohs micrographic surgery, all of which healed. **B:** Tip of toe was disarticulated at the distal interphalangeal joint. **C:** Skin closed. The patient healed well.

ical margins may be deceptively smaller than the histopathologic margins (2). Amputation of the distal phalanx is more radical than may be necessary and results in significant loss of function, especially if the thumb is involved.

The preferred treatment of epidermoid carcinoma of nail tissues is Mohs micrographic surgery, usually left to heal with second intention (1,3,4,6,32), which preserves the maximal amount of normal tissue and function. Amputation of the distal phalanx of the thumb results in 75% impairment of the thumb and 30% hand disability (1). The

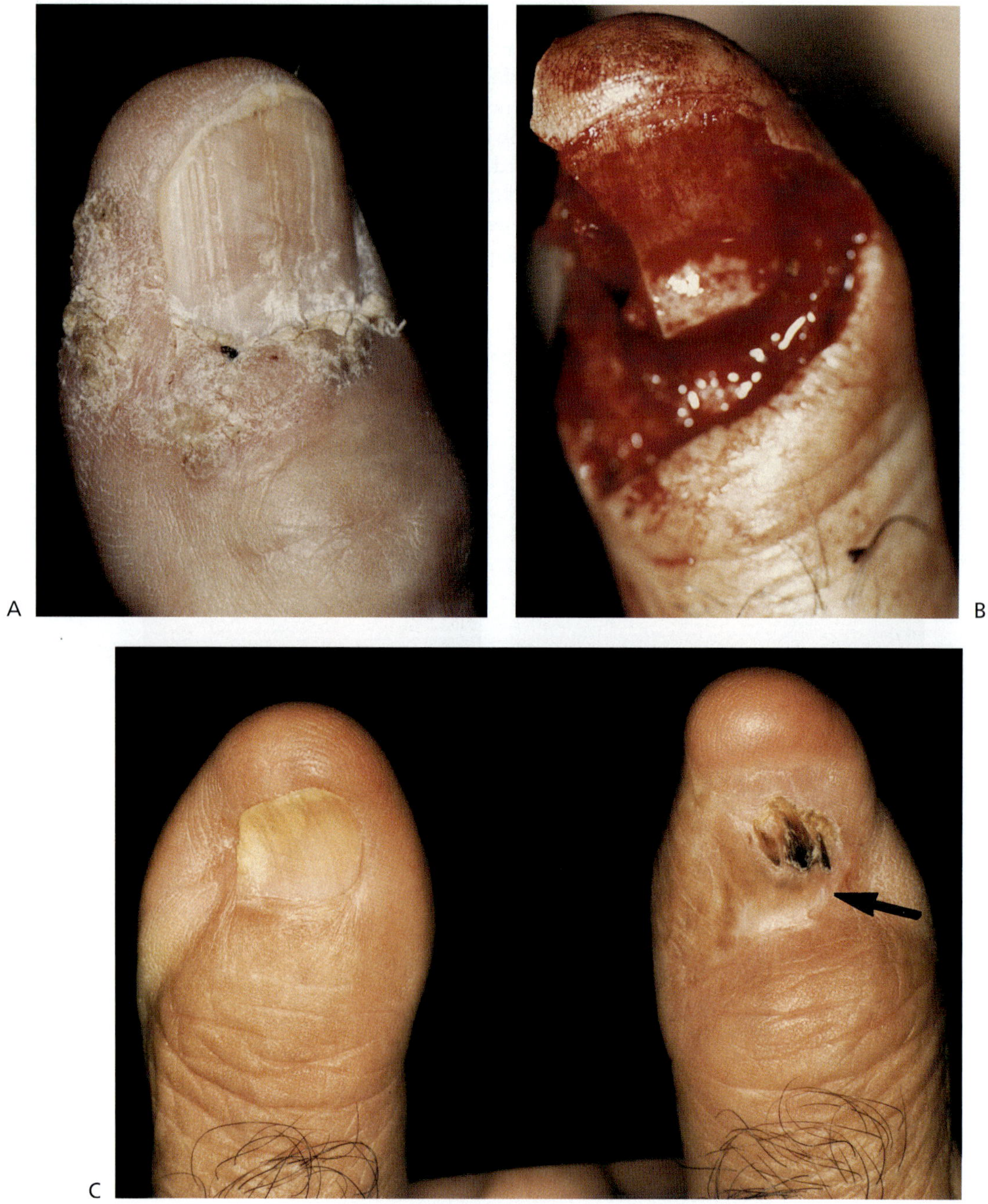

FIGURE 8A.7. A: Extensive Bowen's disease of the right thumb as shown in Fig. 8A.2. **B:** Final appearance after Mohs micrographic surgery of the same thumb. **C:** Appearance after second intention healing (*arrow* points to same thumb as in Figs. 8A, 7A and B). The left thumb also had less extensive Bowen's disease that was treated with Mohs micrographic surgery. Note the amount of tissue and function preserved in both thumbs. There has been no recurrence in 9 years.

cure rates of distal phalangeal amputation and Mohs micrographic surgery are about the same. Mohs micrographic surgery almost always preserves the distal phalanx (Fig. 8A.7) (4). The 5-year cure rate of subungual and periungual epidermoid carcinoma with Mohs micrographic surgery in the series of Goldminz and Bennett (4) was 92%; of the 25 patients in that study, two had recurrences but were treated successfully with repeat Mohs micrographic surgery (4). In a later series by Mikhail (33), the cure rate of epidermoid carcinoma of the nail tissue with Mohs micrographic surgery was 96%; he also had two recurrences that were treated successfully with repeat Mohs micrographic surgery. Because metastasis is of low incidence, approaching 2% even with bone involvement (19–21), preservation of tissue and normal function is a priority in the treatment of epidermal carcinoma of the nail apparatus. Mohs micrographic surgery is recommended as an excellent treatment for epidermoid carcinoma of the periungual and nail tissues and a desirable alternative to amputation (1,4,6).

REFERENCES

1. Mikhail GR. Subungual epidermoid carcinoma. *J Am Acad Dermatol* 1984;11:291–298.
2. Haneke E. Epidermoid carcinoma (Bowen's disease) of the nail simulating acquired ungual fibrokeratoma. *Skin Cancer* 1991;6:217–221.
3. Baran R, Haneke E. Tumors of the nail apparatus and adjacent tissues. In: Baran R, Dawber RPR, eds. *Diseases of the nails and their management,* 2nd ed. London: Blackwell Science, 1994.431–435
4. Goldminz D, Bennett RG. Mohs micrographic surgery of the nail unit. *J Dermatol Surg Oncol* 1992; 18:721–726.
5. Velpeau M. Tumeur cancérereuse de la phalange unquéale de l'indicateur. *Gaz des Hôp* 1850;2:314.
6. Ashinoff R, Li JJ, Jacobson M, et al. Detection of human papilloma virus DNA in squamous cell carcinoma of the nail bed and finger determined by polymerase chain reaction. *Arch Dermatol* 1991; 127:1813–1818.
7. Sau P, McMarlin SL, Sperling LC, et al. Bowen's disease of the nail bed and periungual area. A clinicopathologic analysis of seven cases. *Arch Dermatol* 1994;130:204–209.
8. McHugh RW, Hazen P, Eliezri YD, et al. Metastatic periungual squamous cell carcinoma: detection of human papilloma virus type 35 RNA in the digital tumor and axillary lymph node metastases. *J Am Acad Dermatol* 1996;34:1080–1082.
9. Guitart J, Bergfeld WF, Tuthill RJ, et al. Squamous cell carcinoma of the nail bed: a clinico-pathological study of 12 cases. *Br J Dermatol* 1990;123:215–222.
10. Lai CS, Lin SD, Tsai CW, et al. Squamous cell carcinoma of the nail bed. *Cutis* 1996;57:341–345.
11. Long PI, Espinella J. Squamous cell carcinoma of the nail bed. *JAMA* 1978;239:2154–2155.
12. Baran R, Dupre A, Sayag J, et al. Bowen's disease of the nail apparatus: report of 5 cases and review of 20 cases in the literature. *Ann Dermatol Venereol* 1997;106:227–233.
13. Baran R, Gormley D. Pseudo-acquired fibrokeratoma of the nail apparatus. *J Am Acad Dermatol* 1988;17:201–204.
14. Zaias N. *The nail in health and disease,* 2nd ed. Norwalk, CT: Appleton & Lange, 1990:222.
15. Baran R, Simon C. Longitudinal melanonychia: a symptom of Bowen's disease. *J Am Acad Dermatol* 1988;18:1359–1360.
16. Saijo S, Kato T, Tagami H. Pigmented nail streak associated with Bowen's disease of the nail matrix. *Dermatologica* 1990;181:156–158.
17. Baran R, Eichmann A. Longitudinal melanonychia associated with Bowen's disease. *Dermatology* 1993;186:159–160.
18. Moy RL, Eliezri YD, Nuovo GJ, et al. Human papilloma virus type 16 DNA in periungual squamous cell carcinoma. *JAMA* 1989;261:2669–2673.
19. Campbell GJ, Kloharn T. Squamous cell carcinoma of the nail bed in epidermal dysplasia. *J Bone Joint Surg* 1966;48:92–99.
20. Lombardi LJ, Cleri DJ, Lashinsky AM. Squamous cell carcinoma of the nail bed. *South Med J* 1990; 83:1098–1101.
21. Attiyeh FF, Shah J, Booher RJ, et al. Subungual squamous cell carcinoma. *JAMA* 1979;241: 262–263.
22. Ostrow S, Shaver K, Turnquist S, et al. Human papillomavirus-16 DNA in a cutaneous invasive cancer. *Arch Dermatol* 1989;125:666–669.
23. Kawashima M, Jablonska S, Favre M, et al. Characterization of a new type of human papilloma virus found in a lesion of Bowen's disease. *J Virol* 1986;57:688–692.
24. Rüdlinger R, Grob R, Yu YX, et al. Human papilloma virus-35-positive bowenoid papulosis of the anogenital area and concurrent human papilloma virus-35-positive verruca with bowenoid dysplasia of the periungual area. *Arch Dermatol* 1989;125:655–659.

25. Eliezri YD, Silverstein SJ, Nuovo GJ. Occurrence of human papilloma virus type 16 in cutaneous squamous cell and basal cell carcinoma neoplasms. *J Am Acad Dermatol* 1990;23:836–842.
26. Kettler AH, Rutledge M, Tschen JA, et al. Detection of human papillomavirus in nongenital Bowen's disease by in situ DNA hybridization. *Arch Dermatol* 1990;126:771–781.
27. Carroll RE. Squamous cell carcinoma of the nail bed. *J Hand Surg* 1976;1:92–97.
28. Defiebre BK Jr. Bowen's disease of the nail bed. A case presentation and review of the literature. *J Hand Surg* 1978;3:184–186.
29. Coskey RJ, Mehregan A, Bryan HG. Bowen's disease. Unsuccessful response to topical 5-fluorouracil therapy. *Cutis* 1969;5:1109–1112.
30. Coskey RJ, Mehreghan A, Fosnaugh RP. Bowen's diseases of the nail bed. *Arch Dermatol* 1972;106: 79–80.
31. Dawber RPR, Colver GB, Jackson A. *Cutaneous cryosurgery,* 1st ed. London: Martin Dunitz, 1992: 78–82.
32. deBerker DAR, Dahl MG, Malcolm AJ, et al. Micrographic surgery for subungual squamous cell carcinoma. *Br J Plast Surg* 1996;49:414–419.
33. Mikhail GR. *Mohs micrographic surgery.* Philadelphia: WB Saunders, 1991.Text

8B

MALIGNANT MELANOMA

EDWARD A. KRULL

Subungual melanomas account for 3% of melanomas in white patients and about 20% of melanomas in African-Americans (Table 8B.1) (1–3). The percentage of melanomas in subungual sites is increased in other populations as well: 17% in Chinese (4), 10% in Japanese (5), and 33% of southwestern Native Americans in a limited study (6). However, the overall incidence of melanomas in African-Americans is very low, accounting for only 1% of all melanomas in the United States (7,8). Therefore, the overall number of subungual melanomas in African-Americans is not great.

Subungual melanoma is a disease of the elderly, with the mean age being in the 60s (9–11). It rarely occurs in children (12). Men and women have an almost equal incidence (13). The thumb and first toes were the most common sites of subungual melanomas (5,9–11,14,15). The role of trauma is supported by the occurrence of subungual melanoma and coexistent injuries of the hand or other injuries and trauma (16). Milton et al. (15) concluded that trauma may have been a causative factor in 41% of subungual melanomas. But other authors raise doubt about the role of trauma in the initiation of subungual melanoma (10,11). Subungual melanoma may appear as longitudinal melanonychia (LM) (Figs. 7E.8, 7E.15, 7E.51, and 7E.53). A black or dark brown color, expanding width of the pigmented streak, fuzzy or indistinct edges, variegated colors, especially in the darker LM, and especially nail destruction are suggestive of malignant melanoma change in LM. Hutchinson's sign associated with LM is presumptive of subungual melanoma (17). There is evidence that a significant percentage of subungual melanomas begin with LM (Figs. 7E.8, 7E.15, 7E.51, and 7E.53) (12,18–23).

Many patients with subungual melanomas present with advanced disease because of delays in diagnosis (9,22). A mass under the nail, destruction of the nail plate, a black fungating tumor, and ulceration are some of the late clinical signs of subungual melanoma. From 20% to 25% or more of subungual melanomas are amelanotic (22,24,25). Hutchinson's sign, the spread of pigment to surrounding tissue, is almost pathognomonic of melanoma.

The differential diagnosis of subungual melanoma includes a number of conditions (14,15,22,24). Subungual hematoma is reddish-blue, frequently round or oval shaped, only very infrequently involves the nail plate itself, and migrates with nail growth in most of cases. About 15% of subungual hematomas do not migrate and may make the clinical diagnosis more difficult. However, it is not difficult to confirm the cause of the lesion by removing the piece of nail covering the subungual hematoma and identifying the organized hematoma. The hematoma usually is adherent to the undersurface of the nail or to the underlying tissue (Figs. 8B.1–8B.6) (26). The hematoma can be removed easily by gentle curettage or with forceps. No residual pigment is seen in the nail tissue. The tissue can be examined for the presence of blood. Benzidine stain should be directed toward

E. A. Krull: Department of Dermatology, Henry Ford Hospital, Detroit, Michigan.

TABLE 8B.1. PERCENTAGE OF TOTAL MELANOMAS THAT ARE SUBUNGUAL IN LOCATION

1. 3% of melanomas in whites
2. 20% of melanomas in African-Americans
3. 3% to 11% in Japanese
   A. Possibly higher in Chinese

Twenty percent of subungual melanomas are amelanotic.

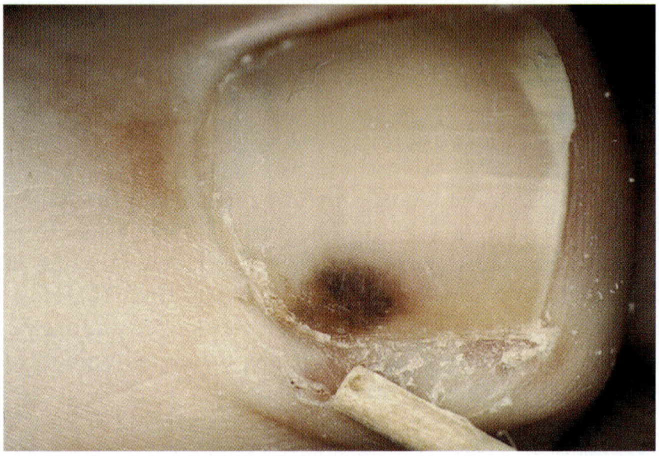

FIGURE 8B.1. Nonmigratory lesion of 6 months' duration. Roundish shape with red-brown color.

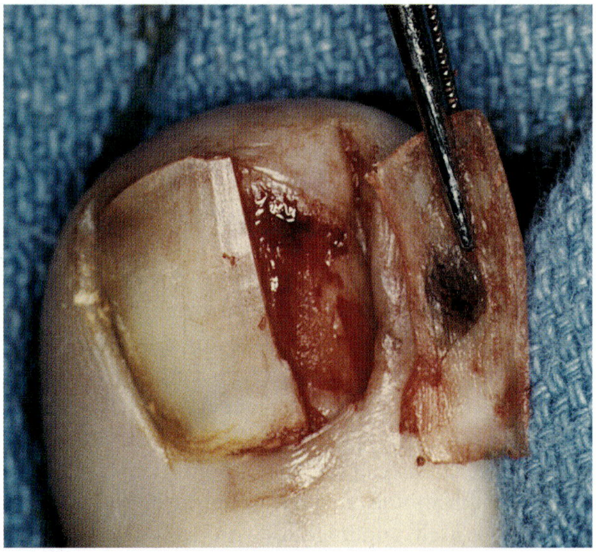

FIGURE 8B.2. Partial nail avulsion of same lesion as shown in Fig. 8B.1. Note the organized blood adherent to the nail plate and the clear nail bed. Benzidine test was positive.

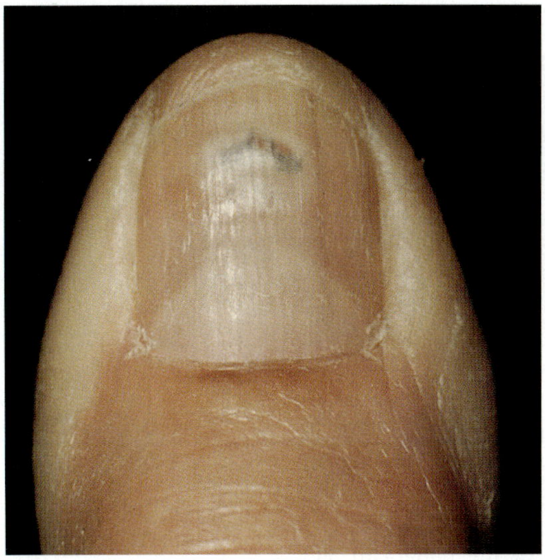

FIGURE 8B.3. Blue lesion of nail bed, present for 7 months.

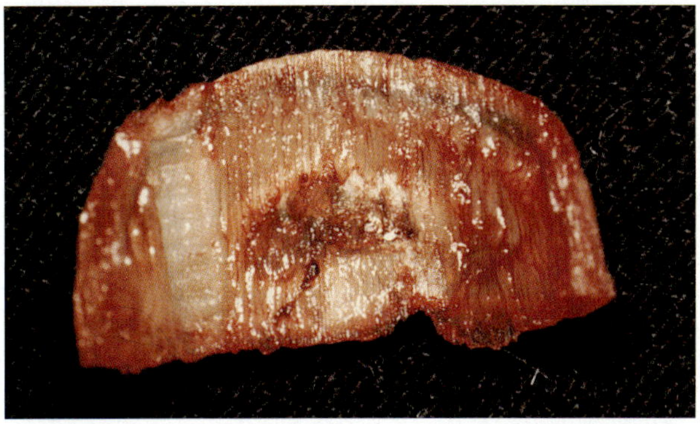

FIGURE 8B.4. Avulsed distal nail plate of patient shown in Fig. 8B.3. Organized hematoma is adherent to nail plate. Benzidine test was positive.

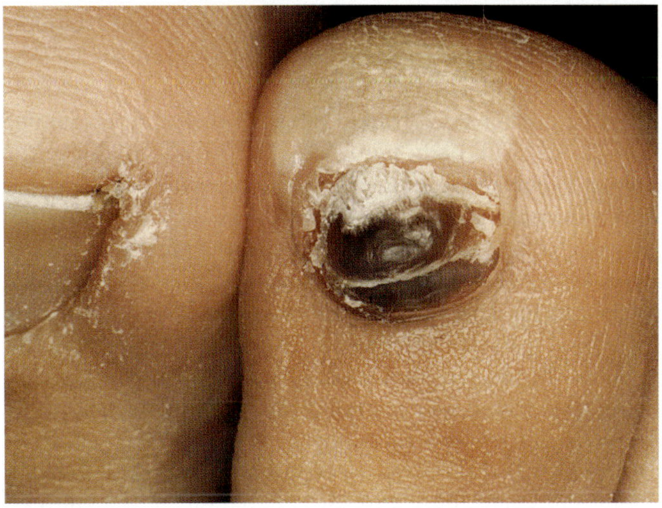

FIGURE 8B.5. Dark red lesion involving most of nail plate for 4 months.

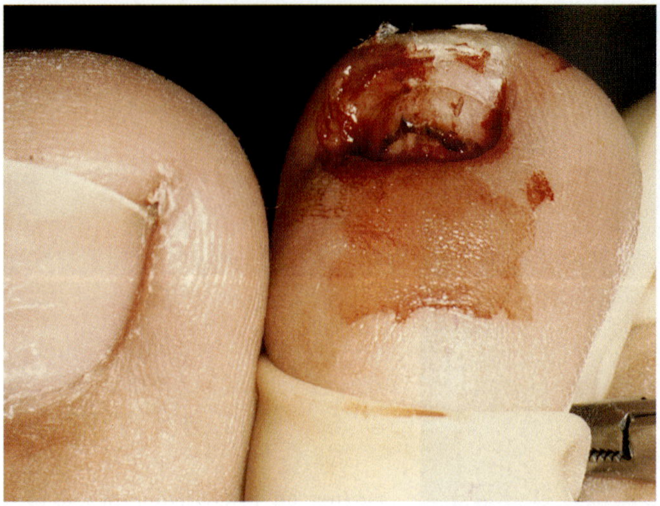

FIGURE 8B.6. Avulsion of the nail plate of the patient in 8B.5 removed most of the organized hematoma.

hemoglobin rather than hemosiderin. The blood deposits are in a space devoid of degrading phagocytes, which induce hemoglobin breakdown to form hemosiderin (26). In cases of subungual hemorrhage into the matrix, the hematoma may be incorporated in the nail plate (27).

Chronic paronychia may mimic subungual melanoma, but acute paronychia usually is not a diagnostic problem. In chronic paronychia, there is a rolling away of the proximal nail fold from the underlying nail plate and often a greenish-black color of the nail plate. When there is pigment spread from the melanoma, differentiation from chronic paronychia is more easily accomplished. However, amelanotic melanoma may be much more easily confused with chronic paronychia. The absence of pigment and the presence of erythema, drainage, and ulceration of amelanotic subungual melanoma can be easily confused with an inflammatory process (14,15,22,23). However, the location of subungual melanoma is usually under the nail plate, differing from chronic paronychial involvement of the nail fold. Occasionally, an amelanotic melanoma involving the nail fold with a subungual component may be confused with chronic paronychia, but in this location the amelanotic melanoma usually spreads beyond the nail fold (22). Therefore, an unexplained or atypical chronic paronychia warrants investigation by biopsy, x-ray study of the bone, and tissue cultures.

Pyogenic granuloma should be differentiated from subungual amelanotic melanoma (Fig. 8B.7). Trauma may be associated with both, but it may be especially minor in the initiation of pyogenic granuloma. Pyogenic granuloma is a reddish papule or tumor that tends to bleed easily and sometimes crusts, raising further concern about subungual amelanotic melanoma. The diagnostic collarette surrounding pyogenic granuloma seen in other sites is not always present in nail lesions. Pyogenic granuloma may affect the nail plate, depending on its location. The granulation tissue response associated with chronic ingrown toe nail may be considered a variant of pyogenic granuloma. Simple excision biopsy should clarify the diagnosis of pyogenic granuloma.

The black-green discoloration associated with microbial infections of the nail, tissue, including *Candida,* chromogenic bacteria, and dermatophytes, may raise concerns about malignant melanoma (Figs. 8B.8–8B.12). That there can be associated pigmented streaks with these lesions can be confusing. Potassium hydroxide preparation, fungal cultures, and exploration and biopsy may be needed to arrive at the correct diagnosis.

Other diagnostic considerations of subungual melanoma include squamous cell carcinoma, Bowen's disease, and glomus tumor (22). A foreign body also may suggest subungual melanoma (Figs. 7E.3–7E.5). The diagnosis of subungual melanoma should be

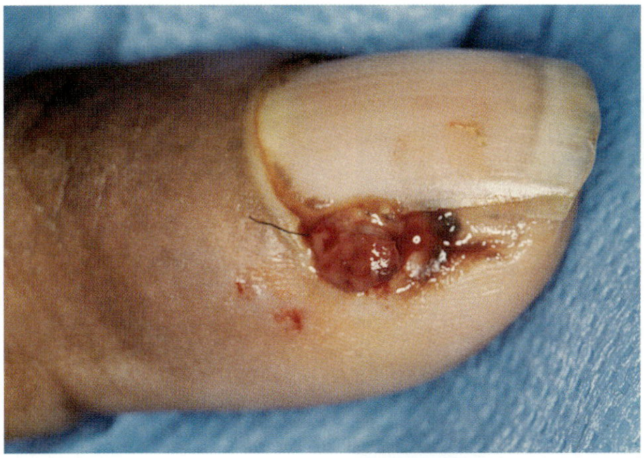

FIGURE 8B.7. Pyogenic granuloma, proven by biopsy and histopathologic examination, affecting the nail plate.

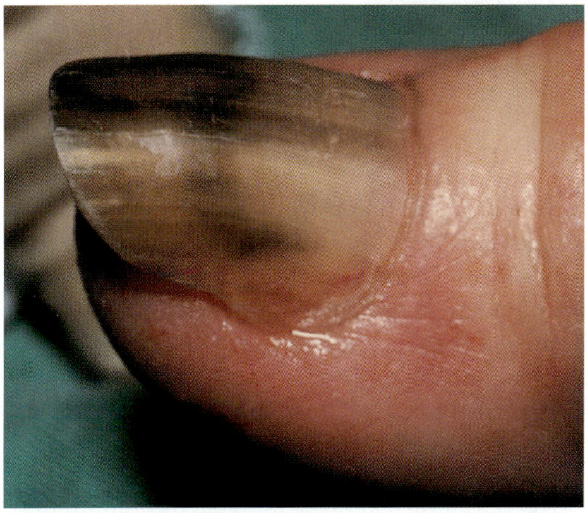

FIGURE 8B.8. Greenish-black discoloration of nail. Onycholysis and subungual debris were present.

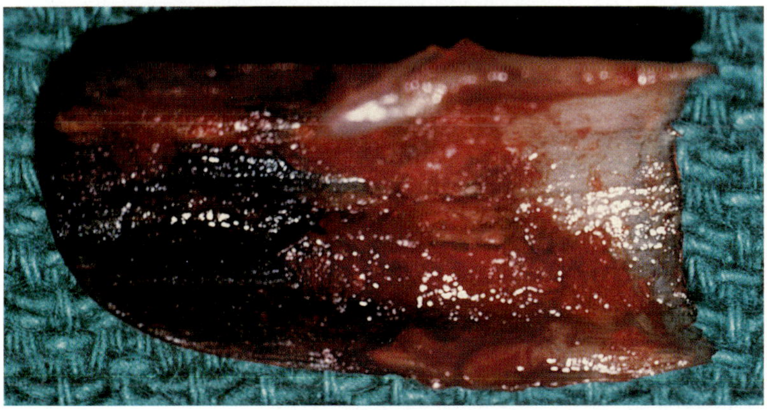

FIGURE 8B.9. Undersurface of avulsed nail plate of patient shown in Fig. 8B.8.

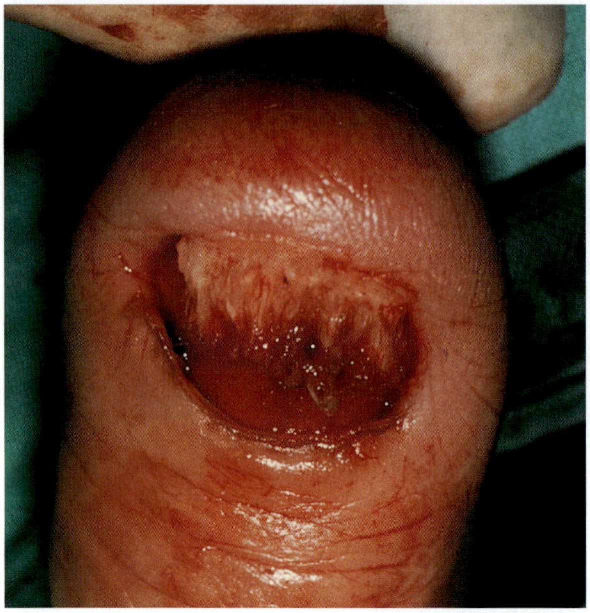

FIGURE 8B.10. Keratotic material under the nail of patient shown in Fig. 8B.8. This occurs under onycholytic nails.

considered for persistent paronychia, chronic granulation tissue, persistent lesions following trauma, unexplained keratotic fissuring changes in the nail fold, and alterations of the nail plates (12). Biopsies should be performed in 6 weeks if there are no signs of healing (28). Exploration, biopsy, and other appropriate studies should be performed to obtain a diagnosis. The biopsy must be adequate and deep enough for histopathologic examination. Often the biopsies are inadequate and improperly performed. As a result, the incorrect diagnosis of a benign lesion leads to a false sense of security and postpones treatment, which can adversely affect outcome (28). When Hutchinson's sign, presumptive of melanoma, is present and other causes are excluded (29) or the clinical evidence for melanoma is formidable, this author prefers to excise the entire nail tissue for serial sections for diag-

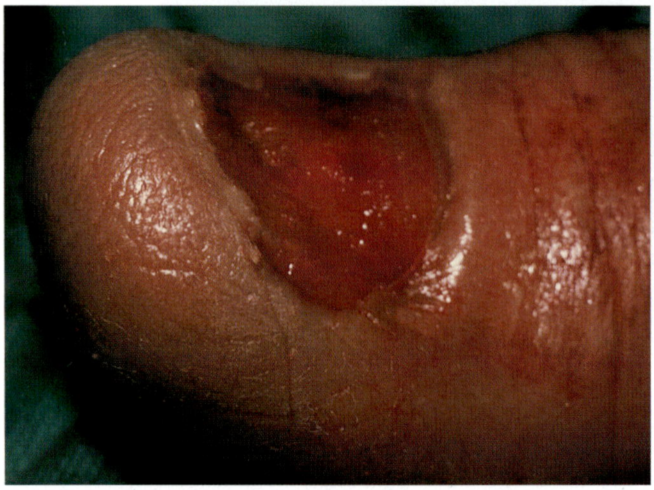

FIGURE 8B.11. Keratotic material removed by curettage of patient shown in Fig. 8B.8. Darker distal nail bed is the result of the curettage.

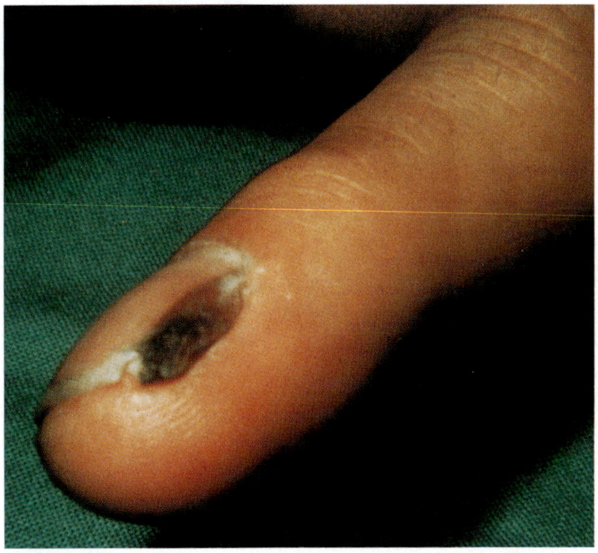

FIGURE 8B.12. Brown-black longitudinal melanonychia and possible black tumor suggesting melanoma was related to chromogenic bacteria and hyperkeratotic material. Cultures for dermatophytes and Candida as well as periodic acid-Schiff stain were negative.

nosis and determination of tumor thickness (Figs. 8B.13–8B.17). In some of my cases associated with Hutchinson's sign, multiple biopsies failed to confirm the diagnosis of subungual melanoma, whereas multiple sections through the entire nail tissue did. In another case, the presence of a pale pink granulating nodule was associated with Hutchinson's sign; all of the pigmented sites were negative for melanoma, but the nonpigmented granulating tumor was melanoma on histopathologic examination (Fig. 8B.18). The excision biopsy of a lesion with Hutchinson's sign includes the pigmented tissue and the entire nail unit. An incision with a margin of 3 to 5 mm is made around the lesion (Fig. 8B.14). The nail unit and surrounding tissue are excised *en bloc* from the underlying bone (Fig. 8B.15). An antibiotic ointment, petrolatum gauze, fine-meshed gauze, Kling dressing, and tubular elastic net are applied. The dressing is changed as needed. Definitive management is determined by the results of the histopathologic examination of the tissue.

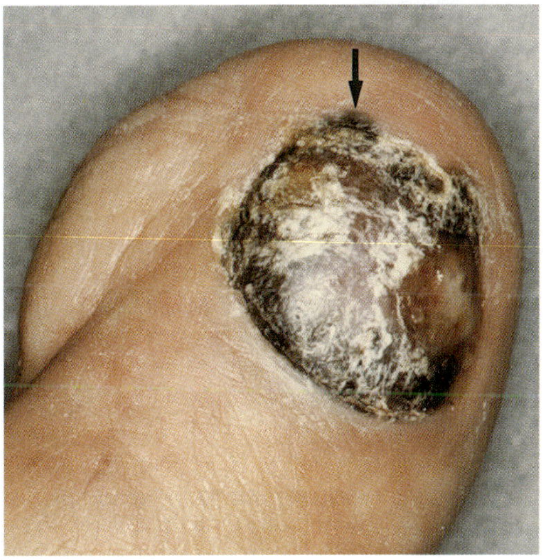

FIGURE 8B.13. Black tumor of entire first toe nail unit with ulceration and early Hutchinson's sign (*arrow*). Lesion is almost certain to be malignant melanoma.

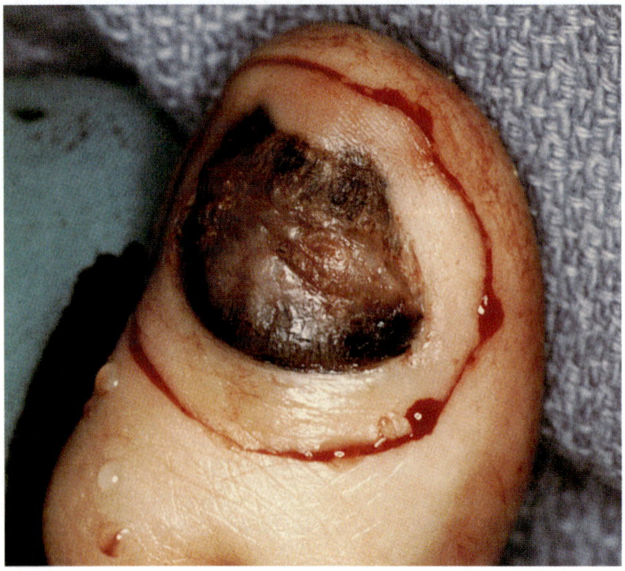

FIGURE 8B.14. Circumferential excision of tumor en bloc with margins.

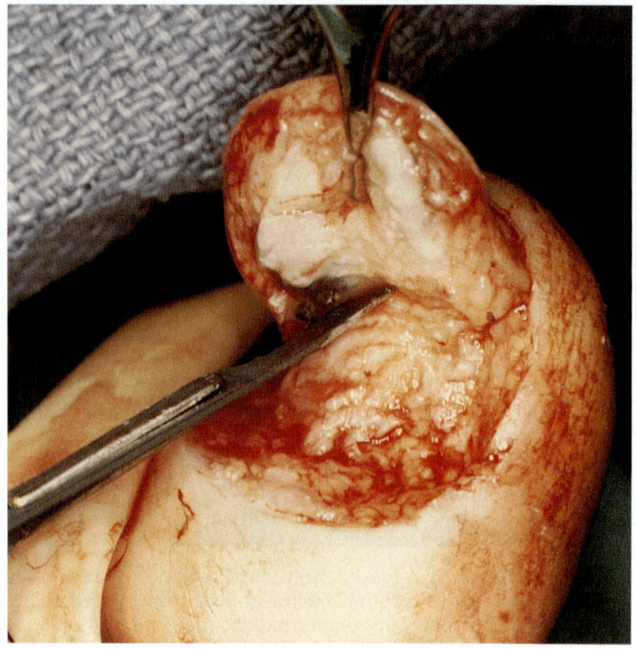

FIGURE 8B.15. Partial excision of tumor down to underlying bone.

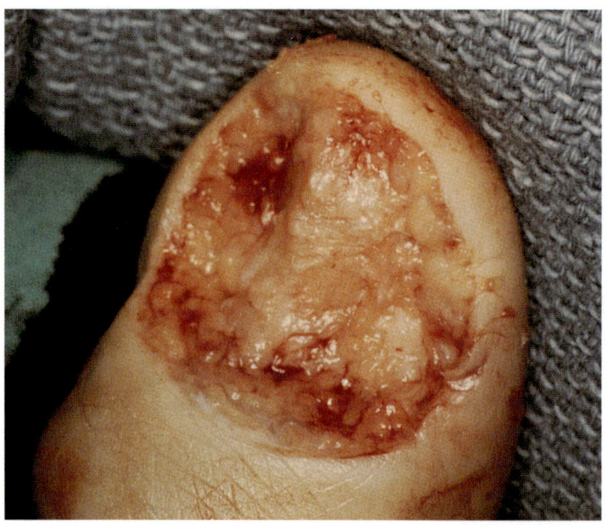

FIGURE 8B.16. Tumor removed. Dressing was applied. Histopathology revealed thick malignant melanoma.

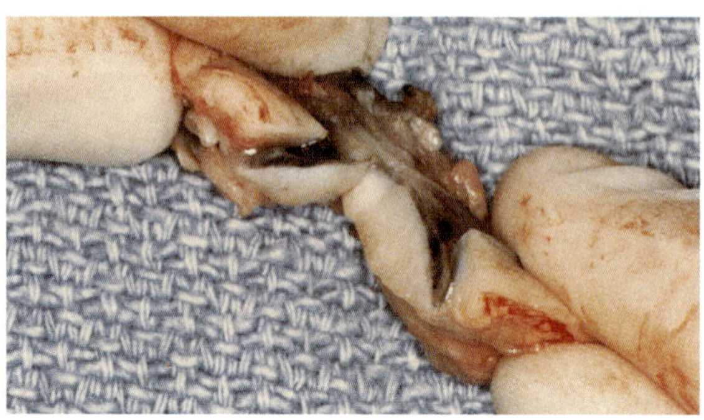

FIGURE 8B.17. Transection of tumor revealing its clinical thickness.

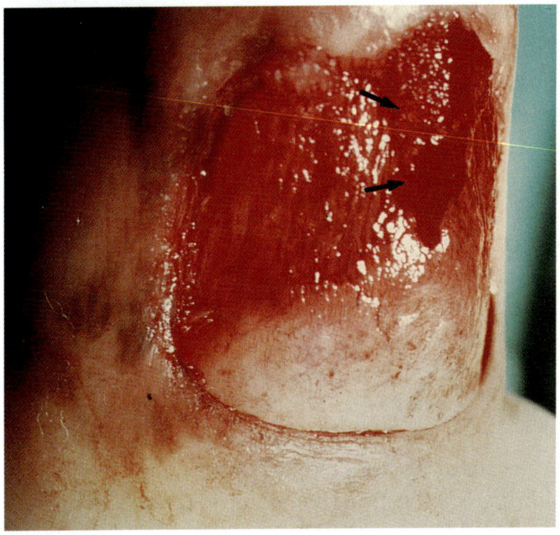

FIGURE 8B.18. Patient had a bluish-black nail with extension of pigment to surrounding tissue, i.e., Hutchinson's sign. Biopsy of brown lesions of soft tissue, lighter brown lesions of the matrix, and random biopsies of nail unit revealed melanocytic hyperplasia. Biopsy of reddish granulating tissue showed thick malignant melanoma (*arrow*).

The histopathology of melanoma is predominantly of the acrolentiginous type, but superficial spreading and nodular melanoma occur (5,9,25,28,30). About 20–25% or more of subungual melanomas are amelanotic (25,31).

The 5-year survival rate of patients with subungual melanoma is poor as compared to other sites. The 5-year survival rate is reported as 16% to 50% (9,15,28). However, some recent reports indicate a significantly higher percentage of 5-year survival, which may be related to earlier diagnosis and adjunctive therapy (30,32,33). The poor prognosis most likely is related to the delay in diagnosis and the resultant thickness of the tumor rather than the anatomic site (28,31). However, others state that there may be a poorer prognosis in subungual melanoma than with melanoma in other sites with similar thickness of the tumor (25,34). Delays in diagnosis of up to 2 years or longer are not uncommon; often lymph node involvement was already present at the time of diagnosis (9,15,28). Most of the melanomas frequently are thick by the Breslau system, with a mean depth of 4.7 mm in one study of 100 cases (9). The two most important 5-year survival predictive factors are thickness of the tumor (35) and the presence of lymphatic disease (36). The c-myc oncogene in melanoma may have potential application as a prognostic marker (34). High levels of the tumor oncogene have been found to be associated with shorter disease-free intervals and lower survival rate.

As the clinical appearance of subungual melanomas becomes better recognized, the 5-year survival results should improve significantly.

The histopathologic diagnosis of subungual acrolentiginous melanoma may be difficult. It is complicated by a number of factors, including inadequate biopsy specimens, distortion of the tissue from previous surgery (28), and the pathologist's lack of experience in the microscopic examination of nail tissue. Serial sections are necessary to identify the hyperplasia of atypical melanocytes with large hyperchromatic pleomorphic nuclei and more prominent nucleoli (31). Atypical melanocytic hyperplasia can be viewed as incipient melanoma *in situ* (31,37). Clusters of tumor cells at the epithelial–dermal junction are indications of progression to melanoma *in situ* (31,37).

Melanocytic hyperplasia that remains benign is not clearly distinguished from similar changes that may be precursors to melanoma. Based on the estimation of number of patients with pigmented nail streaks (Tables 7E.3, 7E.5, and 7E.6), the low incidence of subungual melanoma, and the uncommonness of nail matrix nevi (38), it appears that most patients with LM represent melanocytic hyperplasia and follow a benign course (39).

The management of subungual melanoma has become less radical than in the past (28,33). In early series, amputation often was carried out at or proximal to the metacarpal phalangeal joint (28,33). Tseng et al. (35) recommended that subungual melanomas less than 1.5 mm in thickness be treated surgically by amputation of the distal phalanx and melanomas 1.5 mm or thicker be managed by amputation of the distal phalanx and sentinel node biopsy. Quinn et al. (25) found no statistical difference in local recurrence when comparing proximal and distal interphalangeal amputation. They recommended amputation at the distal interphalangeal joint and the head of the distal phalanx for subungual melanoma of the thumb. Glatt et al. (32) treated melanoma *in situ* with wide excision until negative margins were obtained. Patients with invasive subungual melanoma underwent amputation at the joint proximal to the lesion. Park et al. (28) found that local/distal amputation did not compromise the overall survival compared with those having more proximal amputations.

The role of elective lymph node dissection has been controversial (36). The management of patients has changed over the past 5 years with the development of lymphatic mapping sentinel node biopsies (40,41), which have reduced cost of treatment and morbidity while more accurately assessing the status of the regional lymph nodes (36).

To achieve a better outcome of subungual melanoma, there must be early diagnosis. Further definition of the characteristics of LM that represent precursor lesions of melanoma must be attempted to determine which of the millions of cases of LM should be biopsied. We must raise the index of suspicion for subungual melanoma. There should be less hesitation and more knowledge in performing adequate biopsies of the nail tissue. Pathologists inexperienced in examining nail tissue should use consultants appropriately and work closely with the clinician in orienting and cutting the tissue. Treatment methods of subungual melanoma, including adjunctive therapy, have been better defined, but are frustrated by late diagnoses and thick lesions, which have led to regional and distant spread and adverse 5-year survival rates.

REFERENCES

1. Kopf AW, Waldo E. Melanonychia striata. *Australas J Dermatol* 1980;21:59–70.
2. Baran R, Kechijian P. Longitudinal melanonychia (melanonychia striata): diagnosis and management. *J Am Acad Dermatol* 1989;21:1165–1175.
3. Pack GT, Oropeza R. Subungual melanoma. *Surg Gynecol Obstet* 1967;124:571–582.
4. Collins RJ. Melanoma in the Chinese of Hong Kong. Emphasis on volar and subungual sites. *Cancer* 1984;54:1482–1488.
5. Miura S, Jimbow K. Clinical characteristics of subungual melanomas in Japan: case report and a questionnaire survey of 108 cases. *J Dermatol* 1985;12:393–402.
6. Black, WC, Wiggins C. Melanomas among Southwestern American Indians. *Cancer* 1985;55: 2899–2902.
7. Clark WH Jr., Bernardino EA, Reed RJ, et al. Acral lentiginous melanomas. In: Clark WH, Goldman MI, Mastrangelo MJ eds. *Human malignant melanoma*. New York: Grune & Stratton, 1979: 109–124.
8. Shah JP, Goldsmith HS. Malignant melanoma in the North Anerican Negro. *Surg Gynecol Obstet* 1971;133:437–439.
9. Blessing K, Kernohan NM, Park KGM. Subungual malignant melanoma—clinicopathological features of 100 cases. *Histopathology* 1991;19:425–429.
10. Kopf AW, Bart RS, Rodriguez-Sains RS, et al. *Malignant melanoma*. New York: Masson Pub. USA Inc., 1979.
11. Briggs JC. Subungual melanoma: a review article. *Br J Plast Surg* 1985;38:174–176.
12. Kato, J, Usuba Y, Takematsu H, et al. A rapidly growing pigmented nail streak resulting in diffuse melanosis of the nail. A possible sign of subungual melanoma in situ. *Cancer* 1989;64:2191–2197.
13. Rigby HS, Briggs JC. Subungual melanoma: a clinicopathological study of 24 cases. *Br J Plast Surg* 1992;45:275–278.
14. Leppard B, Sanderson KV, Behan F. Subungual melanoma: difficulty in diagnosis. *Br Med J* 1974; 1:310–312.
15. Milton GW, Shaw HM, McCarthy WH. Subungual melanoma: a disease entity separate from other forms of cutaneous melanoma. *Australas J Dermatol* 1985;26:61–64.
16. Roberts AHN. Subungual melanoma following a single injury. *J Hand Surg* 1984;9B:328–330.
17. Mikhail G. Hutchinson's sign. *J Dermatol Surg Oncol* 1986;12:519–522.

18. Nogaret JM, Andre J, Parent D, et al. Les Melanome des extremities: diagnosis meconnu et traitment delicat. Revue des 20 observations. *Acta Chir Belg* 1986;86:238–244.
19. Retso S, Samman PD. Pigmented streaks in the nail plate due to secondary malignant melanoma. *Br J Dermatol* 1983;108:367–370.
20. Takahashi M, Seiji M. Acral melanoma in Japan. *Pigment Cell* 1983;6:150–166.
21. Patterson R, Helwig EB. Subungual melanoma: a clinical pathological study. *Cancer* 1980;46: 2074–2087.
22. Shukla VK, Hughes LE. Differential diagnosis of subungual melanoma from a surgical point of view. *Br J Surg* 1989;76:1156–1160.
23. Oropeza R. Melanoma of special sites. In: Andrade R, Gumpert SL, Popkin GL, et al., eds. *Cancer of the skin,* vol. 2. Philadelphia: WB Saunders, 1976:974–987.
24. Winslet M, Tejan J. Subungual amelanotic melanoma: a diagnostic pitfall. *Postgrad Med J* 1990;66: 200–202.
25. Quinn MJ, Thompson JE, Crotty K, et al. Subungual melanoma of the hand. *J Hand Surg* 1996; 21A:506–511.
26. Hafner J, Haenseler E, Ossent P, et al. Benzidine stain for histochemical detection of hemoglobin in splinter hemorrhage (subungual hematoma) and black heel. *Am J Dermatopathol* 1995;17: 362–367.
27. Stone OJ, Mullins JF. The distal course of nail matrix hemorrhage. *Arch Dermatol* 1963;88: 186–187.
28. Park KGM, Blessing K, Kernohan NM. Surgical aspects of subungual melanoma. *Ann Surg* 1992; 216:692–695.
29. Baran R, Kechijian P. Hutchinson's sign: a reappraisal. *J Am Acad Dermatol* 1996;34:87–90.
30. Kato T, Suetake T, Sugiyama Y, et al. Epidemiology and prognosis of subungual melanoma in 34 Japanese patients. *Br J Dermatol* 1996;134:383–387.
31. Dawber RPR, Colver GB. The spectrum of malignant melanoma of the nail apparatus. *Semin Dematol* 1991;10:82–88.
32. Metzger S, Ellwanger U, Stroebel W, et al. Extent and consequences of physician delay in the diagnosis of acral melanoma. *Melanoma Res* 1998;8:181–186.
33. Glat PM, Spector JA, Roses DF, et al. The management of pigmented lesions of the nail bed. *Ann Plast Surg* 1996;37:125–134.
34. Grover R, Grobbelaar AO, Hudson DA, et al. The clinical significance of oncogene expression in subungual melanoma. *Br J Plast Surg* 1997;50:15–19.
35. Tseng JF, Tanabe KK, Gadd MA, et al. Surgical management of primary cutaneous melanomas of the hands and feet. *Ann Surg* 1997;225:544–553.
36. Reintgen DS, Rapaport DP, Tanabe KK, et al. Lymphatic mapping and sentinel lymphadenectomy. In: Balch CM, Houghton AN, Sober AJ, et al., eds. *Cutaneous melanoma,* 3rd ed. St. Louis, Missouri: Quality Medical Publishing Inc., 1998:227–244.
37. Baran R, Haneke E. Tumours of the nail apparatus–melanocytic lesions. In: Baran R, Dawber P, eds. *Disease of the nails and their management*, 2nd ed. London: Blackwell Science, 1994:483–497.
38. Tosti A, Baran R. Piraccini BM, et al. Nail matrix nevi: a clinical and histopathologic study of twenty-two patients. *J Am Acad Dermatol* 1996;34:765–771.
39. Duhard E, Calvet C, Mariotte N, et al. Prevalence of longitudinal melanonychia in the white population. *Ann Dermatol Venereol* 1995;122:586–590.
40. Morton DL, Wen DR, Wong JH, et al. Technical details of intraoperative lymphatic mapping for early stage melanoma. *Arch Surg* 1992;127:392–399.
41. Morton DL, Wen DR, Cochran AJ. Management of early stage melanoma by intraoperative lymphatic mapping and selective lymphadenectomy or "watch and wait." *Surg Oncol Clin North Am* 1992;1:247.

SPECIAL TOPICS IN DIAGNOSIS AND SURGERY OF THE NAIL UNIT

9

IMAGING OF TUMORS OF THE NAIL UNIT

JEAN-LUC DRAPÉ
WITH THE COLLABORATION OF
SOPHIE GOETTMANN
AND
JACQUES BITTOUN

Imaging of the nail unit has developed poorly with time due to the superficial location of the nail plate and its access to simple investigations such as transillumination or photograph. Light has long been used as a tool. A simple pocket lamp can diagnose a mucoid cyst by transillumination. Wood light can enhance the modified colors induced by tetracyclines giving a yellow fluorescence. Polarized light may be helpful to examine the undersurface of the nail plates (1). Numerous macrophotographic systems provide high-quality images of the nail unit. Nevertheless, these techniques remain superficial. Tumors of the nail unit may be difficult to diagnose because of the anatomic particulars of this organ. Symptoms, growth, and above all the appearance of the tumors may be modified by the screen produced by the nail plate. The deep lesions that originate close to the nail root are covered by the posterior nail fold and may only be expressed by nail dystrophy. Thus, every suspicious lesion of the nail unit should undergo radiography and biopsy. A complementary imaging modality, such as high-resolution magnetic resonance imaging (MRI), should help in difficult cases by confirming and accurately locating the periungual or subungual mass.

RADIOGRAPHY

In 1895, the first radiograph was that of Mrs. Roentgen's hand. Today, this technique remains the essential complementary imaging modality of the nail unit. Radiographs are essential for study of the distal phalanx, but they usually show very little of the soft tissues of the nail unit. The technique must be adapted to study of the soft tissues with low-voltage radiographs. Progressively, the films without screen have been replaced by high-resolution one-layer breast films. Digital films have a lower spatial resolution despite the imperative use of small field of view, but they provide better contrast resolution. Posteroanterior and lateral views of the involved finger usually are sufficient. Oblique views may be necessary to highlight subtle erosions of the distal phalanx. Some magnification may increase the readability. Most nail dystrophies should benefit from radiography before surgical investigation. These radiographs can show abnormalities of

J-L. Drapé: Department of Radiology B, Hôpital Cochin, 75014 Paris, France; and Centre Interétablissement de Résonance Magnétique (C.I.E.R.M.), Hôpital Bicêtre, Le Kremlin-Bicêtre, France.

S. Goettmann: Department of Dermatology, Hôpital Bichat, Paris, France.

J. Bittoun: Centre Interétablissement de Résonance Magnétique (C.I.E.R.M.), Hôpital Bicêtre, Le Kremlin-Bicêtre, France.

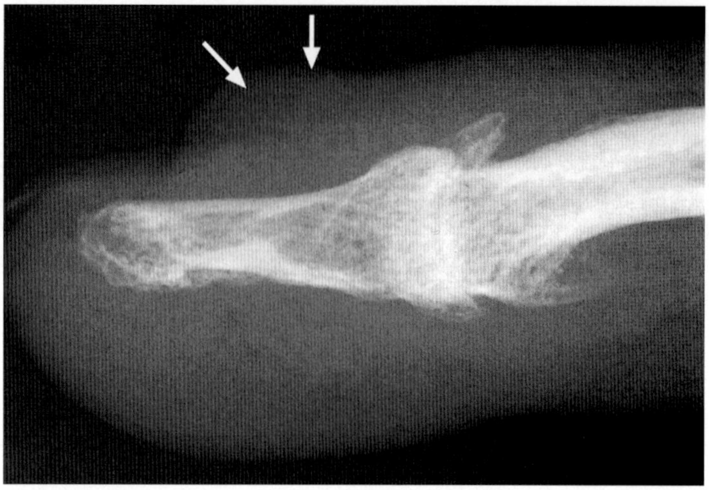

FIGURE 9.1. Mucoid cyst. Lateral radiograph shows thickening of the posterior nail fold *(arrows)* and severe osteoarthritis of the distal interphalangeal joint.

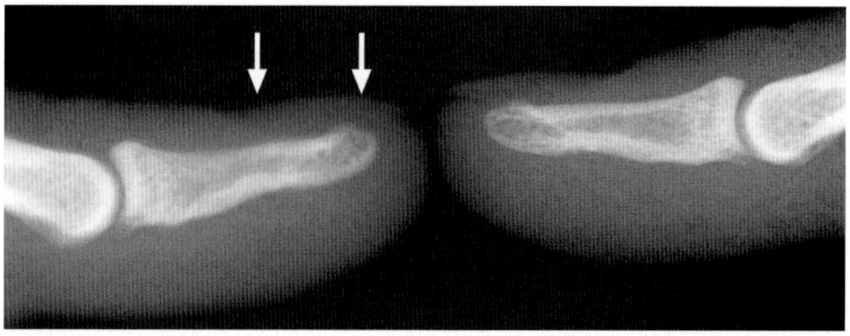

FIGURE 9.2. Subungual glomus tumor. Comparative lateral radiographs show a thickened nail bed on the right side *(arrows)*.

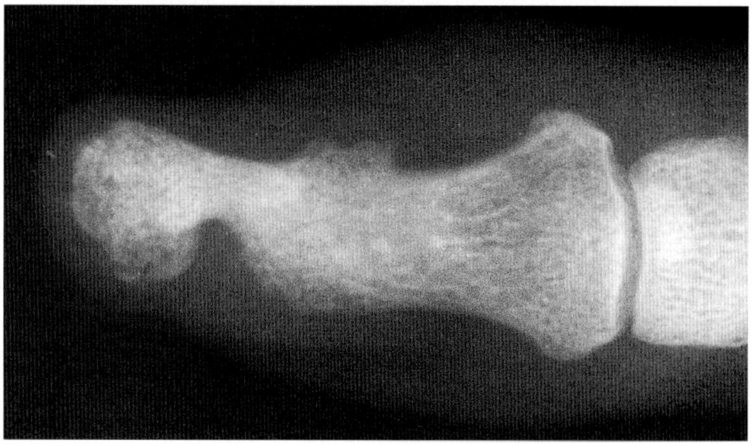

FIGURE 9.3. Subungual exostosis. Posteroanterior view depicts a pedicled ossification emerging from the tuft of the distal phalanx.

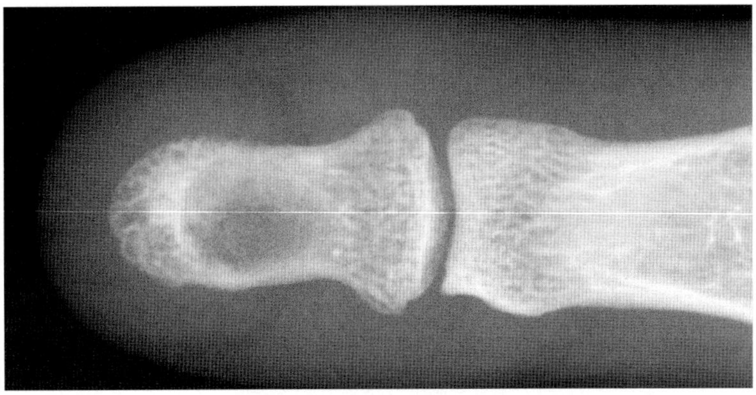

FIGURE 9.4. Epidermoid cyst with an isolated cystic bone lesion.

the soft tissues, such as thickening of the posterior nail fold in case of mucoid cyst (Fig. 9.1). Comparative lateral radiographs can highlight the difference in thickness of subungual soft tissues when a subungual mass is present (Fig. 9.2) (2,3). Radiographs may depict soft tissue calcifications. Accurate study of these calcifications may allow a specific diagnosis to be made. Phleboliths favor indication of an angioma, pedicled ossifications of an exostosis (Fig. 9.3), and mottled calcifications of a chondroma. Radiographs can detect a bone erosion or cyst, isolated (Fig. 9.4) or associated with abnormalities of the soft tissues (Fig. 9.5). Erosion of the distal phalanx may be

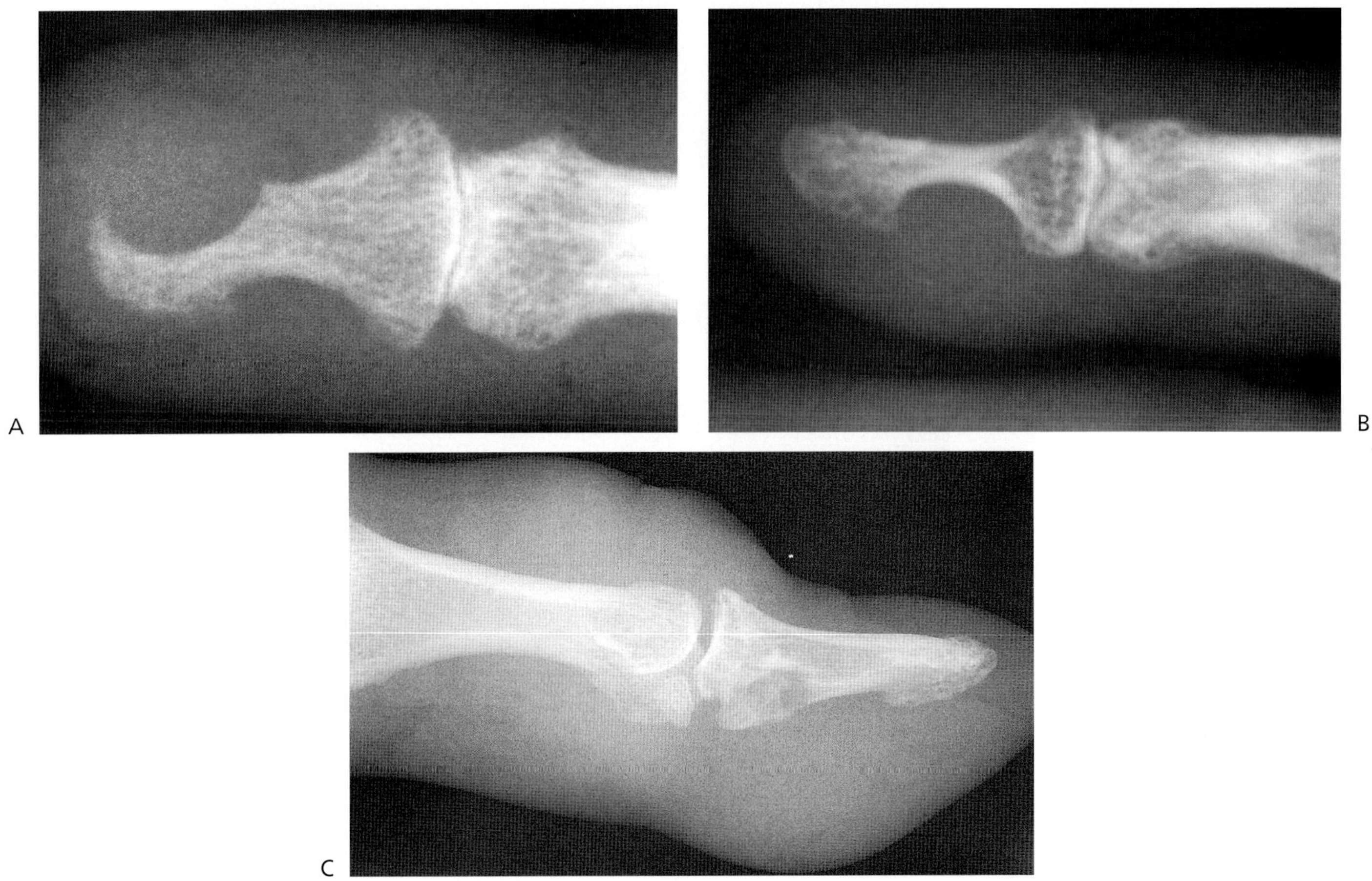

FIGURE 9.5. Bone erosion and mass of the soft tissues. **A:** Keratoacanthoma. **B:** Traumatic arteriovenous aneurysm. **C:** Dorsal and palmar giant cell tumors.

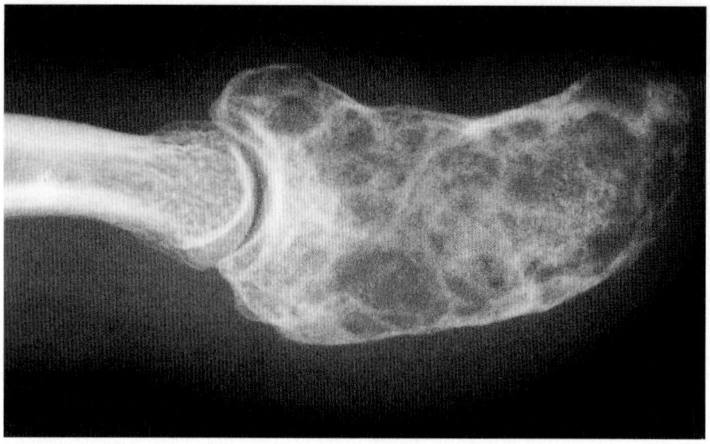

FIGURE 9.6. Expansive enchondroma of the distal phalanx.

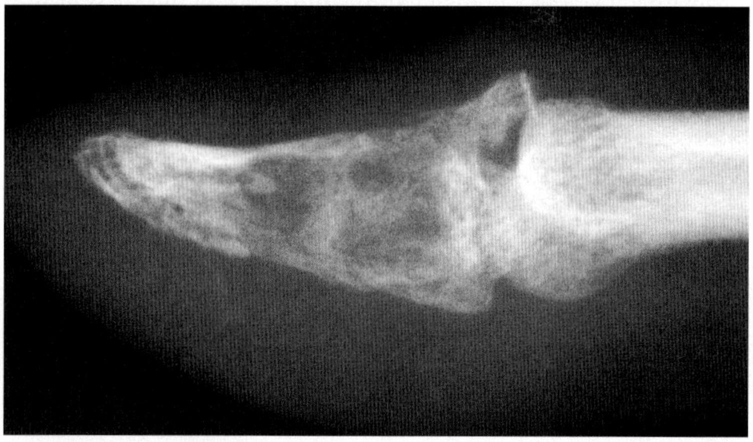

FIGURE 9.7. Enchondroma with a dorsal fracture of the base of the distal phalanx.

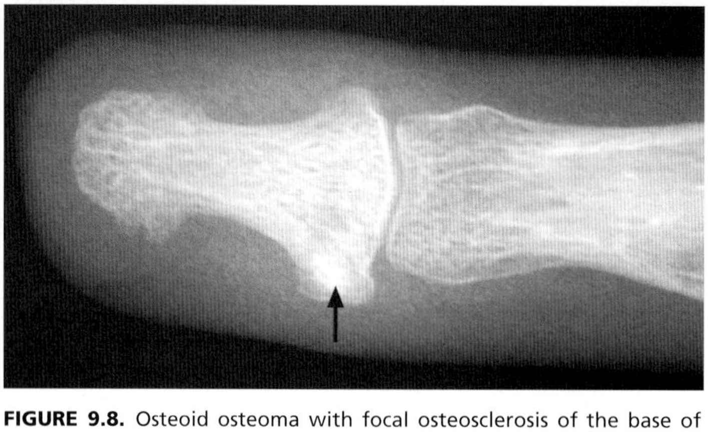

FIGURE 9.8. Osteoid osteoma with focal osteosclerosis of the base of the distal phalanx *(arrow)*.

depicted in several diseases, as in 55% of invasive subungual squamous carcinoma cases (4) or in keratoacanthomas (Fig. 9.5A) or glomus tumors (5). Enchondroma may expand the distal phalanx with mottled calcifications (Fig. 9.6) and be complicated by a pathologic fracture (Fig. 9.7). Radiographs can detect dense bone abnormalities, sometimes secondary to an osteoid osteoma (Fig. 9.8) or to a psoriatic arthropathy with a periosteitis. In case of osteoarthritis of the distal interphalangeal joint, the lateral view is most suitable to detect dorsal osteophytes impinging on the distal band of the extensor tendon (Fig. 9.1). Xeroradiography, another imaging method of the soft tissues, is associated with high irradiation and is no longer used. Arteriography has been proposed for diagnosis of glomus tumors (2) and osteoid osteomas, but it no longer is used since computed tomography and MRI became available (6–8).

COMPUTED TOMOGRAPHY

Exceptionally, computed tomography is indicated in pathologies of the nail unit. However, recent advances in computed tomography with helical acquisitions offer accurate imaging of the distal phalanx. Irradiation remains insignificant at fingertip level. This volumetric approach to the distal phalanx with 0.5- to 1-mm thick slices is necessary to highlight tiny bone abnormalities, such as the nidus of an osteoid osteoma (Fig. 9.9). High-quality three-dimensional remorfated views of the phalanx can be calculated after helical acquisition. However, the study of soft tissues remains poor and requires other imaging techniques such as ultrasonography and MRI.

ULTRASONOGRAPHY

Ultrasonography is rarely indicated in imaging of the nail unit. Its operator-dependent characteristics and the poor experience of radiologists in this anatomic area restrain its development. Technically, a 7.5- to 10-MHz probe dedicated to musculoskeletal imaging is necessary. Use of interposition material is helpful for study of the most superficial layers. The nail bed has a homogeneous hypoechoic appearance, included between the high

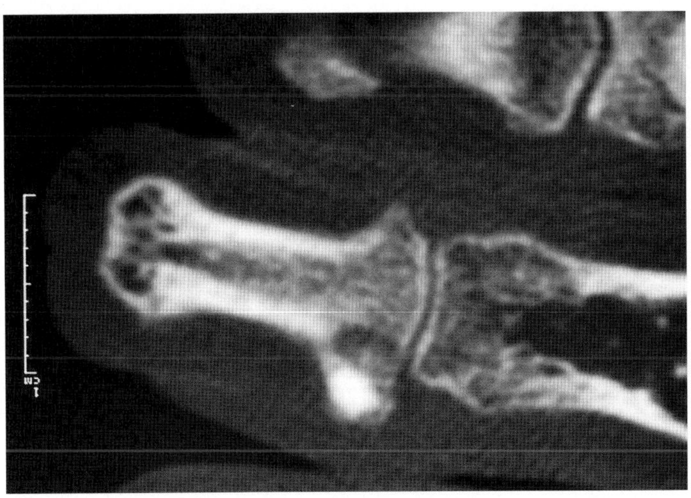

FIGURE 9.9. Osteoid osteoma of the third finger. A 1-mm thick coronal computed tomographic slice of the distal phalanx. The limits of the partially ossified nidus are more accurately defined on the ulnar side of the base of the phalanx.

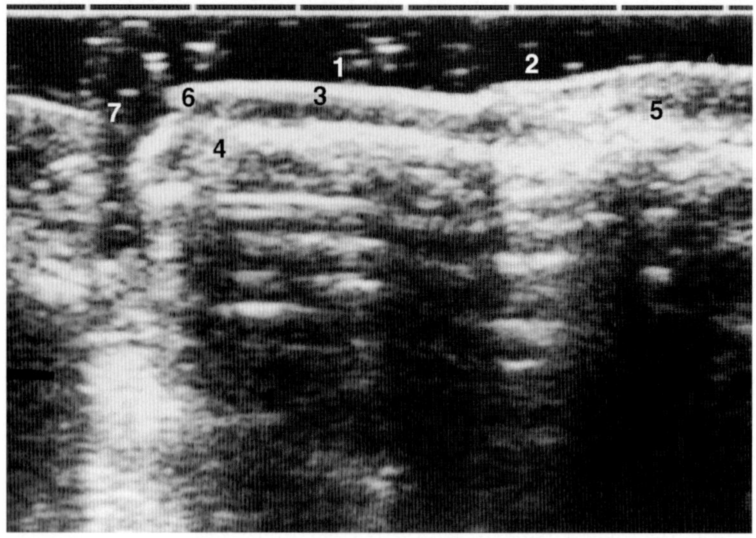

FIGURE 9.10. Sagittal ultrasonographic slice of the nail unit. 1, nail plate; 2, posterior nail fold; 3, nail bed; 4, dorsal cortex of the distal phalanx; 5, extensor tendon; 6, hyponychium; 7, free edge of the nail plate.

echoes of the nail plate and the cortex of the distal phalanx (Fig. 9.10). Artifacts in axial slices are created by the convexity of the nail plate (Fig. 9.11). Ultrasonography was used clinically for diagnosis of glomus tumors. Tumors smaller than 3 mm were hardly visible and the subungual area was difficult to reach (9). The pulp area is easier to access (Fig. 9.12). In case of a cutaneous mass, Doppler imaging can assess the vascular component. Ultrasonography is appropriate for detection of radiotransparent foreign bodies, such as prickles associated with a granulomatous reaction.

Very high-frequency ultrasonography continues to be investigational. Ultrasonography of the skin with high spatial resolution (HSR) was first introduced in 1979 (10). It appeared to be a reliable and noninvasive technique to measure the thickness of the corium (11). A-mode and B-mode ultrasonographic systems dedicated to investigation of

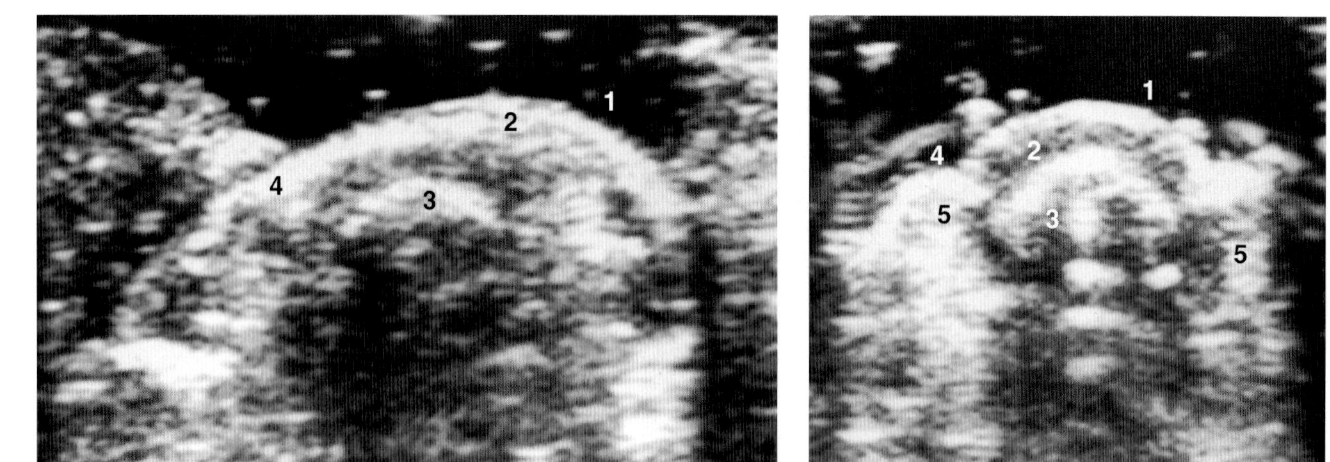

FIGURE 9.11. A: Axial ultrasonographic slice at the level of the matrix. 1, posterior nail fold; 2, area of the matrix (the matrix is not visible); 3, dorsal cortex of the distal phalanx. **B:** Axial slice at the mid part of the nail plate. 1, nail plate; 2, nail bed; 3, dorsal cortex of the distal phalanx; 4, lateral nail fold; 5, artifacts beneath the lateral nail folds.

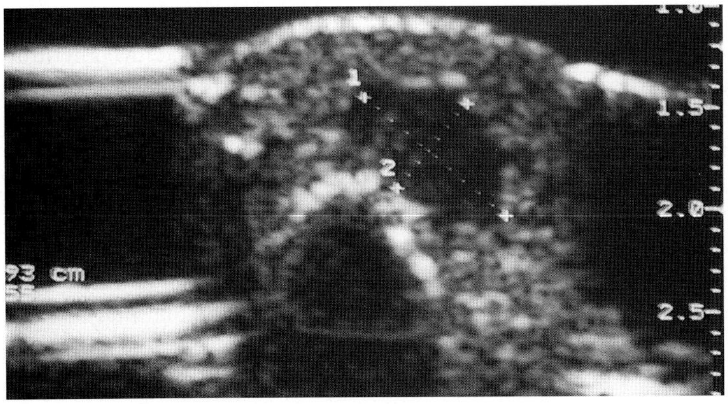

FIGURE 9.12. Histiocytofibroma of the pulp with a hypoechoic mass.

the skin were developed as research tools and as clinical aids, mainly in tumoral and inflammatory diseases (12). M-mode imaging, with high-frequency Doppler and three dimensions, remains in the field of research. The 20-MHz probes are the best compromise between HSR and sufficient depth. The 50-MHz or higher probes can image only the epidermis, with an axial resolution of 37.5 mm and a lateral resolution of 125 μm (13). The applications about the nail unit remain infrequent. Finlay et al. (14,15) introduced study of the thickness of the nail plate proximally and distally with A-mode ultrasonography and a 20-MHz probe. They noted that the thickness of the nail plate varied according to the different fingers, from the thickest thumb to the thinnest fifth finger. They also noted that the nail plate was thicker and more hydrated at the level of the lunula (14,15). Distal conduction speeds (mean 2,470 m/s) were well correlated with measurements of the free edge of the plate obtained with a micrometer. According to those authors, the distal reduction of 8.8% of the transmission time compared to the proximal values should be due to the greater thickness and hydration of the plate at the level of the lunula. On the other hand, Johnson et al. (16,17) demonstrated the contrary. They examined the thickness and the weight of 20 nail plates lost after traumatisms. Their measurements showed that the thickness of the nail plate nearly doubled from the middle part of the lunula to the free edge of the plate, as the weight increased 20%.

Jemec and Serup (18) also studied the sonographic A-mode structure of nail plates of cadaver thumbs *in vitro* and after avulsion. Spatial resolution of 75 μm was obtained with a 20-MHz probe. They noted that the desiccation of the nail plate canceled the correlation between the ultrasonographic measurements and that of the micrometer. This is an important point when it is known that the amount of water in the nail plate may fluctuate about 10%. Unlike Finlay et al., Jemec and Serup noted a dry superficial compartment (conduction speed 3,103 m/s) and a wet deep compartment (conduction speed 2,125 m/s). The conduction speed of the nail bed (1,580 m/s) was close to that of the deep layers of the skin. Nevertheless, the imaging of the whole nail unit was not approached.

MAGNETIC RESONANCE IMAGING

Magnetic resonance imaging is well known for its high contrast of the soft tissues. A few cases of subungual tumors, particularly glomus tumors, were reported in the literature with available commercial coils. However, the spatial resolution was too poor for accurate study of the whole nail unit (6,19–22). On the other hand, the ability to obtain HSR magnetic resonance images has been known since 1986 (23,24). Since 1990, several imaging departments have inserted custom-built high-resolution modules with local gradient surface coils into whole-body MRI units (25–27). The first clinical applications, developed in the Centre Inter-Etablissement de Résonance Magnétique (CIERM) of the Hospital of Bicêtre under the coordination of J. Bittoun, M.D., Ph.D., were used for evaluation of skin imaging (25,28,29). Later, the investigations were expanded to the musculoskeletal system, particularly the wrist and the finger (26,27,30,31), and the eye (31). These developments were based on custom-built local gradient coils or noncommercially available dedicated imaging sequences. However, small surface coils dedicated to wrist or finger imaging have become available recently. When associated with the most powerful whole-body magnetic resonance units with strong gradients (more than 20 mT/m), they provide a clinically suitable image of the nail unit.

Magnetic Resonance Imaging Constraints of the Nail Unit

The nail unit presents a structure with multiple layers on axial images. A voxel height less than 100 μm, about the thickness of the epithelial layer of the nail bed, is necessary in the direction from the surface to the depth to distinguish this layer. Nevertheless, unlike the skin, which is a very superficial structure, the nail unit may require evaluation of the deep layers of the nail bed or even of the pulp when a tumor extends under the lateral interosseous ligament. The thumbs and the big toes may be quite large, and it may be necessary to keep a sufficient signal-to-noise ratio (SNR) about 2 cm or more from the surface coil. Routinely, a field of view of 3 cm is the most adaptable. It provides a whole view of the nail unit while maintaining a sufficient SNR (32). For the small fingers (fifth finger), fields of view of 2 cm may be more appropriate. In case of HSR and strong gradients, the linear distortions are more important, in which case a circumferential coil, such as a solenoid, bird cage, or phased-array surface coil, is preferable.

When using a plane circular surface coil, the nail plate must be placed against the coil to offer the maximum signal close to the superficial layers of the nail unit. The hand is placed supine with the HSR coil fixed on the center of the bed of the magnetic resonance system. Full cooperation of the patient and efficient mechanical support with adhesive bandages are necessary. After several attempts, the most comfortable position for the patient appears to be supine with the arm elevated. Some patients with painful (rotator cuff tears, multiple tendons calcifications) or frozen shoulders cannot maintain this position during the entire examination. However, the failure ratio is very low. For study of the toes, the position of the patient is more comfortable. The patient is prone with the feet in the gantry and the dorsal aspect of the toes against the HSR coil. In all cases, a perfect immobility of the distal phalanx is necessary to avoid movement artifacts, particularly disturbing with HSR. For this reason, children younger than 6 years should not be explored in this manner.

Imaging Sequences

The sequences must be conceived to provide the best compromise between HSR and sufficient SNR. A voxel size of 117 × 234 × 1,000 to 3,000 μm^3 was used with a 3-cm field of view and a 256 × 128 matrix. Routine examination included axial T1-weighted spin-echo (SE) images with a short acquisition time (1 minute 35 seconds), axial T2-weighted SE images, completed with sagittal T1 or T2 SE images, or three-dimensional gradient echo images. The three-dimensional 0.7- to 1-mm thick contiguous images were essential to detect tiny lesions smaller than 2 mm, whereas 3-mm SE images were too thick and presented an interslice gap of 1 mm.

Intravenous injection of 0.1 mmol/kg gadoterate meglumine (Dotarem, André Guerbet, Aulnay-sous-Bois, France) was administered according to the suspected pathology.

Normal Magnetic Resonance Appearance of the Nail Unit

Sagittal Plane

Sagittal images nicely show the nail plate in its full length. However, the *in vivo* magnetic resonance images of the nail plate depict a lack of signal, probably due to the highly organized structure of keratin (33). For collagen, this scleroprotein causes dramatic shortening of the relaxation times (29). If the ventral aspect of the nail plate is well defined by the high signal of the epidermis layer of the matrix and the nail bed, its dorsal aspect is not visible. Its interface with the air, also presenting a lack of signal, is not detectable. Thus, application of Vaseline to the plate before its positioning on the surface coil gives high definition to the dorsal aspect of the nail plate whatever the sequence used (Fig. 9.13). However, the presence of fat causes chemical shift artifact. This artifact underestimates the thickness of the nail plate by about 0.2 mm, which is significant in relation to the thickness of a normal nail plate (0.5 to 1 mm) (34,35). This artifact can be suppressed by using the fat-suppression technique. On three-dimensional gradient-echo images, there is often an artifact of magnetic susceptibility, with an area of low signal around the free edge of the nail plate. The nail plate is very thin close to the nail root, surrounded by the thin high signal of the cul-de-sac of the matrix. The nail plate progressively thickens up to its free edge. This finding is in agreement with ultrasonography reports by Johnson et al. (16,36). However, MRI is not able to distinguish the different layers of the nail plate noted by histologic studies. Whatever the sequence used, the plate always remains without signal through its thickness. Jemec et al. (37) used ultrasonography to highlight two layers of different hydration in the nail plate. But, unlike MRI, ultrasonography was not able to distinguish the epidermis layer of the nail bed.

Sagittal images also accurately depict the distal interphalangeal joint with the insertion of the terminal band of the extensor tendon onto the base of the distal phalanx, articular cartilage, and palmar plate (Fig. 9.14). As shown, the nail root with the surrounding proximal and distal matrix is particularly well defined in this view. It is not possible to highlight a different signal between that of the nail matrix and that of the epidermis layer of the nail bed or the ventral aspect of the proximal nail fold. However, in this plane, the epidermis layer of the nail bed appears thicker than the nail matrix, which seems to be contradictory to the histologic findings (38,39). A partial volume artifact with mixing of the epidermis ridges and the papillary crests due to slice thickness (1 to 3 mm) and curvature of the nail plate could explain this apparent thickening. Nevertheless, the transition between matrix

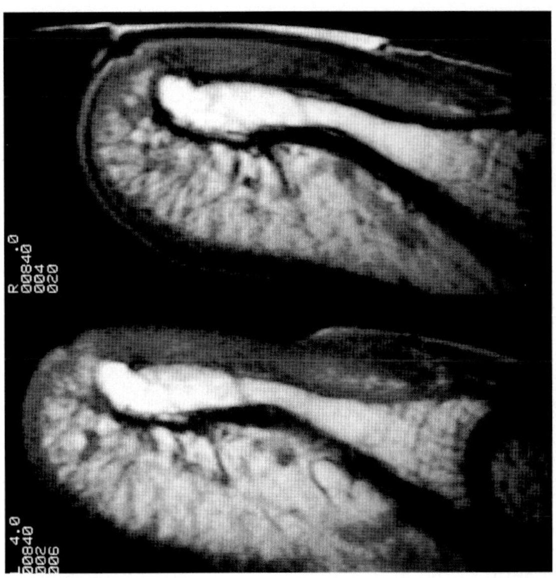

FIGURE 9.13. Sagittal T1-weighted spin-echo images with **(top)** and without **(bottom)** application of Vaseline to the nail plate.

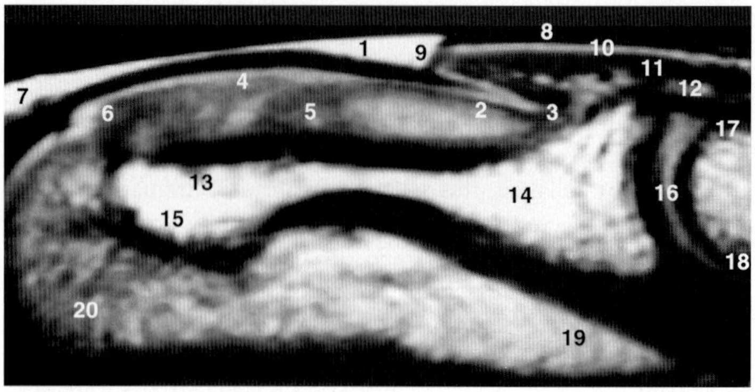

FIGURE 9.14. Sagittal proton density weighted spin-echo image. 1, nail plate; 2, nail matrix; 3, matrix cul-de-sac; 4, epithelium of the nail bed; 5; dermis of the nail bed; 6, hyponychium; 7, free edge of the nail plate (note the round artifact due to magnetic susceptibility); 8, posterior nail fold; 9, cuticle; 10, epidermis; 11, dermis, 12; hypodermis; 13, cortex of the distal phalanx; 14, trabecular bone of the distal phalanx; 15, tuft; 16, distal interphalangeal joint; 17, terminal band of the extensor tendon; 18, palmar plate; 19, flexor digitorum profundus tendon; 20, pulp.

and nail bed is signaled by the steep break of thickness of the epidermis layer. The epidermis layer thickens at the level of the hyponychium. The lack of fat in the nail bed, except for a thin band under the nail root, excludes chemical shift artifact.

T2-weighted sagittal images show an oval-shaped area of high signal in the deep dermis beneath the nail matrix (Fig. 9.15). This structure is related to the lunula, as we demonstrated in 40 healthy volunteers (40). The length of this area is highly correlated to the length of the lunula. When the lunula is lacking, this dermal area still is present but does not pass over the limits of the posterior nail fold. This area presents a strong and homogeneous enhancement after injection of gadolinium, and its limits are well defined. Histologic studies of this area show that the dermis is more uniform, with a looser connective tissue than that of the more distal nail bed. The microvascularization presents a more regular vascular network than in the nail bed (41). The meaning and role of this area are still obscure. According to Baden et al. (42,43), the matrix cells could produce some proteins under the order of the underlying dermis.

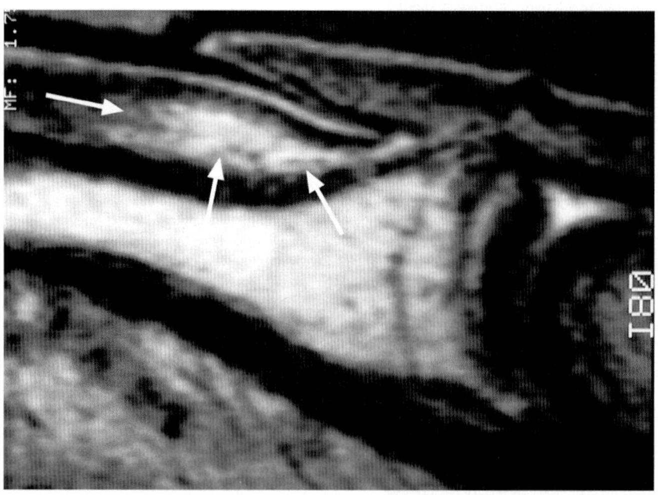

FIGURE 9.15. Sagittal T2-weighted spin-echo image centered at the level of the matrix area. High signal of an oval shaped area beneath the nail matrix *(arrows)*. The distal tip of this area matches with the lunula.

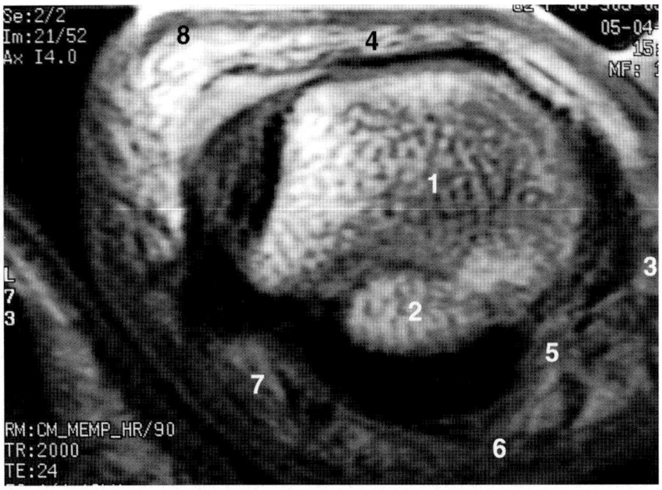

FIGURE 9.16. Axial proton density-weighted spin-echo image at the level of the interphalangeal joint of a thumb. 1, head of the proximal phalanx; 2, base of the distal phalanx; 3, lateral ligaments; 4, terminal band of the extensor tendon; 5, palmar plate; 6, flexor tendon; 7, collateral vessels; 8, proximal nail fold.

Axial Plane

The proximal axial slices easily demonstrate the proximal nail fold, extensor and flexor digitorum tendons, lateral ligaments of the distal interphalangeal joint, and palmar plate (Fig. 9.16). The root of the nail plate appears surrounded by the proximal and distal matrix. The curvature of the matrix is similar to that of the nail plate (Fig. 9.17). The posterolateral borders of the nail plate are contiguous with the matricophalangeal ligaments. At the level of the distal matrix, the slice clearly shows thickening of the epidermis layer close to the median line; the ridges are not visible. The dermis beneath the matrix presents a homogeneous high signal on T2-weighted images and after injection of gadolinium (Fig. 9.17). The lateral nail folds and the rima ungualum, bordered by the Flint lateral interosseous ligaments, are well defined on axial slices of the nail bed (Fig. 9.18) (44). The epidermis layer of the nail bed presents a thin and high signal. The underlying dermis shows a thin superficial layer with low signal and a thick deep layer with a heterogeneous signal. This heterogeneity is reinforced by the injection of gadolinium as result of

A 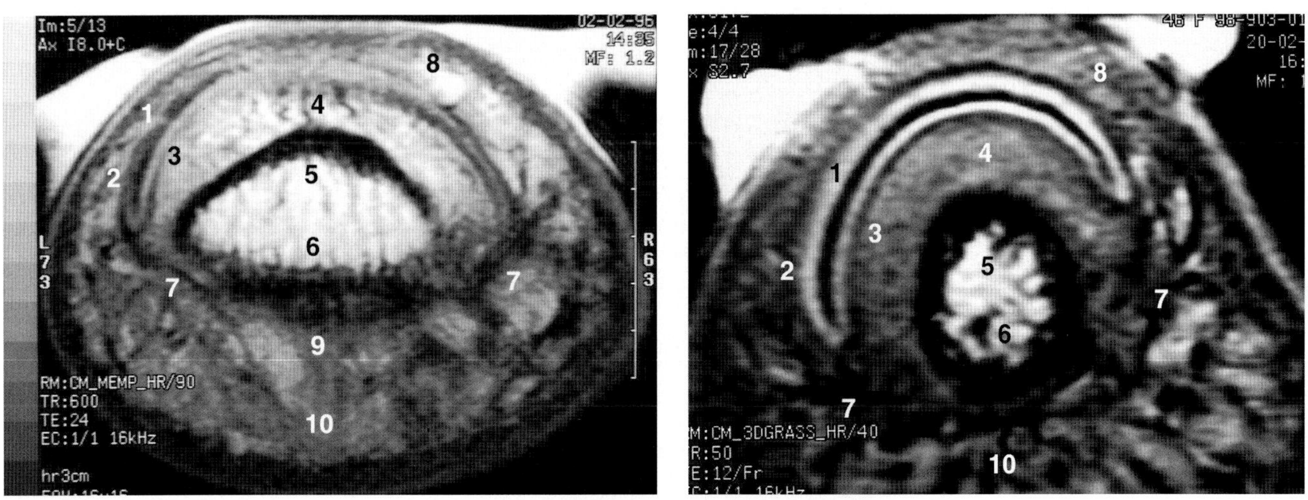 B

FIGURE 9.17. Axial images at the level of the matrix area. **A:** Postenhanced 3-mm thick T1-weighted spin-echo image. **B:** A 0.7-mm thick three-dimensional gradient-echo image. 1, root of the nail plate; 2, proximal nail matrix; 3, distal nail matrix; 4, dermis beneath the nail matrix (note the enhancement of this area and the lack of fat tissue); 5, cortex of the distal phalanx; 6, trabecular bone; 7, dorsal matricophalangeal ligaments; 8, posterior nail fold; 9, insertion of the flexor tendon; 10, pulp.

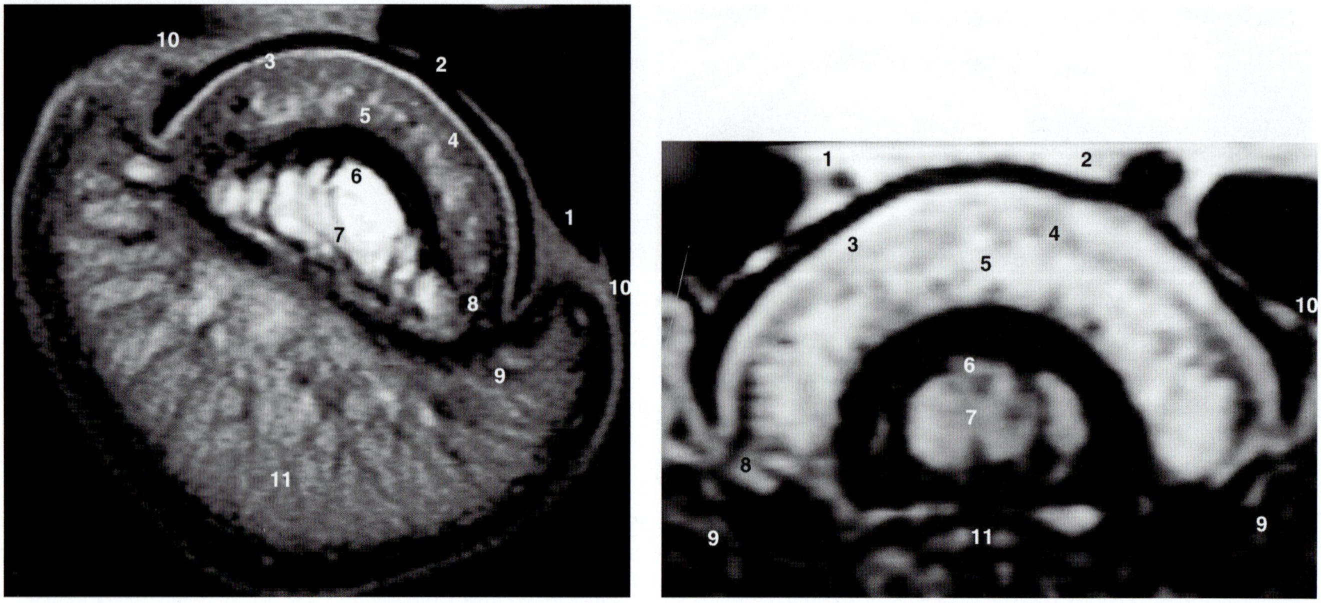

FIGURE 9.18. Axial images at the level of the nail bed. **A:** T2-weighted spin-echo image. **B:** Postenhanced three-dimensional gradient-echo image. 1, Vaseline; 2, nail plate; 3, epidermis of the nail bed; 4, superficial dermis; 5, deep dermis, 6, cortex of the distal phalanx; 7, trabecular bone of the distal phalanx; 8, rima ungualum; 9, lateral interosseous ligament; 10, lateral nail fold; 11, pulp.

the numerous enhanced glomus bodies of the nail bed. These normal glomera are remarkable arteriovenous shunts (Fig. 9.19). They are highly concentrated in the tips of the digits, particularly beneath the nails (45). Each glomus body is a tiny encapsulated oval organ 300 μm long. The nail beds of fingers and toes contain 93 to 501 glomus bodies per square centimeter (46).

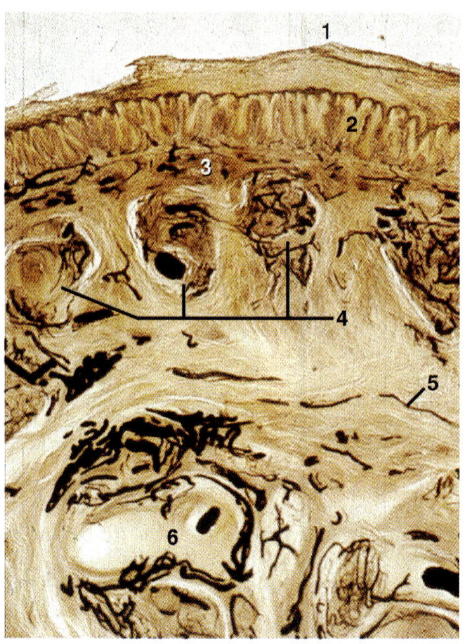

FIGURE 9.19. Vascular networks. Axial section 250 μm thick made transparent by the technique of Spalteholz, vascular injection with gelatinous India ink. 1, nail plate; 2, epidermal ridges and papillary crests; 3, vascular network of superficial dermis; 4, glomus bodies; 5, periosteal vessels; 6, vascular network of trabecular bone. (Courtesy of Pr. R. Wolfram-Gabel, Strasbourg, France.)

Coronal Plane

Coronal slices are not acquired routinely. These slices are disappointing and not adapted to the spatial structure of the nail unit. Its different elements are tangential to the frontal plane and therefore exposed to the partial volume artifact.

Tumors of the Nail Unit

The indications of MRI for pathologies of the nail unit are still under investigation. We do not claim to present an exhaustive MRI atlas of all the tumors of the nail unit in this chapter. This presentation is based on our experience of more than 250 magnetic resonance images of tumors that were confirmed surgically or by their clinical outcome. The main indications were imaging of glomus tumors and mucoid cysts. Numerous other lesions were explored, such as epithelial tumors (warts, epidermoid cysts, onychomatricomas, keratoacanthomas), soft tissue tumors (fibrokeratomas, fibromas, angiomas, giant cell tumors), and osteochondral lesions (exostoses, chondromas, osteoid osteomas, chondrosarcomas). The accurate location of the tumor with magnetic resonance images associated with its signal patterns are important data used for diagnosis.

Glomus Tumors

Glomus tumors early appeared as a main indication for MRI of the nail unit. We collected a large series of 65 tumors. Glomus tumors develop from the glomus bodies, which are particularly numerous in the dermis of the nail bed. High-resolution MRI is able to depict normal glomus bodies with T2-weighted images and after injection of gadolinium. The clinical interest is significant. The classic triad associating pain, painful point, and cold sensitivity is evocative but infrequent. In our series, clinical diagnosis was highly suspected in 68% of cases. The mean diagnostic delay is rather long, varying from 4 to 7 years in the literature (3,21,45,47). Three main findings are well highlighted by MRI: tumor signal, tumor limits, and tumor location.

Tumor Signal

The first cases reported in the literature were pathognomonic: high signal on T2-weighted images and strong and homogeneous enhancement after injection of gadolinium. However, magnetic resonance findings are not so univocal, and some studies in the literature reported conflicting signal behavior. If most of the tumors presented a high signal on T2-weighted images, on T1-weighted images the lesions could present a low signal (6,19,21), a high signal (hemorrhage or vascular component?) (18), or an intermediate (8) or heterogeneous signal with a tumoral core (22). The signal behavior depends on histologic composition. Glomus tumors are the result of hyperplasia of one or several elements of the glomus bodies, and they may be considered hamartomas (45). In 1924, Masson (48) described different histologic variants. They are not routinely mentioned in pathologic reports, as they have no prognostic significance. However, knowledge of them is important to understand their involvement in tumor signal (49).

The vascular type is composed of numerous vascular lumens. Enhancement is very high after injection of gadolinium, and the signal is high on T2-weighted images (Fig. 9.20).

The cellular or solid type mainly presents a proliferation of epithelioid cells (glomus cells) and a relative paucity of vascular lumens. This type of tumor is difficult to detect with MRI. Its signal is close to that of the normal dermis of the nail bed on all sequences. Injection of gadolinium is of little use. Thin three-dimensional contiguous gradient-echo

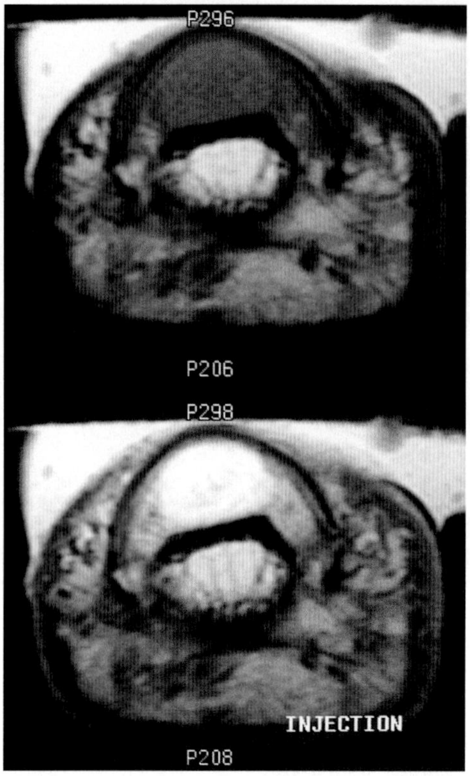

A

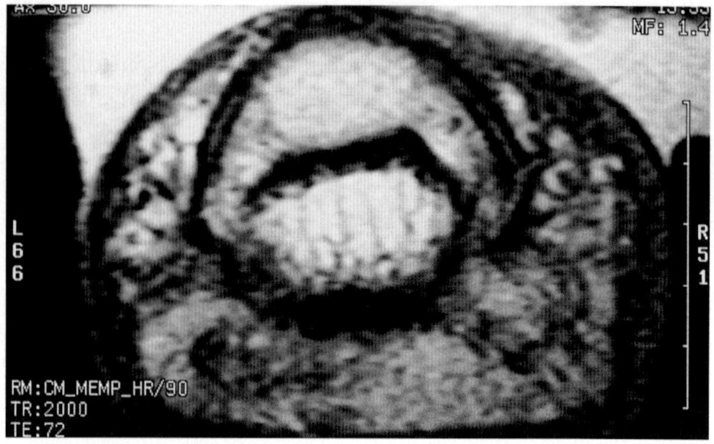

B

FIGURE 9.20. Highly vascularized glomus tumor. **A:** T1-weighted axial spin-echo images before **(top)** and after **(bottom)** injection of gadolinium. High and homogeneous enhancement of a well-defined mass on the midline of the nail bed. Note the bone erosion on the dorsal cortex that was radiologically occult. **B:** T2-weighted axial spin-echo image. The tumor presents a signal faintly higher than that of the nail bed.

A

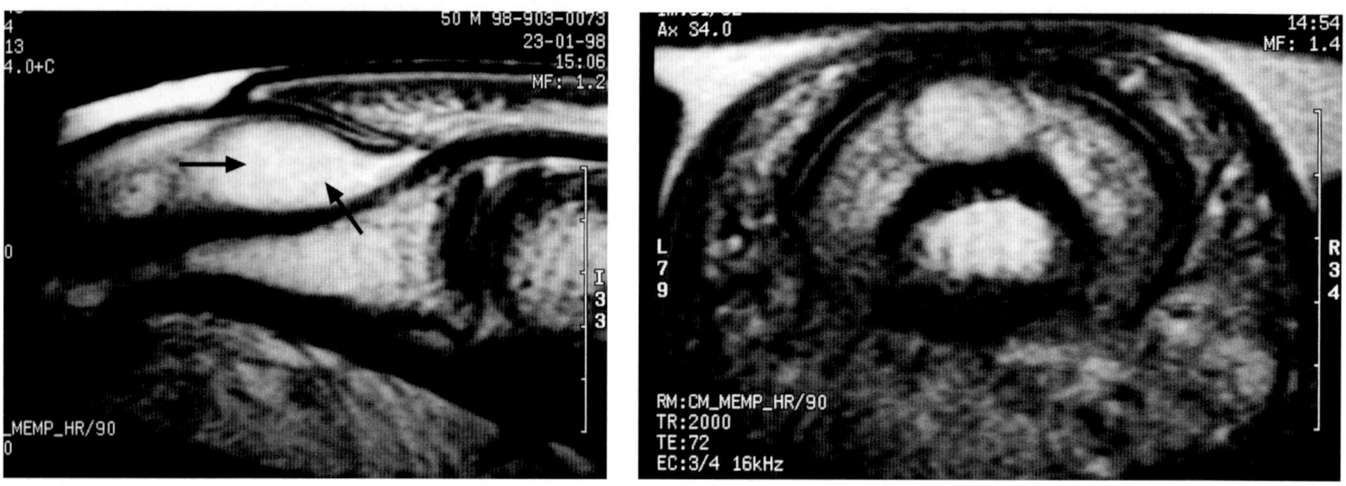

B

FIGURE 9.21. Solid glomus tumor. **A:** Postenhanced sagittal T1-weighted spin-echo image. The tumor *(arrows)* is faintly differentiable from the high signal of the submatrical dermis, even with the injection of contrast media. **B:** Axial T2-weighted spin-echo image.

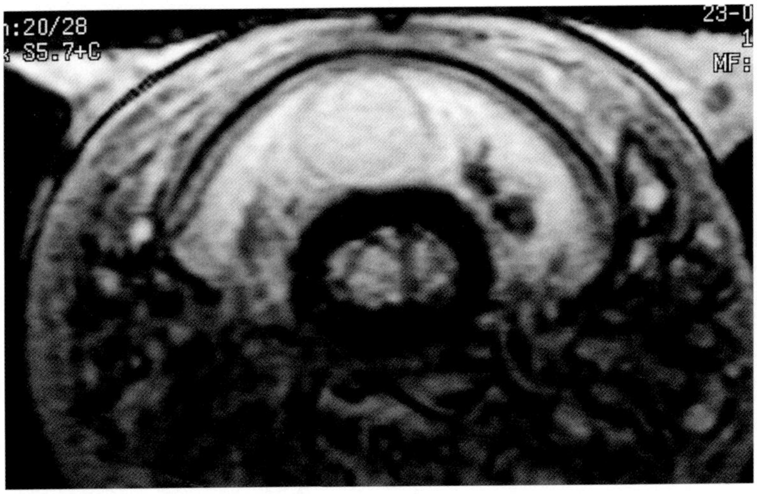

C

FIGURE 9.21 ***(continued).*** **C:** Postenhanced three-dimensional gradient-echo image. The tumor is easier to detect with the low signal of the peripheral capsule.

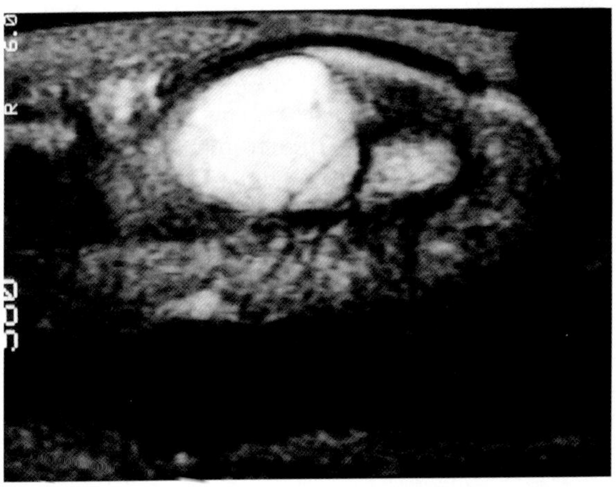

FIGURE 9.22. Mucoid glomus tumor. Sagittal T2-weighted spin-echo image. Note the very high signal of the tumor due to the water content of the mucoid stroma.

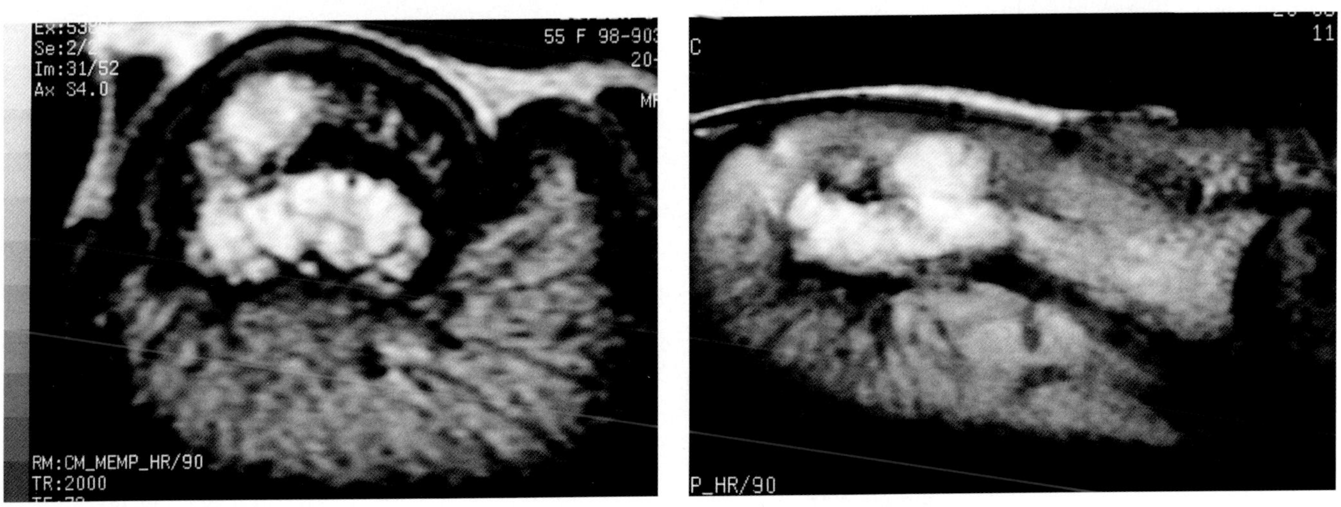

A B

FIGURE 9.23. Glomus tumor. **A:** Axial T2-weighted spin-echo image. **B:** Sagittal postenhanced T1-weighted spin-echo image. The different histologic components of the tumor give an heterogeneous signal of the lesion.

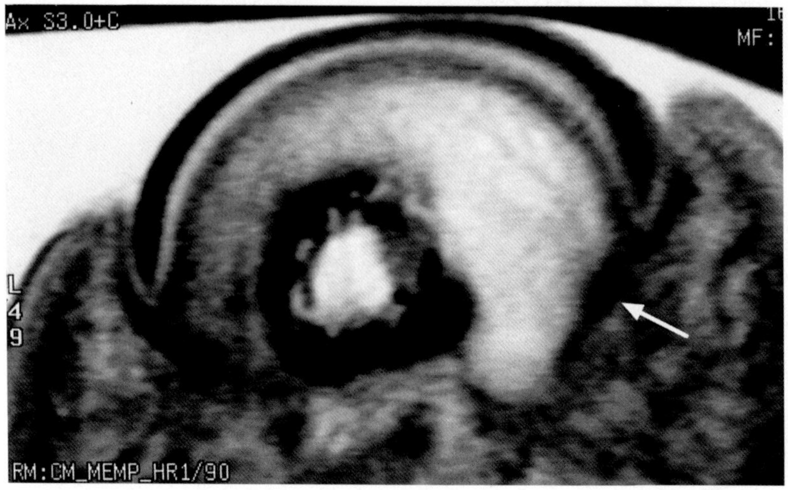

FIGURE 9.24. Ill-defined glomus tumor. Postenhanced axial T1-weighted spin-echo image. The tumor is in the lateral part of the nail bed with an extension toward the pulp via the rima ungualum. The lateral interosseous ligament is pushed by the tumor *(arrow)*. The interface between the tumor and the nail bed is ill defined due to the lack of capsule. Adhesions with the nail bed were surgically noted.

slices are the most helpful. They are the most appropriate to depict a peripheral capsule or a slight bone erosion on the dorsal aspect of the distal phalanx (Fig. 9.21).

The mucoid type with a mucoid degenerescence of the stroma presents a faint enhancement but has a very high signal on T2-weighted images due to the large amount of water in the stroma (Fig. 9.22).

Numerous tumors are a combination of these three elementary types (Fig. 9.23).

Tumor Limits

Most often the tumor limits are well defined with a peripheral pseudocapsule. This capsule is a reactional response of the surrounding connective tissue. The capsule presents a very low signal on all sequences, but it is more visible on T2-weighted images or three-dimensional gradient-echo images (Fig. 9.21). Its analysis is greatly facilitated by high-res-

A 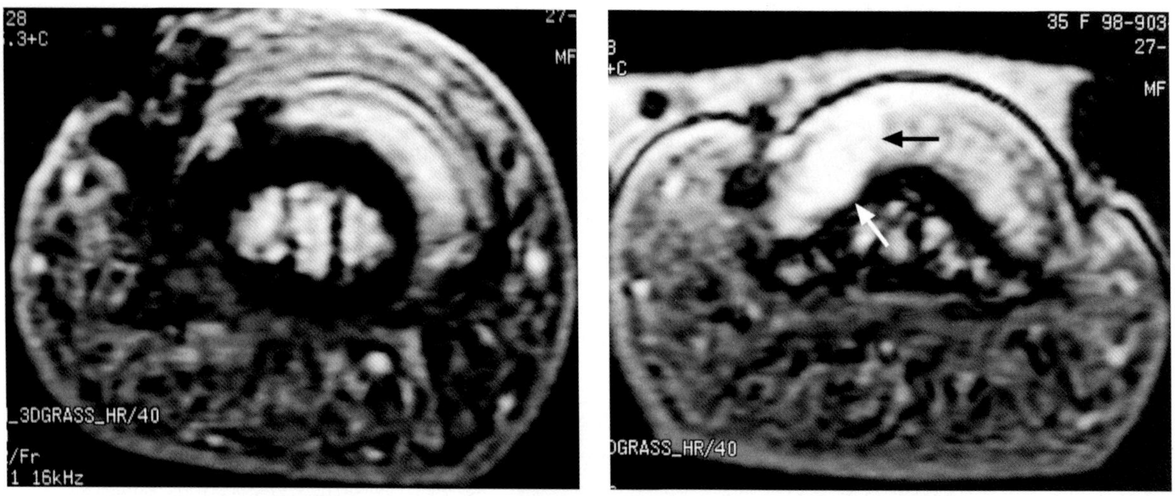 B

FIGURE 9.25. Recurrent glomus tumor 2 years after surgery. Postenhanced axial three-dimensional gradient-echo images. The previous lateral surgical approach is marked by low-signal artifacts on the lateral part of the posterior nail fold, the matrix, and the lateral nail fold. The recurrent glomus tumor is enhanced *(arrows)* on the distal part of the surgical seat.

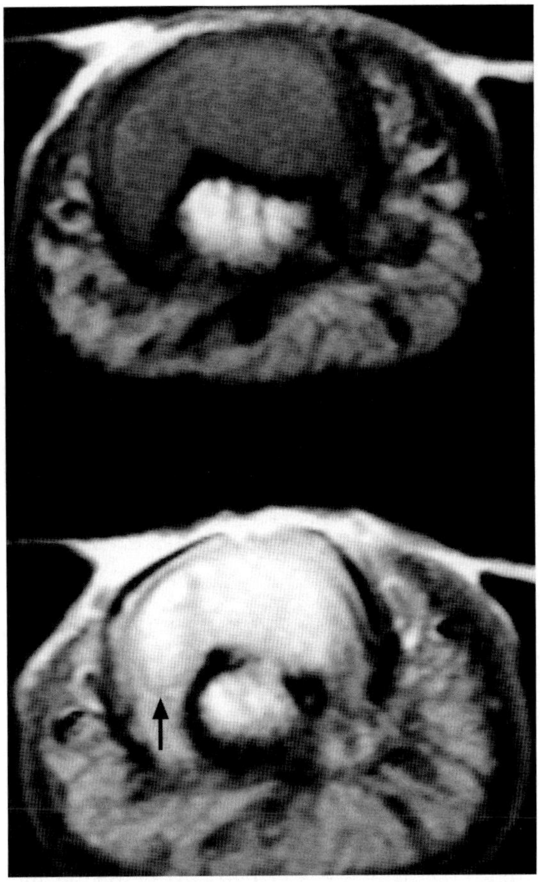

FIGURE 9.26. Multiple glomus tumors. Axial T1-weighted spin-echo images before **(top)** and after **(bottom)** injection of gadolinium. A large enhanced mass on the midline of the nail bed completely erodes the dorsal cortex of the distal phalanx. A second tumor is detected on the lateral part of the nail bed *(arrow)*.

olution MRI (50). In our series, the capsule appears incomplete or absent in 23% of cases. The tumor limits are ill defined, and injection of gadolinium may depict small foci of tumor extending into the nearby nail bed. Often in these cases, some adhesions with the nail bed are noted during surgery (Fig. 9.24). Local invasion of the capsule is debated. It was reported on histologic examinations in 1% to 2% of cases by Kohout and Stout (51) but was not found by Carroll and Berman (45). It is certain that the risk of recurrence is high if some tumoral tissue is left *in situ* during surgery of these ill-defined lesions. The recurrence rate reported in the literature varies from 12% to 24% (45,47,52,53). Magnetic resonance imaging appears to be particularly helpful in cases of recurrent pain after surgery (Fig. 9.25). Magnetic resonance imaging also is able to depict multiple glomus tumors in the same fingertip. In these cases, two tumors usually are close and may be located on each side of the median line (Fig. 9.26). In such cases, MRI is essential to plan the surgical approach.

Tumor Location

In most cases, 66% in our series, the tumor is located in the subungual area, in the supporting tissue of the nail bed or the matrix. These locations are the most difficult to highlight with ultrasonography because the curvature of the nail plate interferes with analysis of the lateral parts of the nail bed (9). Usually, the lesion is deep, close to the periosteum of the underlying phalanx. Often, a cortical erosion is depicted on the axial magnetic resonance images (82% of our cases), although it was occult on radiographs (Fig. 9.20). These axial slices are essential to distinguish the tumors on the median line from those of the lateral part of the nail bed, which sometimes extend into the pulp via the rima ungualum (Fig. 9.24). The surgical approach is planned according to the size and location of the tumor. The lateral type may be excised by a lateral approach and by lifting the nail bed without disruption of the matrix (Fig. 9.27). The median type

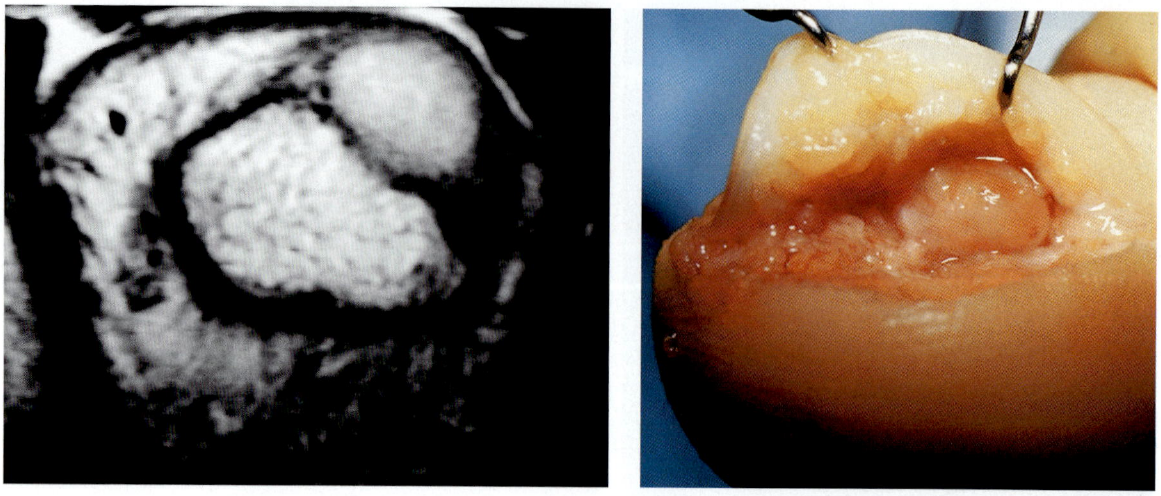

A B

FIGURE 9.27. Lateral glomus tumor beneath the nail matrix cul-de-sac. **A:** Axial postenhanced T1-weighted spin-echo image. **B:** Surgical view. (Courtesy of Dr. S. Goettmann, Paris, France.)

may require a transungual approach, with the resection size adapted to the size of the tumor.

Sagittal slices are essential to determine the relations between the tumor and the nail matrix. Lesions beneath the matrix are difficult to detect (Figs. 9.21 and 9.23). Clinically, they often are unsuspected, although they may lead to a nail fissure by compression of the matrix. As shown, the supporting dermis of the matrix presents specific features with MRI. This oval area with high signal on T2-weighted images and high enhancement after injection of gadolinium decreases the contrast between healthy tissue and tumor on all sequences. In addition, this oval area beneath the matrix may be particularly intense and must not be confused with a glomus tumor. Axial three-dimensional gradient-echo images and injection of gadolinium may help to highlight a faint cortex erosion and a peripheral capsule.

More seldom, the lesions may be located in the pulp or the posterior nail fold. In this case, the contrast healthy tissue/tumor is completely different because of the fat tissue of the hypodermis surrounding the tumor. The low-signal tumor is spontaneously visible on T1-weighted images, surrounded by the high signal of fat. On the other hand, injection of gadolinium blurs the tumor limits by leveling out the signals (Fig. 9.28). Its only interest is to highlight a possible extension into the nail bed. Three-dimensional gradient-echo images with injection of gadolinium yield the best contrast between the low signal of the fat and the enhanced tumor.

Glomus tumors are easily distinguished from other vascular lesions, such as angiomas, with their characteristic blood flow artifacts. Subungual mucoid cysts may mimic glomus tumor, but they only present faint peripheral enhancement after injection of gadolinium.

Mucoid Cysts

The numerous synonyms used to name this lesion are the result of its debated nature: dorsal digital cyst, synovial cyst, recurrent myxoid cyst, cutaneous mucoid cyst (54,55), dorsal ganglion of the distal interphalangeal joint (56), mucoid digital pseudocyst (57), focal myxoid degenerescence (58), and mucoid cyst (59). Whereas some authors regard it as a synovial cyst, others consider it a degenerative lesion. Interest in a complementary imaging modality may seem poor because the clinical diagnosis is easy; most of the cysts originate from the accessible posterior nail fold. However, the high recurrence rate despite numerous proposed treatments may increase the interest in accurate preoperative imaging. High-resolution MRI will accurately analyze the relations between the cyst and the distal interphalangeal joint.

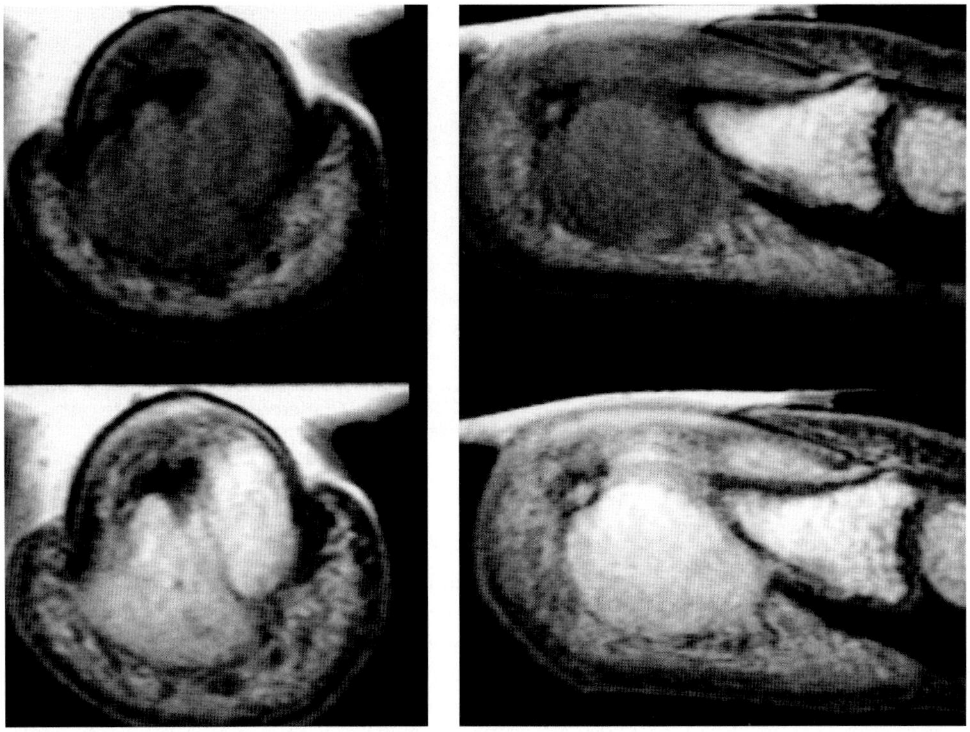

A B

FIGURE 9.28. Glomus tumor of the nail bed with invasion of the distal phalanx and the pulp. Axial **(A)** and sagittal **(B)** T1-weighted images before **(top)** and after **(bottom)** injection of gadolinium. The nail bed component is better detected after injection of contrast media, whereas the pulp component is better seen spontaneously with the natural contrast of the high signal of the surrounding fat tissue.

Magnetic Resonance Imaging Findings

These findings are based on a review of 23 magnetic resonance images of mucoid cysts (60). Most of the cysts are solitary and located on the proximal nail fold. Their appearance is specific: thin regular walls, low signal on T1-weighted images, and very high signal on T2-weighted images (Fig. 9.29). Intracystic septa are best seen on T2-weighted

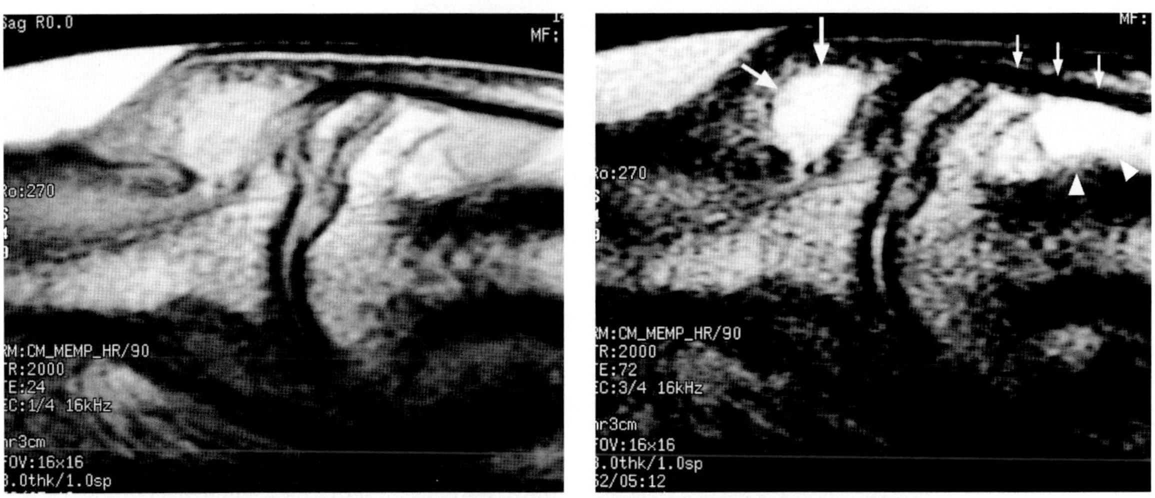

A B

FIGURE 9.29. Mucoid cyst of the posterior nail fold. Sagittal proton-density **(A)** and T2-weighted **(B)** spin-echo images. The cyst presents a very high signal on the T2-weighted image and sharp limits *(arrow)*. Note the large dorsal osteophytes of the distal interphalangeal joint lifting up the extensor tendon *(small arrows)*. Some fluid is depicted in the dorsal recess of the joint beneath the extensor tendon *(arrowheads)*.

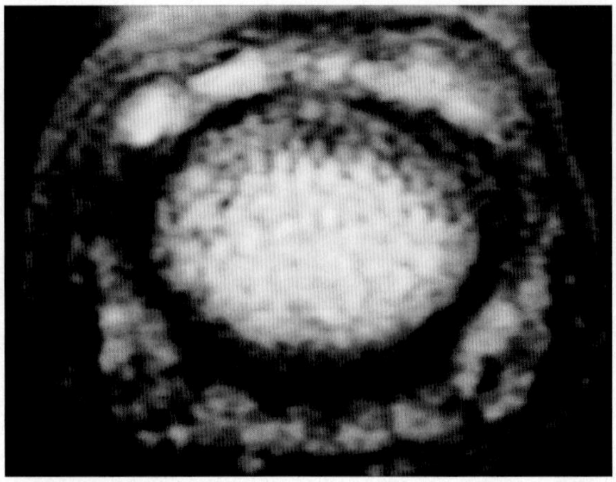

FIGURE 9.30. Flattened mucoid cyst of the posterior nail fold. Axial T2-weighted spin-echo image. Numerous septa separate cystic lobules.

images in 39% of cases (Fig. 9.30). Injection of gadolinium shows early faint peripheral enhancement that is different from that of glomus tumors (Fig. 9.31). With time, the enhancement moves toward the center of the cyst. This diffusion of contrast media may be compared to the intraarticular diffusion of gadolinium at the level of the knee through the synovium after intravenous injection of gadolinium (61). However, a true synovial membrane was not found in digital cysts, apart from a possible peduncle (57). Magnetic resonance imaging is able to highlight satellite cysts (Figs. 9.32 and 9.33) or sagging multiloculated cysts (Fig. 9.30). These latter forms may be difficult to detect clinically, unless the typical signs of swelling and discharge of a thick fluid from the proximal nail fold are found. In these infrequent forms (22%), we could not detect a connection with the distal interphalangeal joint. These cysts may develop independently from the underlying

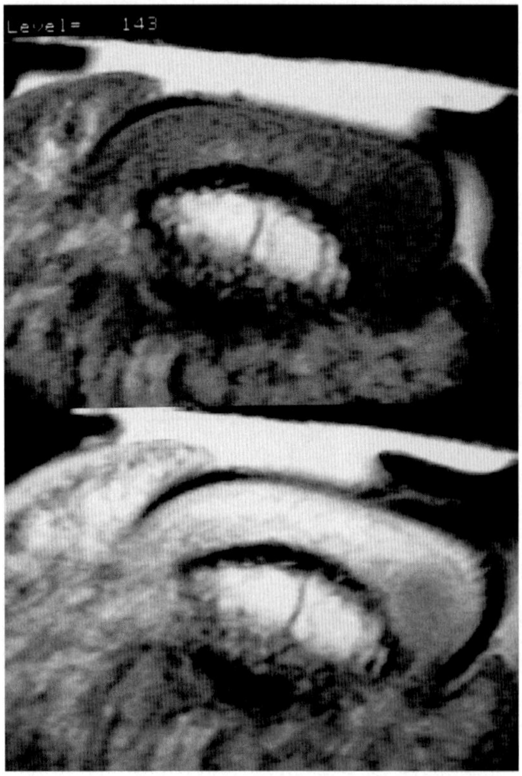

FIGURE 9.31. Painful subungual mucoid cyst. Axial T1-weighted spin-echo images before **(top)** and after **(bottom)** injection of gadolinium. The mass in the lateral part of the nail bed with a slight bone erosion may spontaneously evoke a glomus tumor. The thin and faint peripheral enhancement eliminates this hypothesis.

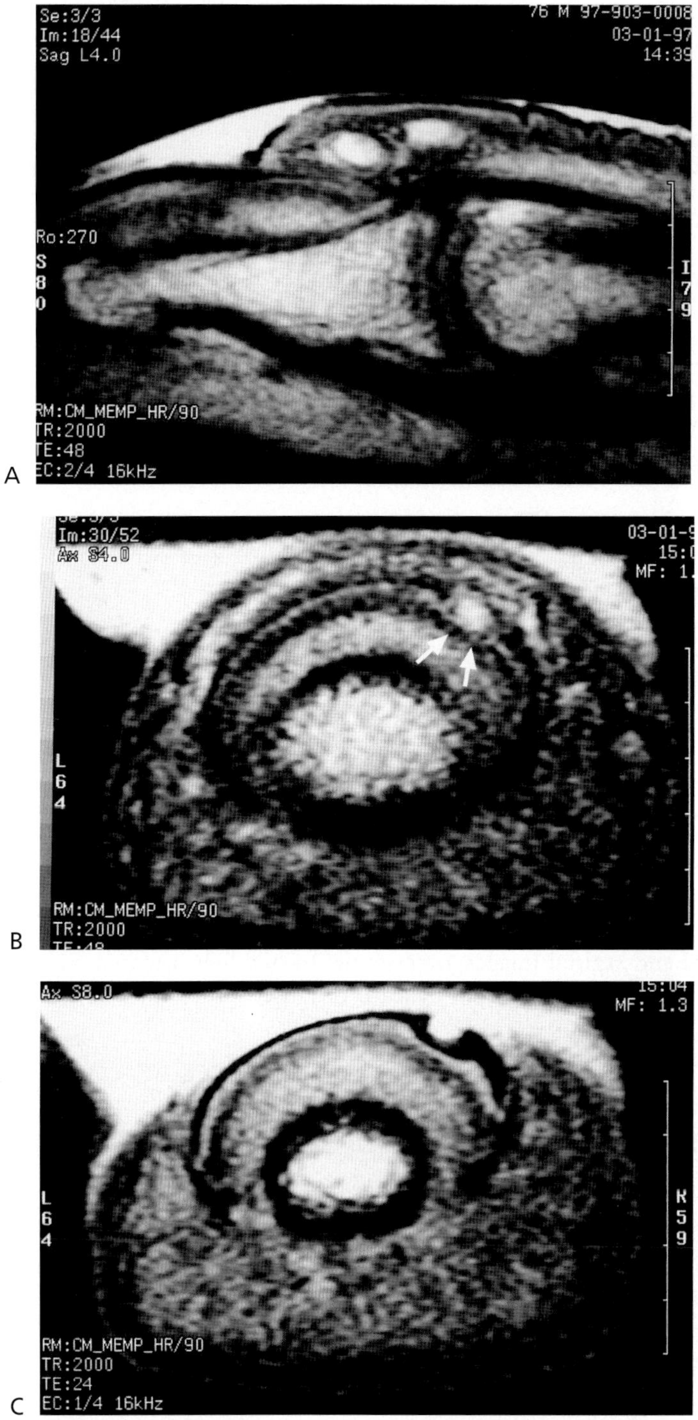

FIGURE 9.32. Multifocal mucoid cysts. **A:** Sagittal T2-weighted spin-echo image. Two cysts are depicted in the proximal nail fold close to the insertion of the extensor tendon. **B:** Axial T2-weighted spin-echo image shows the deformity of the nail matrix by the distal cyst *(arrow)*. **C:** A more distal axial image depicts the distal fissure of the nail plate in front of the matrix compression.

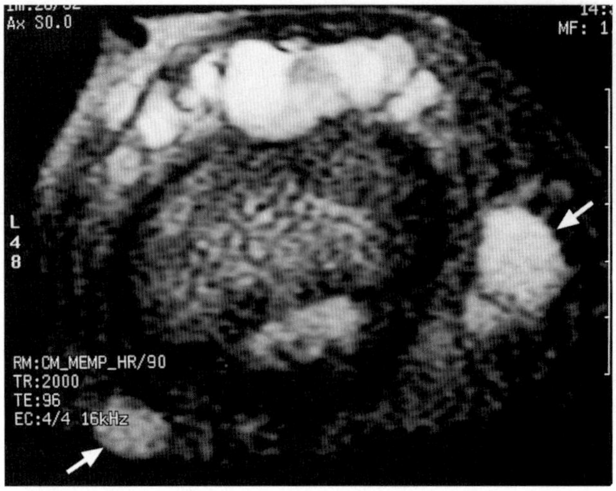

FIGURE 9.33. Multifocal mucoid cysts. Axial T2-weighted spin-echo image at the level of the distal interphalangeal joint. There are numerous cysts in the proximal nail fold. Note other cysts on the lateral parts of the pulp *(arrows).*

joint and result from increased production of hyaluronic acid due to the metaplasia of fibroblasts. This process may be compared to the cutaneous myxomas with a focal storage of mucoid material in the dermis (62,63,54).

Mucoid Cysts and Osteoarthritis

A peduncle connecting the cyst and the distal interphalangeal joint is visible by MRI in most cases (82%). Axial T2-weighted images and injection of gadolinium (peduncle enhancement) on axial and sagittal images are essential to highlight the peduncle (Fig. 9.34). In all cases, the peduncle is lateral, beneath the insertion of the extensor

A

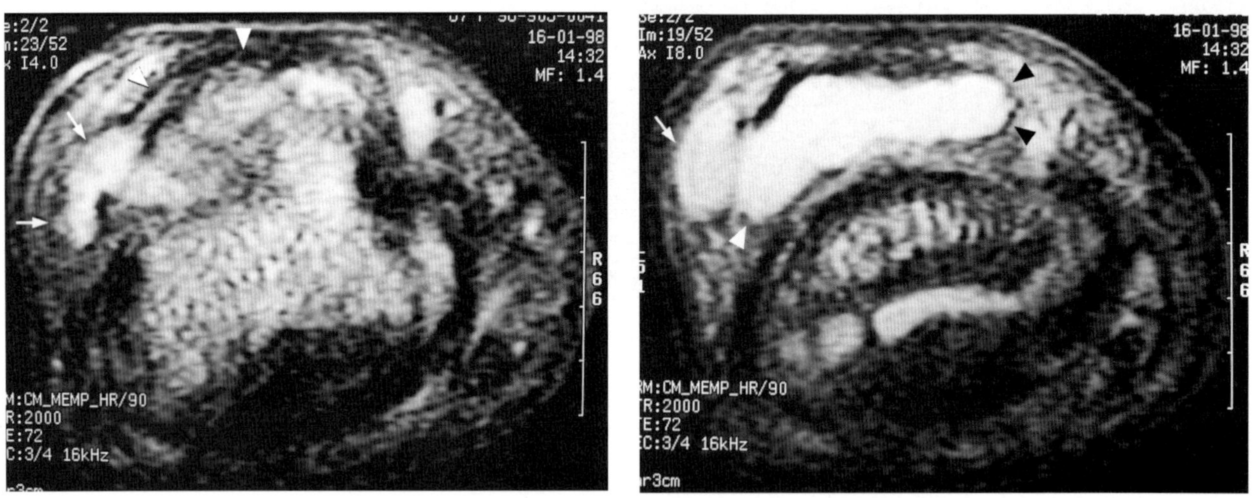

B

FIGURE 9.34. Peduncled mucoid cyst. **A:** Axial T2-weighted spin-echo image at the level of the head of the middle phalanx. A lateral peduncle *(arrow)* is depicted beneath the extensor tendon lifted by dorsal osteophytes *(arrowheads).* **B:** A more proximal slice shows a distended dorsal articular recess *(arrowheads)* and the nearby mucoid cyst *(arrow).*

digitorum tendon on the base of the distal phalanx. The cysts may be compared to true synovial cysts. When the surgical decision is made, the peduncle must be detected and tied up or removed to avoid frequent recurrences. The peroperative injection of methyl blue mixed with hydrogen peroxide into the palmar aspect of the distal interphalangeal joint has been proposed to color the peduncle, but its realization is not always easy (56).

Osteoarthritis of the distal interphalangeal joint is noted on radiographs in 70% of cases (50). Dorsal osteophytosis of the head of the middle phalanx may injure the insertion of the extensor digitorum tendon (Figs. 9.29 and 9.34). These cases present with a mucoid cyst with a peduncle. Even if osteoarthritis of the distal interphalangeal joint seems to be the initial lesion, mucoid cysts with peduncles may be noted without detectable osteoarthritis. The dorsal osteophyte must be removed at the same time as the cyst and the peduncle to avoid recurrence. Thickening and enhancement of the synovium of the distal interphalangeal joint is seen in half of the cases. This hyperplasia of the synovium associated with the osteophytes may contribute to development of Heberden's nodes (55).

Subungual Mucoid Cysts

Mucoid cysts extend into the nail bed in 30% of cases (50). This location has been poorly studied and rarely mentioned in the literature. High-resolution MRI alone is able to detect this type of cyst (Figs. 9.31 and 9.35). Symptoms may lead to misdiagnosis of glomus tumor when the cyst is painful. We previously showed the differential magnetic resonance findings. When the cyst is large, erosion of the cortex of the underlying phalanx may occur in the confined space of the nail bed (Fig. 9.36). The cyst is in the dermis beneath the nail matrix, close to the distal interphalangeal joint. Matrix compression may induce a fissure of the nail plate with a claw deformity. Most often, the cyst is bilobated with a component in the proximal nail fold (Fig. 9.37) and more rarely in the pulp, associated with the nail bed component (Fig. 9.38). The submatrical extension may be clinically occult and responsible for recurrence. Detection of a peduncle is crucial, because its resection may be enough to collapse the cyst and avoid direct access to the matrix.

A 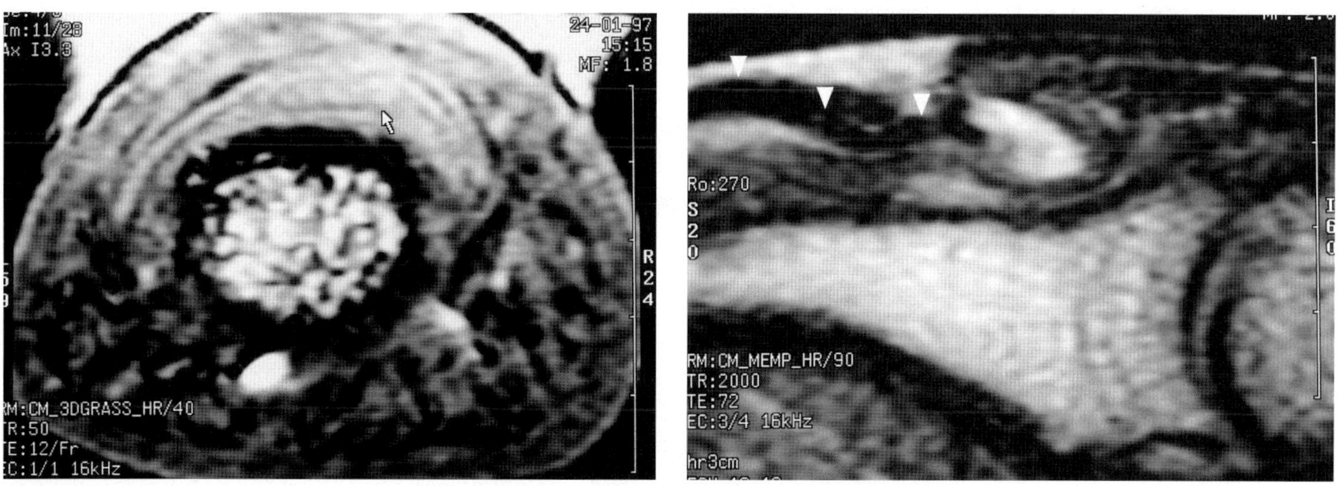 B

FIGURE 9.35. Small mucoid cyst of the nail matrix area. **A:** Axial three-dimensional gradient-echo image depicts a mass (*arrow*) on the midline of the nail matrix. The contrast with the surrounding tissues is poor. **B:** Sagittal T2-weighted spin-echo image shows the high signal of the mucoid cyst emerging from the nail matrix cul-de-sac. Note the distal nail dystrophy (*arrowheads*).

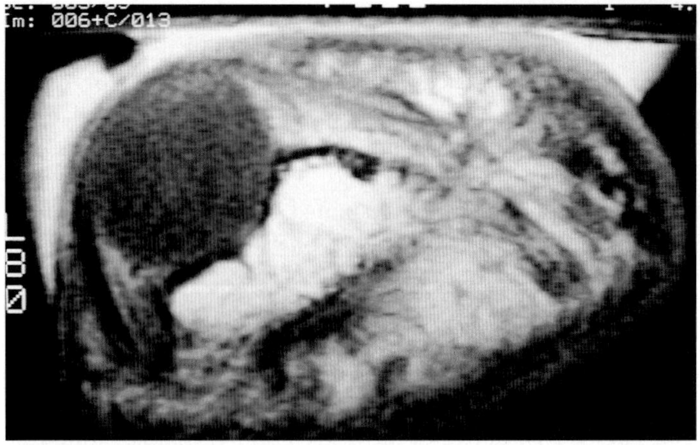

FIGURE 9.36. Subungual mucoid cyst. Axial T1-weighted spin-echo image shows a large cyst in the nail bed with deep bone erosion of the underlying phalanx.

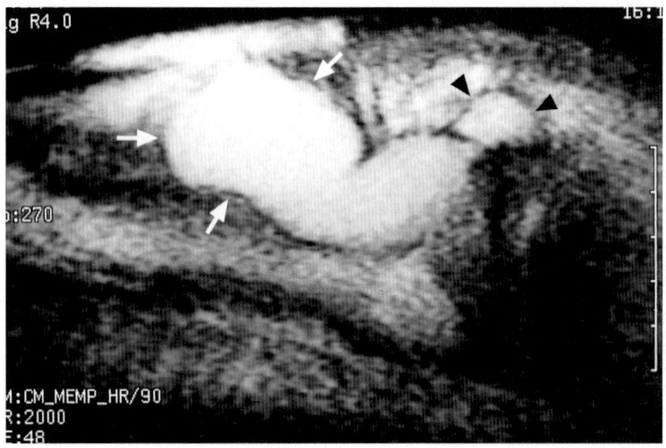

FIGURE 9.37. Bilobate mucoid cyst. Parasagittal T2-weighted image shows the continuity between the part of the cyst beneath the nail matrix *(arrows)* and the part in the posterior nail fold *(arrowheads)*.

A 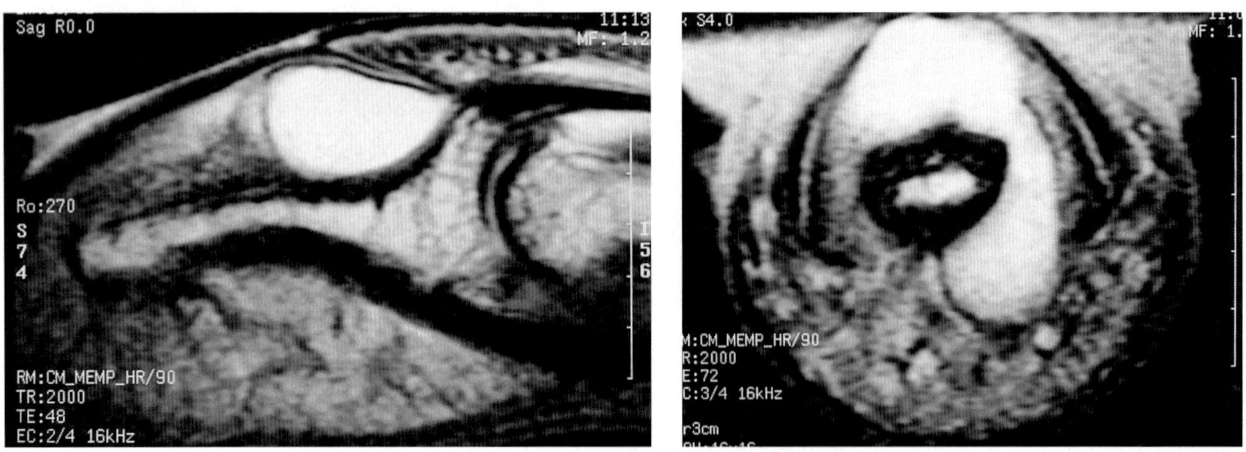 B

FIGURE 9.38. Bilobate mucoid cyst. **A:** Sagittal T2-weighted spin-echo image shows a large cyst beneath the lifted nail matrix. **B:** Axial T2-weighted spin-echo image depicts an extension of the subungual cyst toward the pulp via the rima ungualum.

Epithelial Tumors

Among the epithelial tumors, we only explored benign tumors such as periungual warts, epidermoid cysts, keratoacanthomas, and onychomatricomas.

The *warts* are induced by human papillomaviruses. They are benign fibroepithelial tumors, poorly contagious, with an irregular keratotic surface. Typically, the warts are periungual. The subungual locations, when they are painful, may mimic a glomus tumor, mainly because the long-drawn forms sometimes induce bone erosion (64). Warts of the hyponychium may be subtle and go unnoticed. Histologic examination may be necessary to distinguish it from verrucous Bowen's disease and early squamous cell carcinoma.

Magnetic resonance imaging easily estimates the thickening of the epidermis in depth and surface, particularly in the subungual area and at the level of the hyponychium. The signal is identical to that of normal epidermis on all sequences. A thick superficial layer with lack of signal may cover the thickened epidermis and corresponds to the corneous layer or to aftereffects (cryotherapy, laser?) (Fig. 9.39).

Keratoacanthoma is a rare, benign, but rapidly growing tumor located in the most distal part of the nail bed (65). The lesion may start as a small and painful keratotic nodule beneath the free edge.

Magnetic resonance imaging shows a dome-shaped nodule with a homogeneous signal (intermediate T1, hyper T2) and strong enhancement. A central area of low signal may indicate a central plug of horny material filling the crater, but this is inconstant (Fig.

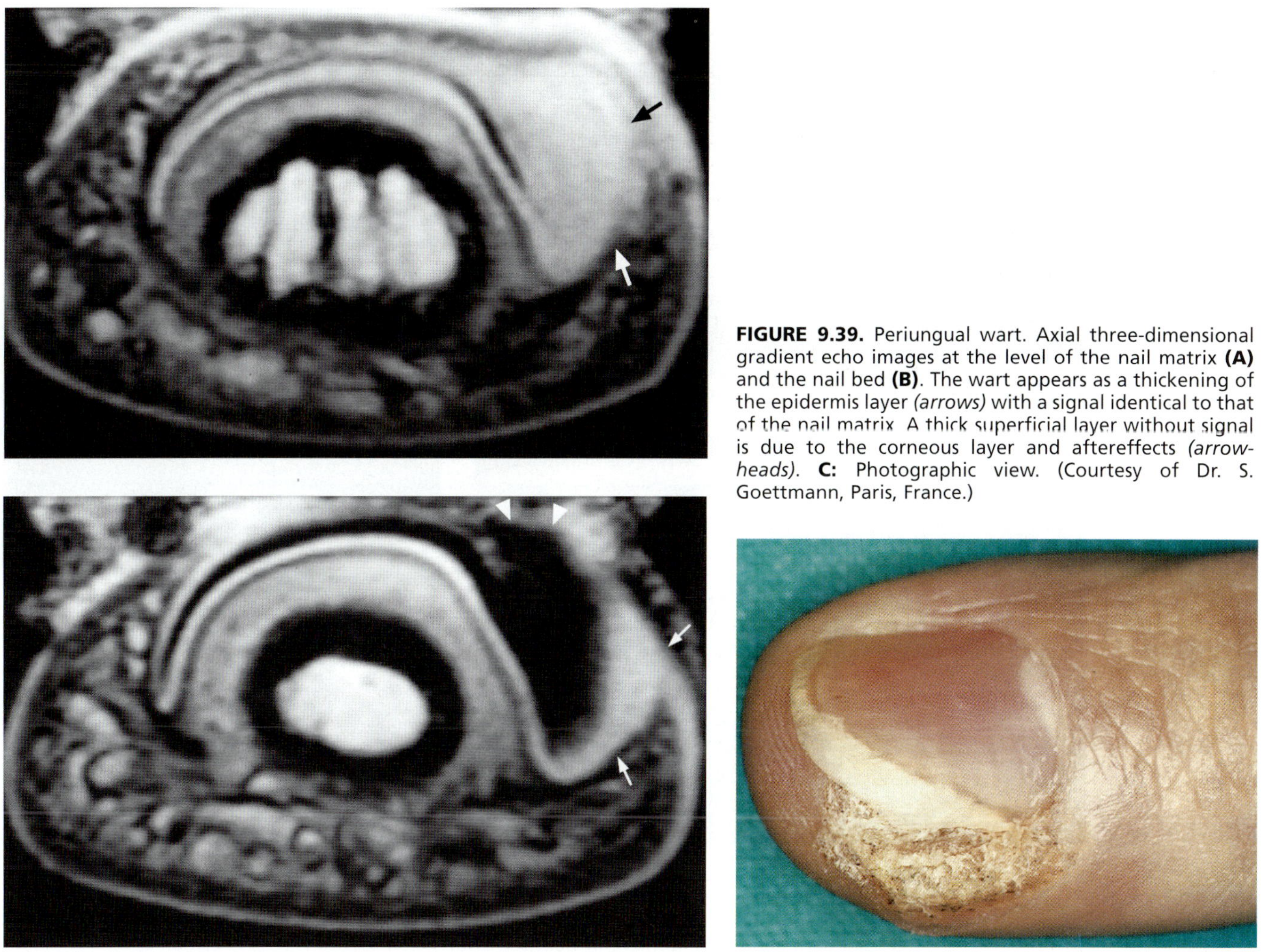

A

B

C

FIGURE 9.39. Periungual wart. Axial three-dimensional gradient echo images at the level of the nail matrix **(A)** and the nail bed **(B)**. The wart appears as a thickening of the epidermis layer *(arrows)* with a signal identical to that of the nail matrix. A thick superficial layer without signal is due to the corneous layer and aftereffects *(arrowheads)*. **C:** Photographic view. (Courtesy of Dr. S. Goettmann, Paris, France.)

9.40). The limits may be ill defined due to edema in the surrounding tissues. Magnetic resonance images show a deep infiltrating lesion and detect more accurately than radiographs a frequent bone erosion of the distal phalanx (Fig. 9.5A).

Epidermoid cysts of the distal phalanx are rare, usually secondary to a trauma with implantation of epidermis into subcutaneous tissue or even into the bone. An old trauma often goes unnoticed. The cyst may develop on a scar after surgery (66). The phalanx progressively expands and clubbing becomes obvious. The pain is of late onset, sometimes on the occasion of a pathologic fracture. Histology shows a simple epidermoid cyst filled with orthokeratin and lined with a thin layer of epidermis.

Radiographs depict a round, accurately rimmed erosion of the distal phalanx without septa or peripheral sclerosis (67). Early, the bone erosion is absent or subtle and is not visible on radiographs. Magnetic resonance imaging shows a regular mass with slightly heterogeneous content and intermediate signal on T1-weighted and T2-weighted images. Heterogeneous enhancement is noted after injection of gadolinium (Fig. 9.41). The thin epidermal layer is depicted as a regular rim with high signal identical to that of normal epidermis (Fig. 9.42). Bone erosions, even subtle, are highly detected on axial images. The area of an old penetrating injury may be marked by dark artifacts on gradient-echo images (Fig. 9. 43). In contrast, subungual epidermoid inclusions develop from the epidermis ridges and are not visible with MRI, as they usually remain microscopic (58,68).

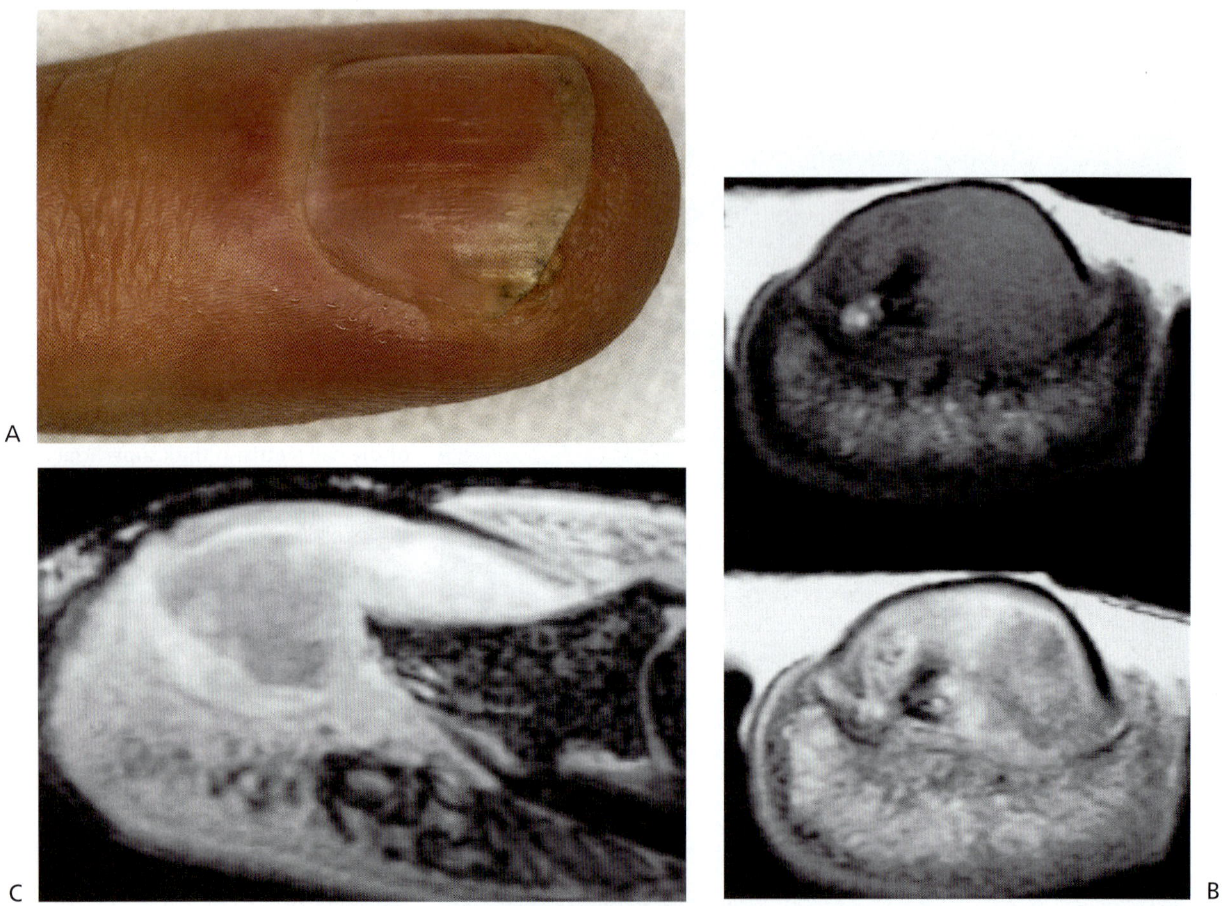

FIGURE 9.40. Keratoacanthoma. **A:** Photographic view. **B:** Axial T1-weighted spin-echo images before **(top)** and after **(bottom)** injection of gadolinium. **C:** Postenhanced sagittal three-dimensional gradient-echo image. A distal mass with a central core of low signal and a high and irregular peripheral enhancement invades the distal phalanx with extensive osteolysis.

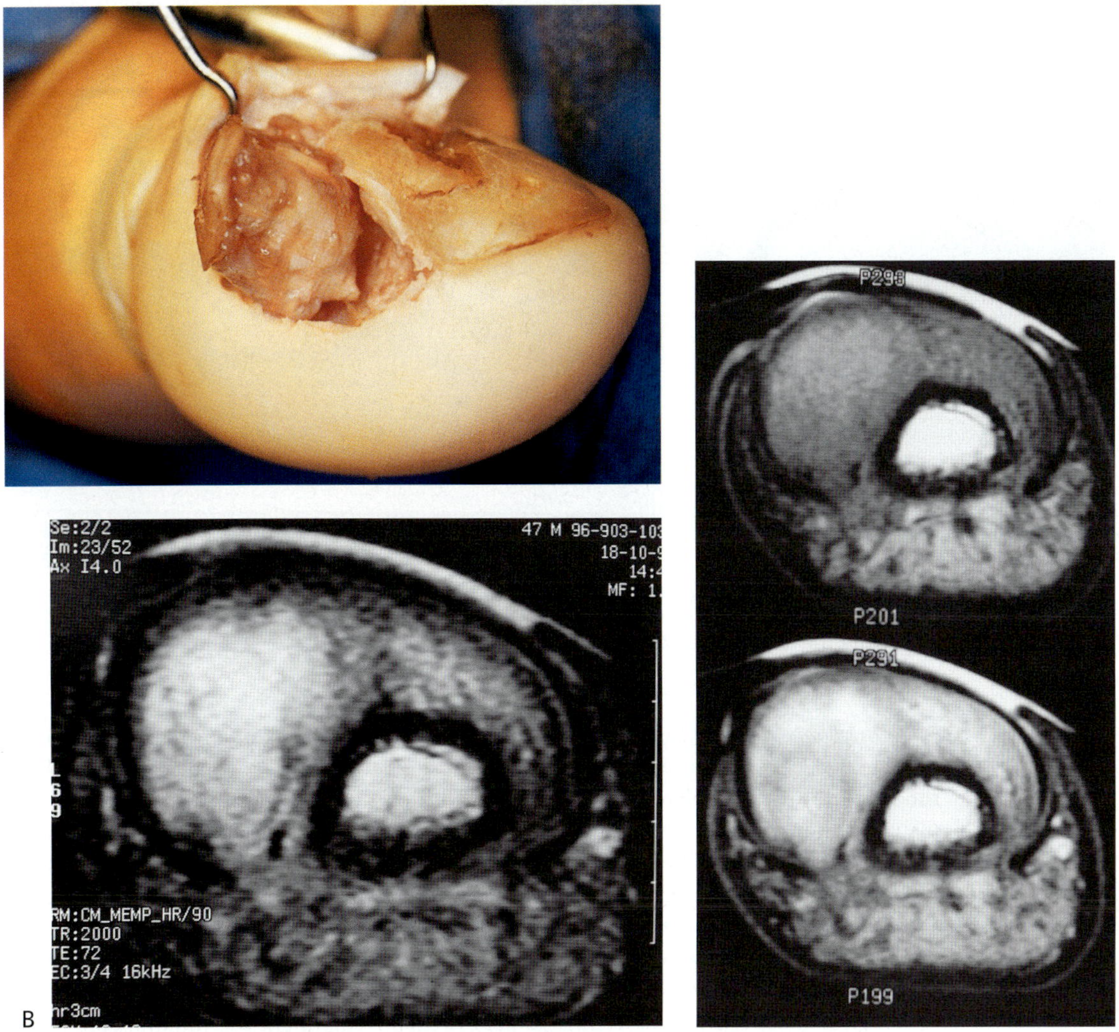

FIGURE 9.41. Epidermoid cyst. **A:** Surgical view. (Courtesy of Dr. S. Goettmann, Paris, France.) **B:** Axial T2-weighted spin-echo image. The cysts has a signal faintly higher than that of the nail bed. **C:** Axial T1-weighted spin-echo images before **(top)** and after **(bottom)** injection of gadolinium. The cyst locates in the lateral part of the nail bed and presents an heterogeneous peripheral and central enhancement.

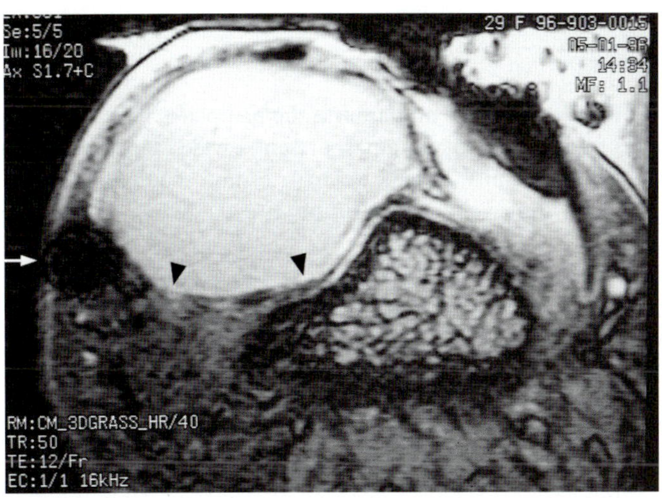

FIGURE 9.42. Epidermoid cyst. Axial postenhanced three-dimensional gradient-echo image. A large cyst lifts up the nail matrix. A thin epidermal layer surrounds the cyst *(arrowheads)*. Note the bone erosion and the artifacts of a previous penetrating trauma *(arrow)*.

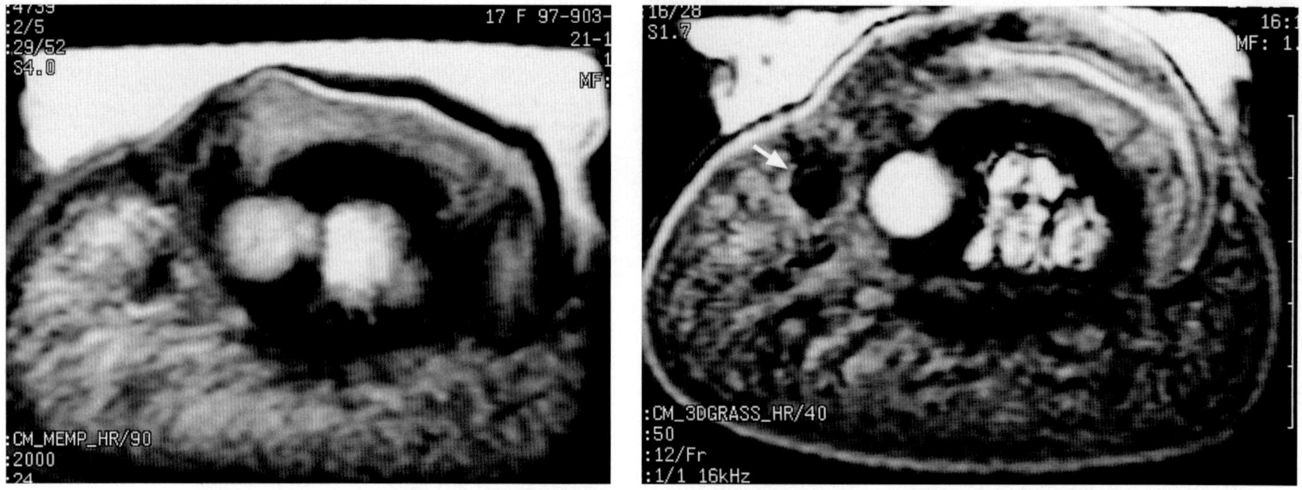

A B

FIGURE 9.43. Epidermoid cyst. Axial T1-weighted spin image **(A)** and axial three-dimensional gradient-echo image **(B)** show a well-defined round cyst in the lateral cortex of the distal phalanx. Note the artifacts due to a previous surgery *(arrow)*.

Onychomatricomas (filamentous tufted tumors in the matrix of a funnel-shaped nail) must be suspected based on four main clinical signs (69): (i) yellow coloration all along the length of the nail plate with a variable width, (ii) prominent appearance of the nail plate, (iii) transverse overcurvature of the plate, and (iv) tumor emerging from the matrix visible after avulsion of the plate. The nail plate appears as a flattened tunnel in which the filamentous expansions grow. Histology shows epithelial proliferation of the matrix or surrounding epidermis. The lobules are delimited by normal basal cells and are composed of keratinocytes identical to those of the matrix. After removal of these parakeratotic cells, it remains as an invagination resembling the infundibulum of a hair follicle.

Sagittal magnetic resonance images are essential to highlight the tumoral core in the matrical area and the invagination of the lesion into the funnel-shaped nail plate (Fig. 9.44A). The center shows a low signal on all images, with a peripheral rim with a signal identical to that of normal epidermis (Fig. 9.44B). The distal part with the filamentous extensions presents a higher signal on T2-weighted images due to a mucoid stroma. Axial slices accurately show the holes in the substance of the nail plate, filled with the filamentous extensions (Fig. 9.44C).

Tumors of the Soft Tissues

Numerous fibrous tumors can develop beneath and around the nail plate. These tumors, ranging from dermatofibromas to digital fibrokeratomas, present a wide range of clinical patterns despite relative histologic uniformity. In fact, Koenen's tumor, acquired fibrokeratoma, and dermatofibroma present some "clinical continuity."

Acquired Periungual Fibrokeratomas

In our series, we mainly explored acquired periungual fibrokeratomas. These asymptomatic benign nodules spontaneously develop with a hyperkeratotic tip and a narrow base in the periungual area or elsewhere on the fingers. They may be multiple and located beneath the nail plate, but most of them emerge from the proximal nail fold and grow in a splint of the nail plate (70). A traumatic origin often is mentioned. Biopsy is the rule, as Bowen's disease may simulate a fibrokeratoma (71). Three histologic types may present different magnetic resonance findings: one type is composed of thick and dense bundles of collagen, a second type includes an increased amount of fibroblasts in the dermis, and the third type has an edematous component and a poor amount of cells. Fibrokeratoma is thought to consist of newly formed collagen.

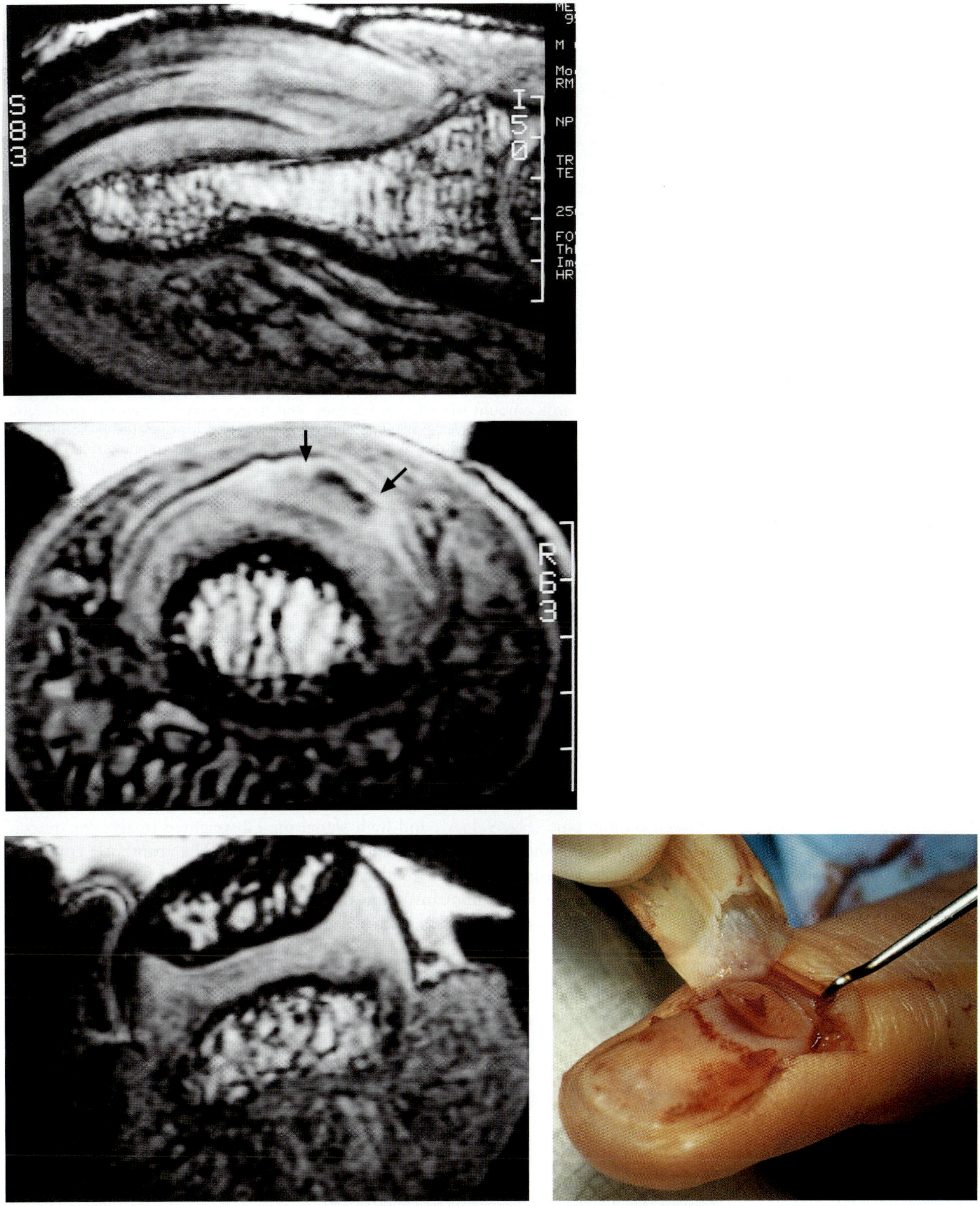

A B C D

FIGURE 9.44. Onychomatricoma. **A:** Sagittal three-dimensional gradient-echo image shows a tumor emerging from the nail matrix and invaginating into a funnel-shaped nail plate. **B:** Axial three-dimensional gradient-echo image. The core of the tumor *(arrows)* is inserted in the nail matrix with a central low signal. **C:** More distal axial three-dimensional gradient-echo image. High-signal filamentous expansions into the substance of the nail plate. **D:** Surgical view. (Courtesy of Dr. S. Goettmann, Paris, France.)

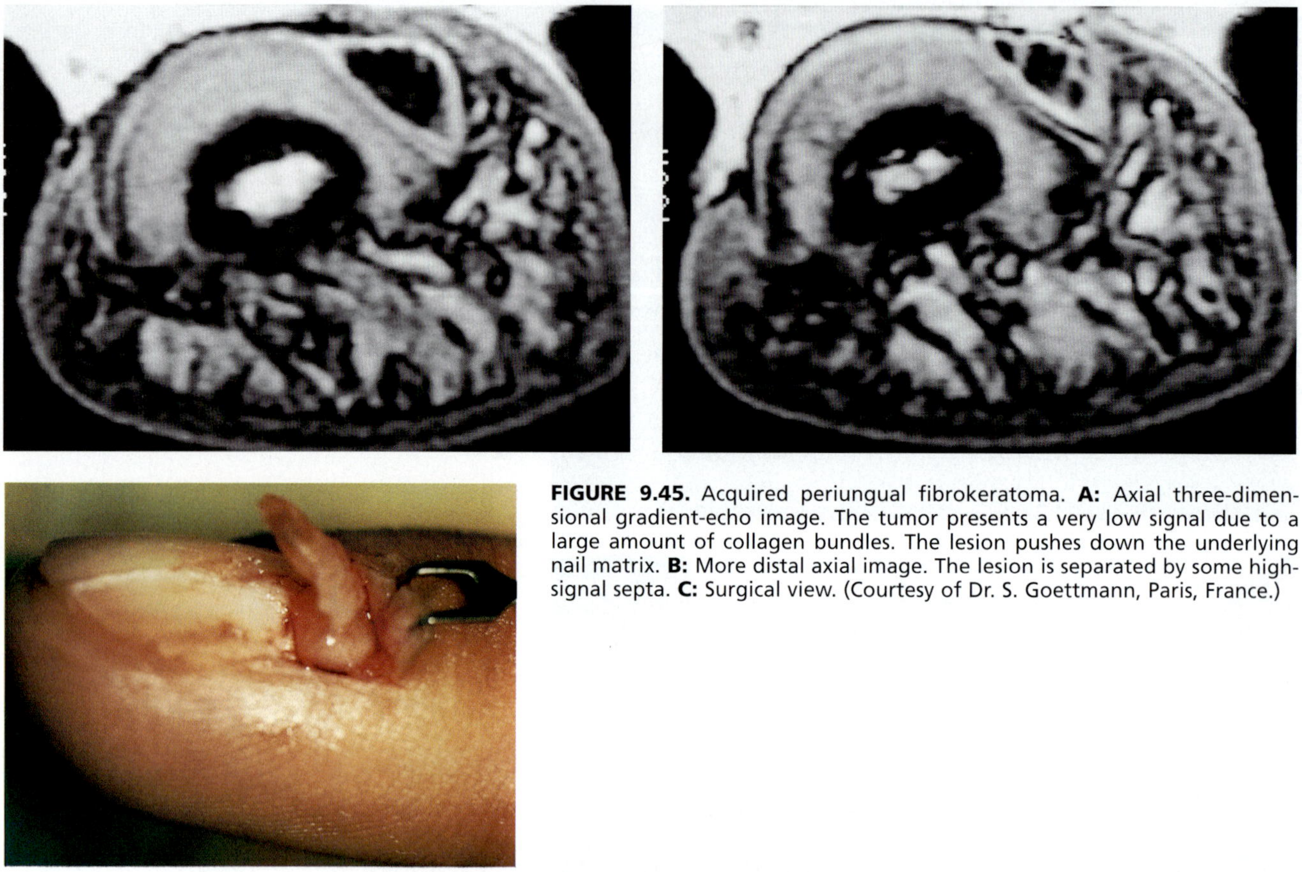

A B C

FIGURE 9.45. Acquired periungual fibrokeratoma. **A:** Axial three-dimensional gradient-echo image. The tumor presents a very low signal due to a large amount of collagen bundles. The lesion pushes down the underlying nail matrix. **B:** More distal axial image. The lesion is separated by some high-signal septa. **C:** Surgical view. (Courtesy of Dr. S. Goettmann, Paris, France.)

Magnetic resonance imaging accurately depicts the component emerging from the proximal nail fold in the splint of the nail plate. Overall, magnetic resonance images highlight the deep implantation close to the nail root. The signal of the tumor depends on the different histologic types: very low signal on all sequences for the dense and numerous collagen bundle type (Fig. 9.45A), and high signal on T2-weighted images for mucoid stroma (Fig. 9.46). The lesion may be septated (Fig. 9.45B). The acanthotic epidermic coverage presents a signal identical to that of normal epidermis. Magnetic resonance imaging is able to depict a tumor involving the ventral aspect of the proximal nail fold with an epithelial invagination. This invagination acts as an incidental matrix and produces a pseudonail of collagen.

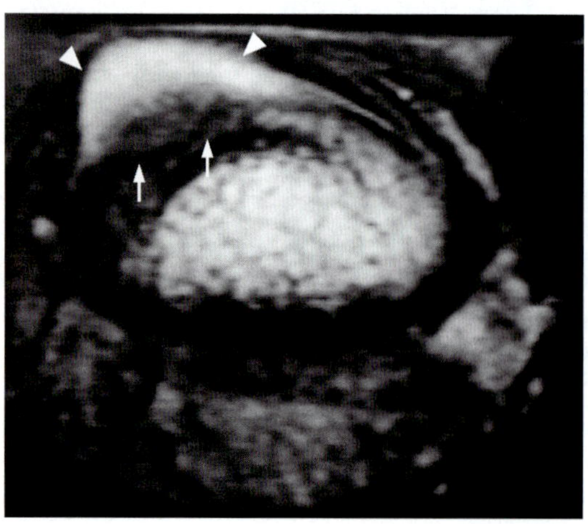

FIGURE 9.46. Acquired periungual fibrokeratoma. Axial T2-weighted spin-echo image. The low-signal deep component *(arrows)* bends the nail matrix and is associated with a high-signal superficial component of mucoid stroma *(arrowheads)*.

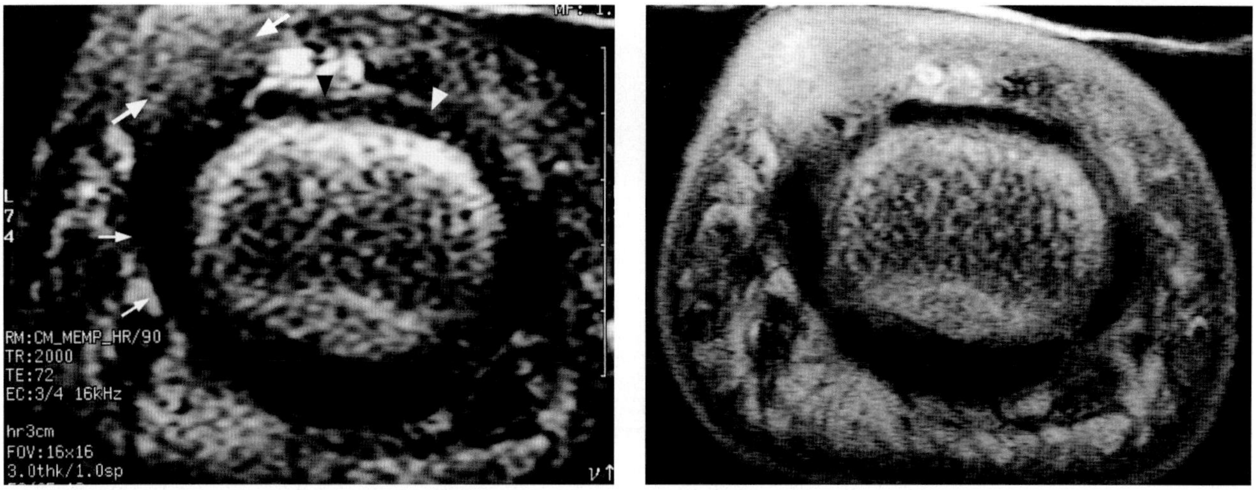

A B

FIGURE 9.47. Knuckle pad. **A:** Axial T2-weighted spin-echo image. Low-signal nodule *(arrows)* on the dorsolateral aspect of the interphalangeal joint. The lesion is close to the extensor tendon *(arrowheads)* and the ulnar lateral ligament *(small arrows)*. **B:** Postenhanced axial T1-weighted spin-echo image. Heterogeneous enhancement of the lesion with a faintly enhanced medial part.

Knuckle Pads

Knuckle pads are asymptomatic, persistent, and keratotic nodular plaques that involve the dorsal aspect of the interphalangeal joints. Histology reveals hyperkeratosis associated with thickened collagen bundles.

Magnetic resonance images accurately show the relations between the extensor apparatus and the plaque. The signal behavior is not specific, but is common to other fibrous lesions rich in collagen, with rather low and heterogeneous signal on T1-weighted and T2-weighted images (Figs. 9.47 and 9.48).

Fibromas

Fibromas are painless slow-growing nodules that can develop in any epidermal structure of the nail unit. Clinical features vary according to the anatomic site of the lesion. The nail may be lifted in the distal subungual area. Histology shows very dense connective tissue bundles with elastic fibers.

Radiographs may depict bone erosion and thickening of the soft tissues. There are no calcifications. Findings on MRI are similar to those of knuckle pads, with a low signal lesion on all sequences and haphazard very dark areas of extremely dense connective bundles. These patterns easily differentiate them from acquired fibrokeratomas.

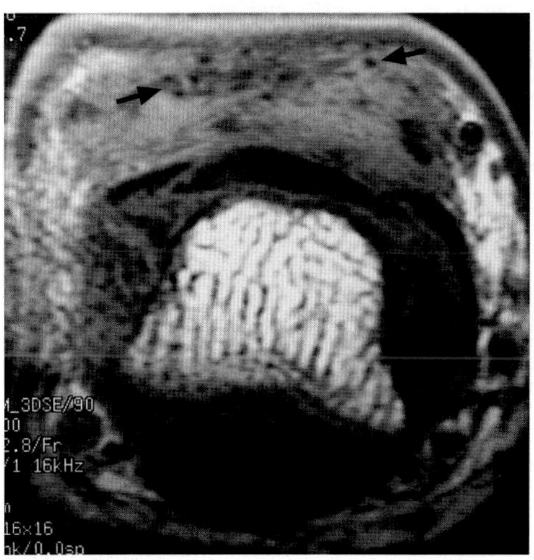

FIGURE 9.48. Knuckle pad. Axial postenhanced T1-weighted spin-echo image. Note mottled areas of low signal *(arrows)* in the dorsal part of the tumor due to the thickened collagen bundles.

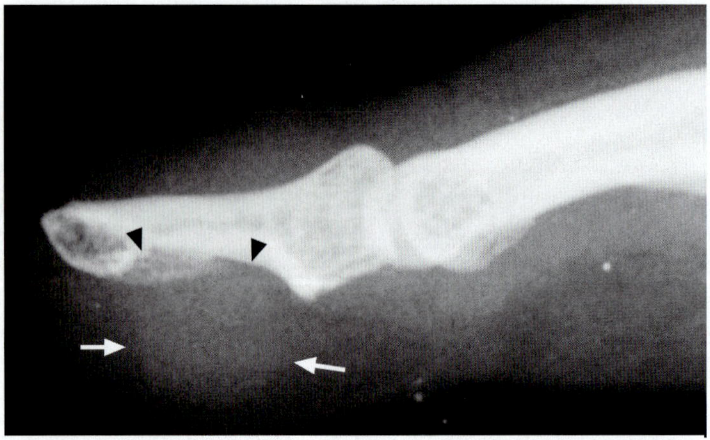

FIGURE 9.49. Cavernous hemangioma. Lateral radiograph. Note round mass in the palmar soft tissues *(arrows)* with a bone erosion of the palmar aspect of the distal phalanx *(arrowheads)*.

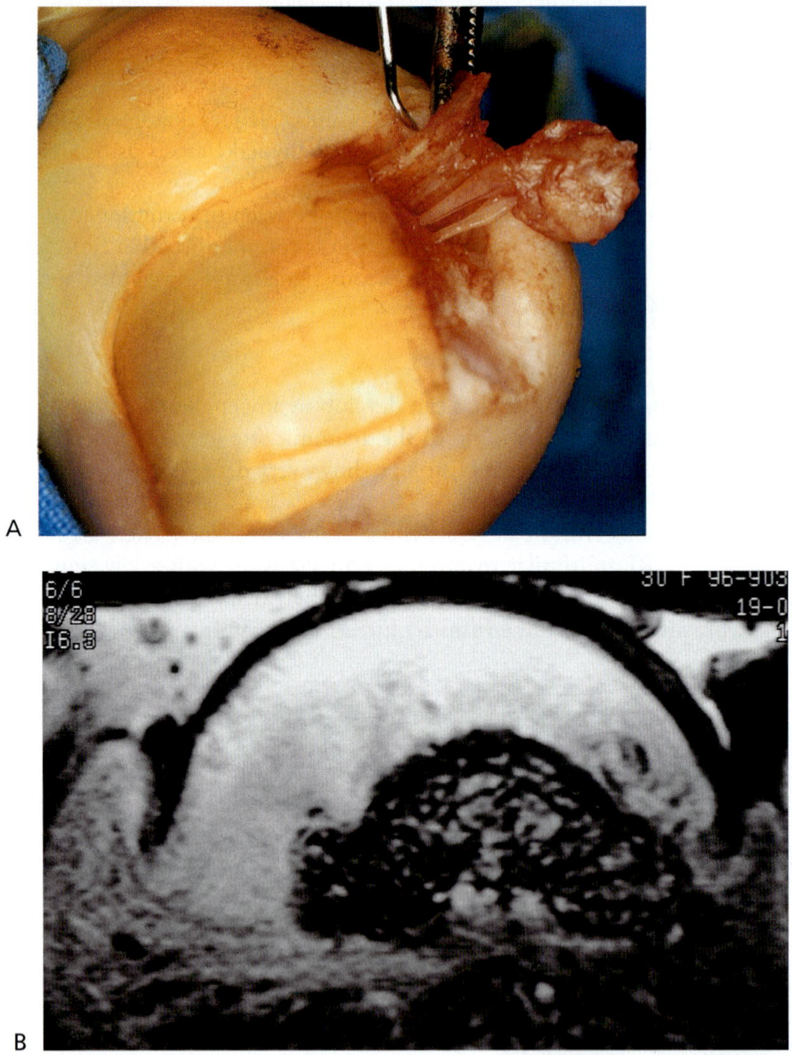

FIGURE 9.50. Botriomycoma. **A:** Surgical view. (Courtesy of Dr. S. Goettmann, Paris, France.) **B:** Axial postenhanced gradient-echo image shows a ill-defined thickening of the lateral part of the nail bed without a true mass. The inflammatory reaction blurs the limits of the lesion.

Vascular Tumors

As we have seen, the vascular lesions are mainly represented by glomus tumors. Most of the vascular tumors that involve the nail unit are benign, except for Kaposi's sarcoma. The benign vascular tumors include numerous histologic types, such as the exceptional hemangiomas of the nail bed and the capillary malformations present since birth. Radiographs can depict a mass in the soft tissues, phleboliths, and even bone erosion with venous and arteriovenous malformations, or an epithelioid hemangioendothelioma (Fig. 9.49). The bone may be primarily involved by an hemangioma (linear striations parallel to the shaft of the phalanx) or an aneurysmal bone cyst (expansive osteolytic lesion of the phalanx). The botriomycoma, little inflammatory angiomatous tumor, is the most frequent vascular lesion. It is caused by an excoriation or a prick and may be confused with an achromic melanoma.

Magnetic resonance imaging is not able to differentiate these different types of vascular tumors and may be misleading without injection of contrast media (Fig. 9.50). On the other hand, the vascular nature often is obvious (flow void artifacts, signal of hemorrhage, and high enhancement after injection of gadolinium) (Fig. 9.51). Magnetic resonance imaging also assesses the extension of the lesion into the soft tissues and the relations with the digital collateral vessels (Fig. 9.52). The original location, soft tissues or bone, is easily depicted on by MRI. Cystic blood-filled spaces with horizontal levels may be encountered in numerous lesions, particularly in aneurysmal bone cyst.

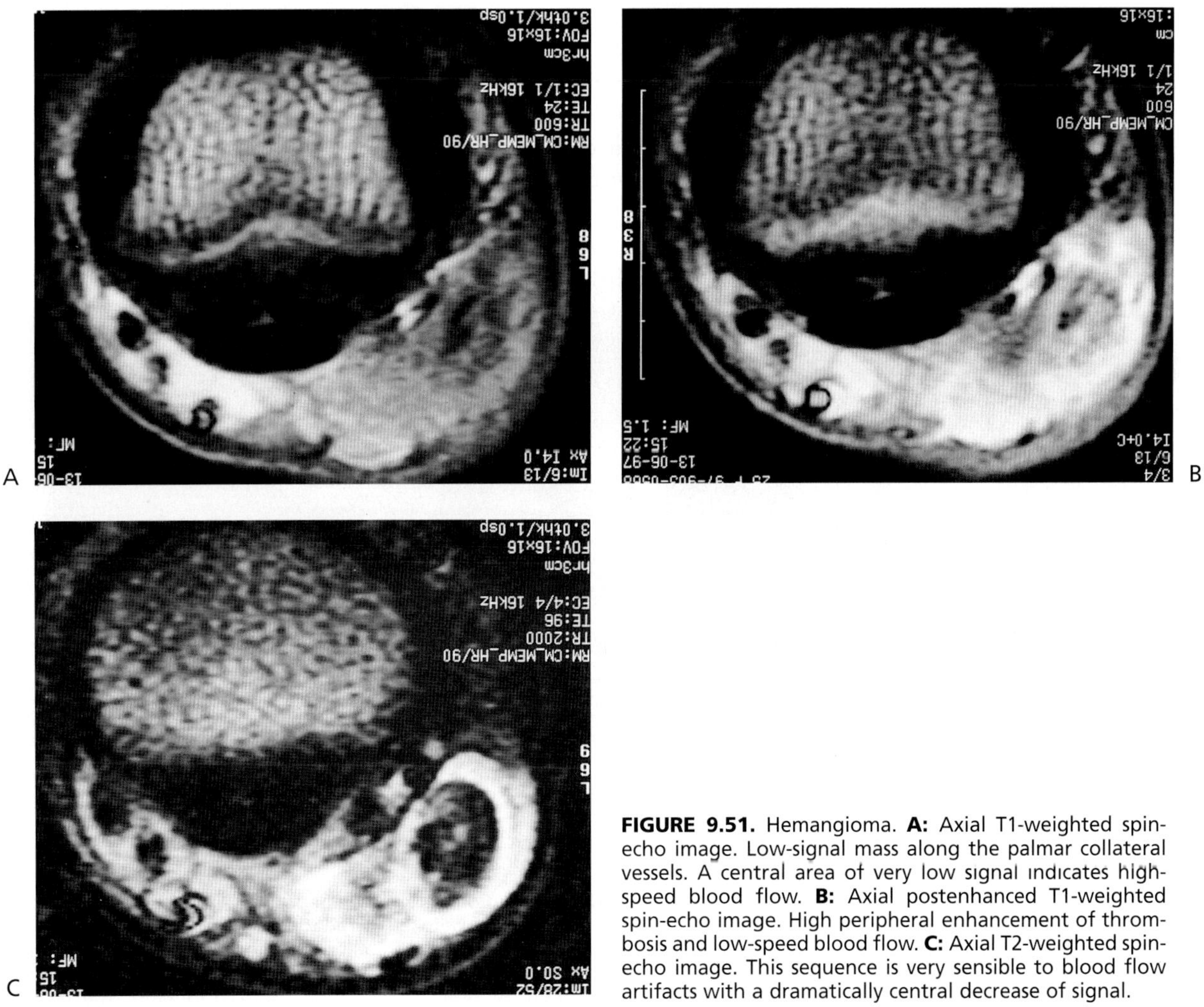

FIGURE 9.51. Hemangioma. **A:** Axial T1-weighted spin-echo image. Low-signal mass along the palmar collateral vessels. A central area of very low signal indicates high-speed blood flow. **B:** Axial postenhanced T1-weighted spin-echo image. High peripheral enhancement of thrombosis and low-speed blood flow. **C:** Axial T2-weighted spin-echo image. This sequence is very sensible to blood flow artifacts with a dramatically central decrease of signal.

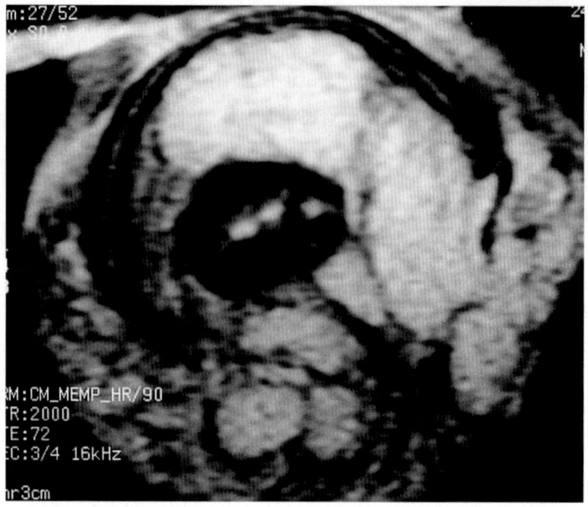

FIGURE 9.52. Hemangioma of the nail bed. Axial T2-weighted spin-echo image. Note numerous high-signal blood-filled cavities in the nail bed, extending toward the lateral nail fold and the pulp via the rima ungualum.

Neuromas

Neuromas develop from the numerous nerve fibers in the nail bed and in the pulp after an injury or repeated microtraumas.

Magnetic resonance imaging can detect a small, well-circumscribed nodule in the nail bed or pulp. The signal of the lesion is low to intermediate on T1-weighted and T2-weighted images. Usually there is no or only faint enhancement after injection of gadolinium. This enhancement is increased by fat suppression (Fig. 9.53).

Schwannomas

Schwannomas rarely occur in the subungual area. The lesions have sharp limits and show strong enhancement of the whole or a part of the tumor. A cystic component is possible. The signal is very high on T2-weighted images (Fig. 9.54).

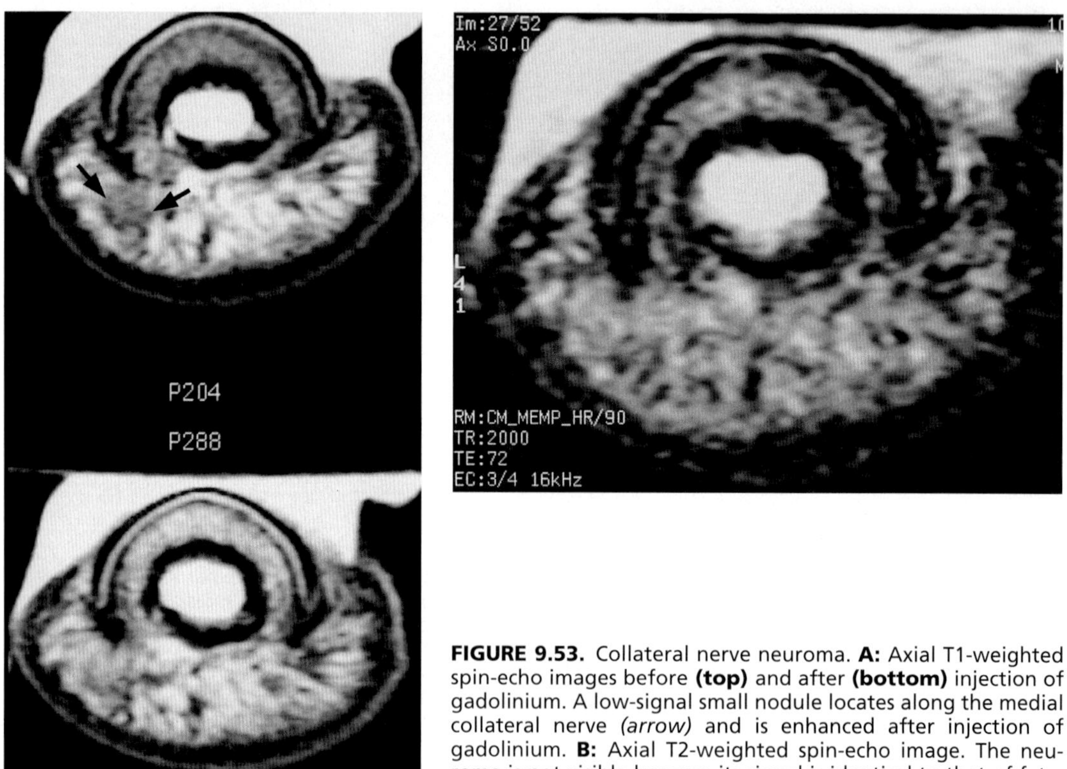

A B

FIGURE 9.53. Collateral nerve neuroma. **A:** Axial T1-weighted spin-echo images before **(top)** and after **(bottom)** injection of gadolinium. A low-signal small nodule locates along the medial collateral nerve *(arrow)* and is enhanced after injection of gadolinium. **B:** Axial T2-weighted spin-echo image. The neuroma is not visible because its signal is identical to that of fat.

A 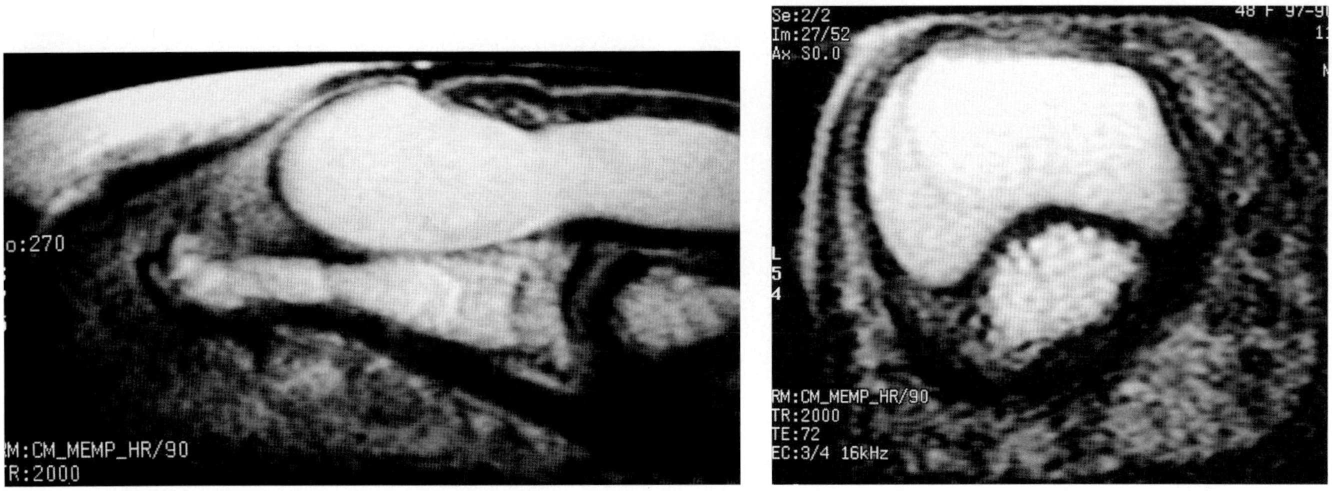 B

FIGURE 9.54. Schwannoma. Sagittal **(A)** and axial **(B)** T2-weighted spin-echo images. The large lesion is well defined and presents a high signal on T2-weighted images. The tumor is located in the posterior nail fold and in the nail bed.

Giant Cell Tumors

Giant cell tumors originate from the tendon sheaths or the articular synovium. They are the second most frequent tumors of the soft tissues in the hand. This lesion occurs frequently in women. At the level of the fingers, the tumor most often is solitary and on the dorsal aspect of the interphalangeal joint. These tumors rarely involve the nail unit. There are no calcifications, as opposed to synovial sarcomas. Histology shows the patterns of villar pigmented synovitis with histiocytes and fibroblasts, and a various amount of giant cells.

Giant cell tumors may extend into the distal phalanx, with bone erosion visible on radiographs and even more so on MRI (Fig. 9.55). The osseous lesion is expansive in one-third of cases, sometimes with a pathologic fracture (Fig. 9.56). The lesions are multiple at the level of the hand in 18% of cases (72). Hemosiderin deposits can lead to a brownish coloration of the tumor and typical signal void artifacts on T2-weighted images. The hemosiderin deposits are less often visible in the fingers than in the knee (Fig. 9.56). However, an intermediate-to-low signal of a part of the tumor is very frequent and must

A 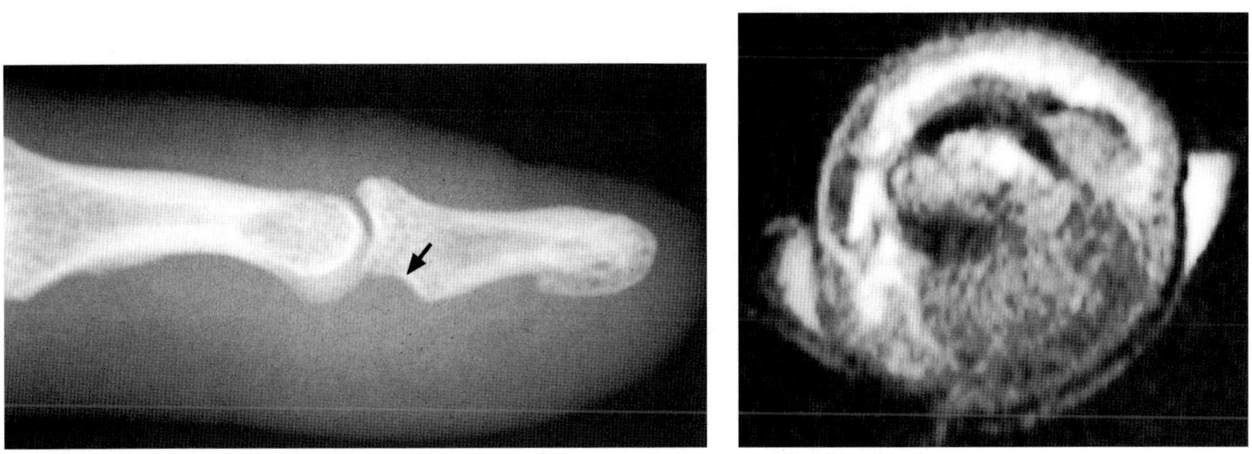 B

FIGURE 9.55. Giant cell tumor of the tendon sheaths. **A:** Lateral radiograph. There is thickening of the palmar soft tissues and bone erosion of the palmar aspect of the base of the phalanx *(arrow)*. **B:** Axial T1-weighted spin-echo image. Note intermediate signal mass along the insertion of the flexor tendon. A second nodule is depicted along the lateral insertion of the extensor tendon.

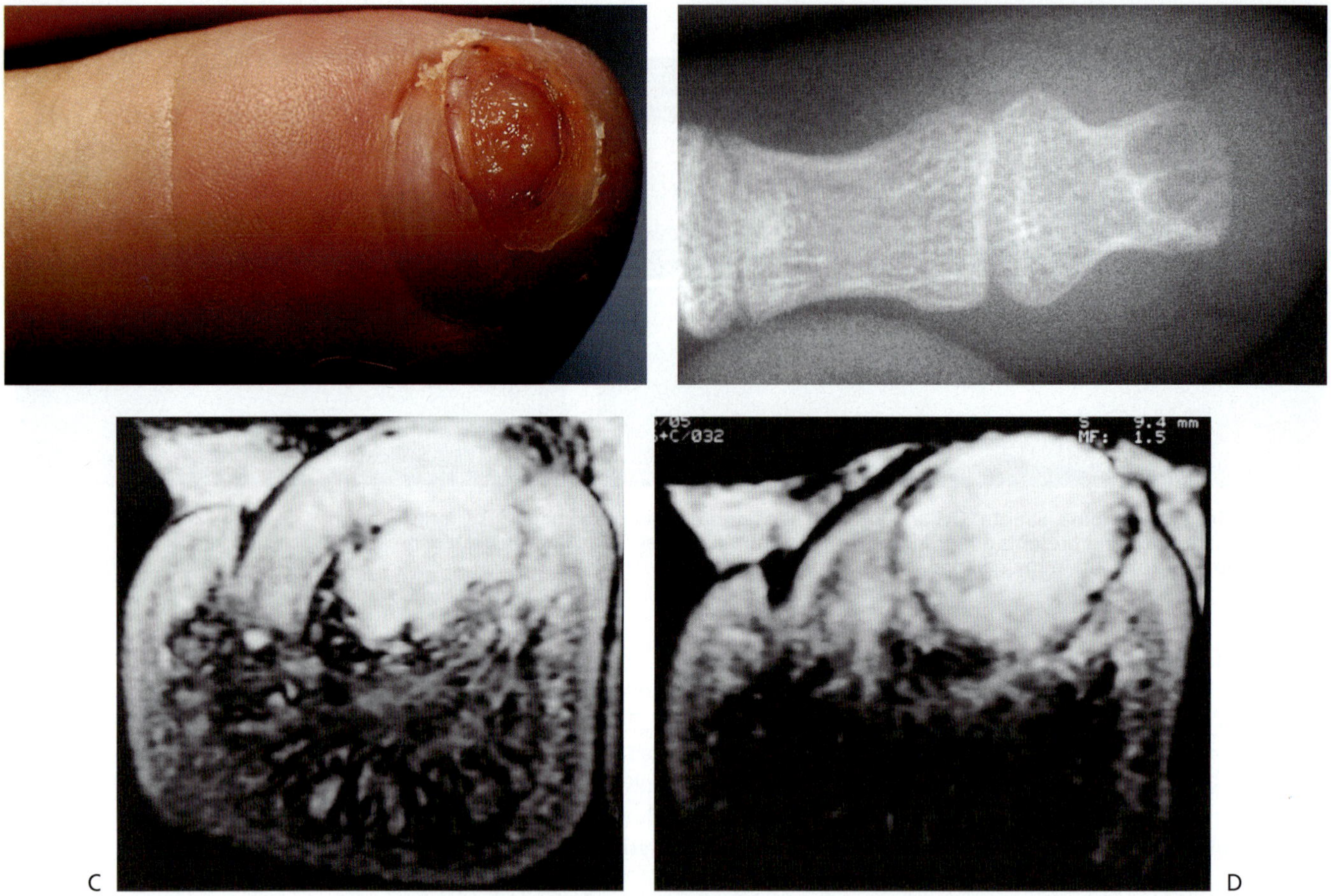

FIGURE 9.56. Giant cell tumor of the distal phalanx. **A:** Surgical view. (Courtesy of Dr. S. Goettmann, Paris, France.) **B:** Posteroanterior radiograph shows a well-circumscribed lytic bone lesion with a sclerotic rim and septa. The distal cortex is ruptured. **C:** Axial postenhanced three-dimensional gradient-echo image. High-signal tumor emerging from the bone and invading the nail bed. **D:** More distal axial slice. Peripheral areas of low signal are compatible with hemosiderin deposits.

evoke the diagnosis. The close relations of the lesion with a tendon sheath or an articular synovium is the second main MRI pattern of giant cell tumor (Fig. 9.55). In contrast, the signal enhancement after the injection of gadolinium is not specific.

Osteocartilaginous Tumors

Exostoses and Osteochondromas

Benign tumors by far are the most frequent and include mainly subungual exostoses and osteochondromas. The distinction between these two lesions is vague and debated. According to Norton (73), exostoses have a peduncle and a fibrocartilaginous cap, whereas osteochondromas are sessile with a hyaline cartilage cap. Lemont and Christman (74) suggest a new classification of the subungual exostoses into genetic and acquired types. This classification is based on histology, radiology, tumor location, and patient age. Genetic subungual exostoses of type I appear in patients between ages 20 to 40 of years. They often present with paronychia and enlargement limited to the medial side of the nail bed. Acquired subungual exostoses of type II are noted in patients between 50 and 70 of years. They show middorsal blistering of the distal nail bed, without a cartilaginous cap. The nail plate presents an inverted U shape. Initially, the exostosis is composed of fibrous tissue in the nail bed with a cartilaginous metaplasia, followed by ossification and secondary linking with the distal phalanx. This type II lesion is comparable to subungual hyperostosis, which is noted often in claw deformities. Traumatisms are an essential etiopathologic condition. These lesions mainly involve the big toenail (75). They present as a pink mass with a fibrous or fibrokeratotic appearance. The nail plate is lifted up with a secondary onycholysis. The triad of pain, nail deformity, and radiographic patterns are sufficient for diagnosis, but the radiographic criteria distinguishing exostosis from osteochondroma are not well established.

Magnetic resonance imaging can easily distinguish these two populations, as the cartilaginous cap is directly visible. Proton density-weighted images or gradient-echo images are the most accurate to depict the cap. The cap presents a high signal, when it is composed of hyaline cartilage (Fig. 9.57), or presents a much lower signal in case of fibrocartilage (Fig. 9.58). In some cases, the two components are associated (Fig. 9.59). The thickness of the cap can be measured. The thicker the cap, the higher the risk of degenerescence. This complication is exceptional at the level of the extremities. The acquired exostoses described by Lemont and Christman (74) present no cap (Fig. 60), much like ossified hematomas (Fig. 9.61). Magnetic resonance imaging easily depicts the implantation of exostoses and osteochondromas. The trabecular bone is nicely visible on gradient-echo images, where the trabecular network is highlighted by magnetic susceptibility artifacts. Continuity of the cortex of osteochondromas is well demonstrated on axial magnetic resonance images. These slices also accurately show the asymmetric deformity of the nail bed (Fig. 9.62).

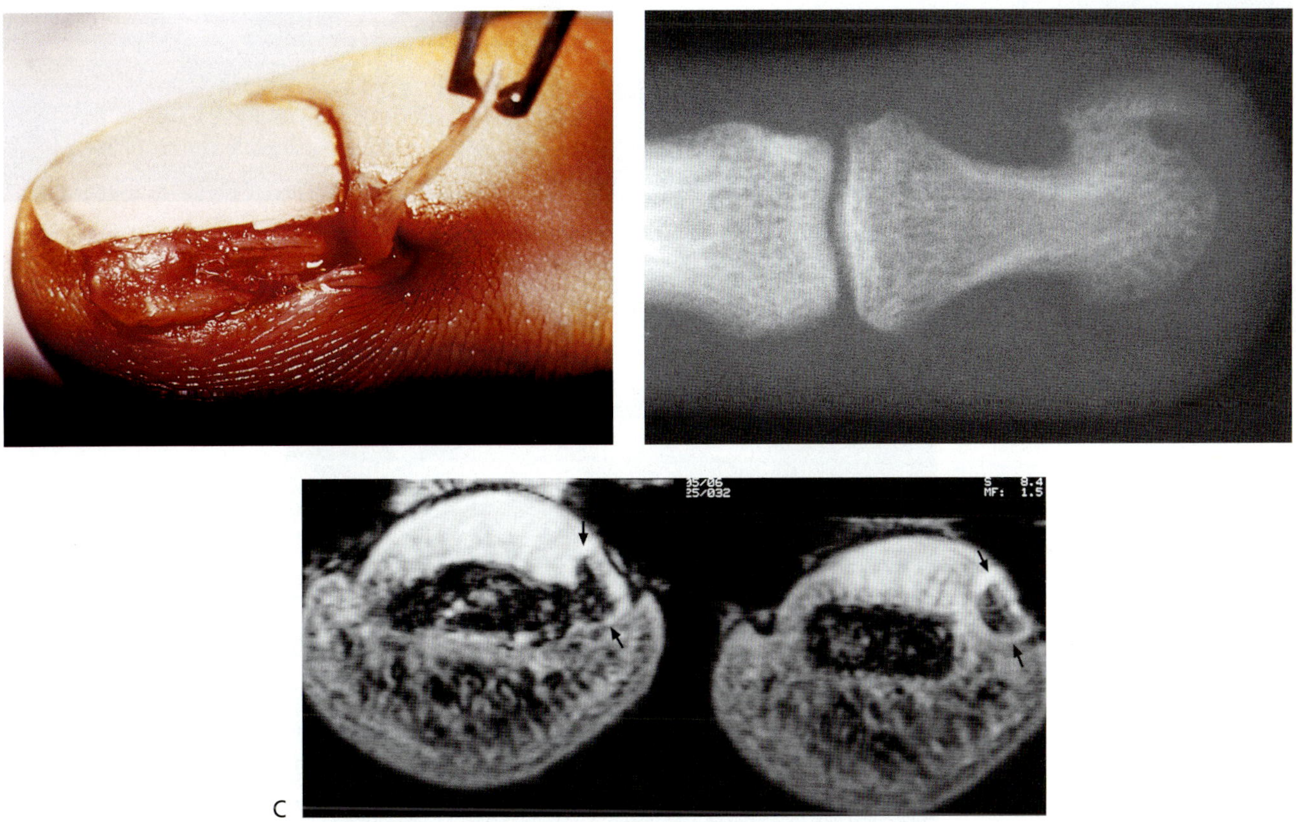

A B C

FIGURE 9.57. Subungual osteochondroma. **A:** Surgical view. (Courtesy of Dr. S. Goettmann, Paris, France.) **B:** Posteroanterior radiograph depicts a lateral ossification peduncled on the tuft. **C:** Axial three-dimensional gradient-echo image. There is a continuity between the trabecular bone of the phalanx and the osteochondroma. Note the high signal of the hyaline cartilaginous cap *(arrows)*.

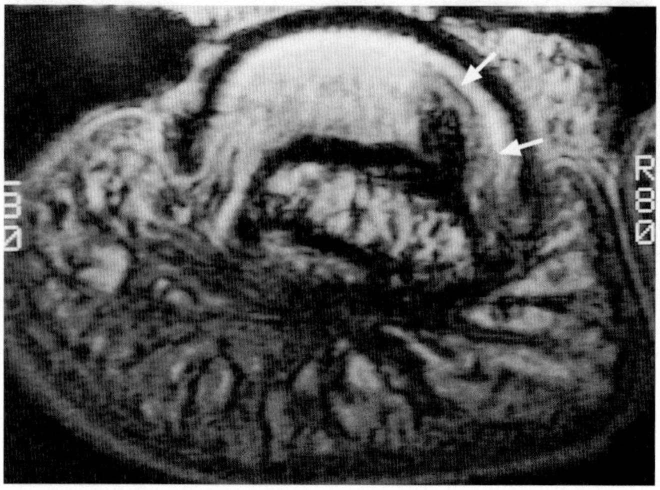

FIGURE 9.58. Subungual osteochondroma. Axial three-dimensional gradient-echo image. The fibrocartilaginous cap *(arrows)* presents an intermediate signal that is less intense than in the previous case.

A 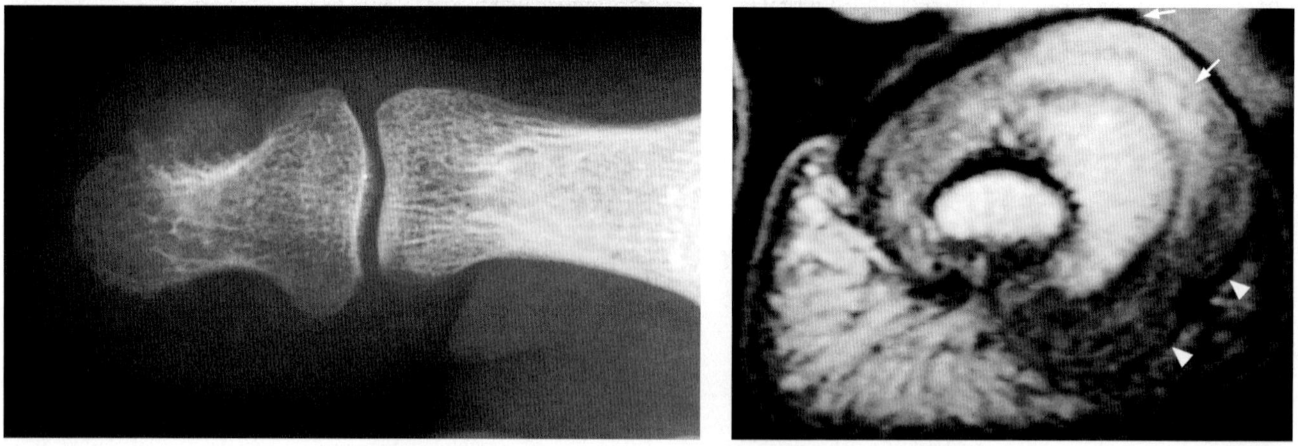 B

FIGURE 9.59. Subungual osteochondroma. **A:** Posteroanterior radiograph. Lateral sessile ossification. **B:** Axial T2-weighted spin-echo image. Thick cartilaginous cap with high signal dorsal hyaline cartilage *(arrows)* and low-signal fibrocartilage *(arrowheads)*. Note the lateral implantation of the osteochondroma and the asymmetric deformity of the nail bed.

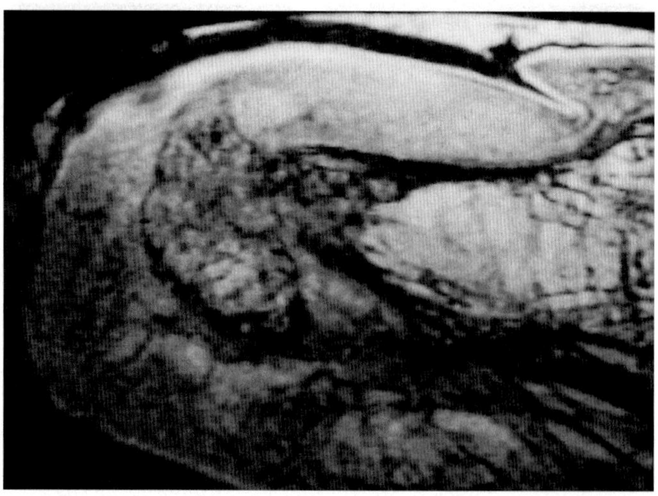

FIGURE 9.60. Subungual exostosis. Sagittal three-dimensional gradient-echo image. Distal acquired exostosis without detectable cartilaginous cap.

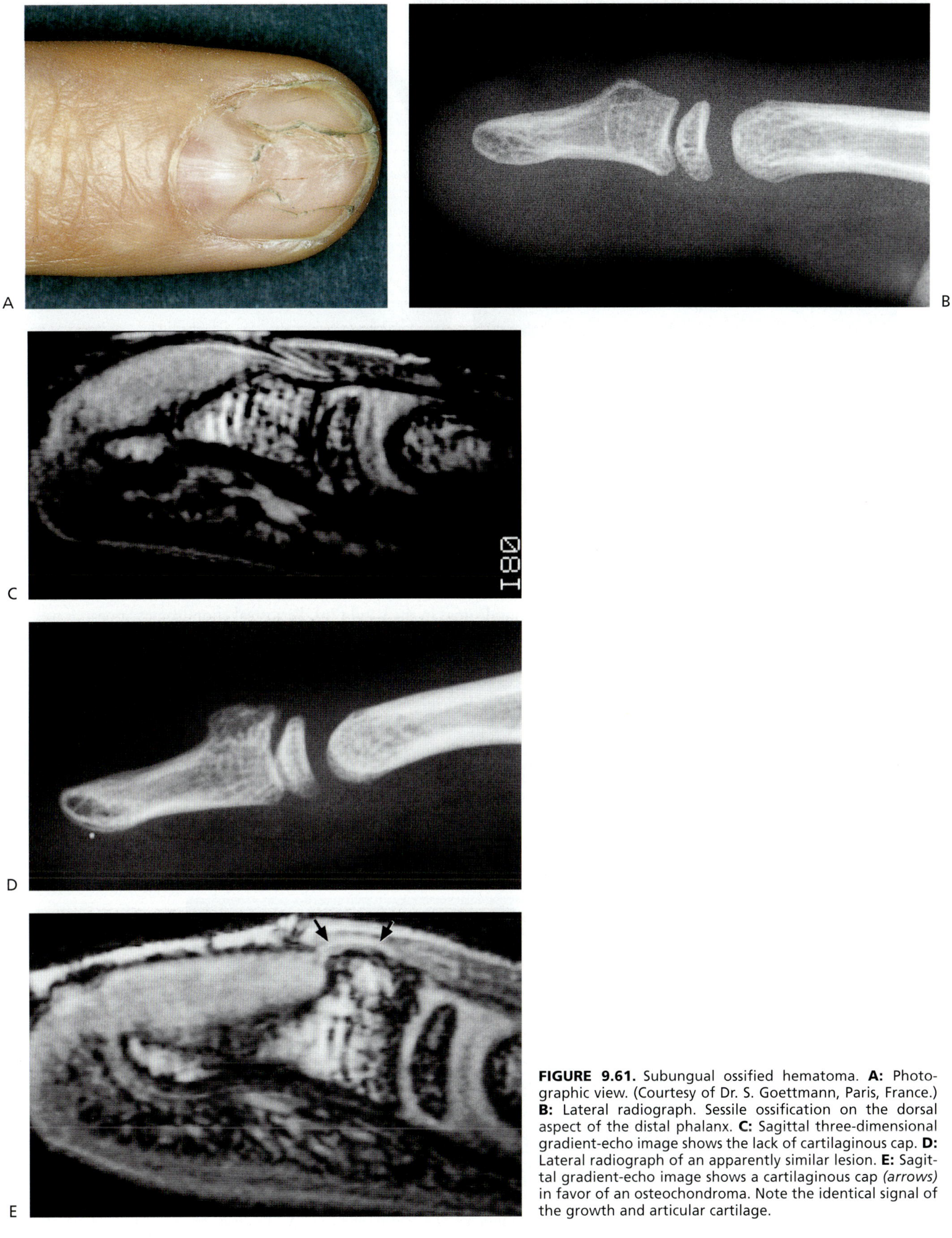

FIGURE 9.61. Subungual ossified hematoma. **A:** Photographic view. (Courtesy of Dr. S. Goettmann, Paris, France.) **B:** Lateral radiograph. Sessile ossification on the dorsal aspect of the distal phalanx. **C:** Sagittal three-dimensional gradient-echo image shows the lack of cartilaginous cap. **D:** Lateral radiograph of an apparently similar lesion. **E:** Sagittal gradient-echo image shows a cartilaginous cap *(arrows)* in favor of an osteochondroma. Note the identical signal of the growth and articular cartilage.

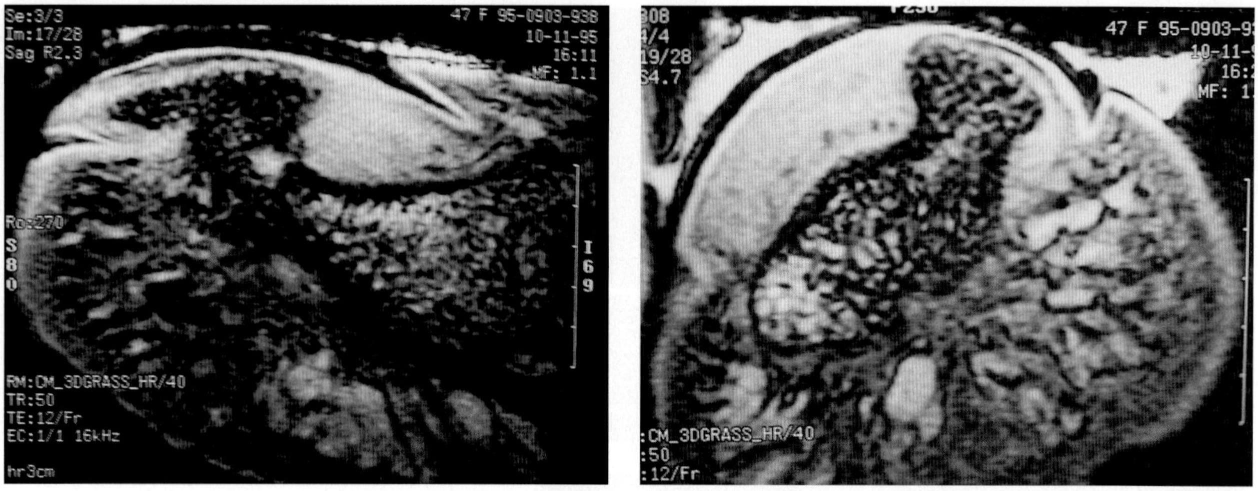

A B

FIGURE 9.62. Subungual osteochondroma. Sagittal **(A)** and axial **(B)** three-dimensional gradient-echo images. Note with this sequence the high visibility of the trabecular bone and the continuity of the cortex. The osteochondroma has a high-signal cartilaginous cap. The lateral implantation and the asymmetric deformity of the nail bed are depicted in the axial view.

Chondromas

Chondromas are the most frequent bone tumors of the hand, accounting for almost 50%. The MRI study of enchondromas is of poor interest, as the radiographs most often are specific with lobulated radiolucent defect, bone expansion, and flecks of calcifications (Figs. 9.6 and 9.7). However, MRI nicely shows the cartilaginous content in atypical cases and in the very rare extraosseous chondromas (Fig. 9.63). These cases account for only 4.6% of all chondromas in the hand (76). Periosteal chondromas appear as well-circumscribed intracortical radiolucencies (77). These patterns may be very subtle on radiographs (Fig. 9.64). High-resolution MRI more accurately depicts these abnormalities on axial gradient-echo images and reveals the cartilaginous signal of the tumor, with a low signal on T1-weighted images and a high signal on T2-weighted images, identical to that of water, and a faint peripheral enhancement without visible septa (Fig. 9.64). Usually, this cartilaginous component is less important in parosteal osteochondromatous proliferations (78). Chondromas may be located only in the soft tissues, such as in the nail bed (79). The nail bed is enlarged without abnormalities of the underlying phalanx (Fig. 9.65). Calcifications are helpful for diagnosis but are present in only 60% of cases (80).

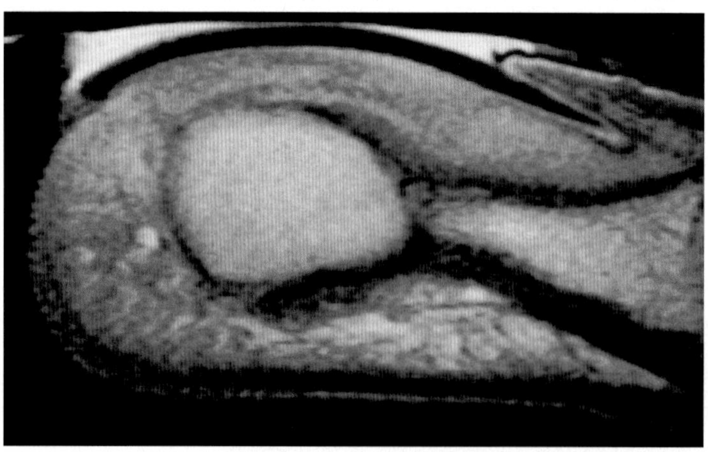

A

FIGURE 9.63. Enchondroma of the distal phalanx. **A:** Sagittal T1-weighted spin-echo image.

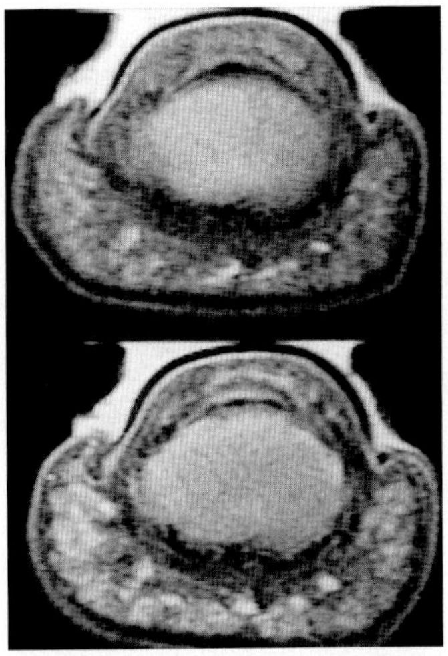

B

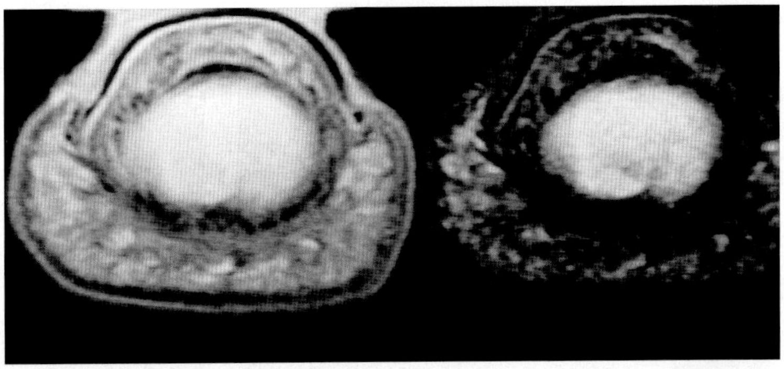

C

FIGURE 9.63 ***(continued).*** **B:** Axial T1-weighted spin-echo images before **(top)** and after **(bottom)** injection of gadolinium. The cartilaginous content is homogeneous with a low or intermediate signal. Note the thin and faint peripheral enhancement. **C:** Axial proton density **(left)** and T2-weighted **(right)** spin-echo image. Chondroma has a very high signal on T2-weighted images.

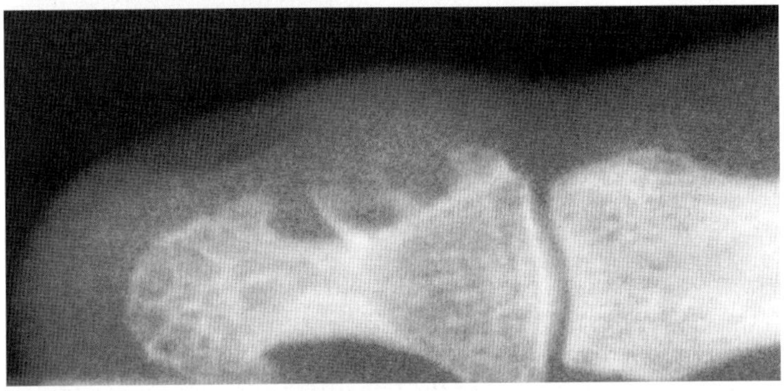

A

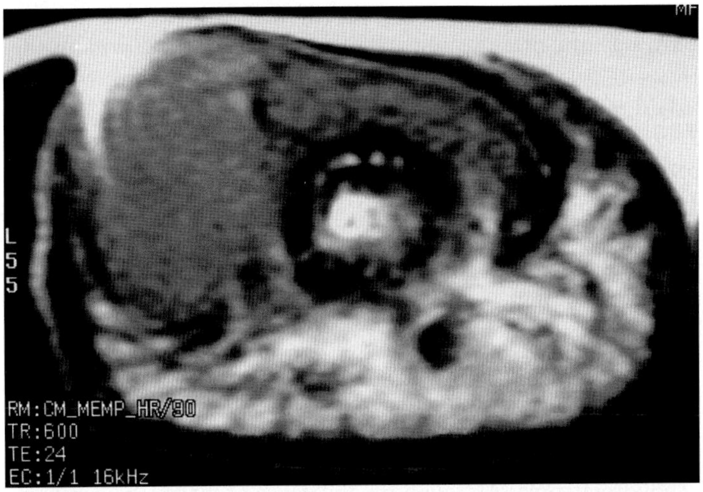

B

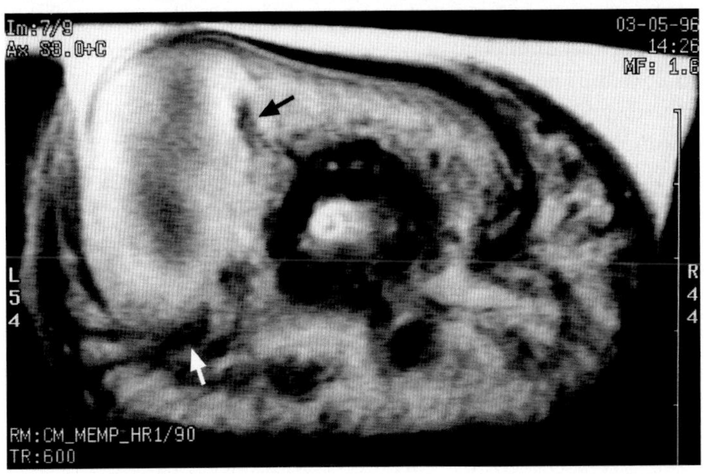

C

FIGURE 9.64. Periosteal chondroma. **A:** Posteroanterior radiograph. Lateral soft tissue mass with juxtacortical peripheral ossifications. Axial T1-weighted spin-echo images before **(B)** and after **(C)** injection of gadolinium. The nail bed is distorted by a lateral mass presenting a peripheral enhancement. The lesion is close to the lateral cortex of the phalanx with periosteal ossifications *(arrows)*. There is no invasion of the trabecular bone.

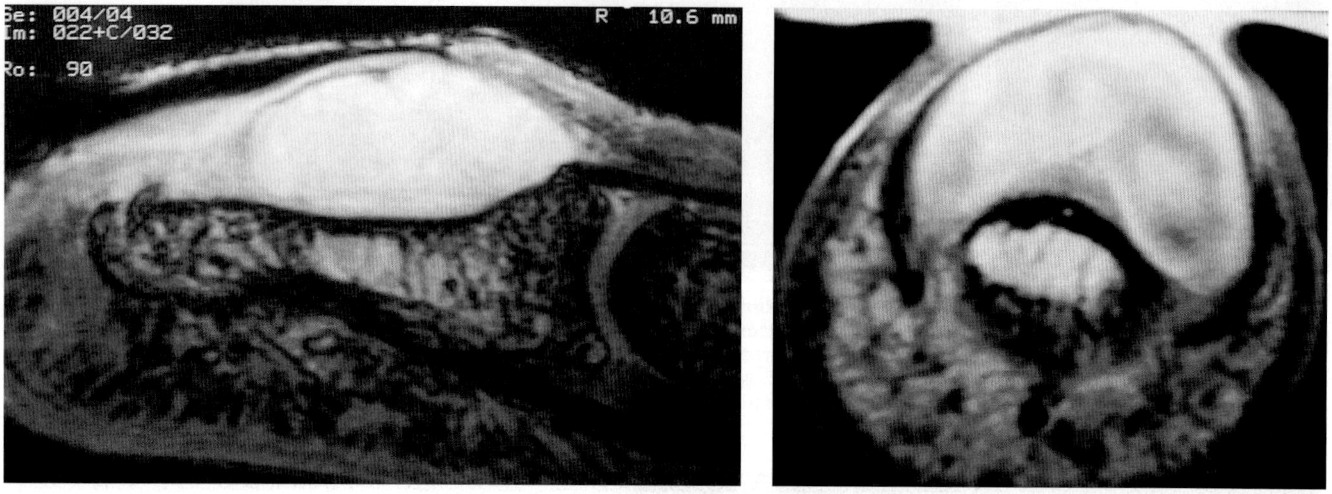

FIGURE 9.65. Extraosseous chondroma. **A:** Sagittal three-dimensional gradient-echo image. Well-defined mass in the nail bed with an homogeneous high signal. The posterior nail fold and the matrix are lifted up. Note the lack of bone abnormalities of the distal phalanx. **B:** Axial postenhanced T1-weighted spin-echo image. The whole nail bed is occupied by the lesion. There is peripheral and central enhancement.

Osteoid Osteomas

Osteoid osteomas are rare at the level of the distal phalanx. About 8% of osteoid osteomas involve the phalanges. They cause swelling of the fingertip, with widening and thickening of the nail plate. The skin may present a normal or slightly purplish coloration. Probing with a pin locates a painful pressure point. Radiographs may depict a nidus (Fig. 9.8) or only osteosclerosis or periostitis (Fig. 9.66). Computed tomography with thin contiguous slices is the best imaging technique to detect the nidus, eventually guided by a previous technetium 99 (^{99}Tc) scintigraphy (Figs. 9.9 and 9.66). As in other locations, the nidus may be subperiosteal, cortical, or intramedullary. High-resolution MRI is as accurate as computed tomography to detect the tiny nidus if contiguous 1-mm thick slices are acquired (three-dimensional sequence). The osteoid tissue is intense on gradient-echo images and enhances (Fig. 9.67). The large inflammatory reaction of the nail bed is very difficult to see with computed tomography, although it is obvious with MRI as the associated medullary edema of the distal phalanx (Fig. 9.66). Synovitis of the distal interphalangeal joint may be associated.

Chondrosarcomas

The occurrence of pain is rare with enchondromas, except in cases of pathologic fracture; pain is common with chondrosarcoma. Although the incidence of malignant degenerescence of solitary enchondromas is less than 1%, it can increase to 50% in Ollier's disease and Maffucci's syndrome. The risk of malignant transformation in multiple exostoses disease ranges from 1% to 25%, but no case was reported at the level of the distal phalanx. On early radiographs, the diagnosis may be difficult. Focal areas of ill-defined osteolysis of the cortex and calcifications in the soft tissues must be located (Fig. 9.68). Magnetic resonance imaging is helpful and shows abnormal patterns for simple enchondroma. Extension to the nail bed may be large, with lifting of the nail plate. Some areas show signal behavior typical of hyaline cartilage, but other areas present an unusual mottled enhancement (Fig. 9.68).

In conclusion, despite the high value of clinical examination of the nail unit, radiographs remain necessary if a deep mass beneath the nail plate is suspected. High-frequency ultrasonography may be helpful in experienced hands, but high-resolution MRI offers the most promising developments for imaging of the nail unit.

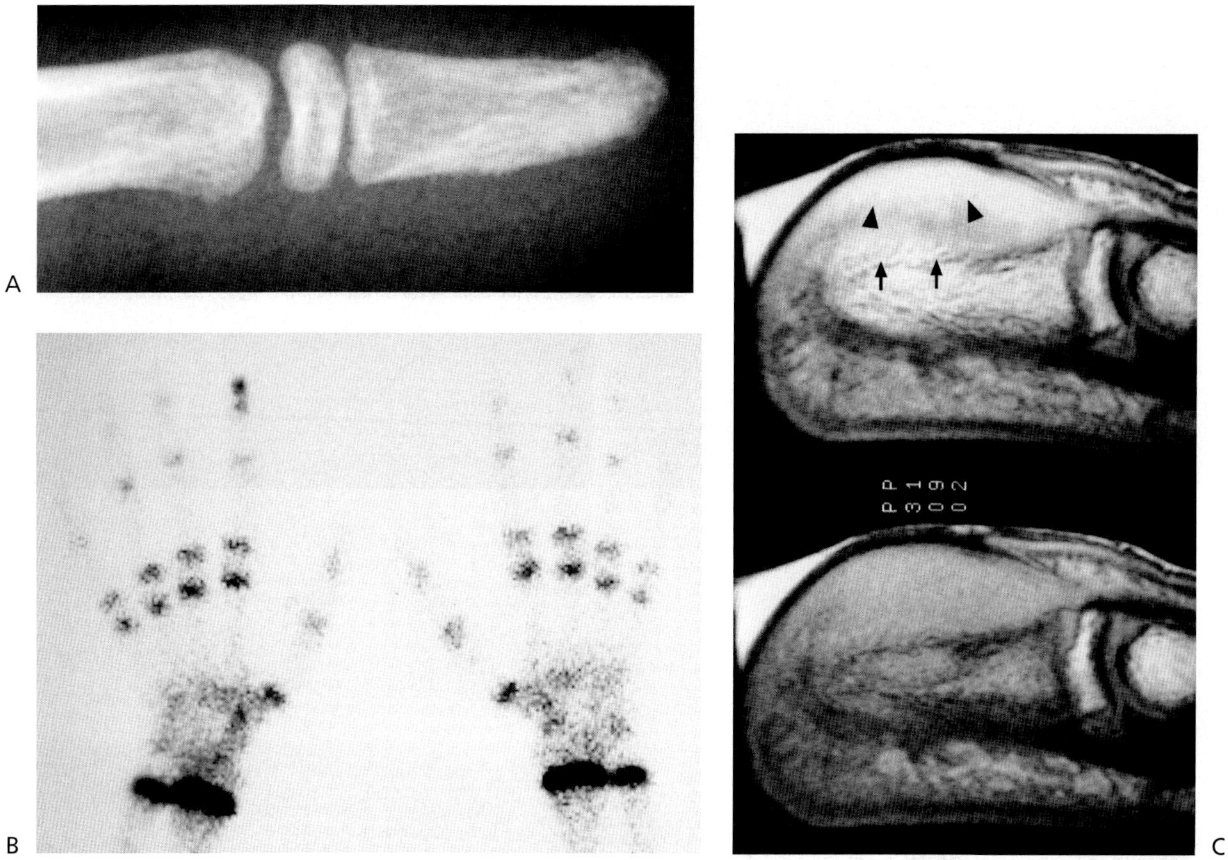

FIGURE 9.66. Osteoid osteoma of the distal phalanx. **A:** Lateral radiograph. There is osteosclerosis and diffuse periosteal reaction of the distal phalanx. There is no visible bone nidus. **B:** Spotty enhancement of the distal phalanx on the bone scan. **C:** Sagittal T1-weighted spin-echo images before **(bottom)** and after **(top)** injection of gadolinium. Note the low signal of a diffuse bone edema **(bottom)** and the strong enhancement of the periosteal reaction *(arrow)* and the inflammatory nail bed *(arrowheads)*. There is secondary sagittal hypercurvature of the nail plate.

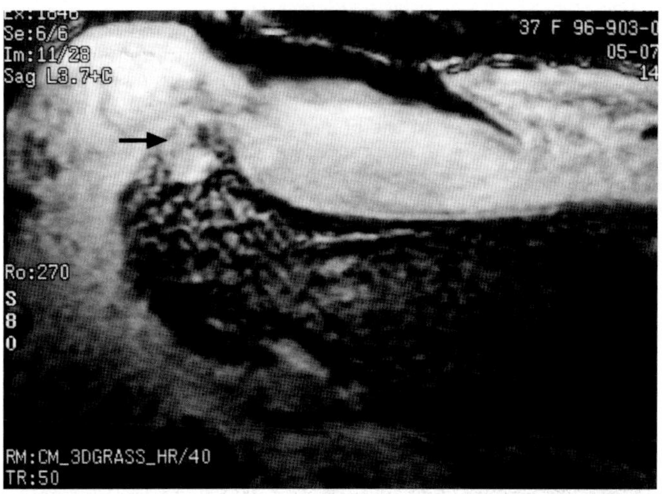

FIGURE 9.67. Osteoid osteoma of the distal phalanx. Sagittal postenhanced gradient-echo image. A round distal nidus of the tuft presents a high signal *(arrow)*.

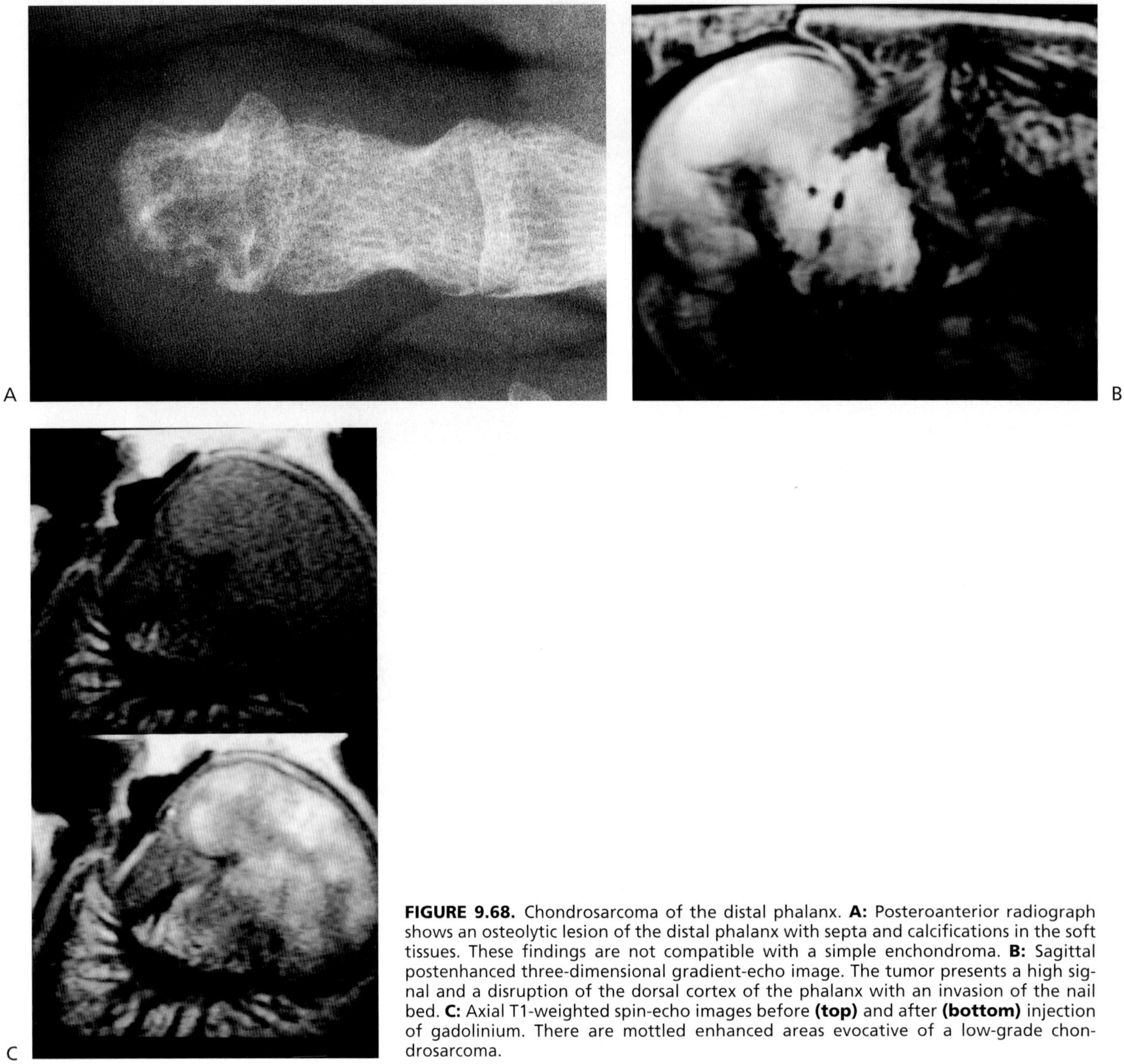

FIGURE 9.68. Chondrosarcoma of the distal phalanx. **A:** Posteroanterior radiograph shows an osteolytic lesion of the distal phalanx with septa and calcifications in the soft tissues. These findings are not compatible with a simple enchondroma. **B:** Sagittal postenhanced three-dimensional gradient-echo image. The tumor presents a high signal and a disruption of the dorsal cortex of the phalanx with an invasion of the nail bed. **C:** Axial T1-weighted spin-echo images before **(top)** and after **(bottom)** injection of gadolinium. There are mottled enhanced areas evocative of a low-grade chondrosarcoma.

REFERENCES

1. Apolinar E, Rowe WF. Examination of human fingernail ridges by means of polarized light. *J Forensic Sci* 1980;25:154–161.
2. Camirand P, Giroud JM. Subungual glomus tumour. Radiological manifestations. *Arch Dermatol* 1970;102:677–679.
3. Gandon F, Legaillard P, Brueton R, et al. Forty-eight glomus tumors of the hand: retrospective study and four-year follow-up. *Ann Hand Surg* 1992;11:401–405.
4. Lumpkin LR, Rosen T, Tschen JA. Subungual squamous cell carcinoma. *J Am Acad Dermatol* 1984; 11:735–738.
5. Mathis WH, Schulz MD. Roentgen diagnosis of glomus tumours. *Radiology* 1948;51:71–76.
6. Jablon M, Horowitz A, Bernstein DA. Magnetic resonance imaging of a glomus tumor of the finger tip. *J Hand Surg* 1990;15A:507–509.

7. Malghem J, Maldague B. X-ray computed tomography of phalangeal osteoid osteoma. *Rev Rhum Mal Osteosarc*, 1988;55:719.
8. Holzberg M. Glomus tumor of the nail. *Arch Dermatol* 1992;128:160–162.
9. Fornage BD. Glomus tumours in the fingers: diagnosis with ultrasound. *Radiology* 1988;167:183–185.
10. Alexander H, Miller DL. Determining skin thickness with pulsed ultrasound. *J Invest Dermatol* 1979;72:17–19.
11. Tan CY, Marks R, Payne P. Comparison of xeroradiographic and ultrasound detection of corticosteroid induced dermal thinning. *J Invest Dermatol* 1981;76:126–128.
12. Serup J. Ten years' experience with high-frequency ultrasound examination of the skin: development and refinement of technique and equipment. In: Altmeyer P, El-Gammal S, Hoffmann K, eds. *Ultrasound in dermatology.* Berlin: Springer-Verlag, 1991:41–54.
13. El-Gammal S, Hoffmann K, Auer T, et al. A 50-MHz high-resolution ultrasound imaging system for dermatology. In: Altmeyer P, El-Gammal S, Hoffmann K, eds. *Ultrasound in dermatology.* Berlin: Springer-Verlag, 1991:297–322.
14. Finlay AY, Moseley H, Duggan TC. Ultrasound transmission time: an in vivo guide to nail thickness. *Br J Dermatol* 1987;117:765–770.
15. Finlay AY, Western B, Edwards C. Ultrasound velocity in human fingernail and effects of hydration: validation of in vivo nail thickness measurement techniques. *Br J Dermatol* 1990;123:365–373.
16. Johnson M, Comaish JS, Shuster S. Nail is produced by the normal bed: a controversy resolved. *Br J Dermatol* 1991;125:27–29.
17. Johnson M, Shuster S. Continuous formation of nail along the bed. *Br J Dermatol* 1993;128:277–280.
18. Jemec GBE, Serup J. Ultrasound structure of the human nail plate. *Arch Dermatol* 1989;125:643–646.
19. Kneeland JB, Middelton WD, Matloub HS, et al. High resolution MR imaging of glomus tumor. *J Comput Assist Tomogr* 1987;11:351–352.
20. Schneider LH, Bachow TB. Magnetic resonance imaging of glomus tumor. *Orthop Rev* 1991;20: 255–256.
21. Matloub HS, Muoneke VN, Prevel CD, et al. Glomus tumor imaging: use of MRI for localization of occult lesions. *J Hand Surg [Am]* 1992;17:472–475.
22. Hou SM, Shih TTF, Lin MC. Magnetic resonance imaging of an obscure glomus tumour in the fingertip. *J Hand Surg [Br]* 1993;18:482–483.
23. Aguado JB, Blackband SJ, Schoeniger J, et al. Nuclear magnetic resonance imaging of a single cell. *Nature* 1986;322:190–191.
24. Johnson GA, Thompson MB, Gewalt SL, et al. Nuclear magnetic resonance imaging at microscopic resolution. *J Magn Reson* 1986;68:129–137.
25. Bittoun J, Saint-Jalmes H, Querleux BG, et al. In vivo high resolution imaging of the skin in a wholebody MR system at 1.5T. *Radiology* 1990;176:457–460.
26. Wong EC, Jesmanowicz A, Hyde JC. High-resolution short echo time MR imaging of the fingers and wrists with a local gradient coil. *Radiology* 1991;181:393–397.
27. Blackband SJ, Chakrabarti I, Gibbs P, et al. Fingers: three-dimensional MR imaging and angiography with a local gradient coil. *Radiology* 1994;190:895–899.
28. Querleux B, Yassine MM, Darrasse L, et al. Magnetic resonance imaging of the skin. A comparison with ultrasonic technique. *Bioeng Skin* 1988;4:1–14.
29. Richard S, Querleux B, Bittoun J, et al. In vivo proton relaxation times analysis of the skin layers by magnetic resonance imaging. *J Invest Dermatol* 1991;97:120–125.
30. Dion E, Idy-Perretti I, Bellin MF, et al. MR imaging of the carpal tunnel with a specific high-resolution coil. *Radiology* 1991;181(P):106.
31. Foo TK, Shellock FG, Hayes CE, et al. High resolution MR imaging of the wrist and eye with short TR, short TE, and partial echo acquisition. *Radiology* 1992;183:277–281.
32. Bittoun J, Leroy-Willig A, Idy I, et al. Relation entre rapport signal-sur-bruit et paramètres d'acquisition en IRM pour un contraste donné. *Ann Radiol* 1987;30:5–13.
33. Fleckman P. Basic science of the nail unit. In: Scher RK, Daniel CR III, eds. *Nails: therapy, diagnosis, surgery.* Philadelphia: WB Saunders, 1990:36–51.
34. Zook EG, Van Beek AL, Russell RC, et al. Anatomy and physiology of the perionychium: a review of the literature and anatomic study. *J Hand Surg* 1980;5:528–536.
35. Zook EG. The perionychium: anatomy, physiology, and care of injuries. *Clin Plast Surg* 1981;8:21–31.
36. Johnson M, Shuster S. Continuous formation of nail along the bed. *Br J Dermatol* 1993;128:277–280.
37. Jemec GBE, Agner T, Serup J. Transonychial water loss: relation to sex, age and nail plate thickness. *Br J Dermatol* 1989;121:441–446.
38. Lewin K. The normal fingernail. *Br J Dermatol* 1965;77:421–430.
39. Parent D, Achten G, Stouffs-Vanhoof F. Ultrastructure of the normal human nail. *Am J Dermatopathol* 1985;7:529–535.
40. Drapé JL, Wolfram-Gabel R, Idy-Peretti I, et al. The lunula: resonance imaging approach to the subnail matrix area. *J Invest Dermatol* 1996;106:1081–1085.
41. Wolfram-Gabel R, Sick H. Vascular networks of the periphery of the finger nail. *J Hand Surg* 1995; 20B:488–492.

42. Baden HP, Kubilus J. Fibrous proteins of bovine hoof. *J Invest Dermatol* 1983;81:220–224.
43. Baden HP. Structure, composition, and physiology of nails. In: Baden HP, ed. *Diseases of the hair and nails.* Chicago: Year Book Medical Publishers, 1987:3–11.
44. Guéro S, Guichard S, Fraitag SR. Ligamentary structure of the base of the nail. *Surg Radiol Anat* 1994;16:47–52.
45. Carroll RE, Berman AT. Glomus tumors of the hand: review of the literature and report of 28 cases. *J Bone Joint Surg* 1972;54A:691–703.
46. Dawber RPR, Baran R. Science of the nail apparatus. In: Baran R, Dawber RPR, eds. *Diseases of the nails and their management.* Oxford: Blackwell Science, 1994:1–34.
47. Rettig AC, Strickland JW. Glomus tumors of the digits. *J Hand Surg* 1977;2:261–265.
48. Masson P. Le glomus neuromyo-artériel des régions tactiles et ses tumeurs. *Lyon Chir* 1924;21: 256–280.
49. Drapé JL, Idy-Peretti I, Goettmann S, et al. MR imaging of subungual glomus tumors. *Radiology* 1995;195:507–515.
50. Drapé JL, Idy-Peretti I, Goettmann S, et al. MR imaging of digital mucoid cysts. *Radiology* 1996;200:531–536.
51. Kohout E, Stout AP. The glomus tumor in children. *Cancer* 1961;14:555–556.
52. Varian J, Cleak DK. Glomus tumors in the hand. *Hand* 1980;12:293–299.
53. Davis TS, Graham WP III, Blomain EW. A ten-year experience with glomus tumors. *Ann Plast Surg* 1986;6:297–299.
54. Johnson WC, Helwig EB. Cutaneous focal mucinosis: a clinicopathological and histochemical study. *Arch Dermatol* 1966;93:13–20.
55. Sonnex TS. Digital myxoid cysts: a review. *Cutis* 1986;37:89–94.
56. Newmeyer WL, Kilgore ES, Graham WP. Mucous cyst: the dorsal distal interphalangeal joint ganglion. *Plast Reconstr Surg* 1974;53:313–315.
57. Goldman L. Transillumination of the fingertip as aid in examination of nail changes. *Arch Dermatol Chicago* 1962;85:644.
58. Zaïas N. *The nail in health and disease*, 2nd ed. Norwalk, CT: Appleton & Lange, 1990.
59. Armijo M. Mucoid cysts of the fingers: differential diagnosis, ultrastructure, and surgical treatment. *J Dermatol Surg Oncol* 1981;7:317–332.
60. Drapé JL, Idy-Peretti I, Goettmann S, et al. Standard and high resolution magnetic resonance imaging of glomus tumors of toes and fingertips. *J Am Acad Dermatol* 1996;35:550–555.
61. Drapé JL, Thelen P, Gay-Depassier P, et al. Intraarticular diffusion of Gd-DPTA after intravenous injection in the knee: MR imaging evaluation. *Radiology* 1993;188:227–234.
62. Jensen D. Ganglia and synovial cysts: their pathogenesis and treatment. *Ann Surg* 1937;105:592–601.
63. Isaacson NH, McCarthy DD. Recurring myxomatous cutaneous cysts. *Surgery* 1954;35:621–623.
64. Kumar B, Shazma SC, Kaur S. Phalangeal erosions with subungual warts. *Ind J Dermatol Venereol* 1980;46:166–168.
65. Keeney GL, Banks PM, Linscheild RL. Subungual keratoacanthoma. Report of a case and review of the literature. *Arch Dermatol* 1988;124:1074–1076.
66. Baran R, Kechijian P. Longitudinal melanonychia: diagnosis and management. *J Am Acad Dermatol* 1989;21:1165–1175.
67. Drewers J, Günther D, Nolden HH. Intraossäre epidermiszysten der finger und zehen. *Akt Chir* 1985;20:171.
68. Lewin K. Subungual epidermoid inclusion. *Br J Dermatol* 1969;81:671–675.
69. Baran R, Klint A. Onychomatrixoma. *Br J Dermatol* 1992;126:510–515.
70. Goettmann S, Drapé JL, Idy-Peretti I, et al. Magnetic resonance imaging: a new tool in the diagnosis of nail tumors. *Br J Dermatol* 1994;130:701–710.
71. Haneke E. Fungal infections of the nail. *Semin Dermatol* 1991;10:41–53.
72. Averill RM, Smith RJ, Campbell CJ. Giant-cell tumors of the bones of the hand. *J Hand Surg* 1980;5:39–50.
73. Norton LA. Disorders of the nails. In: Moschella SL, Hurley HJ, eds. *Dermatology,* 2nd ed. Philadelphia: WB Saunders, 1985:1398–1400.
74. Lemont H, Christman RA. Subungual exostosis and nail disease and radiologic aspects. In: Scher R, Daniel CR, eds. *Nails: therapy, diagnosis, surgery.* Philadelphia: WB Saunders, 1990:250–257.
75. Kato T, Nakagawa T, Suji T, et al. Subungual exostosis clinicopathological and ultrastructural studies of 3 cases. *Clin Exp Dermatol* 1990;15:429–432.
76. Besser E, Roessner A, Brug E, et al. Bone tumors of the hand. *Arch Orthop Trauma Surg* 1987; 106:241–247.
77. Feldman F. Primary bone tumors of the hand and carpus. *Hand Clin* 1987;3:269–289.
78. Derrick EK, Darley CR, Tanner B. Bizarre parosteal osteochondromatous proliferations of the tubular bones of the hands and feet. *Clin Exp Dermatol* 1994;19:53–55.
79. Dumontier C, Abimelec P, Drapé JL. Soft-tissue chondroma of the nail bed. *J Hand Surg* 1997;22B:474–475.
80. Dahlin DC, Salvador AH. Cartilaginous tumors of the soft tissues of the hands and feet. *Mayo Clin Proc* 1974;49:721–726.

10

CONGENITAL AND/OR HEREDITARY CONDITIONS

ROBERT BARAN
ECKART HANEKE

RACQUET THUMBS

A short, broad terminal phalanx of the thumb (Fig. 10.1) results in a short and wide nail plate that usually lacks lateral nail folds. The esthetic appearance may be improved by narrowing the nail plate and creating lateral nail folds (1,2).

Nail and tissue excision are performed on both sides of the thumb as for a lateral longitudinal nail biopsy (Fig. 10.2). The lateral soft aspects of the distal phalanx are dissected from the bone, and back stitches are used to create lateral nail grooves (Fig. 10.3). The needle is run into the lateral aspect about 2 to 3 mm volar to the plane of the nail bed bone interface, through the nail bed and plate, and back again through the lateral thumb skin, which on knotting will be elevated, thus forming a lateral nail fold. Nail groove epithelium develops by secondary intention.

CONGENITAL NAIL MALALIGNMENT OF THE BIG TOENAIL

Congenital malalignment of the big toenail may present with gross nail deformation, greenish-gray discoloration, thickening, and oyster shell-like appearance (Fig. 10.4) (3,4). The deviation almost always is to the lateral side, and the nail usually grows faster on its medial side, adding to the lateral deviation. The obliquely positioned nail usually has sharply downward bent lateral margins often digging into the lateral nail fold and distal nail bed. There is no attachment to the nail bed because there is no counterpressure to the forces acting on the tip of the toe during gait; the tip and pulp tissue are dislocated dorsally to form a distal nail wall. The lack of attachment to the nail bed epithelium can be shown readily by placing a blunt-tipped probe under the nail. The ridged skin of the pulp of the toe is constantly dislocated under the free end of the nail, obscuring the hyponychium. The thickened nail plate easily breaks into slightly curved pieces. It is only adherent to the matrix, and repeated trauma from inadequate shearing forces to the matrix most probably are the cause for the oyster shell-like nail plate growth. The proximal nail fold commonly covers just a short portion of the proximal nail plate, which, however, has lateral matrix horns reaching deep proximally.

About 50% of patients experience spontaneous improvement leading to an almost normal looking yet still malaligned nail with good nail bed attachment before the age of 10 years. Guéro et al. (5) suggested the existence of a ligamentous structure corresponding to a dorsal expansion of the lateral ligament of the distal interphalangeal joint, and

R. Baran: Nail Disease Centre, 42, Rue des Serbes, 06400 Cannes, France.
E. Haneke: Klinikk Bunæs, Lokkeåsveien, 1337 Sandvika, Oslo, Norway.

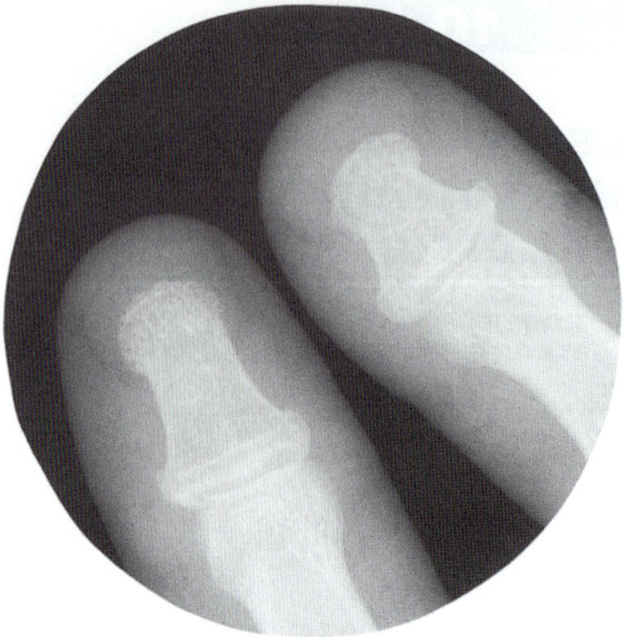

FIGURE 10.1. Broad distal bony phalanx.

this "bone-nail ligament" appears to be connected with the nail matrix (Fig. 10.5). Its physical properties may change with the passage of time and therefore explain spontaneous cure.

If the appearance of the nail is extreme, surgery is chosen and best performed before 2 years of age, as this diminishes the risk of permanent dystrophy.

Realignment requires rotation of a bulky nail unit flap, including the entire nail, nail bed, and matrix. This requires an external Burow's triangle. An eccentric crescentic excision undermines the nail unit with the maximum width located on the internal side of the foot, corresponding to the side to which the nail needs to be redirected (Fig. 10.6). This crescent ends on each side 3 to 4 mm behind the most proximal level of the proxi-

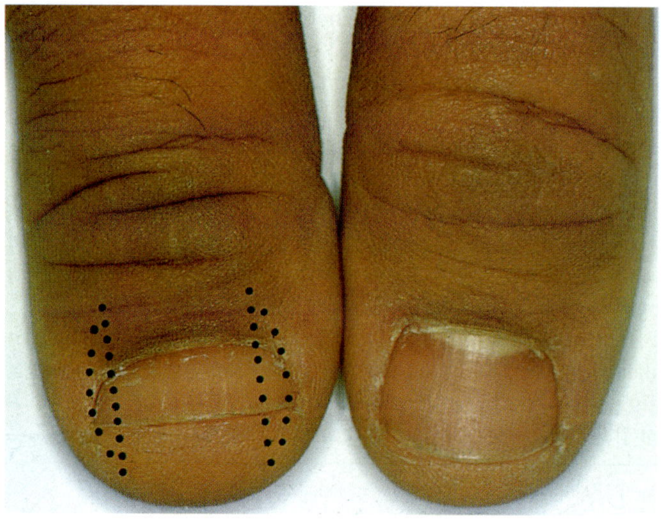

FIGURE 10.2. Wide nail plate with narrow lateral nail folds.

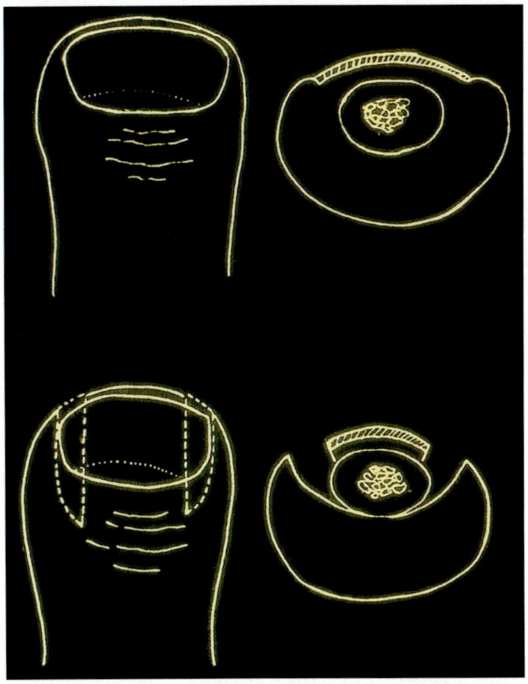

FIGURE 10.3. Back stitches to are used to create lateral nail grooves.

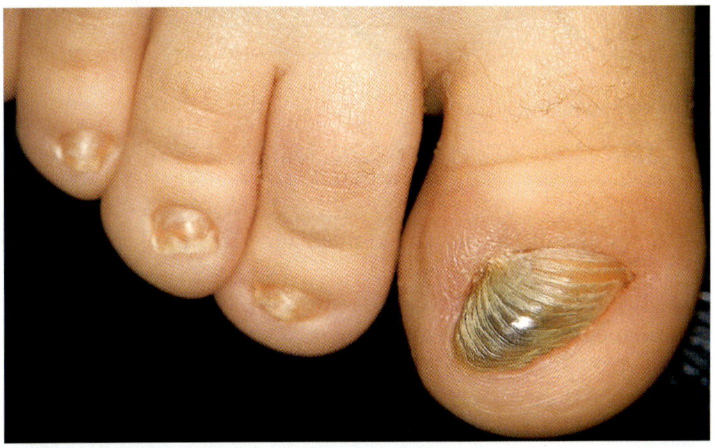

FIGURE 10.4. Congenital malalignment of the big toe nail.

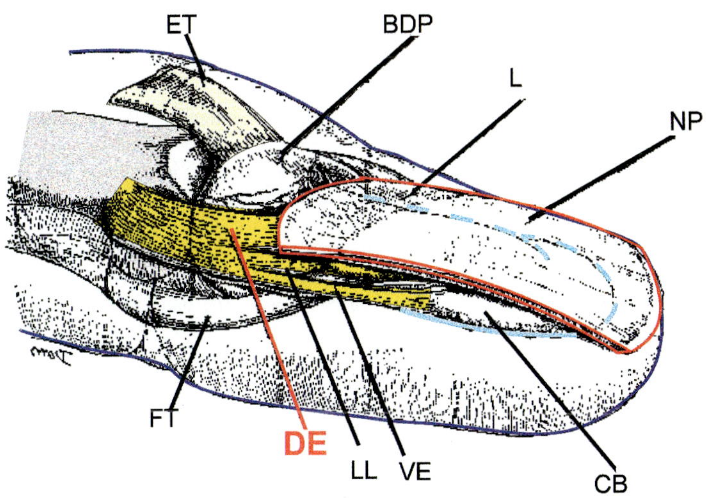

FIGURE 10.5. Dorsal expansion of the lateral ligament of the distal interphalangeal joint pulling laterally the nail plate. (Courtesy of S. Guéro, Paris, France.)

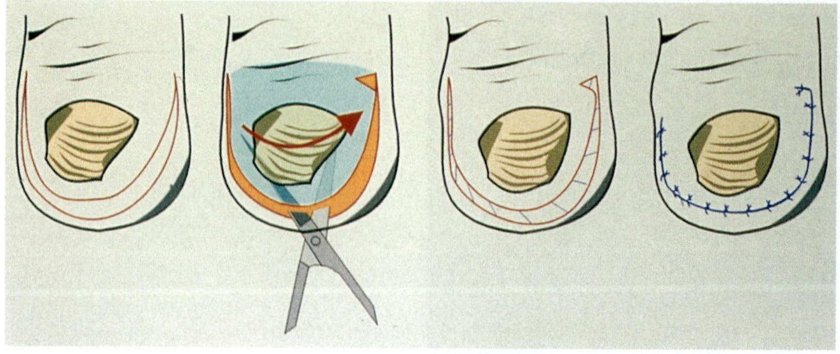

FIGURE 10.6. Surgery of congenital malalignment.

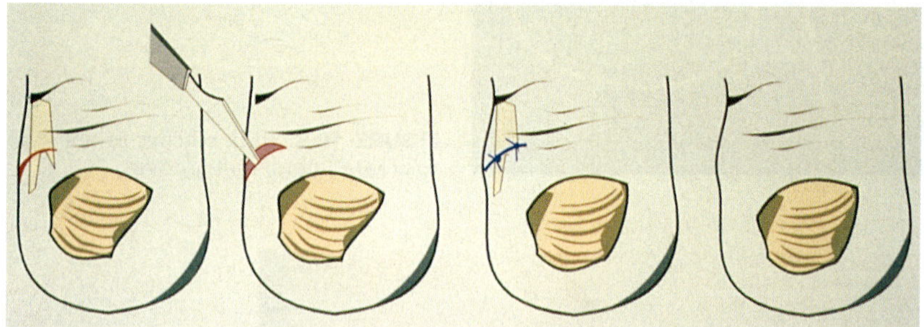

FIGURE 10.7. Section of Guéro's ligament.

mal nail fold. The nail bed and the matrix are undermined and lifted until the fibers of the extensor tendon are visible on its bony insertion.

Suturing together the edges of the excised triangle reduces the loss of cutaneous substance. The nail unit is realigned to the inside as the maximum cutaneous resection is mostly distal and medial.

Taking into account the dorsal expansion of the lateral ligament of Guéro, we have proposed its section (Fig. 10.7) as a treatment for congenital malalignment of the great toenail, but we believe it should be performed only as a supplement to the previous surgery.

REFERENCES

1. Haneke E. Behandlung einiger Nagelfehlbildungen. In: Wolff HH, Schmeller W, eds. *Fehlbildungen Nävi, Melanome.* Berlin: Springer-Verlag, 1985:71–77.
2. Haneke E. Reconstruction of the lateral nail fold after lateral longitudinal nail biopsy. In: Robins P, ed. *Surgical gems in dermatology.* New York: Journal Publ. Group, 1988:91–93.
3. Baran R, Bureau H. Congenital malalignment of the big toenail as a cause of ingrowing toenail in infancy. Pathology and treatment (a study of 30 cases). *Clin Exp Dermatol* 1983;8:619–623.
4. Baran R, Haneke E. Etiology and treatment of nail malalignment. *J Dermatol Surg* 1998;24:719–721.
5. Guéro S, Guichard S, Fraitag SR. Ligamentary structure of the base of the nail. *Surg Radiol Anat* 1994;16:47–62.

11

PERIUNGUAL AND DIGITAL LASER SURGERY

TIMOTHY ROSIO

> The hands are a sort of feet,
> which serve us in our passage towards Heaven,
> curiously distinguished into joints and fingers,
> and fit to be applied to any thing
> which reason can imagine or desire.
> *Meditations on the Six Days of Creation* (1717)

The nail and its environs constitute a superbly functional and aesthetic area of high sensitivity and vascularity, and one that is rich in anatomic complexity. This chapter will discuss how periungual neurovascular and superficial three-dimensional structures lend particular advantages to lasers when cases are properly selected and techniques correctly applied.

Most literature on laser periungual surgery mentions only the carbon dioxide (CO_2) laser and inclusively cites many controversial applications that few experienced laser users would advise. I will review the utility of multiple laser wavelengths and laser techniques that I believe may be recommended for serious consideration in the treatment of conditions, disorders, and diseases of the nail and periungual region. I will list specific advantages of laser therapy and relative indications, and offer criteria used in my decision-making process for when and how to choose and use particular lasers for periungual laser surgery. Additionally, I group the current plethora of lasers into procedural groups by major mechanism of action.

In the final analysis, the surgeon familiar with both laser and nonlaser techniques can choose the approach best suited to the individual patient and problem.

ADVANTAGES OF LASER THERAPY

Eight advantages of laser applications will be discussed and are listed in Table 11.1.

1. Hemostatic visibility derives from the coagulating ability of excisional and ablational lasers to seal blood vessels as they remove tissue. Although the CO_2 laser is the most commonly used for the skin in this manner, the erbium-YAG and other far infrared lasers share this property. The bloodless field allows one to distinguish surgical depth and subtle differences in tissue targeted for removal. An example is disordered dermatoglyphs in warts (Fig. 11.1). Laser ablation provides high visibility without the need to place many tourniquets as would be required for multiple fingers. Hemostasis from laser coagulation is superior to that achievable with chemical hemostatics, yet laser is more uniform and less

T. Rosio: Fulton Skin Institute of Newport Beach, Newport Beach, California 92660.

TABLE 11.1. ADVANTAGES OF LASER THERAPY ON THE DIGITS

1. Hemostatic visibility
2. Visualization (differential tissue effects of CO_2)
3. Nontouch, less traumatic, noninoculation of infectious material
4. Precision in amount of tissue affected, limited injury depth compared to mechanical or electrosurgical methods
5. Uniform distribution (line of sight)
6. Sterilization and sealing of tissue against bacterial or viral infection
7. Speed
8. Specific absorption by tissues, cells, or subcellular chromophores below intact surface (selective photothermolysis)

deeply injurious to tissue than electrocoagulation. Furthermore, postoperative hemostasis is excellent with minimal dressings.

Laser choice and method of use are both important when precise effects are needed to preserve adjacent tissue. A standard YAG is hemostatic but damages tissue imprecisely; therefore, more disadvantage than advantage accrues to its use in periungual excisional procedures.

2. Visualization and identification of various tissue types and depths are aided by differential laser effects and not just by improved hemostasis. CO_2 ablational laser surgery depends on highly efficient water absorption of the 10,600-nm CO_2 wavelength. Epidermis and wart-infected tissue contain more water than, for example, dermis, fibromas, scar tissue, and fat. Therefore, wart tissue vaporizes more readily and notably "bubbles' compared to other tissue types (Fig. 11.2).

In contrast to the "bubbling" of epidermal and wart tissue, collagen and muscle contracts or "shrinks" much more than warts or the other tissues. Dermal collagen contraction becomes much more obvious when one reaches the reticular dermis (Fig. 11.3).

Another exciting example is the noninvasive identification of viable tissue. Ascertaining adequate circulation is difficult in burn patients, yet is the key criterion used to avoid amputations or more debridement than necessary. Vascular identification is possible by 800-nm diode laser illumination, resulting in 840-nm fluorescence from systemically injected indocyanine green. If vascularity is sufficient, the fluorescence yields an infrared photographic image.

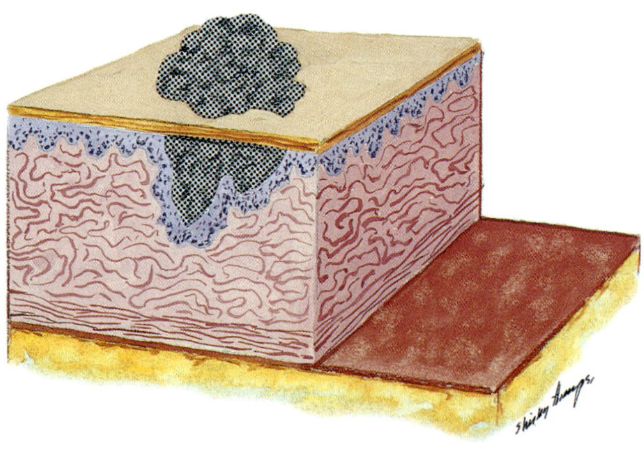

FIGURE 11.1. Wart-infected tissue lacks normal dermatoglyphs. Precision removal of thin tissue layers by CO_2 laser results in a bloodless field while allowing one to distinguish surgical depth and normal from abnormal tissue targeted for removal. Landmark of disordered dermatoglyphs in warts is shown.

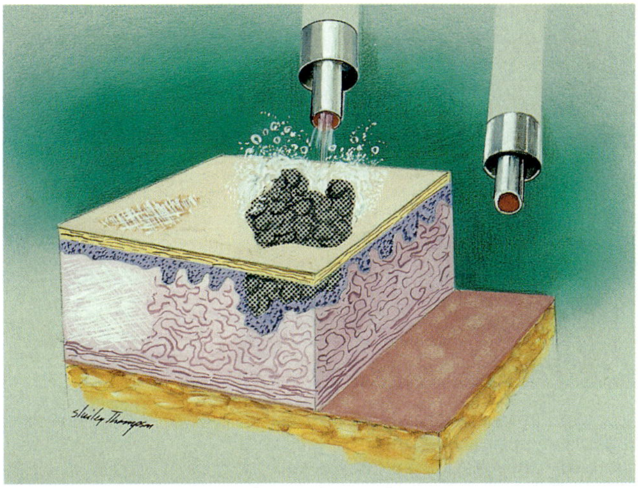

FIGURE 11.2. Differential laser tissue effects. Wart-infected epidermal tissue vaporizes more readily and notably, which is visualized as "bubbling," compared to normal dermis, fibromas, scar tissue, virally infected collagen, or fat.

These visual observations in laser surgery help us distinguish pathology, tissue types, and depth in a fashion unmatched by other modalities.

3. Laser therapy allows tissue removal without mechanical injury. When nail splitters, hemostats, curettes, and other instruments are applied, mechanical injury occurs to surrounding tissues, and infective visions are carried along the instrument track. Infection may be transplanted from a more easily treatable area, such as the hyponychium and lateral nail fold, to the proximal nail fold (Fig. 11.4). The nail plate may be laser incised or vaporized with minimal effects to the subjacent bed.

4. Lasers may be preset to deliver set fluences and specific electromechanically timed pulses or scan patterns that remove very small, precise layers of tissue, now in the 50-μm range. No other method currently allows for this degree of control. Improved functional and cosmetic outcomes may be achieved due to tissue preservation, which otherwise would have required full-thickness removal or greater destruction thickness. Less scarring, smoother less obvious scar outlines from feathered wound edges, and less recurrence of

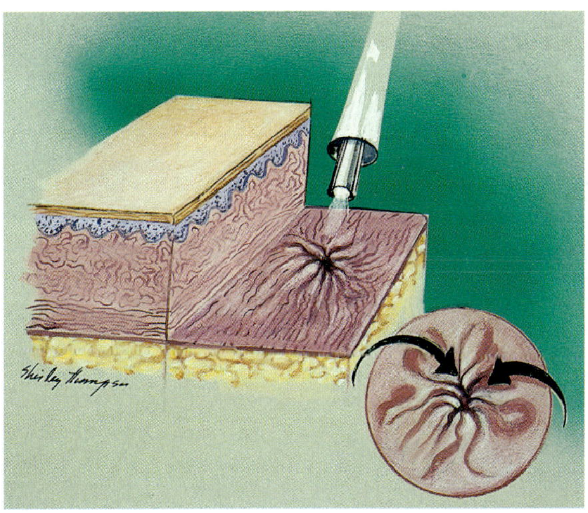

FIGURE 11.3. Laser thermal retraction on reticular dermis "pucker sign." Collagen contracts or "shrinks" much more than warts or epidermal tissues. This sign becomes more obvious when one reaches the reticular dermis.

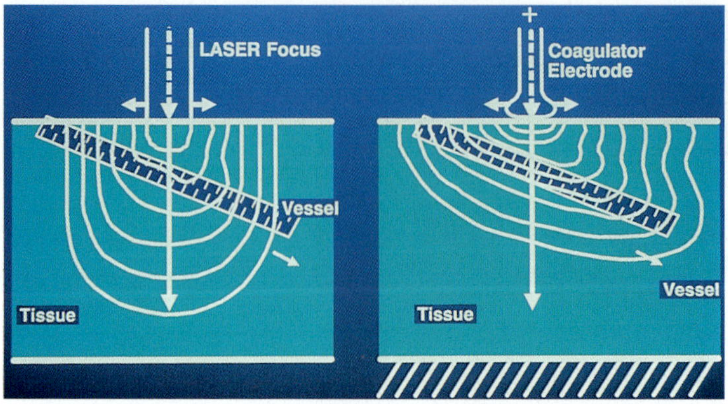

FIGURE 11.4. Laser versus electrocoagulator isotherms. Lasers travel in a highly linear fashion. Electrosurgical current may be preferentially conducted to nearby nontarget tissue and concentrated below the surface tissue at neurovascular bundles. This concentration can cause unseen damage that may result in prolonged or permanent nerve injury and paresthesias.

malignant or infectious conditions are all examples of how laser precision translates to better outcomes.

5. Lasers travel in a highly directional (collimated) beam. Lasers' line of sight targeting contrasts with electrosurgery, which may be preferentially conducted to nearby nontarget superficial tissue. Dispersed electrosurgical current also may concentrate below the surface tissue at neurovascular bundles. This concentration can cause unseen damage that may result in prolonged or permanent nerve injury and paresthesias (Fig. 11.5).

6. Sterilization of tissue and sealing by laser treatment prevents reinfection of treated sites and seeding of bacteria or viruses to adjacent tissue. This is particularly important when treating warts where viral count and infectivity are high.

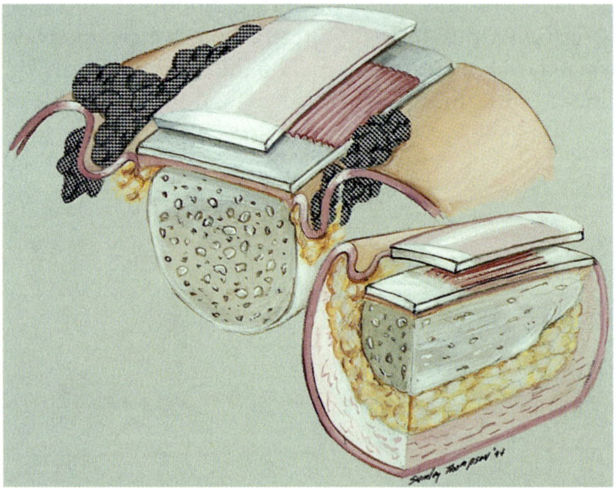

FIGURE 11.5. Periungual and subungual wart infection. Laser therapy allows precise nontouch tissue removal without mechanical injury or viral particle transfer caused by nail splitters, hemostats, curettes, and other instruments. Infective virions carried along by these manual instrument tracks may be transplanted deeper and more extensively than the original infection. Laser sterilizes and seals tissue, preventing reinfection of treated sites and inhibiting seeding of bacteria or viruses to adjacent tissue.

7. Speed is definitely enhanced by improved visibility and visualization, and by avoiding the application and removal of finger tourniquets, blotting the surgical site, placement of bulkier dressings, and occasional returns of patients a few hours after surgery to replace bloody saturated dressings.

8. Lasers can perform selective tissue or other target destruction. Vascular tissue, such as hemangiomas and malformations, as well as pigmented melanocytic lesions or tattoos on the digits, may be treated more specifically and elegantly with lasers than any other method. Using selective photothermolysis, a specific tissue, cell, or subcellular target is destroyed while sparing surrounding tissue from injury and scarring.

Recently, some patients with smaller, limited numbers of verrucae were observed to have clinical cures from several flashlamp excited dye laser (FEDL; pulsed dye laser) treatments with less morbidity than often associated with ablational CO_2 laser therapy (1). Although this phenomenon needs further verification of efficacy and longevity, some patients have benefited from this approach. My early impression is that approximately one-third of patients with smaller, nonperiungual warts on the extensor aspect of the digits will show some significant response with two to three treatments, with a little less discomfort than liquid nitrogen, but less efficacy as well. Presently for the digits, I reserve trials of the FEDL for small initial cases that are superficial but refuse routine therapies, or for treatment failures as one more treatment approach to the patient with resistant warts.

HOW DOES ONE CHOOSE LASER THERAPY?

One chooses laser therapy when the advantages to the patient or surgeon weighed against disadvantages indicate a net benefit. The eight laser therapy advantages listed in Table 11.1 are discussed in the section on "Advantages of Laser Surgery" along with diagnostic examples.

Clinical judgment considers that potential advantages accruing to laser utilization often are relative to the severity and location as well as the particular patient's diagnosis. For example, a small singular wart on the dorsum of the finger may as well be treated with liquid nitrogen, but tens of lesions in a similar location or a singular lesion in the periungual fold growing under the nail plate are strong indications for CO_2 laser (2–5). Procedures that cannot be performed without the laser, or those with clearly superior outcomes, often shorter physician time, or easier postoperative course, warrant a strong indication for laser therapy.

Table 11.2 lists the author's ranking of relative indications for laser therapy into "strong," "selective," and "no" indication by listing diagnoses. Each diagnosis is further

TABLE 11.2. LIST OF DIAGNOSTIC PROBLEMS AND PROCEDURE BY DEGREE OF LASER INDICATION

Strong Indication	Selective Indication	No Indication
Warts (A?V)	Nail ablation (A)	Myxoid cyst (A)
Arteriovenous malformation (V)	Matricectomy (A)	Onychomycosis (A)
Hemangioma (V)	Periungual fibroma (A)	Subungual hematoma (A)
Pyogenic granuloma (A/V)	Ingrown nail and granulation tissue (A)	
Benign melanocytic lesions (P/V)	Superficial skin cancer (e.g., Bowen's disease, multifocal basal cell carcinoma) (A)	
Decorative and traumatic tattoo removal (P)		
Proximal and lateral fold plasty (A)		

A, ablation; P, photomechanical technique; V, vascular.

TABLE 11.3. PROCEDURAL LASER CATEGORIES

1. Ablation or incision of periungual solid tissues, lesions, or tumors: CO_2, erbium-YAG, holmium-YAG
2. Vascular lesion photocoagulation: flashlamp excited dye laser ("pulsed dye"), tunable dye, argon
3. Pigment destruction: PLDL, argon, KTP, Q-switched (ruby, alexandrite, KTP)
   A. Decorative and traumatic tattoos: on back of fingers, pencil lead, bike/car accident
   B. Melanin, e.g., lentigines, Peutz-Jeghers syndrome

alphabetically coded to indicate the category(ies) of laser(s) recommended for consideration: "A" for ablation, "V" for vascular lasers, and "P" for pigment lasers.

Table 11.3 organizes lasers into primary use groups: solid tissue ablation, vascular photothermolysis, and pigmented lesion photomechanical destruction. The categories listed in Table 11.3 are discussed in the section on "Procedural Laser Categories and Applications."

Disadvantages of laser approaches are relative equipment and operational and training costs when simpler, cheaper methods that accomplish the same task as effectively without other drawbacks are available. Setup time for some laser systems is another cost factor. When one already owns a laser system, the operational cost factors may be overlooked in cost-to-benefit analysis.

PROCEDURAL LASER CATEGORIES AND APPLICATIONS:

The principles, indications, equipment, methods of treatment, and aftercare will be summarized for the three laser categories: ablation, vascular photothermolysis, and pigmented lesion photomechanical destruction.

Ablation or Incision of Solid Tissue Lesions or Tumors with Infrared Lasers: Carbon Dioxide, Erbium-YAG, and Holmium-YAG Lasers

Principles, Equipment, and Methods of Treatment

Basic Principles

Bloodless and thin-layer laser ablation of solid tumors and lesions, or their incision is accomplished using CO_2, erbium-YAG, and holmium-YAG lasers. These infrared lasers all emit wavelengths between 1,200 and 10,600 nm that are efficiently absorbed by water. Vaporization of water occurs at 100°C with energy densities of approximately 4 J/cm^2. Due to shallow tissue penetration of these wavelengths, at vaporization threshold, wounds show minimal peripheral injury. Measurements with CO_2 laser result in less than 50 μm if exposure time is kept close to 1 ms, i.e., the epidermal thermal relaxation time. Either short pulses or rapidly moving intense beams via a scanner can meet requisite energy fluences and exposure times to avoid injuring sensitive tissues.

The tissue between the nail bed and the periosteum is only 1 to several millimeters; thus, laser precision offers superior control in tissue removal compared to manually controlled excision or electrodesiccation and minimizes the risk of matriceal damage or osteomyelitis.

The CO_2, erbium-YAG, and holmium-YAG are all infrared lasers. As the erbium-YAG is not yet widely available and the holmium-YAG is similar to both the CO_2 and erbium-YAG but has less precision, discussion will center on the CO_2 laser.

Indications

Therapeutic indications for infrared laser ablation or incision include wart removal, nail ablation, matricectomy, periungual fibroma, pyogenic granuloma, ingrown nail with granulation tissue, superficial skin cancer (e.g., Bowen's disease), proximal and lateral fold plasty (6). Biopsy should be performed in cases of uncertain etiology because a variety of malignancies can masquerade as benign lesions (7–9).

CO_2 laser ablation of warts exemplifies the "nontouch" benefits of one of the most clearly advantageous laser techniques, especially for periungual warts. The recurrence rate of periungual warts is higher with other methods due to the deep sequestration of wart infected tissue under the proximal nail fold, nail bed, and deep sulcus of the lateral nail fold (Fig. 11.6). Perisurgical new site infection rates are increased due to lack of sterilization of viral particles and inoculation of surgically traumatized epithelium. In contrast, cure rates with proper CO_2 laser technique are 80% for first laser treatment of previously untreated warts and 73% for residual or recurrent warts after previous CO_2 laser treatment (Fig. 11.7) (2). In addition to the lower cure rates of other methods, patients note they are more painful (bleomycin, liquid nitrogen) and, in the case of bleomycin, can cause permanent nail dystrophies, nail loss, and Raynaud's disease (10–15).

A 1-mm curette is a valuable tool to probe residual dermal thickness and integrity, and to debride and explore the texture of unknown tissue foci (Fig. 11.8). Figure 11.8 illustrates the use of a curette to "feel" an incipient periungual fat herniation. No further tissue is removed. Healing will occur with less scarring and contraction when the dermal-subcutaneous boundary can be maintained intact, which is not always possible with many deep periungual and digital warts.

A more selective indication is in order for laser nail ablation, with or without laser matricectomy, which may be performed for therapy-resistant painful dystrophic nails. Laser matricectomy may be helpful if previous surgical attempts were unsuccessful with chemical or surgical methods. In my experience, standard chemical matricectomy is successful and superior for the majority of cases.

I have seen three patients who had surgical matricectomies with subsequent deep abscess formation due to buried matrix remnants, usually of the lateral horns. In these patients, the excellent visibility provided by combined laser incision and ablation provided a safe and effective means to remove the remaining matrix and nail spicule abscesses without recurrence.

Periungual fibromas have a selective indication for laser therapy. They may have multiple causes and may be treated palliatively or for cure (16–20). Biopsy is recommended if diagnosis is not obvious, as in tuberous sclerosis (Koenen's tumors). In the case of Koenen's tumors, larger lesions are more difficult to treat and may require matricectomy for large and central lesions. Recurrences are not uncommon with conservative technique designed to minimize injury to the matrix with resultant dystrophy. Smaller acquired fibromas may originate from proximal nail folds and can be resolved easily without permanent nail damage.

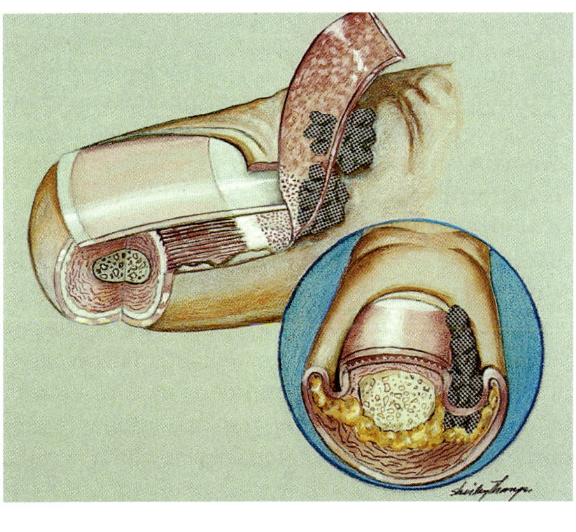

FIGURE 11.6. Sequestration of wart within lateral nail fold. Deep sequestration of wart within lateral nail fold precipitates recurrences. There is a narrow distance between the nail bed, periungual epidermal sulcus, and periosteum.

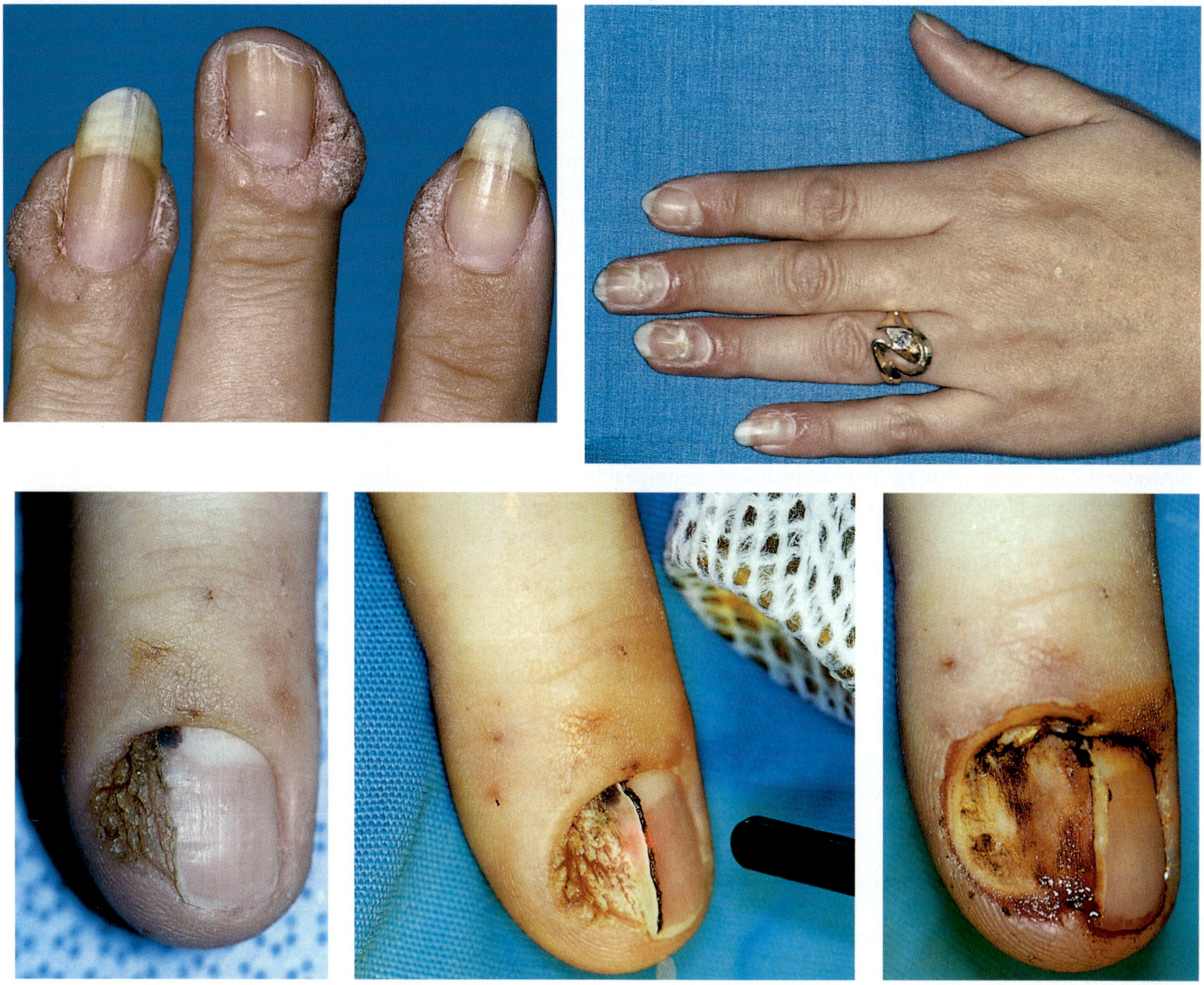

A B C D,E

FIGURE 11.7. Periungual warts. The recurrence rate of periungual warts is higher with nonlaser methods due to the deep sequestration of wart-infected tissue under the proximal nail fold, nail bed, and deep sulcus of the lateral nail fold. The CO_2 laser vaporizes tissue precisely while maintaining excellent visualization of the tissue landmarks. **A:** Preoperative view of woman with extensive periungual warts. **B:** Postoperative results after CO_2 laser excision. **C:** Preoperative view of male with lateral nail fold and subungual wart. **D:** Intraoperative view of male having CO_2 division of nail plate. **E:** Postoperative view of nail fold and bed after deep wart vaporization.

Most ingrown nails are treatable with small conventional instruments and partial lateral chemical matricectomy, if needed. A laser may be used in selected cases to aid the surgeon with speed and hemostasis in debulking and sealing bleeding granulation tissue at the lateral folds. Otherwise, there is no particular benefit.

Pyogenic granuloma has been recognized as the third most common tumor of the nail. These benign proliferative vascular tumors may be excised or ablated. Excision technique of polypoid lesions benefits greatly from a laser backstop, which may provide tension as well as protecting normal tissue that is left behind (Fig. 11.9). Some practitioners have noted less recurrence and satellite pyogenic granuloma formation after laser removal than with other methods. It is postulated that the sealing of excision margin tissue with minimal adjacent injury decreases the abnormal wound healing response leading to new pyogenic granulomas. Preliminary conventional biopsy or excisional laser technique at the

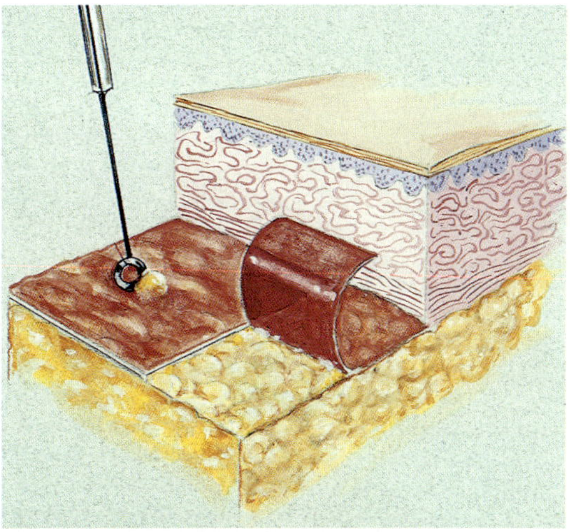

FIGURE 11.8. Use of small curette probe to delineate dermal depth and texture of incipient digital fat herniation. A 1-mm curette is a valuable probe to ascertain residual reticular dermal thickness and integrity, and to debride and explore the texture of unknown tissue foci. Here an incipient periungual fat herniation is detected, and no further tissue is removed. Healing will occur with less scarring and contraction.

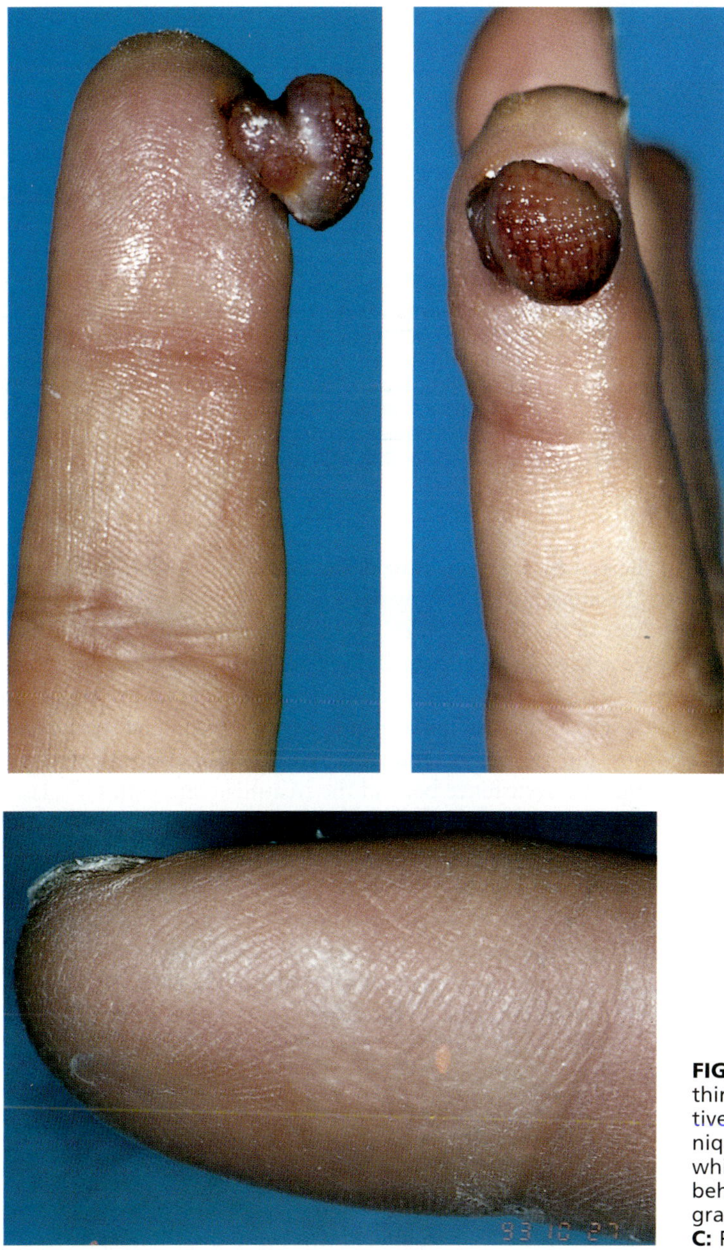

A B C

FIGURE 11.9. Pyogenic granuloma has been recognized as the third most common tumor of the nail. These benign proliferative vascular tumors may be excised or ablated. Excision technique of polypoid lesions benefits greatly from a laser backstop, which may provide tension as well as protect normal tissue left behind. **A:** Preoperative palmar view of fingertip pyogenic granuloma. **B:** Preoperative radial view of pyogenic granuloma. **C:** Postoperative results after CO_2 laser excision.

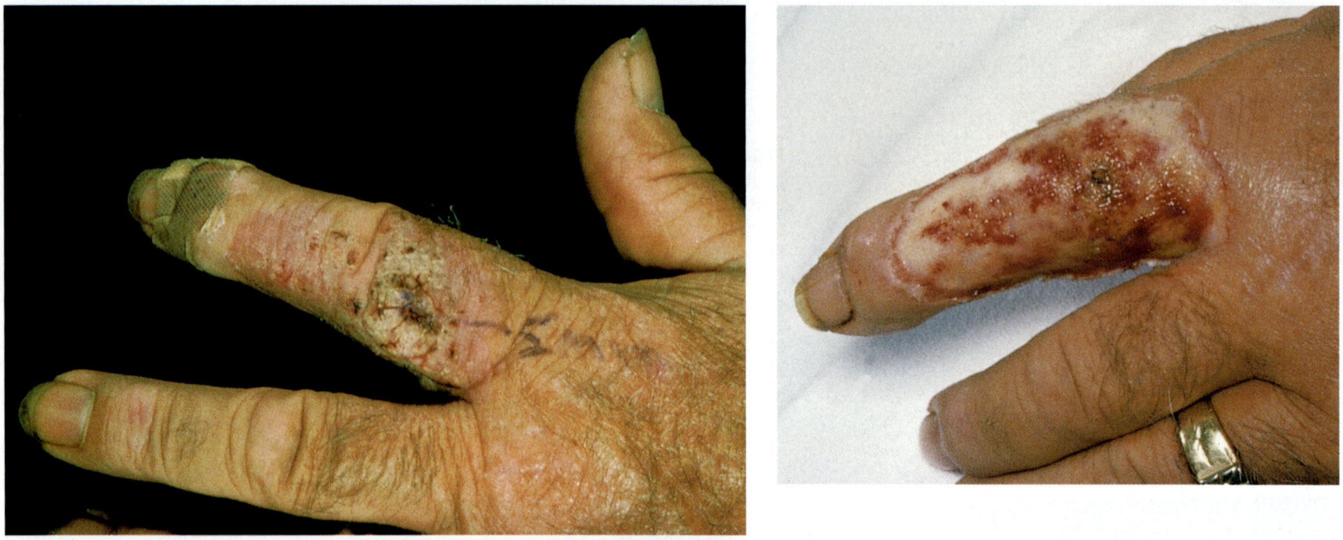

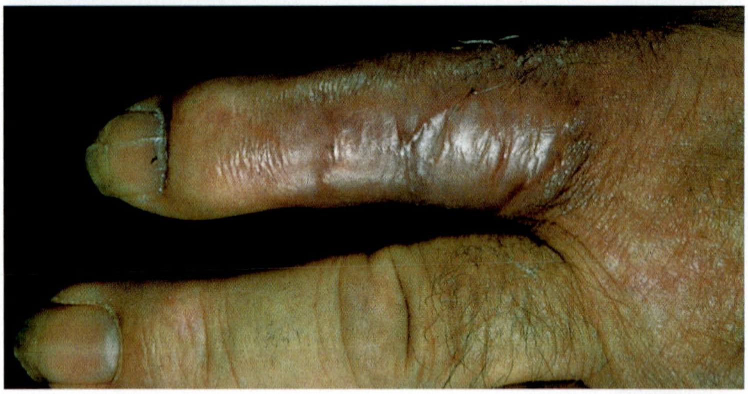

FIGURE 11.10. Bowen's disease of the finger. Functional gains and shorter rehabilitation times are benefits of CO_2 laser therapy, together with less risk of peripheral recurrence beyond the treatment site. (Courtesy of R. Roenigk, M.D.) **A:** Preoperative patient slated for amputation. **B:** Intraoperative patient treated with ablational CO_2 removal of Bowen's disease. **C:** Postoperative acceptable scarring with salvage of digit.

lesional base as depicted, followed by ablation, is recommended; this ensures pathologic confirmation of the clinical diagnosis if there is any doubt.

Superficial skin cancer, i.e., of the Bowen's type, or extensive superficial and multifocal basal cell carcinomas have a selective treatment indication for laser therapy (6). Full-thickness removal and deeper destructive ablation, i.e., with electrocoagulation, are prone to excessive scarring and attendant functional impairment, even with skin graft placement (Fig. 11.10). Functional gains and shorter rehabilitation times are additional benefits, together with less risk of peripheral recurrence beyond the treatment site with CO_2 laser treatment. These advantages of laser treatment are balanced against the possibility of residual skin cancer tracking down a follicle that could be missed by less than full-thickness destruction or removal (21).

Proximal and lateral fold plasty earns a strong indication for the ease performed with laser technique when asymmetry of thickness or proximal nail fold length needs adjustment. The ability to remove thin layers bloodlessly and to sculpt the thickness of the folds is carried through to easily blend the transition to the adjacent periungual areas.

Laser and Ancillary Equipment

Equipment for basic CO_2 laser surgery includes a 20-W or higher laser possessing electromechanical shuttered pulsing at 0.05 seconds or shorter, a focusing handpiece with focal length of 50 to 125 mm, a basin of sterile water or saline, small and medium curettes (2 and 5 mm), gauze, cotton applicators, and toothed forceps. More advanced equipment includes a high-energy pulsed laser or a scanning attachment for a low-energy laser.

Anesthesia, Operative Settings, and Techniques

Anesthesia for most ablative and excisional periungual surgery is by local injection or preferably by nerve blocks on both sides of the digit with lidocaine (Xylocaine), mepivacaine (Carbocaine), or bupivacaine (Marcaine) according to standard techniques (22).

Operative laser settings correspond to five general techniques and their most frequent application in specific conditions. For simplicity, these five main techniques and settings may be summarized as follows:

1. Slow superficial incisions with laser settings of 3 to 5 W, repeat mode at 0.05 seconds, focused beam at 0.1- to 0.2-mm spot size. The laser impulses are repeated automatically and laid down contiguously with the handpiece held steady during laser exposures, then advanced between exposures. This technique is useful for incising the nail plate or fold, or anywhere slow delicate incisions are desired. Alternatively, one may use the low-power, 3- to 5-W setting in continuous mode, moving the handpiece quickly and continuously to make very shallow, rapid incisions, for example, through the nail plate for partial nail plate removal or avulsion. This technique may be used when a tumor is growing partially under the nail plate and better access is desired for ablation, biopsy, or excision.
2. Rapid deeper incisions with laser settings of 10 to 20 W, continuous mode, focused beam of 0.1- to 0.2-mm spot size. The laser fires continuously with steady quick advancement of the handpiece during exposures, usually with a laser-absorbing backstop behind the tissue. This technique is useful for excising or debulking thicker tissue or anywhere rapid deeper incisions are desired (e.g., large warty masses, pyogenic granuloma, nail fold abscess, or granulation tissue).
3. Slow superficial ablations with laser settings of 15 to 50 W, repeat mode at 0.05 seconds, defocused beam of 2- to 5-mm spot size. The laser impulses are repeatedly superimposed on small targets, laid down contiguously on broad areas to be ablated, or feathered into surrounding zones. Target zones are wiped clean of charred residual tissue before subsequent exposures. The handpiece is held steady during laser exposures, then advanced between exposures. This technique is useful for ablating the nail plate, bed, or thin or deeper areas, e.g., the base of the nail folds or matrix, or anywhere delicate thin ablations are called for. Note that the slow, controlled working speed is achieved at high powers via selection of the short exposure duration of 0.05 seconds, resulting in low average power but minimal charring and tissue damage.
4. Rapid deeper ablations with laser settings of 15 to 50 W, continuous mode, defocused beam of 2- to 4-mm spot size. The laser impulses are continuous with footpedal control. Rapid movement of the handpiece minimizes tissue dwell time of the laser beam, preventing excess thermal conduction to adjacent tissue. Continuous movement of the handpiece during exposures is necessary to achieve uniform tissue removal without grooving. Target zones are wiped clean of charred residual tissue before subsequent exposures. This technique is useful for planing or debulking thicker tissue, or anywhere rapid ablation is desired (e.g., warts, pyogenic granuloma, abscess, or granulation tissue).
5. Coagulation of small superficial blood vessels is performed with laser settings of 5 to 10 W, continuous mode with intermittent manual laser activation for 1 to 3 seconds, defocused beam of 3- to 5-mm spot size. The handpiece is moved continuously just until gray-white blanching coagulation of tissue seals vessels 0.5 mm and smaller. Larger vessels may require compression during coagulation to be sealed. Overlapping and repeated beam exposures to the tissue are discouraged, as injury depth and potential scarring are increased.

Aftercare and Follow-up

Antibiotic ointment and a dressing such as nonadherent gauze or Vaseline-impregnated gauze are applied until reepithelialization is complete. Superficial wound care is followed by use of over-the-counter analgesics, whereas deeper wounds are treated with prescription pain relievers. The patient is seen as needed and at intervals of 4 and 8 weeks.

Vascular Lesion Photothermolysis with Visible and Infrared Light Lasers: Flashlamp Excited Dye, Tunable Dye, Metal Vapor, and KTP-YAG Lasers

Infants occasionally suffer rapidly growing hemangiomas on the fingers, where maceration and mechanical injury from sucking can cause tissue breakdown and bleeding. Early treatment with FEDL at 585 nm or other wavelengths near hemoglobin's alpha or beta absorption peaks (540, 577) can shrink hemangiomas and impede further growth, especially of the predominantly superficial type. Higher power output systems, such as the potassium-titanyl-phosphate (KTP) laser at 532 nm, tunable dye at 577 to 585 nm, and copper-bromide laser in the range of 3 to 10 W, are valuable not only in the treatment of hemangiomas, but also for superficial dermal arteriovenous malformations. Although less precise, neodymium-YAG lasers may be helpful for coagulation of larger and deeper vascular malformations, if some scarring is acceptable.

Principles, Equipment, and Methods of Treatment

Basic Principles

Vascular photothermolysis is accomplished using flashlamp excited dye, tunable dye, metal vapor, and KTP-YAG lasers. These lasers emit visible wavelengths between 488 and 585 nm, except for the YAG wavelength, which emits at 1,060 nm. These wavelengths exhibit preferential absorption by blood compared to the surrounding tissue, but competition with melanocytic pigments is significant. Coagulation of appropriately selected vascular targets (not too large and not too deep) allows selective destruction of abnormal vessels. The visible wavelengths are characterized by shallow penetration and limited scattering, whereas YAG photons penetrate and scatter by more than 10-fold.

The mechanism of visible light lasers to treat warts is unknown. Postulated mechanisms include destruction of necessary vessels for verrucal nutrition or traumatic injury to the wart causing increased immune recognition and susceptibility.

Indications

Diagnostic indications for vascular destruction include proliferative vascular lesions (hemangiomas), vascular malformations (port-wine stains, arteriovenous malformations, telangiectasias), and warts.

Hemangiomas on the digits of an infant are subject to minor trauma and maceration, which may lead to bleeding, ulceration, and scarring. Selective photothermolysis treatment of hemangiomas in their growth phase has demonstrated value in that the superficial proliferative tumors or problems related to their rapid growth are usually resolved in one to three treatments REF.

Vascular malformations, i.e., port-wine stains, arteriovenous malformations (spider angiomas, as well as larger vessels), and telangiectasias, may be treated for their undesirable appearance or their potential for present or future enlargement or discomfort.

Laser and Ancillary Equipment

Equipment for vascular lesion laser surgery includes a broad range of laser types and wavelengths: FEDL (pulsed dye), continuous tunable dye, quasi-continuous copper vapor, and KTP-YAG lasers. Adjunctive equipment includes various handpieces, spot sizes, and scanning attachments for continuous and quasi-continuous wave lasers. Specifics will be repeated within the operative techniques description for each type.

The FEDLs typically use 585 nm, 450-μs pulses, with beam diameters between 5 and 10 mm, and fluence of 5 to 10 J for treating superficial microsize malformations.

Alternatively, continuous visible or infrared lasers may be used. The YAG (1,060 nm) with its deep penetration may be the last resort for larger or deeper vascular problems if significant fibrosis is acceptable and perhaps necessary.

Tracing uses lower- to medium-power continuous wave lasers (including "quasi-continuous" lasers) such as KTP-YAG, metal vapor, tunable dye, and argon. Tracing is performed with 100 μm- to 2-mm beam spot sizes and power settings of 1 to 10 W, or 8 to 20 J. I favor the smaller beam diameters for vessels up to 1.5 mm because beam spread begins immediately at the epidermis and scatters to a radius of 1 mm. The laser impulses are shuttered at 0.05 to 0.1 seconds and laid down in repeat or single exposure mode contiguously 1 to 2 mm apart.

Pulsed drilling also uses continuous wave (or quasi-continuous wave) lasers and is helpful for vascular malformations on the fingers 1 mm or deeper, or malformations with a higher flow rate and diameter. Pulsed drilling is performed with medium to high powers of 3 to 20 W of visible wavelengths, shuttered exposures of 0.1 to 0.5 seconds, and 0.1- to 1.0-mm spot size.

Anesthesia, Operative Laser Settings, and Techniques

Anesthesia needs vary by the type of vascular laser treatment and the depth of the lesion. No anesthesia is necessary for selective photothermolysis of superficial hemangiomas and telangiectasias of the digits. Treatment of warts generally is performed without anesthesia on the extremities, or with EMLA cream.

Moderate anesthesia of extensive port-wine stain malformations on the digits with EMLA cream is successful a majority of the time after waiting 1 to 1.5 hours after cream application and plastic wrap. Treatment of port-wine stain is young children is greatly facilitated by heavy sedation with a combination of meperidine (Demerol), promethazine (Phenergan), and chlorpromazine (Thorazine). Sedation is helpful due to children's high degree of anxiety about restraint or even mild discomfort.

Complete local anesthesia with lidocaine 1% or 2% usually is needed for shuttered or pulsed laser microdrilling photocoagulation of arteriovenous malformations.

Operative laser settings correspond to three general techniques for vascular lesions: defocused contiguous spots; shuttered tracing; and pulsed drilling.

These three techniques will be described, along with their most frequent application in specific conditions. The immediate therapeutic endpoint depends on the type of laser and the problem treated. The endpoint for malformations treated with continuous and quasi-continuous lasers is disappearance of the vessel with the least amount of peripheral heat damage. Peripheral heat damage is indicated by gray-white "blanching." Some vessels will show shrinkage or darkening color before disappearance. When FEDLs are used, purpura is required for effectiveness.

The immediate treatment endpoint for hemangiomas and verruca is a slight blanching with continuous lasers, or purpura with FEDL treatment.

1. Defocused contiguous spots generally are used to treat port-wine stain and diffuse matte telangiectatic vascular malformations, and occasionally warts and proliferative hemangiomas to stimulate regression.

The FEDLs typically use 585 nm, 450-μs pulses, with beam diameters between 5 and 10 mm and fluence of 5 to 10 J for treating superficial microsize malformations and proliferative hemangiomas (Fig. 11.11).

Alternatively, continuous visible or infrared lasers may be used. Brief exposures in the millisecond range and energy densities in the 5- to 20-J range are necessary to minimize nonspecific damage and scarring. The YAG (1,060 nm) with its deep penetration may be the last resort for larger or deeper vascular problems if significant fibrosis is acceptable and perhaps necessary.

The 1-mm laser impulses may be manually directed or passed through a robotic scanner, such as the StarPulse or Hexascanner, to distribute a uniform array of more than 100 or as few as several laser exposures, to match large or small areas. The laser impulses are laid down automatically and contiguously until the area is covered like a checkerboard. The handpiece is held steady during laser scan times averaging 30 seconds; otherwise,

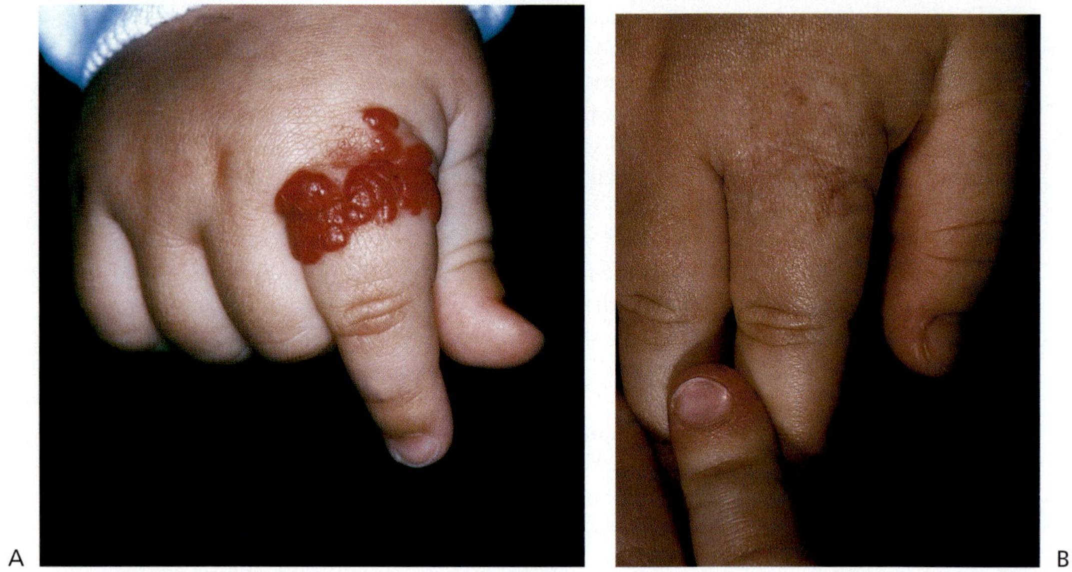

FIGURE 11.11. Proliferative hemangioma, flashlamp excited dye laser (FEDL) method. Rapidly growing hemangiomas may be halted or caused to regress or resolve by FEDL treatment. The laser can be supplemented with intralesional steroid injection if needed for deeper (transcutaneous) lesions. Hands and digits of infants and young children are constantly being traumatized by crawling and sucking. **A:** Preoperative proliferative hemangioma. **B:** Resolved proliferative hemangioma.

some exposures may be superimposed while other spots are left untreated. The handpiece is advanced between exposures.

Warts have recently been treated with close to maximum fluence by pulsed dye lasers. Double or triple exposure to the same spot often is performed, especially in resistant or thicker lesions. Pretreatment paring is vital to aid penetration of laser light through thick callus, e.g., on the sole. Continuous lasers usually perform mild coagulation of the wart with mild-to-moderate coagulative blanching of the wart visible. Patients return for at least one to three retreatments at 2- to 4-week intervals with either FEDL or continuous lasers.

A key to successful hemangioma treatment is early treatment during the rapid growth phase, preferably before the thickness attained exceeds the penetration depth of the vascular laser. Generally, hemangiomas that are several millimeters thick still respond well. Thicker lesions and especially those developing a deep subcutaneous or transcutaneous component will need intralesional steroid treatment. Along with a laser treatment, I inject a small dose of equal parts of triamcinolone 40 mg/mL and betamethasone 6 mg/mL every 1 to 3 weeks, until regression is rapidly proceeding. Doses average a few tenths of a milliliter. After a single treatment, ulcerations usually begin rapid healing, and surrounding hemangioma tissue begins to fade and flatten, and reveals gray involutional changes.

2. Shuttered tracing is used for small, superficial linear vessels that are uncommonly seen on the fingers. Tracing uses lower- to medium-power continuous wave lasers (including "quasi-continuous" lasers) such as the KTP-YAG, metal vapor, tunable dye, and argon. Tracing is performed with 100-μm to 2-mm beam spot sizes and power settings of 1 to 10 W, or 8 to 20 J. I favor the smaller beam diameters for vessels up to 1.5 mm because beam spread begins immediately at the epidermis and scatters to a radius of 1 mm.

The laser impulses are shuttered at 0.05 to 0.1 seconds and laid down in repeat or single exposure mode contiguously 1 to 2 mm apart.

3. Pulsed drilling is helpful for vascular malformations on the fingers 1 mm or deeper, or malformations with a higher flow rate and diameter. Pulsed drilling is per-

formed with medium to high powers of 3 to 20 W of visible wavelengths, shuttered exposures of 0.1 to 0.5 seconds, 0.1- to 1.0-mm spot size.

The high-intensity small beam vaporizes the tissue in a narrow shaft. A wet cotton applicator is used to remove small amounts of char from the opening, and the aiming beam is repositioned exactly in line with the shaft for repeated exposures until the arteriovenous malformation is photocoagulated (Fig. 11.12). Another arteriovenous malformation case demonstrates that occasionally several laser drill shafts are required due to

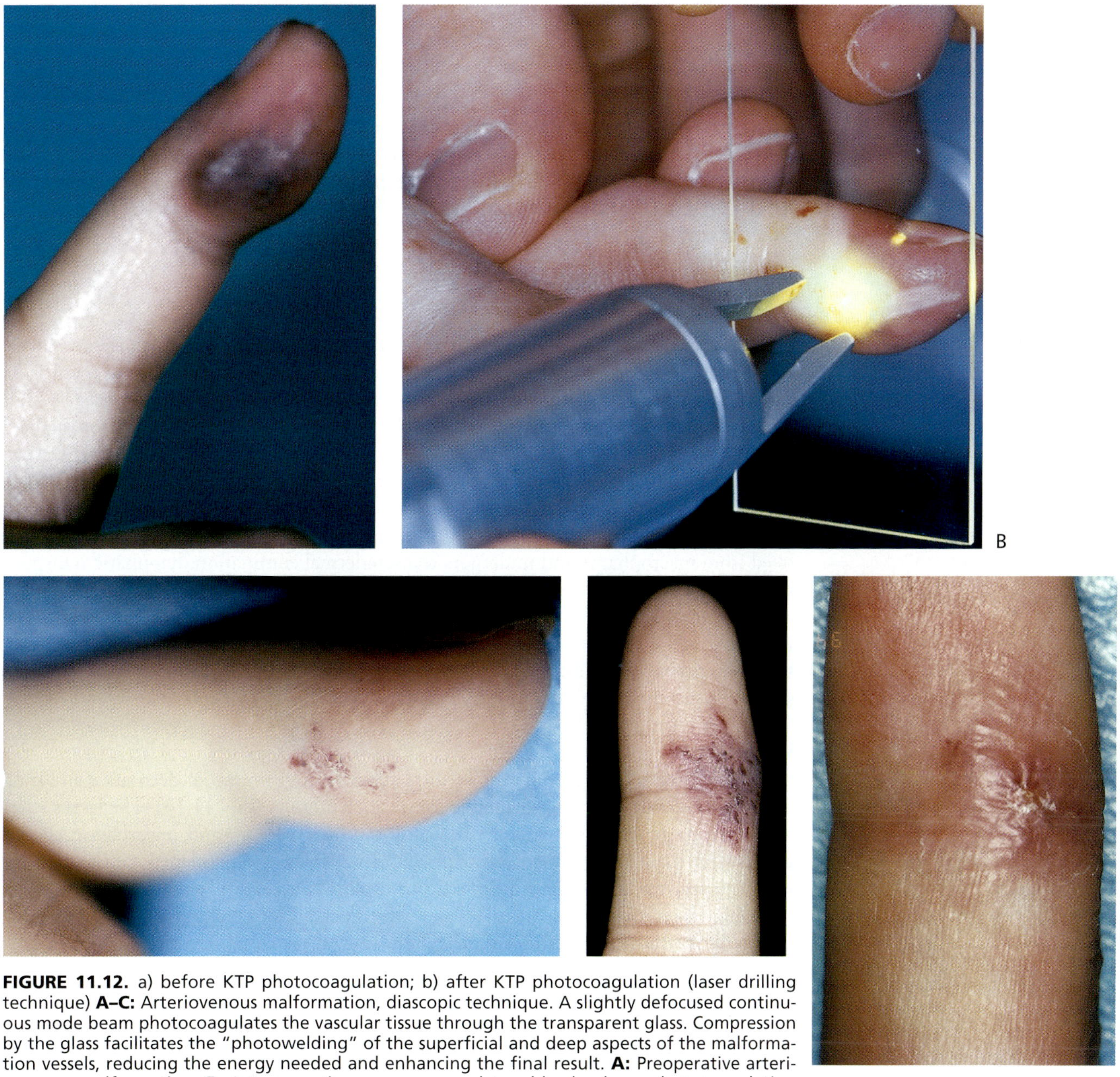

A B C D,E

FIGURE 11.12. a) before KTP photocoagulation; b) after KTP photocoagulation (laser drilling technique) **A–C:** Arteriovenous malformation, diascopic technique. A slightly defocused continuous mode beam photocoagulates the vascular tissue through the transparent glass. Compression by the glass facilitates the "photowelding" of the superficial and deep aspects of the malformation vessels, reducing the energy needed and enhancing the final result. **A:** Preoperative arteriovenous malformation. **B:** Intraoperative argon-pumped tunable dye laser photocoagulation with transilluminating diascopic technique. **C:** After tunable dye laser photocoagulation (diascopic technique). **D,E:** Arteriovenous malformation, pulsed drilling technique. Deeper malformations may require pulsed laser drilling, in contrast to the diascopic technique. The high-intensity small beam vaporizes the tissue in a narrow shaft. A wet cotton applicator is used to remove small amounts of char from the opening, and the aiming beam is repositioned exactly in line with the shaft for repeated exposures until the arteriovenous malformation is photocoagulated. **D:** Preoperative. **E:** Postoperative.

multiple vascular connections. Superficial thermal spread from multiple drill shafts may allow separation of the epidermis that facilitates repositioning of the aiming beam (Fig. 11.11).

Aftercare and Follow-up

Antibiotic ointment and a dressing such as nonadherent gauze or Vaseline-impregnated gauze are applied until reepithelialization is complete. Superficial wound care is followed by use of over-the-counter analgesics, whereas deeper wounds are treated with prescription pain killers. The patient is seen as needed and at intervals of 4 and 8 weeks. The patient must be prepared for the possibility of minor bleeding episodes, which could result from breakdown of the coagulated vascular tissue in the first week after treatment. In general, a small pressure dressing and elevation are all that may be necessary.

Pigmented Lesion Photomechanical Destruction with Q-switched Far Red and Infrared Lasers (Q-switched Ruby, Alexandrite, and YAG), Flashlamp Excited 510-nm Dye Laser, and Continuous-wave Lasers in the Visible (Yellow-Green) Spectrum

Principles, Equipment, and Methods of Treatment

Basic Principles

The basic principles for pigmented lesion selective removal are either photomechanical disintegration (for Q-switched systems) or semiselective photocoagulation (depending on which laser is used). Q-switched lasers are represented in the infrared and green visible ranges; there is one flashlamp dye laser in the green light range (510) used for pigment removal.

Q-switching produces extremely brief (nanosecond) laser pulses with very high energy (gigawatt peak powers). Wavelengths matching only the target's absorption allows selective destruction at the cellular or subcellular level. Pigment thus reduced in size is freed from macrophages and undergoes transepidermal elimination, lymphatic transport, and dispersal in the dermis, and gradually disappears clinically. Tissue surrounding the target is spared due to the narrow spatial confinement of the photomechanical disruption (23). Individual microphage destruction in tattoos or melanosomes in pigmented cells illustrates the most precise functional category of laser treatment. Examples of these lasers are the Q-switched ruby, alexandrite, and YAG lasers.

Indications

Q-switched lasers have become the medical standard for removal of decorative and traumatic tattoos and benign pigmented lesions, e.g., lentigines (Fig. 11.13). The laser's selective targeting of pigment, ease of use, comfort to the patient, and unsurpassed clinical results have made most other treatments obsolete, unless the pigment is resistant to the laser. Organic carbon pigment from road or explosion accidents, penetrating injuries, or tattooing is removed by sequential laser treatments. More treatments on average are required for treatment of tattoos on the digits than at proximal skin sites.

Laser and Ancillary Equipment

Equipment for tattoos and melanocytic pigmented lesion treatment includes one or more of the following Q-switched far red and infrared/green lasers: Q-switched ruby (694 nm), alexandrite (755 nm), YAG (1064, 532 nm), and FEDL (510 nm).

Anesthesia, Operative Settings, and Techniques

Anesthesia for most pigmented lesion surgery is unnecessary, or it may be performed with brief icing or EMLA cream application 1 hour before treatment. When more anesthesia is desired, local anesthetic injection or preferably nerve blocks on both sides of the digit

A 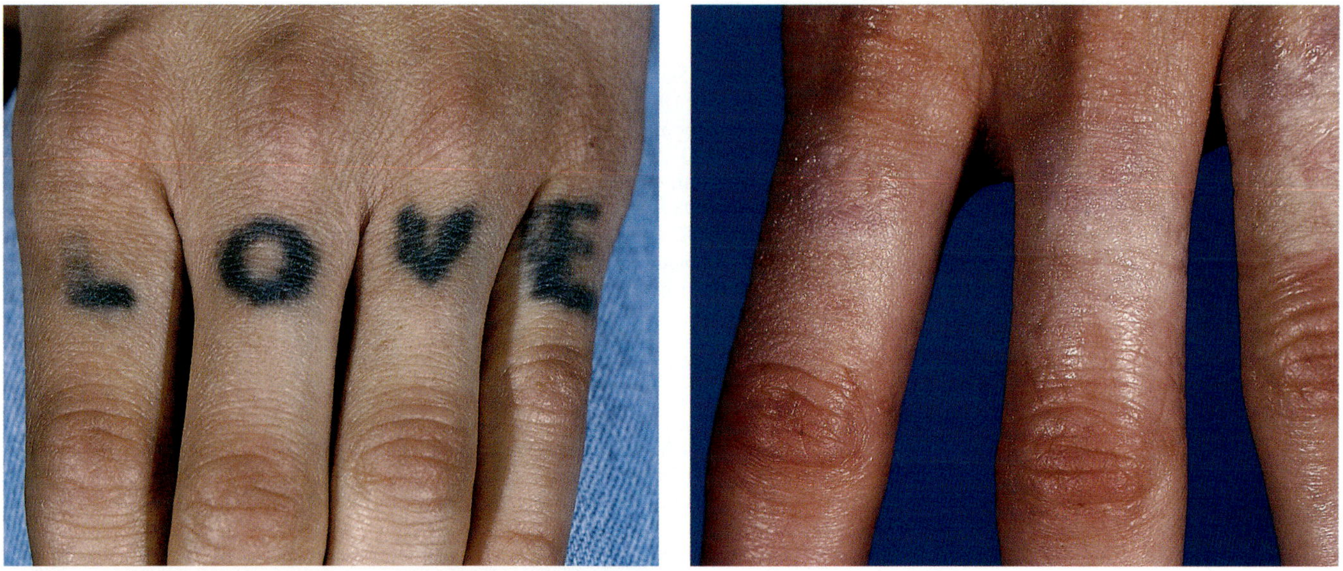 B

C 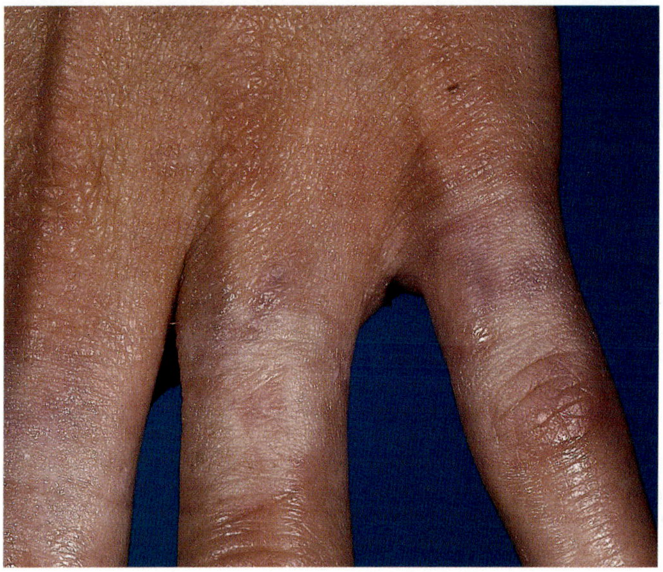

FIGURE 11.13. Finger tattoos. Traumatic tattoos of the fingers from accidents, lentigines, or decorative tattoos can be removed without scarring using Q-switched lasers. Treatment performed with Q-switched ruby laser. **A:** Preoperative view of decorative finger tattoos. **B:** Postoperative results after Q-switched ruby laser treatment. **C:** Postoperative results after Q-switched ruby laser treatment.

may be administered with lidocaine, mepivacaine, or bupivacaine according to standard techniques.

Operative laser settings correspond to the specific laser and pigment type and depth involved. Usually energy densities range from 2 to 10 J/cm^2. Laser technique consists of administration of impulses with the defocused spots laid down contiguously with minimal to no overlap. Energy densities may be adjusted based on the desired clinical endpoint of seeing transient (minutes) white macules or wheals at the laser impact site. These are interpreted as cavitation bubbles within cellular and extracellular fluid. Alternatively, with the Q-switched YAG superficial tissue disruption with bleeding points or pigmented dye laser may be used where a fine ash is seen.

Aftercare and Follow-up

Antibiotic ointment is applied to any eroded areas until reepithelialization is complete, usually in 5 to 7 days. A dressing such as nonadherent gauze or Vaseline-impregnated gauze is optional. The patient is seen after 4 to 8 weeks to evaluate the need for repeat treatments. The patient is advised to avoid trauma to the treated area for 1 week.

REFERENCES

1. Tan OT, Hurwitz RM, Stafford TJ. Pulsed dye laser treatment of recalcitrant verrucae: a preliminary report. *Lasers Surg Med* 1993;13:127–137.
2. Lim JT, Goh CL. Carbon dioxide laser treatment of periungual and subungual viral warts. *Aust J Dermatol* 1992;33:87–91.
3. Logan RA, Zachary CB. Outcome of carbon dioxide laser therapy for persistent cutaneous viral warts. *Br J Dermatol* 1989;121:99–105.
4. Street ML, Roenigk RK. Recalcitrant periungual verrucae: the role of carbon dioxide laser vaporization. *J Am Acad Dermatol* 1990;23:115–120.
5. Taylor MB. Successful treatment of warts. Choosing the best method for each situation. *Postgrad Med* 1988;84:126–128.
6. Gordon KB, Robinson J. Carbon dioxide laser vaporization for Bowen's disease of the finger. *Arch Dermatol* 1994;130:1250-111252.
7. Goldminz D, Bennett RG. Mohs micrographic surgery of the nail unit. *J Dermatol Surg Oncol* 1992; 18:721–726.
8. Goldenhersh MA, et al. Merkel cell tumor masquerading as granulation tissue on a teenager's toe. *Am J Dermatopathol* 1992;14:560–563.
9. Ishihara Y, et al. Detection of early lesions of "ungual" malignant melanoma. *Int J Dermatol* 1993; 32:44–47.
10. Urbina GF, et al. Cutaneous toxicity of intralesional bleomycin administration in the treatment of periungual warts. *Arch Dermatol* 1986;122:974–975.
11. Shumer SM, O'Keefe EJ. Bleomycin in the treatment of recalcitrant warts. *J Am Acad Dermatol* 1983;9:91–96.
12. Miller RA. Nail dystrophy following intralesional injections of bleomycin for a periungual wart. *Arch Dermatol* 1984;120:963–964.
13. Czarnecki D. Bleomycin and periungual warts [Letter]. *Med J Aust* 1984;141.
14. Amer M, et al. Therapeutic evaluation for intralesional injection of bleomycin sulfate in 143 resistant warts. *J Am Acad Dermatol* 1988;18:1313—1316.
15. Abbott L. Bleomycin and periungual warts [Letter]. *Med J Aust* 1984;141.
16. Chin FE, McCarthy DJ. The cytological and biochemical implications of periungual fibroma. *J Foot Surg* 1992;31:486–497.
17. Goldman EP, Boike AM, Snyder AJ. An unusual presentation of a periungual fibroma. *J Am Podiatr Med Assoc* 1992;82:215–217.
18. Kojima T, Nagano T, Uchida M. Periungual fibroma. *J Hand Surg [Am]* 1987;12:465–470.
19. Pasyk KA, Argenta LC. Argon laser surgery of skin lesions in tuberous sclerosis. *Ann Plast Surg* 1988;20:426–433.
20. Shelley WB, Phillips E. Recurring accessory "fingernail": periungual fibrokeratoma. *Cutis* 1985;35: 451–454.
21. Rudlinger R, et al. Human papillomavirus-35-positive bowenoid papulosis of the anogenital area and concurrent human papillomavirus-35-positive verruca with bowenoid dysplasia of the periungual area. *Arch Dermatol* 1989;125:655–659.
22. Epstein RH, Halmi B, Lask GP. Anesthesia for cutaneous laser therapy. *Clin Dermatol* 1995;13: 21–24.
23. Levins PC, Anderson RR. Q-switched ruby laser for the treatment of pigmented lesions and tattoos. *Clin Dermatol* 1995;13:75–79.

12

CRYOSURGERY

RODNEY P.R. DAWBER

During the past 30 years, cryosurgery has developed a high profile in dermatologic surgery. For many benign, premalignant, and malignant skin lesions, cryosurgery can have cure rates as high as many of the excisional and radiotherapeutic methods, and it has the advantages of being inexpensive, readily available, and easy to perform if sufficient background, scientific knowledge and clinical training are obtained (1–3). This latter statement is, of course, basic to all clinical medical and surgical practice, but because of its relative simplicity, cryosurgery often is performed enthusiastically by practitioners with insufficient knowledge. It is important to note that accuracy of technique is almost more crucial for nonexcisional methods of surgical "killing" if high cure rates are to be obtained, which in many ways is similar to radiotherapy.

Before considering applications of cryosurgery specific to the nail apparatus, it is important to consider the basic general principles of cryosurgery and its scientific background (3) and to show variations from the "norm" as they occur in the nail apparatus. Because the nail performs many subtle functions and has great esthetic significance, many applications will not be appropriate to this anatomic compartment because of risk of scarring that may be of no significance or may be quite acceptable at other sites.

MECHANISM OF DAMAGE DUE TO COLD INJURY

Cryosurgery can be loosely defined as the deliberate destruction of diseased tissue by freezing in a controlled manner. The controversy regarding whether cryotherapy would be a more representative term can safely be left to the "semantologists" (3). Many well-recognized events follow the rapid lowering of temperature in biologic systems, and all contribute to the necessary cellular injuries required for successful treatments.

Extracellular Ice

When cryosurgery was first introduced, it was thought that the main cause of cell death was the formation of ice in the tissues and the mechanical damage produced as the ice crystals formed. Observation has demonstrated that if cells are frozen *in vitro,* extracellular ice forms first, which gradually squeezes the cells together. It has been difficult to demonstrate actual disruption of membranes by this means; however, volume changes in the extracellular and intracellular compartments have been measured during freezing and thawing. It was concluded that disruption of cell membranes must occur. It appears that extracellular ice formation alone is not sufficient to kill cells. The temperature changes during cryosurgery are so rapid that there is inevitable formation of intracellular ice, which is physically very destructive.

R. P. R. Dawber: Department of Dermatology, Churchill Hospital, Oxford, OX3 7LJ, United Kingdom.

Hypertonic Damage

When extracellular ice is formed in association with cell suspension *in vitro,* the amount of extracellular water decreases, which causes an increase in concentration of solutes in the remaining fluid. Changing osmotic gradients between cells and extracellular fluids are produced, which lead to passage of electrolytes out of the cells and subsequent decrease in cell volume and disruption of cell membranes. It has been shown that when a certain concentration is reached, normally intracellular components, such as hemoglobin in red blood cells, pass out of the cell, causing irreversible damage. Rapid electrolyte transfer also has been implicated as the cause of damage to cell proteins and enzyme systems. It is generally accepted that the temperature gradients produced during cryosurgery cause damage by this means, especially during the reverse process of thawing.

Sensitization

In experiments using red blood cells, it has been shown that gross cell damage is produced even if hypertonic conditions necessary for disruption are not achieved. Therefore, it has been suggested that this "sensitization damage" is the result of disruption of phospholipids in cell membranes. It may be that events occurring during thawing, such as reversed osmotic gradients, may give rise to this form of damage.

Intracellular Ice

Extracellular ice formation and sensitization damage can only occur when freezing is slow, i.e., when differential freezing in different parts of a system is allowed to occur. When very rapid freezing takes place, intracellular ice formation occurs. It is thought that this intracellular ice formation causes cellular death even though cells assume a remarkably normal appearance immediately after thawing. Damage to cell organelles, such as mitochondria and endoplasmic reticulum, has been postulated as the cause of injury from intracellular ice. The size of the ice crystals probably is important; the larger the crystals, the greater the damage. Most of the evidence supports the general principle that intracellular ice is lethal. It is probable that recrystallization of ice during the slow thaw time after cryosurgery is responsible for tissue destruction. This process is as important as the initial freeze in causing cell death.

Circulatory Changes

Many authorities believe that other factors are as important as the initial freeze/thaw events in causing cell and tissue death: (a) circulatory (capillary and lymphatic) malfunction associated with early endothelial damage and edema; and (b) capillary and venous occlusion seen several days after treatment, leading to anoxia, further cell death, and tissue necrosis. Of interest with regard to the nail apparatus is that even though cold injury may cause severe necrosis to the nail folds, it is very rare for subungual necrosis to occur, possibly because the many glomus bodies in these sites are resistant to cold injury.

Immunologic Events

Cryosurgery has been shown to excite immunologic reaction against tumor cells. Lymphocytes and serum from tumor-bearing animals receiving a single cryosurgical dose show greater cytotoxicity to an identical tumor in a syngeneic animal than lymphocytes and serum from an untreated animal. This response is tissue specific and stimulated by the release of tumor-specific antigen either during or after freezing. Other workers observed the effects of an immune reaction after cryosurgery to treat pathology such as prostate carcinoma. Whether such events are important in human cutaneous cryosurgery remains to be proven.

EQUIPMENT AND METHODS

The most common equipment used in routine clinical practice is designed to utilize liquid nitrogen as a refrigerant because it is almost universally available and inexpensive. It is essential for the treatment of premalignant and malignant skin lesions, and it provides the most consistent "cell killing " (4).

The simplest technique is the dipstick method (3). This long-used method employs either a cotton-wool swab or copper disk with insulated handles. The swab or disk is dipped into a robust metal Dewar flask containing liquid nitrogen and then applied to the area to be treated. The time of application depends on the size and nature of the lesions. To maintain relatively long periods of tissue freezing, repeat dipping and reapplication may be necessary. It is very difficult to standardize this technique in view of the many variables, which include ambient temperature, pressure applied, distance the dip instrument travels from the Dewar flask to the lesion, and "dripping" of liquid nitrogen. This technique is considered suitable only for small, benign, very superficial conditions. "Artistry" and experience are essential if consistently good cure rates are to be achieved with this method. The technique has been used successfully by dermatologists for many decades. If premalignant and malignant integumentary lesions are to be treated, modern standardized equipment is required.

For routine outpatient cryosurgery, most dermatologists prefer a small hand-held spray unit or a compact tabletop unit capable of either spray or cryoprobe application. The former is by far the most common used in clinical practice. For details of the various commercially produced units available, the reader is referred to more detailed text (1,3,5). Most units allow for variation in the "width" of the spray and have probe attachments of different sizes to correspond with the size of the area to be treated. The probes generally are cylindrical and preferably have flat contact surfaces. They are particularly useful when pressure is needed, for example, with vascular lesions. Very small lesions can be treated with pointed probes, as spray techniques result in too wide an icefield and therefore greater morbidity. Instruments that use carbon dioxide snow or nitrous oxide cooling (Joule-Thomson effect) for dermatologic lesions will not be considered here, because these instruments are no longer in widespread use.

Various items of auxiliary equipment are important if the full range of cryosurgical techniques is to be used. Truncated nonconducting cones frequently are used to limit surface application of the spray. For small lesions, if carefully localized spray is required, auroscope cones can be used (Fig. 12.1). This method gives a very rapid rate of temperature decrease that is probably more destructive than the open spray technique. Some recently introduced machines have an attachable "closed cone" that sprays liquid nitrogen into the cone, which is pressed onto the skin.

Monitoring Devices

If only benign, relatively flat, and small premalignant and malignant lesions are to be treated and if liquid nitrogen is the refrigerant, then monitoring equipment is unnecessary because no physical instrument can measure uniform, "adequate" cell death. Treatment of deep or large tumors requires careful "depth dose" monitoring equipment, most commonly a pyrometer-thermocouple combination. Some methods use electrical impedance or tissue resistance-measuring devices that, in principle, have the advantage of measuring actual freezing. Only thin, inexpensive electrodes are needed, and many areas of the tumor can be monitored using a single probe (1,6).

In an attempt to standardize the treatment used for different lesions at varying sites and so that we can further our knowledge of the relative sensitivity or resistance of different lesions to cryosurgery, we have adopted a "spot" freeze technique that enables medical personnel with varying degrees of experience to obtain the same results (3,7).

The spot freeze technique involves first defining the size of the field to be treated (as with radiotherapy) and then inducing ice formation within that field by liquid nitrogen

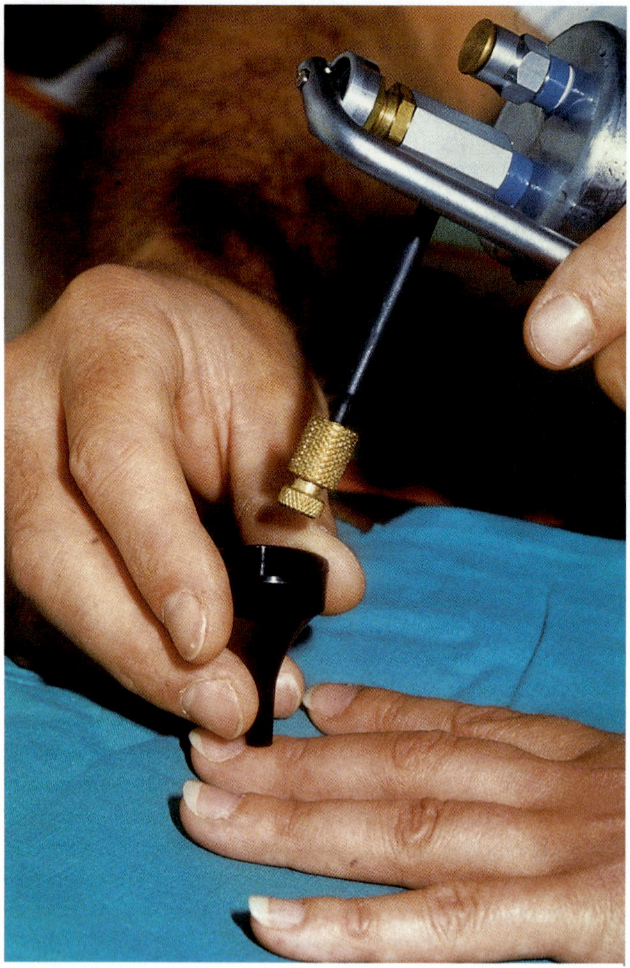

FIGURE 12.1. Spraying into an auroscope earpiece to localize the liquid nitrogen is more accurate than using the open spray method.

spray. Large lesions are divided into overlapping circles of 2-cm diameter using a skin marker. The liquid nitrogen spray (e.g., the "B" spray of CryAC units, Brymill Corp.) is held approximately 1 cm from the skin surface in the center of a 2-cm circle and spraying is commenced. The white "iceline" is allowed to extend outward until it fills the circle. This icefield is "held" for a measured time by continuing the spray with sufficient jet pressure to maintain the iceline. The measured time depends entirely on the nature of the lesion. Once the time has passed, spraying is stopped and thawing commenced. Each 2-cm circle is treated similarly.

A single freeze and thaw is termed a freeze/thaw cycle (FTC). Malignant lesions usually receive two FTCs, and the intervening thaw time is at least three times the duration of the initial freeze. Evidently, treatment fields smaller than 2-cm diameter do not need to be divided.

Determining the time to add after icefield formation must be learned by experience, but it varies according to the size, site, and type of pathology. The record in the hospital notes is usually made, for example, as follows:

$$LN_2 : \text{Single icefield} : 1 \times 15 \text{ seconds}$$

$$\text{Liquid nitrogen} : \text{Single} <2 \text{ cm} : 1 \text{ FTC} \times \text{time after ice formed}$$

This schedule is typical of that used for a small flat plaque of Bowen's disease of the skin. Small viral warts may require as few as 4 to 5 seconds, whereas malignant lesions need up to 30 seconds. Times of less than 30 seconds do not usually cause connective tissue distortion and scarring (3,5).

We tested the adequacy of the spot freeze method in experiments on the flank skin of pigs and showed consistent cell killing and satisfactory temperature levels within the field of treatment. The method is exactly repeatable from patient to patient and from site to site. If the treated lesion is not cured following a particular schedule, the time of treatment can be lengthened. Also, for the physician who handles the case at follow-up, the exact treatment procedure is known and can be specifically modified. This technique can quickly be taught to inexperienced medical staff and is exactly reproducible.

Many other techniques are available that involve varying spray methods for treating large lesions and liquid nitrogen probe methods. The important factor is to gain experience with whatever techniques are to be learned by (a) experimenting on the skin of inexpensive meats or other animal tissue, e.g., pigs' feet or hock available from most meat retailers; (b) working with an animal "model" whenever possible; and (c) observing experienced operators in practice. This is particularly important if malignant lesions are to be treated, because much of the therapeutic failure with small basal and squamous epitheliomas can be ascribed to poor technique.

As with all surgical treatments, accurate diagnosis is essential before cryosurgery is performed. Evidently this does not mean biopsy of all lesions before freezing, because clinical diagnosis often will be adequate, particularly with benign lesions and many superficial basal cell carcinomas. As with radiotherapy, misdiagnosis can lead to "blurring" of physical signs, for example, with melanotic lesions, and may facilitate tumor spread before the correct diagnosis is made.

Treatment Schedules

Most of the benign, premalignant, and malignant lesions that affect the skin may affect the nail apparatus and be amenable to cryosurgery as for other sites (3,4,7,8). This particularly applies to periungual tissues, including the lateral nail "walls," proximal nail folds, and distal subungual skin. For lesions at these sites, treatment schedules will be similar in principle to those used at other sites. Only those lesions specific to the nail area, or those that behave differently and require modified treatment, will be considered here. Table 12.1 lists some lesions for which cryosurgery may be the treatment of choice or useful under certain circumstances.

TABLE 12.1. NAIL APPARATUS LESIONS THAT MAY RESPOND TO CRYOSURGERY (1,3,7)

Ingrowing toenail (Fig. 12.2)
Subungual corn (heloma)
Myxoid cyst (Fig. 12.3)
Granuloma annulare
Pyogenic granuloma
Viral warts: periungual or distal subungual lesions (Figs. 12.4 and 12.5)
Molluscum contagiosum
Malignant lesions
Bowen's disease (carcinoma *in situ*)
Squamous carcinoma
Basal cell carcinoma
Lentigo maligna
Lentigo maligna melanoma
Kaposi's sarcoma lesions
Palliation

Ingrowing Toenail

Of the whole range of ingrowing toenail types, cryosurgery is particularly useful for those lesions that are (a) very painful, because of its anesthetic effects; (b) show hypertrophic lateral nail walls; and (c) have prominent granulation tissue, whether epithelialized or not (Fig. 12.2) (9). A single FTC up to 30 seconds is used. The FTC is quite painful, and EMLA cream applied under occlusion 2 hours before treatment or injection or plain 1% lidocaine may be needed. Sonnex and Dawber (9) reported a 64% cure rate for this regime, which is useful given its simplicity. *Pyogenic granuloma* freezing requires no more than a single FTC of 15 seconds with liquid nitrogen spray after ice formation, as does *subungual heloma.*

Myxoid Cyst

Myxoid cysts that occur in relation to the proximal nail fold, typically in fingers with distal interphalangeal osteoarthritis and Heberden's nodes (Fig. 12.3), are very difficult to cure (10). Many techniques are available. Dawber et al. (10) obtained a cure rate of 86% for small, clearly visible cystic lesions. For those patients who do not respond, more complex surgical excision can be performed.

Viral Warts

Periungual warts are notoriously resistant to treatment and may last for many years (3,4,7,8). Cryosurgery is considered appropriate, providing reasonable cure rates. The treatment consists of four to five courses given at 3- to 4-week intervals, using the spot freeze method, FTC up to 20 seconds freeze time, and EMLA cream applied 2 hours before treatment (Figs. 12.4 and 12.5). This treatment schedule should allow healing without any scarring. Molluscum contagiosum lesions of periungual tissue, which are rare at this site unless immunosuppression is present, usually respond to a single FTC treatment with maximum 10 seconds of freeze time.

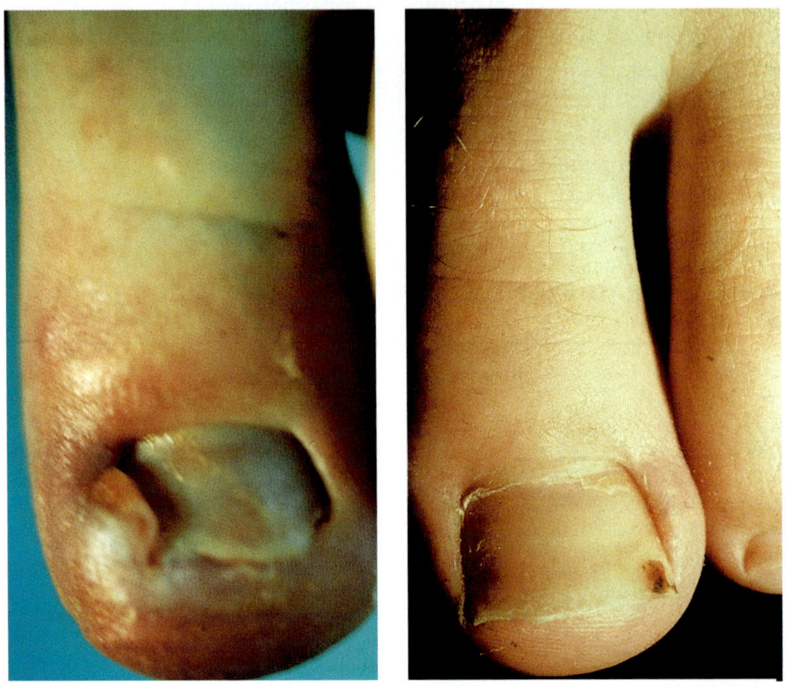

A B

FIGURE 12.2. Ingrowing toenail. **A:** Considerable epithelialized vascular overgrowth. **B:** Good result from cryosurgery. Liquid nitrogen spray of 20 seconds after ice formation. Slight temporary onycholysis and subungual hemorrhage.

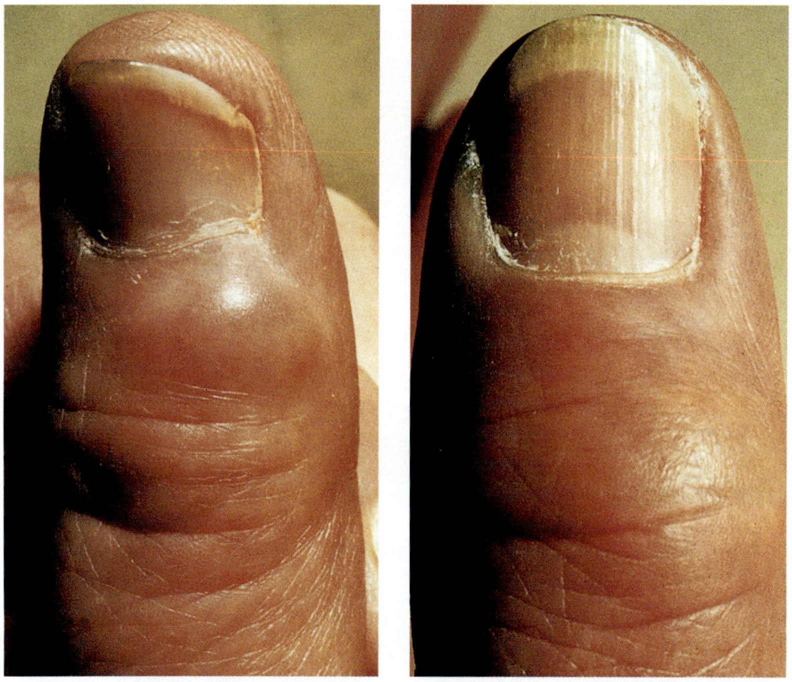

A B

FIGURE 12.3. A: Myxoid cyst of thumb overlying nail matrix area. **B:** Three months after cryosurgery.

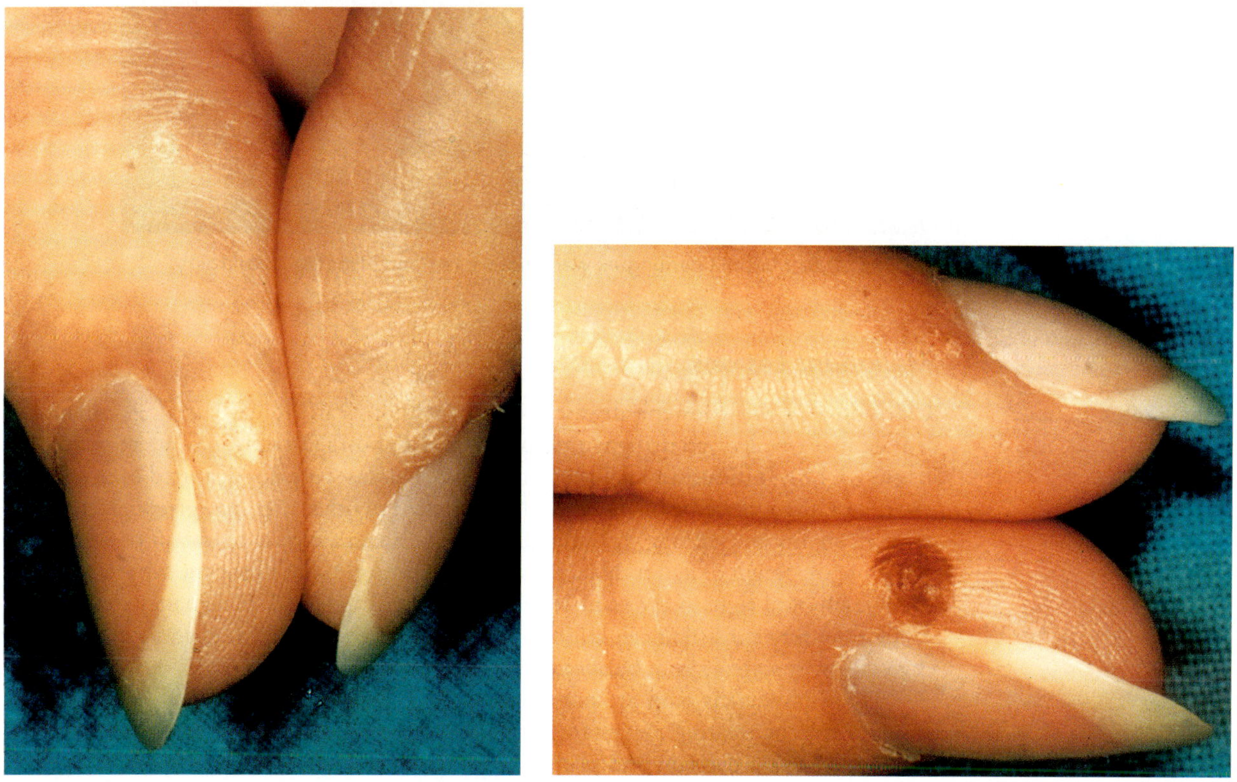

A B

FIGURE 12.4 A and B. Periungual warts. Before **(A)** and after **(B)** treatment.

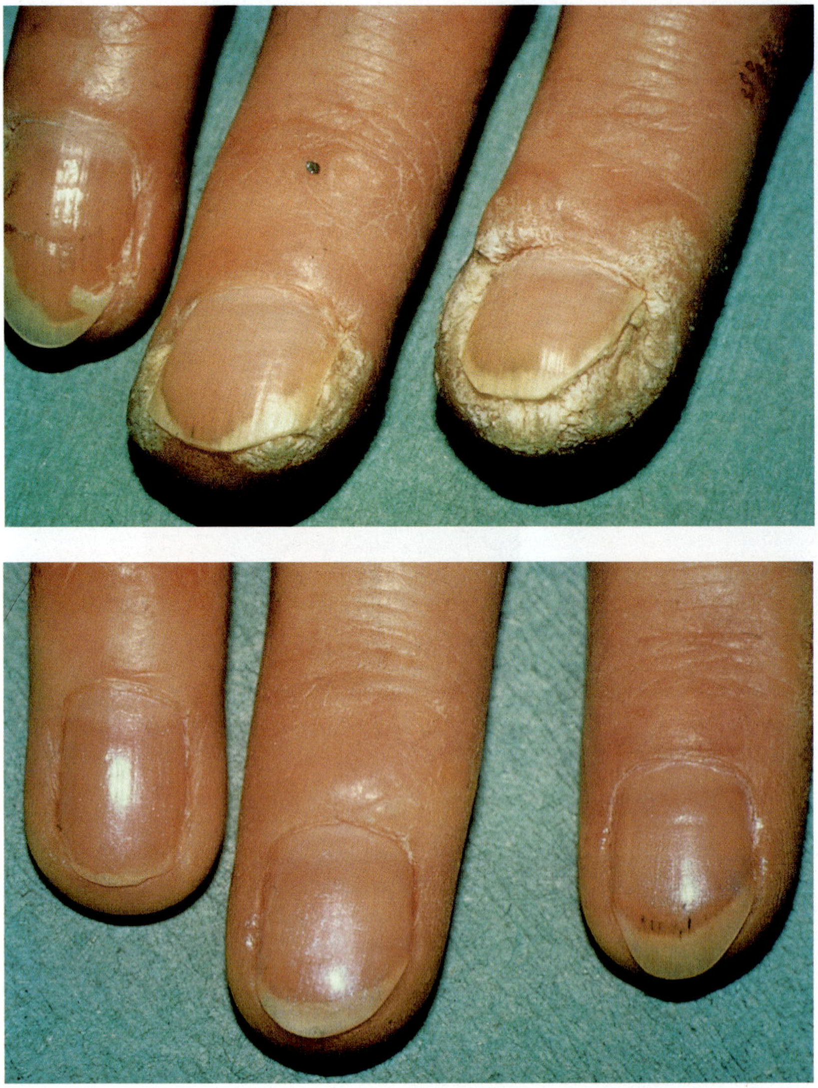

A

B

FIGURE 12.5. Multiple periungual warts before **(A)** and after **(B)** cryosurgery.

Malignant Lesions

Carcinoma *in situ* (Bowen's disease), small squamous carcinoma, basal cell carcinoma, lentigo maligna, lentigo maligna melanoma, and Kaposi's sarcoma may affect periungual tissues (3,4,8). If the central nail and subungual structures are not distorted or invaded by the tumor, then the same freeze schedules as applied to other sites may give high cure rates without significant scarring. It is mandatory to perform a biopsy to obtain tissue diagnosis before treatment of nail tumors, because these tumors do not have the same easily diagnosable physical signs as the same lesions elsewhere. In general, malignancy requires two FTCs to ensure consistent cell killing and good success rates. This schedule limits tumor cryosurgery if scarring is to be avoided.

Those physicians who are only familiar with using cryosurgery for benign lesions should begin by treating malignant tumors that might otherwise undergo curettage and cautery or simple excision and primary closure. Until greater skill is acquired, sites that have the highest recurrence rates, whatever the method of treatment, except Mohs surgery, are best avoided. Sites included with the nail apparatus are the inner canthus, nasolabial folds, and periauricular lesions. Basal cell and squamous carcinomas due to previous x-irradiation and in Gorlin's (basal cell nevus) syndrome are susceptible to cold killing and require only one FTC of 30 seconds in periungual sites. With careful case selection, cure rates higher than 97% can be obtained.

It is the author's opinion that cryosurgery is the treatment of choice for carcinoma *in situ* (Bowen's disease) of the skin and of the penis (erythroplasia of Queyrat). Periungual lesions fit into this broad group and can be treated successfully by cryosurgery. In Bowen's disease, distorting central nail structures almost always have some invasion, thus the term "epidermoid carcinoma." Such lesions are better treated by excisional surgery. Even though good cure rates can be obtained if two FTCs are used, poor cosmetic results and scarring are common. The clinical signs of squamous carcinoma in the early stages are less clear-cut than those of basal cell carcinoma; therefore, tissue diagnosis is crucial before treating squamous cell carcinoma. This leads to the conclusion that very small lesions are better treated by excision biopsy, as primary closure often is possible even in the nail apparatus.

LENTIGO MALIGNA AND LENTIGO MALIGNA MELANOMA

Dermatologists have used various freezing techniques in clinical practice for many decades to treat lentigo maligna and lentigo maligna melanoma (6,11). Anecdotal reports have always been favorable. It is important to note that cryobiologic research confirmed the ease with which normal melanocytes are killed by short freezes as used in clinical practice (12), which fully justifies pilot studies of "good prognosis" malignant melanoma such as biopsy-proven early lentigo maligna. Studies of other types of melanoma are ongoing (13). These lesions, which distort the nail apparatus, always require biopsy followed by adequate excision.

Palliation

Many authorities showed the value of freezing temperatures applied to surface malignant lesions: the size of primary and secondary (often fungating) malignancies in the skin was decreased; pain caused by such tumors often was decreased; and any chronic bacterial infection usually was improved or cured (3,14). In parts of the world where surgical facilities and radiotherapy are not available, such palliative methods still are useful, because liquid nitrogen is a better "killing" refrigerant than 19th-century salt/ice mixtures. Kuflik (14) showed that it is still a useful principle to use for many "incurable" malignant skin lesions. Liquid nitrogen is so widely available around the world in hospitals, veterinary and other biologic units (mostly for tissue preservation), and other industries that it usually can be obtained even in "third world" countries.

ADVANTAGES, SIDE EFFECTS, AND DISADVANTAGES

The patient usually feels a burning sensation during freezing and thawing (3,15). Any pain experienced usually is transient due to the anesthetizing effect of freezing. Local anesthesia is not required for short freeze times, but it may be indicated when treating malignant lesions or for patients with a low pain threshold. Periungual treatment produces relatively greater discomfort than other digital sites. Pretreatment with EMLA cream may help, but injection of local anesthetic is more likely to be required.

Some degree of erythema and edema is to be expected with cryosurgery treatments. In areas where the skin is lax in the nail area, although edema may be minimal, blisters may form even after short freeze times. These blisters may cause pain and severe inflammation and may slow healing.

Because this acute inflammation was thought to be unnecessary to obtain good cure rates, for many years the author advocated pretreatment and posttreatment (3 to 5 days) antiinflammatory therapy with soluble aspirin 300 to 600 mg up to four times daily and clobetasol propionate (Dermovate) cream daily to the treated area. The value of this regime was confirmed by objective assessment. Some authorities recommend systemic corticosteroids to minimize acute inflammation. EMLA cream applied three times daily may help to minimize any prolonged pain (rare!) after treatment.

Many of the conditions often considered curable by cryosurgery also are amenable to other surgical methods. The modality chosen often depends on the skills available in the department to which the patient has been referred. Compared to all other modes of treatment, cryosurgery has the advantages of being quick, inexpensive, and easy to learn and perform. Sterile surgical facilities usually are not required, and treatment can be initiated even in the presence of bacterial infection, e.g., ingrowing toenail. That posttreatment connective tissue distortion generally does not occur makes cryosurgery advantageous where scarring would be progressively troublesome (e.g., some digital myxoid cysts) over joints where a full range of movement can be expected to be retained even after treatments using two FTCs. Proximal nail fold cryosurgery may cause a transverse white band, a Beau's line, onychomadesis, and, rarely, permanent atrophy (Figs. 12.6 and 12.7).

Anything but the shortest freeze schedules will cause pigment changes in the treatment area (12), i.e., hypopigmentation with very short freezes and at the edge of more aggressively treated areas. Hyperpigmentation occurs after prolonged freezing (more than

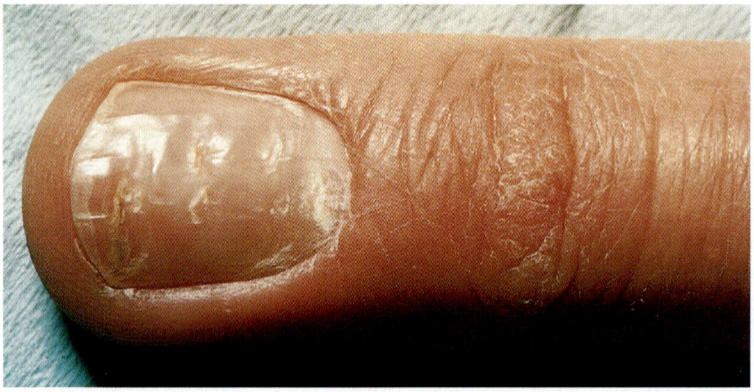

FIGURE 12.6. Temporary nail plate distortion after cryosurgery to proximal nail fold warts.

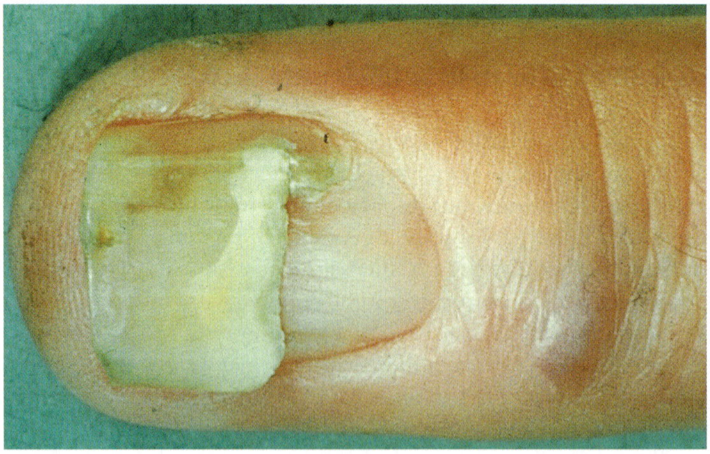

FIGURE 12.7. Nail shedding, usually temporary, after aggressive cryosurgery for a digital myxoid cyst. Note the absence of proximal nail fold scarring, which is a good prognostic sign.

5 seconds after icefield formation) and may be permanent; therefore, cryosurgery is less valuable in patients with racially dark skin, e.g., Asians and black Africans, and for "exposed part" lesions in white individuals who tan darkly on sun exposure. Because normal pigmentation is limited in periungual tissue even in black-skinned individuals, this is not usually an esthetic problem in nail tissues.

Temporary impairment of sensation in the treatment area is common after freezing (16). Only rarely is the patient aware of this. Such nerve ending damage can be expected to disappear within a few months, except after "malignancy regimes" using two FTCs. At sites such as the finger pulp, lateral nail walls, and distal subungual skin, permanent sensory loss may cause important functional impairment. Although nerve trunk damage and "distant" sensory and motor loss have been recorded, they are very rare and reversible, usually within a few months.

A rare side effect of cryosurgery is delayed bleeding, which may be due to granulation tissue formation as in pyogenic granuloma and chronic ingrowing toenail, or from erosion of a small artery. The former may require no more than pressure to abort it, chemical treatment (e.g., silver nitrate), or electrocautery. "Patent" bleeding arteries require tying off with an appropriate suture.

It now is evident that cryosurgical equipment and skill are essential in all dermatology and surgical departments that regularly treat both benign and malignant lesions. There are many skin and nail conditions in which excision, radiotherapy, and cryosurgery may be alternative modes of treatment under consideration; clearly for many of these lesions, cryosurgery is now the treatment of choice.

REFERENCES

1. Zacarian SA. *Cryosurgery for skin cancer and cutaneous disorders.* St. Louis: CV Mosby, 1985.
2. Dawber RPR. Cold kills! *Clin Exp Dermatol* 1985;13:137–150.
3. Dawber RPR, Colver GB, Jackson A. *Cutaneous cryosurgery,* 2nd ed. London: Martin Dunitz, 1997.
4. Kuflik EG. Cryosurgery updated. *J Am Acad Dermatol* 1994;31:925–944.
5. Breitbart EW. Cryosurgery: method and results. *Hautarzt* 1983;34:612–619.
6. Zacarian SA. Cryosurgery in the treatment of skin cancer. In: Friedman RJ, Rigel DS, Dopf AW, eds. *Cancer of the skin.* Philadelphia: WB Saunders, 1991:451–465.
7. Thai KE, Sinclair RD. Cryosurgery for benign skin lesions. *Australas J Dermatol* 1999;40:175–186.
8. Drake LA. Guidelines of care for cryosurgery. *J Am Acad Dermatol* 1994;31:648–653.
9. Sonnex TS, Dawber RPR. Treatment of ingrowing toenails with liquid nitrogen spray cryotherapy. *Br Med J* 1985;291:173–175.
10. Dawber RPR, Sonnex T, Leonard J, et al. Myxoid cysts of the finger: treatment by liquid nitrogen cryosurgery. *Clin Exp Dermatol* 1982;8:153–156.
11. Kuflik EG, Gage AA. The five-year cure rates achieved by cryosurgery for skin cancer. *J Am Acad Dermatol* 1991;24:1002-1004.
12. Burge SM, Bristol M, Millard PR, et al. Pigment changes in human skin after cryosurgery. *Cryobiology* 1986;23:111–116.
13. Sinclair R, Dawber RPR. Cryosurgery of malignant and premalignant diseases of the skin: a simple approach. *Australas J Dermatol* 1995;36:135–142.
14. Kuflik EG. Cryosurgery for palliation. *J Dermatol Surg Oncol* 1985;11:867–869.
15. Dawber RPR. Cryosurgery: complications & contraindications. In: Breibart EW, Dachow-Siwiec E, eds. *Clinics in dermatology 8.* New York, Elsevier, 1990:108–114.
16. Sonnex TS, Jones RL, Weddell AG, et al. Long term effects of cryosurgery on cutaneous sensation. *Br Med J* 1985;290:188–190.

13

THE PAINFUL NAIL

ECKART HANEKE
ROBERT BARAN

DIAGNOSIS

Pain is a nonspecific symptom of many nail conditions, such as trauma, infection, and tumor. It is often the reason for consulting a physician for alteration of the nail apparatus or distal phalanx that otherwise was neglected by the patient and/or physician.

The unique anatomic relations of the nail plate as a firm cover, the nail bed dermis as a very tough fibrous structure, and the terminal phalanx as a sensory structure receiving about 60% of the axons of the digital nerves all are responsible for the sometimes excruciating pain that an otherwise bland alteration would cause (1).

Diagnosis requires an accurate detailed history that includes the following information:

- Onset: sudden, slow, insidious, unnoticed
- Course: continuous, repeated, day and/or night
- Nature: continuous, lancinating, throbbing, nagging, dull, sharp
- Intensity
- Response to physical alteration: intensification or relief on raising the corresponding extremity; on cold or heat; on applying a tourniquet
- Response to specific drugs.

EXPLORATION

The clinical examination focuses on the nail, periungual soft and distal phalanx, and entire digit and compares them with their contralateral homologues for size, shape, circumference, color, and temperature. The nail must be examined in a relaxed state, under forced extension and flexion, as well as when the fingertips are pressed against a hard base. These maneuvers provide additional information about possible differences in blood circulation. A magnifier lens often is of great help.

The next step is careful palpation of the nail region, probing of the nail plate and perionychium, and comparison of the consistency of the periungual tissue.

Swellings may be examined using diascopy (glass spatulum pressed on the lesion will make the lesion's proper color appear) and diaphanoscopy (a light held against the swollen part will make the lesion shine if it is cystic and filled with a clear fluid). Probing is mandatory to locate the point of maximum pain.

Epi-illumination microscopy is particularly helpful in the clinical diagnosis of pigmented lesions under and around the nail. Specifically, dense minute bleedings of the dig-

E. Haneke: Klinikk Bunæs, Lokkeåsveien, 1337 Sandvika, Oslo, Norway.
R. Baran: Nail Disease Centre, 42, Rue des Serbes, 06400 Cannes, France.

ital tip that tend to become eliminated transepidermally can easily be distinguished from melanin pigmentation.

X-ray investigation is compulsory if pain after a single trauma or inflammation does not subside in due course. We have found xerography with magnification to be particularly valuable because it permits evaluation of the bone and provides outlines of the nail plate and soft tissues.

Magnetic resonance tomography recently proved to be useful for the diagnosis and exact localization of glomus tumors, as did arteriography.

Biopsy often is mandatory and the ultimate tool to make an exact diagnosis.

A number of highly sophisticated techniques, such as magnetic resonance imaging, are available for research purposes and, under certain circumstances, can be useful for examination of a painful nail.

DIFFERENTIAL DIAGNOSIS

Being so unspecific, subjective, and inconstant, pain as a symptom is not a reliable feature of many nail conditions. Some patients complain of irresistible pain even in normally painless conditions; other patients are extremely indolent and pretend not to feel pain.

For clarity, a few hints to possible differential diagnoses are given according to clinical features and pathology in Table 13.1.

Whenever the nail apparatus is painful, the possible mechanism should be considered. Edema due to infection, other inflammations, or trauma cause pain that usually is alleviated by raising the affected extremity as high as possible. Infection is commonly associated with pulsating pain. Pain that intensifies over several days despite surgical intervention of an acute subungual felon may indicate osteitis of the terminal phalanx. Differential diagnosis of periungual pyogenic infection is erysipeloid. Chronic candidal paronychia is characterized by episodic acute flares during which some pain may occur. Herpetic whitlow must be considered, particularly when lymphangitis accompanies the lesions from their beginning and appears to be extraordinarily intense and painful. Ingrown nails are frequent in adolescents and mainly occur on the great toe; whenever a seemingly normal nail grows in on a digit other than the big toe, a tumor of the lateral nail groove or wall must be considered. However, malignant tumors also may develop on the great toe and mimic painful ingrown toenail.

All space-occupying processes, such as rapidly growing tumors, can be painful. Keratoacanthoma is a typical example: it elicits pain in contrast to the slow-growing subungual squamous cell (epidermoid) carcinoma. Most other benign tumors and pseudotumorous lesions, such as dorsal myxoid pseudocysts, usually remain asymptomatic.

The best known subungual lesion is the glomus tumor. It is an inherently painful tumor and may cause spontaneous, sometimes pulsating pain that often is precipitated by slight trauma or cold. The pain characteristically radiates up to the shoulder. Application of a tourniquet will relieve the pain. The tumor often is seen through the nail plate as a reddish-violaceous spot 3 to 10 mm in diameter. A reddish stripe may appear in the nail bed, and there may be a slight longitudinal elevation in the nail plate. Probing allows further localization of the tumor. Approximately half of the glomus tumors cause a depression in the distal phalanx visible on x-ray films.

TREATMENT

Any therapy requires an exact diagnosis. Nail avulsion must never be done without (suspicion of) a diagnosis or even as a substitute for treatment. Although the nail plate usually obscures the clinical picture, it may provide invaluable hints as to the history of the disease and its exact localization.

Treatment options for painful nail lesions are summarized in Table 13.2.

TABLE 13.1. CLINICAL SIGNS IN THE PAINFUL NAIL (1)

Lesion	Key Symptoms and Examination
Clinically normal-appearing nail	
Cold injury	Color of digit, pain on rewarming
Raynaud's phenomenon	History of Raynaud's attacks and digiti mortui
Some chemical burns	Contact with alkali, caustic substances
Subungual glomus tumor	Excruciating pain radiating into arm and triggered by cold, mechanical shock, probing
Osteoid osteoma	Nagging pain, poorly localized, more intense at night, x-ray
Gout	Pain attacks, tophi on fingertip, uric acid determination
Neuroma	Biopsy
Enchondroma	X-ray, biopsy
Osteoma	X-ray
Visible nail changes, trauma, inflammation	
Hammer blow, crush injury	Personal history
Subungual hematoma	Inspection
Splinter	Inspection, history
Acute paronychia	Inspection
Subungual felon	Inspection, bacterial culture
Herpetic whitlow	Close inspection, history of recurrences, lymphangitis, Tzanck test
Ingrown toenail	Inspection, probing
Pincer nail (1,2)	Inspection
Retinoid nails	History
Different types of hereditary epidermolysis bullosa	History
Pterygium unguis inversum	Inspection
Pulpitis sicca and periungual eczema	Inspection, history of recurrences
Subungual cracks	Mainly during cold season
Dermatomyositis	Inspection, nail fold capillary microscopy
Systemic lupus erythematosus	Inspection, autoimmune phenomena
Occlusive vascular disease (3)	Decreased pulses and blood pressure on the affected extremity
Visible nail changes, tumors	
Subungual warts	Inspection (cut away overlying nail plate)
Ungual fibrokeratoma	Inspection
Exostosis	Bone-hard tumor, x-ray
Osteochondroma	Bone-hard tumor, x-ray
Glomus tumor (4)	Probing
Leiomyoma	Biopsy
Pyogenic granuloma	Inspection, biopsy
Keratoacanthoma	Inspection, dermatopathology
Maffucci's syndrome	X-ray, history
Aneurysmal bone cyst	Bulbous enlargement of phalanx, x-ray
Epidermoid cyst	Inspection (history), x-ray

TABLE 13.2. PAINFUL NAIL CONDITIONS AND THEIR TREATMENT

Condition	Treatment
1. Infections	
Herpetic whitlow	Conservative, oral acyclovir 5 × 200 mg daily
Periungual and subungual felon	Drainage, splinting, systemic antibiotics
Distal phalangeal osteitis	Wide incisions, drainage, splinting, systemic antibiotics
Acroosteolysis	Only painful when infected
Atypical myobacteriosis	Excision, oral chemotherapy
Tuberculosis cutis verrucosa	Excision, oral tuberculostatic treatment
Acute flares of candida paronychia	Topical broad-spectrum or anticandida antifungals, oral ketoconazole or itraconazole
Acute flares of allergic paronychia	Topical corticosteroids (in association with antifungals)
2. Inflammation	
Cutaneous gout	During attack: colchicine, nonsteroidal antiinflammatory drugs; in interval: allopurinol
Sarcoidosis	Systemic steroids
Raynaud's phenomenon	Slow warming, avoid cold
3. Trauma	
Cold injury	Slow careful rewarming, calcium channel blocker
Foreign bodies	Removal, prevention of infection
Acute injury, squeeze, crush, hammer blow	Rule out bone fracture as well as nail bed and matrix laceration, immobilization
Windsurfer's toe, tennis toe, ski-boot toe	Reduce sports activities, change habit and footwear
4. Characteristic nail lesions	
Ingrown toenail	Selective matrix horn dissection/cautery
Pincer nail (2)	Nail plate and bed plasty (see Chapter 5D)
Onychogryphosis	Cut softened nail, nail ablation
Onychophosis	Soften nail groove hyperkeratosis with urea or salicylic acid ointment
Pterygium inversum	Consistent application of emollient, excision and grafting
Retinoid nail	Reduce dose
Painful nail splits in lichen planus unguium	Emollients, intralesional steroid injection
5. Benign tumors	
Aneurysmal bone cyst	Ligate arteriovenous fistula
Cirsoid angioma	Extirpation
Subungual corn	Treat underlying foot deformity
Enchondroma	Enucleation
Eccrine angiomatous hamartoma	Excision
Epidermoid cyst	Extirpation
Exostosis, osteochondroma	Excision
Fibroma, fibrokeratoma	Removal
Glomus tumor	Extirpation, primary multiple glomus tumors may occur and cause pseudorecurrence
Keratoacanthoma	Excision, etretinate or acitretin for multiple keratoacanthomas
Leiomyoma	Excision
Osteoid osteoma	Naproxen, *en bloc* resection
Subungual tumors of incontinentia pigmenti	Curettage
Warts	Aggressive keratolysis, cryosurgery, laser ablation
6. Malignant tumors	
Bowen's disease	Excision, Mohs' surgery
Squamous cell carcinoma	Excision, Mohs' surgery
Malignant melanoma (painful only with secondary infection)	Excision with safety margin
Sarcomas	Removal, amputation

REFERENCES

1. Mathes U, Hartwig R. Der empfindliche Nagel. Z. Hautler 1993;68:351–358.
2. Baran P. Pincer and trumpet nails. *Arch Dermatol* 1974;110:639.
3. Kerdel FA, Baden HP. Subclavian occlusive disease presenting as a painful nail. *J Am Acad Dermatol* 1984;10:523–524.
4. Sheils WC, Becton JL, Christian JD. Subungual glomus tumor: a cause of pain beneath the finger nail. *J Med Assoc Georgia* 1972;64:268–270.

SUBJECT INDEX

SUBJECT INDEX

Page numbers followed by f *indicate figures; those followed by* t *indicate tables.*

R

S